THE KEPI

Table of Contents

THE KEPI Volume I Number 1

Table of Contents

THE KEPI Volume I Number 2

Table of Contents

THE KEPI Volume I Number 3

Table of Contents

THE KEPI Volume I Number 4

Table of Contents

THE KEPI Volume I Number 5

Table of Contents

THE KEPI Volume II Number 1

Table of Contents

THE KEPI Volume II Number 2

Table of Contents

THE KEPI Volume II Number 3

Table of Contents

THE KEPI Volume II Number 4

Table of Contents

THE KEPI Volume II Number 5

Table of Contents

THE KEPI Volume II Number 6

Library of Congress Cataloguing-in-Publication Data

ISBN # 978-1-945594-80-9 ebook
978-1-945594-81-6 paperback

The KEPI Volume I and II
/ S.L. Kotar and J.E. Gessler

Manufactured in the United States of America

Ahead of The Press Publishing
St. Louis, Missouri

AUTHOR'S NOTE

Putting together *The Kepi* Magazine

In speaking with devotees of the Civil War it has often struck me that those deeply immersed in studying the conflict, whether they refer to it as, "The War of Northern Aggression," the "Blue and the Gray," or the "War Between the States," have a life-long fascination with it. When people ask me how and why I got interested, my answer is always the same: I was born to it. As early as I can remember, I was drawn to the Civil War. Nothing was too obscure or settled for me: I wanted to know everything about the battles and leaders, but more specifically I craved an understanding of the times, because to truly understand the past, you have to view it from the perspective of those who lived it. Believing that others felt as I did, and that they weren't satisfied with coffee table publications specializing in repeating the tried and the true without daring to reinterpret a battle or a strategy or even a personality, my partner and I decided we would publish a magazine dealing not only with fresh (and sometimes uncomfortable) views, but one that offered a more thorough examination of the mid-19th century. With an eye toward the iconoclastic, the odd and the unexplored, we began publishing in the time before personal computers were commonplace.

To present a professional magazine meant using a professional printer and typesetter. Starting out with grand ideas and a limited budget we quickly learned how to do a paste-up which is actually more complex than it sounds. We had to teach ourselves about various typefaces, column width and how to efficiently use space, where and how to insert illustrations and how to be judicious in the use of lines. The description "paste-up" comes from the fact that pages were literally pasted-up. After deciding how to arrange a page, the typesetting (which came on a long roll), was cut to size and blocked out on a piece of blue graph paper (the camera didn't pick up the color blue). Then the back of the typesetting was covered with hot wax and placed on the graph paper, using the lines and squares to make the column even. When the wax dried, the page was "set" and ready to be "shot."

As you read the issues from start to finish (the printer did the paste-up on the premiere issue and we did subsequent issues) I think you'll see how we experimented with typefaces and lines, getting more and more adept as we went along. Ironically, this turned out to be the easiest part of the equation. What we anticipated would be the easiest actually turned out to be the most difficult: finding authors who were capable of writing about the War without falling into the trite and the rehashed. We didn't want the same old story of Gettysburg, or a romanticized version of Stonewall in the Valley. We sought hard, gritty stories, or ones with new interpretations of cannon. And believe me, when you talk about the Civil War, there is most definitely "cannon." And by that I mean, *This is the way it is.*

Perhaps there's a comfort in the thought that you know a particular battle inside out. Perhaps that makes it easy to discuss when everyone is on the same page. But when you're afraid to delve into probing questions, then you're not an historian, you're a fan. Fans were not our audience. That said, it proved extremely difficult to find articles worth publishing. We wanted to shake people out of their complacency but to do that we needed writers capable of producing thought-provoking material. We assumed they would beat a path to our door (we even paid a small fee) but they didn't. Those articles we received might have been all right for a weekend reenactment, but it wasn't interesting enough for us. So we ended up writing a number of articles ourselves, primarily on "Stonewall" Jackson, who happened to be our specialty. In re-reading these articles, as fresh to me now as when I wrote them decades ago, I hope you find some new insights and some thought-provoking interpretations. To add to our in-house staff we recruited several regular writers who had interests in medicine and religion. And I have to mention a gentleman who became a semi-regular writer for us, Thomas J. Evans. Although it's been a long time since we communicated, Tom was a first rate investigator and we were honored to publish his material.

This is the background on *The Kepi.* After we ceased publishing we devoted our energies to fiction writing. We have quite a number of books in the *R.B. Saga* that I intend to publish in the near future that deal specifically with the War. Our trilogy, *The Kansas Pirates,* available on Amazon, is set pre-War and deals with a number of uncomfortable issues in "Bleeding Kansas."

Thank you for choosing THE KEPI and good reading!

SLK

GSFE@aol.com

Dedication to THE KEPI, Volume I & II

If it can be said the past makes us what we are today, then a great deal of what I am was formed by what I absorbed from early television: that which was my Golden Age, the mid-1950s through the decade of the 60's. Integral to this time was an actor named Robert Horton. Not only was Flint McCullough, scout to Major Adam's *Wagon Train,* an inspiration to me, *A Man Called Shenandoah* captured a sense of the Old West in ways no other series ever did. But, it wasn't the characters as much as it was the actor who gifted them with life. There was an honesty and an intensity about Robert Horton that reached me on so many levels. He was *real* in ways that touched my soul. There was a wildness and a sense of dreams unfulfilled that challenged me to strive beyond what the eyes beheld. Just as the Civil War represents a scope wider than its individual parts, it has always summoned me to look beyond the Battles and Leaders to the people who were so intimately involved. Thus, through the War and into the Wild West days I rode – and am still riding – with those who lived and those who instilled characters with their own uniqueness, adding depth and dimensions to my imagination and the core of my being. It gives me great honor, therefore, to dedicate this first volume of THE KEPI to Robert Horton as a way of thanking him for letting me ride along, harboring both questions and answers as to who any of us really are and where we're going.

S. L. Kotar
And JEG, always

THE KEPI

APRIL, 1983 **FIRST ISSUE** **$3.00**

IN MEMORIAM
RICHARD ROWLAND KIRKLAND
CO. G, 2ND SOUTH CAROLINA VOLUNTEERS
C. S. A.
AT THE RISK OF HIS LIFE THIS AMERICAN
SOLDIER OF SUBLIME COMPASSION BROUGHT
WATER TO HIS WOUNDED FOES AT
FREDERICKSBURG THE FIGHTING MEN ON
BOTH SIDES OF THE LINE CALLED HIM
"THE ANGEL OF MARYE'S HEIGHTS"

APRIL, 1983 VOLUME I, NUMBER 1

The Articles

The Irregulars

Editors-in-General S. L. Kotar and J. E. Gessler. THE KEPI, 12115 Magnolia Boulevard, P. O. Box 107, North Hollywood, California 91607
Return postage must accompany all inquiries and submissions. THE KEPI reserves the right to edit, and copyright all letters.

Generally Speaking

We would like to take the opportunity in this, the first issue of THE KEPI, to explain the theme and goals of our publication.

Its title, THE KEPI, was chosen because the cap itself symbolizes the Civil War. Worn by both North and South, officers and privates, the kepi is a non-partisan capsulization of the War Between the States. This is exactly what we hope to achieve with our magazine: an unbiased representation of the war years.

Those who participated in the conflict necessarily lived before it, and so were inescapably exposed to the culture of their time. Therefore, a complete study of the Civil War should include an examination of the times preceding the outbreak of hostilities. Thus, THE KEPI will present articles dealing with all aspects of mid-19th Century life: religious thoughts and diversities that influenced daily living; practices of medicine, caught between taboos, superstitions and ignorance on one hand, and the trends of the day, such as hydrotherapy, phrenology and tonics on the other; the developing science of war, tactics and ordnance, from the experience young officers gained at West Point, the Mexican War and life on the frontier; politics, now facing the immensely complex and emotional issues of slavery; the growth of railroads and communications systems; agriculture; education; humor; fashion and food of the people who were caught up in these tumultuous years. There will also be a unique feature highlighting poetry and fiction from our contemporaries and the works of their age.

From accounts of the times to modern works, we shall provide a greater knowledge of daily life, making it familiar rather than unusual; enabling us to suspend our hindsight and view events from a new perspective. With our studies of soldiers and battles, we hope to stimulate new thoughts and points of view, as well as offer little-known or newly researched information.

There is no limit to the length of an article. If a subject needs to be discussed in-depth, it will be presented that way. We hope also to provide a forum for readers' opinions and questions, allowing you freedom to explore your own ideas, speculations and grudges.

In future issues, we will supply information on museums, battlefields, battle re-enactments and Round Tables to complete the overall view of the Civil War.

We welcome readers' comments, suggestions and questions. We are accepting contributions that deal with the war or mid-19th Century life. In this issue, we are including two columns which will appear semi-regularly; The Bugaboo Corner, *featuring fictional stories with a supernatural theme; and* Nostrums and Notions, *which will deal with medical or health practices in a humorous vein. On the serious side,* Medical Manual *will present discussions on the science of healing as it was understood by the 19th Century practitioners.* Tracts *will deal with religious topics;* Mess Call *with food and recipes;* The Gentleman Agronomist *with crop economy and technique; and* Missing in Action *with great people or ideas that either went awry or were never heard from again. You are welcome to write for any of these topics, or submit ideas of your own.*

We also want to encourage the submission of war-related articles. In the following pages we are presenting a study of the court-martial of Richard B. Garnett, because we feel the entire issue has been misrepresented and treated too lightly by historians. Kearnstown was General Jackson's first major battle to command from inception to retreat. His expectations, actions and determinations are crucial for an understanding of this great man. To read any biography on Jackson, and find only the vaguest references to Garnett's court-martial totally ignores the military and the psychological consequences. It also ignores the inner workings of the Confederate high command, and the important testimony given at the actual trial. Supplementing this is General Garnett's report on Kearnstown that was never printed in the Official Records. This is the type of war-related material we are seeking-thought-provoking, informative, comprehensive. If a subject has been ignored by historians, if it needs to be investigated or new material presented, THE KEPI will be the forum for it.

When submitting articles on the war or the times, we require that they be well-researched, and that footnotes be included. We are reproducing this documentation after every article to encourage our readers to investigate the arguments and the proofs for themselves.

We feel THE KEPI will help make the war period come alive by complementing the war with its times. With continued interest and support, THE KEPI will expand, both in pictures and content to satisfy the needs of our readers.

S. L. Kotar J. E. Gessler
Editors-in-General

THE KEPI

A New Concept in Civil War Reporting

The first Civil War Magazine to deal with all aspects of soldiers' and civilians' lives.

Have you ever wondered what were the exact charges against Brig. Gen'l. R. B. Garnett, or why they were eventually dismissed? Has the Phrenological Head with all its subdivisions and numbers mystified you? Ever wondered what was in those religious tracts passed out to soldiers, or how a Jewish Synagogue was consecrated in 1860? What was going on in the rest of the world while the States were preparing for deadly conflict? Read all this and more in THE KEPI. Published bi-monthly.

Next issue due out in June. Subscriptions now being accepted: $18.00 a year or $3.00 per single copy.

Subscribe early!

ZULLO AND VAN SICKLE BOOKS

OFFERING Civil War Books, rare editions, personalized service and extensive title search. Over 7,000 volumes in stock. Inquiries welcome. We are also beginning a new endeavor by reprinting Civil War books in our BUTTERNUT PRESS. Five reprints available soon are: "Mosby's Men," "In Camp and Battle With the Washington Artillery," "Bedford's Light Artillery," "The History of the Nottoway Grays," and "Camp Life of a Confederate Boy in Bratton's Brigade, Longstreet's Corps." THE BUTTERNUT PRESS will also offer some new, original work in the future.

Write Zullo & Van Sickle Books, 12137 Darnestown Rd., Gaithersburg, MD. 20878.

Newspapers, Harper's Weeklies, CSA documents, prints, Autographs, books. $1.00 for new list. Deborah Lambert, 1945 Lorraine Ave., McLean, Va. 22101.

Civil War Newspapers, books, documents, letters, images, stereoviews. Comprehensive Catalog $1.00. Gordon Totty, Scarce Paper Americana, 576K Massachusetts Ave., Lunenburg, MA 01462.

The Battle Behind Kearnstown*

By

S. L. KOTAR and J. E. GESSLER

On August 6, 1862, the court-martial of Richard Brooke Garnett, second commander of the Stonewall Brigade, took place at Liberty Mills, Virginia. The charge was Neglect of Duty with seven specifications, which was brought against him by Major-General T.J. Jackson for Garnett's actions during the battle near Kearnstown, Virginia, March 23, 1862. The court adjourned without reaching a verdict, and Garnett's subsequent reinstatemen to command seemed to preclude further investigation into the matter: history has judged Garnett innocent while cashiering Jackson for his motives and actions. Garnett's name has been irrevocably associated with that of Jackson's. He is seldom, if ever, mentioned without an additional comment on how he was "pilloried"[1] by Stonewall, or how badly this "unfortunate,"[2] "as brave as the bravest,"[3] "useful officer"[4] was treated by the "rigid"[5] "unrelenting"[6] Jackson. A careful consideration of the facts and events, as well as the assumptions precipitated by this incident, demand a reevaluation.

Who was Richard Brooke Garnett? What were the events leading to his court-martial? Born in 1819 on his father's estate in Essex County to the "best"[7] of Tidewater families, he was a member of the Virginia aristocracy. With all the benefits of a wealthy family and a respected name, he and his cousin, Robert S. Garnett, entered West Point. Both graduated in 1841, after 4 years of cadetship.[8] The system of assigning officers a position in the army was based on class standing. Robert ranked high enough to merit a position in the artillery, but Richard, graduating 29th, was only entitled to enter the infantry. Assigned to the 6th Infantry as a brevet 2nd lieutenant, he was stationed in Florida from 1841-42. Then after a tour of frontier duty, he served as aide-de-camp in Louisiana and Texas from 1845-51. Consequently, he saw no service in the Mexican War, where most West Point officers, including Thomas J. Jackson, garnered valuable experience in command and technique which they would bring to bear during the Civil War. From 1852-54, Garnett was assigned to recruiting service. On May 9, 1855, after 14 years of continuous service, he was promoted to captain.[9]

On May 17, 1861, Garnett resigned from the U.S. Army[10] and offered his services to the Confederacy, where he was commissioned a major of artillery.[11] On November 14th, he was promoted to brigadier-general, P.A.C.S.,[12] and on November 30th was directed to report to General J.E. Johnston, commanding at Manassas.[13] Early in December, he was ordered to Winchester to assume command of the Stonewall Brigade.[14] Garnett replaced Colonels J.T.L. Preston and James Allen, who had acted as temporary commanders after Jackson's promotion to major-general. During this time of stern discipline and training, while Jackson strove to mold militia and recruits into an efficient army, Garnett made himself a favorite with the men while gaining the reputation of pampering the Brigade.[15]

On January 1, 1862, Gen. Jackson marched the combined troops of the Valley Army and the Army of the Northwest out of Winchester on his Romney expedition. Mild weather turned foul, and the wagons were not able to keep pace with the marching troops. It was not until the morning of January 3rd that supplies came up; distributing and cooking rations meant a long delay. Coupled with the bitter weather and the realization that a postponement now could adversely affect the discipline he was maintaining, as well as delaying his campaign, Jackson gave the order to push on. Gen. Garnett took it upon himself to countermand this order and allowed his men to fall out of ranks to cook their food. By disobeying the commanding general's orders, Garnett displayed a blatent disregard for the overall strategy of the campaign, instead placing the comfort of one Brigade over that of the entire army. In so doing, Garnett revealed a character trait that would reappear at Kearnstown. When Gen. Jackson rode up and demanded to know who had given permission for the halt, Garnett replied it was he. Jackson explained there was no time for that, but Garnett claimed it was impos-

* In this article, Kearnstown is spelled with a "a" as this was its spelling in original documents of the time.

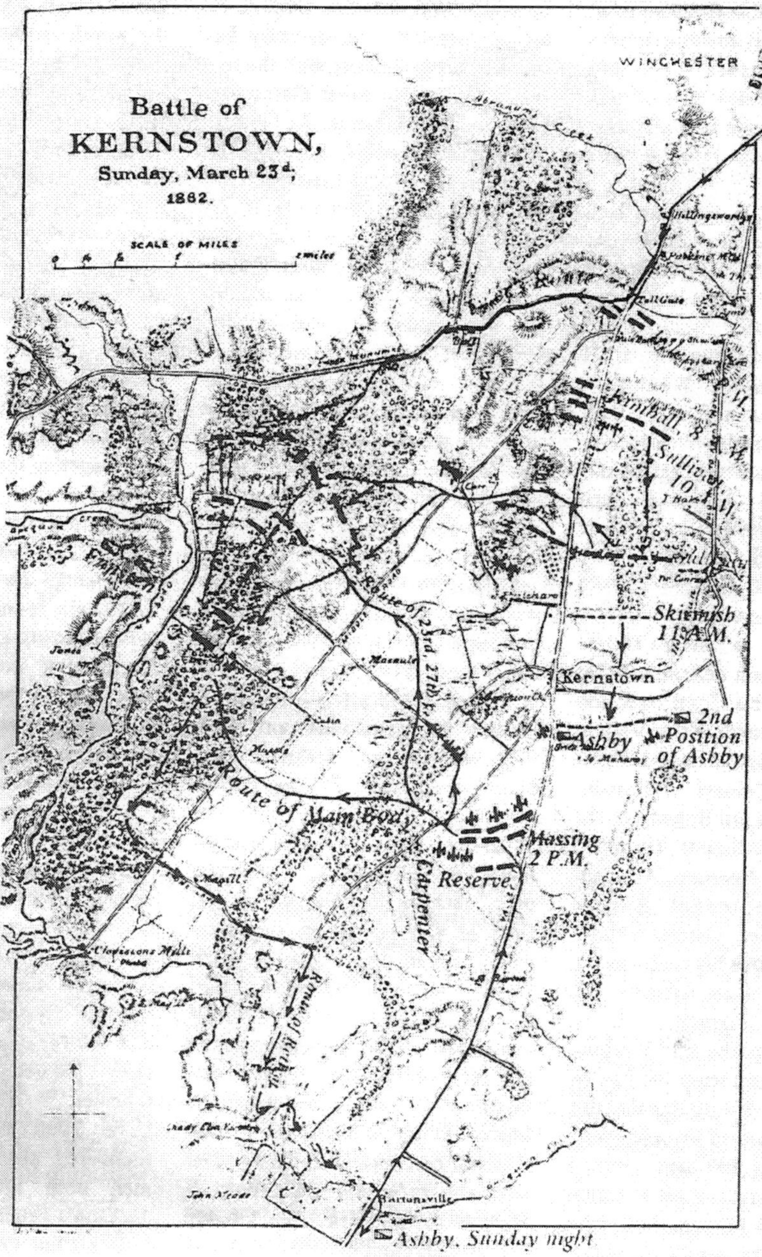

From William Allan's "History of the Campaign of Gen. T. J. (Stonewall) Jackson in the Shenandoah Valley."

sible for the men to march further without eating. Jackson stated firmly, "*I* never found anything impossible with this Brigade!"[16] This may have been the incident that prompted Jackson to write to Richmond, which Garnett later refers to as a "covert attack."[17]

On March 21st, Gen. Jackson received information from his cavalry commander, Turner Ashby, that Federal forces were leaving the Shenandoah Valley, and only a small force amounting to three regiments remained at Winchester. Jackson, whose duty it was to keep these Union troops from reinforcing those at Fredericksburg, seized this opportunity and ordered his men back down the Valley. They arrived in the vicinity of Kearnstown at midday of March 23rd, where Ashby was skirmishing with the enemy. After a brief reconnaissance, Jackson determined that the enemy in his front was too strong to be forced, but believed it could be done by turning his right. He ordered Colonel Fulkerson, commanding the 3rd Brigade, to the left of the Winchester Turnpike, then sent his Assistant Adjutant General, Lieutenant Sandie Pendleton, to Gen. Garnett with instructions to move his command in support of Fulkerson. After moving some distance, Garnett, with his leading regiment, the 33rd Virginia, became separated from the rest of the 1st Brigade. Calculating that this regiment and the 3rd Brigade would amount to only 800 men, Garnett informed Fulkerson of his intention to withdraw his support. Fulkerson protested that this would necessitate his falling back and that he was waiting for orders. Garnett, however, withdrew the 33rd, abandoning Fulkerson, forcing his withdrawal.[18] Meanwhile, Gen. Jackson, at the front, desired to bring up infantry support for the artillery, which was already engaged. Major Frank B. Jones, who was detached from the 2nd Brigade to act as Jackson's aide-de-camp because of his familiarity with the terrain, was sent to order Garnett to bring up his regiments. As Garnett could not be found, Jackson was forced to give orders directly to Colonel Echols of the 27th Va. Since no other regiments of the 1st Brigade were in sight, Jackson directed Colonel Patten of the 21st Va. to move forward and support Col. Echols. These two regiments opened the infantry attack. Fulkerson and Garnett, hearing musketry fire, moved up, their troops intermingling. The 3rd Brigade formed the left flank, while Garnett held the center.

It was soon apparent that Colonel Ashby's information had been grossly incorrect; the Confederates were facing largely superior numbers. Concluding that a victory under these circumstances could not be expected, and safety consisted in holding their positions until dark then withdrawing, Jackson made plans accordingly. The key to an orderly retreat lay in the protection of an open space over which the infantry, in falling back, would pass over. Jackson left Garnett on the field as the second in command while he went to the rear to hurry reinforcements. Jackson believed Garnett could and would hold the center, despite the enemy's pressure and the Confederates' depleted ammunition. Garnett, however, felt the line could not be held, and without Jackson or Fulkerson's knowledge, gave the order for the Stonewall Brigade to fall back. This left the 3rd Brigade completely without protection.[19]

Jackson found Colonel Harman of the 5th Va., which constituted part of his reserves, and ordered him to move forward to support the troops engaged. After ordering up the 42nd and 48th regiments, Jackson returned to the front, where he found Col. Harman, by order of Gen. Garnett, filing his regiment to the woods in the rear of the open space. Jackson tried to remedy this retrograde movement, but it was too late. Upon entering the woods, Jackson discovered Garnett and his troops in full retreat. The General, who was unaware of Garnett's order to fall back, approached him and demanded to know why he was not attempting to rally his men. Garnett claimed he had done so until he was hoarse, possibly to sustain Gen. Jackson's impression that his men had retreated solely on their own. Jackson ordered Garnett to rally his men; together they held a number of the troops until the 42nd Va. had come up. After a brief stand, the retreat continued, Federal pursuit being halted due to darkness.

Daylight found the Valley Army moving slowly up the Shenandoah. Skirmishing continued, but the belief that Jackson had been heavily reinforced stayed the Union army from pressing their advantage,[20] allowing Jackson valuable time to reposition himself in a more defensible area.

On the morning after the battle, Jackson remarked, "Had I been able to bring up 2000 more men, I should have beaten them."[21] Even with the knowledge of Federal reserves, his opinion was firm. "If I had put the men engaged to flight, they would all have gone together."[22] Although he did not have 2000 men, his reserves numbered about 1000; had these men been brought into battle, Jackson's original determination to hold his line until dark would have been realized.

With his army near Newtown on March 24th, Jackson addressed a letter to Gen. J. E. Johnston, informing him of the battle, "As the enemy had been sending off troops from the district, and from what I could learn were still doing so, and knowing of your great desire to prevent it, and having a prospect of success, I engaged him yesterday about 3 PM near Winchester and fought until dusk, but his forces were so superior to mine that he repulsed me with the loss of valuable officers and men killed and wounded; but from the obstinacy with which our troops fought and from their advantageous position I am of the opinion that his loss was greater than mine in troops, but I lost one piece of artillery and three caissons.... This fight will probably delay, if not prevent, their leaving, and I hope will retain others."[23]

An order was sent out to the field officers to submit their reports on the battle of March 23rd. Colonel Allen of the 2nd Va., Lieutenant-Colonel Ronald of the 4th, Col. Harman of the 5th, Colonel Grigsby of the 27th, and Colonel Cummings of the 33rd submitted their reports to Captain R. J. Wingate, Gen. Garnett's Assistant Adjutant General, by March 27th; Gen. Garnett's own report was drafted March 30th.[24]

On April 1st, Lt. Pendleton was ordered to place Gen. Garnett under arrest for "improper conduct on the battle-field."[25] He was promptly sent up the Valley to Harrisonburg, and Brigadier-General Charles S. Winder was sent to replace him. Winder proved to be a very strict disciplinarian, which caused the men of the 1st Brigade to more deeply miss Garnett.

On April 7th, the charge and specifications preferred by Gen. Jackson against Richard Garnett were sent to Gen. Johnston's headquarters. They were as follows:

Charge: Neglect of Duty

Specification 1st: In this that Brig. Gen'l R. B. Garnett after having been directed by Gen'l T. J. Jackson to support Col. S. V. Fulkerson on the day of the battle near Kearnstown, Va. on the 23d of March, 1862, did after advancing with four regiments for some distance, leave three of them in rear, and continued to advance with only one, and finally notified Col. S. V. Fulkerson, he would move back his regiment, thus leaving him with out any support.

Specification 2d: In this that Brig. Gen'l. R. B. Garnett dis so separate himself from his command that he could not be found by Major F. B. Jones, who was directed to deliver an order to him.

Specification 3d: In this that Brig. Gen'l. R. B. Garnett did neglect to be with his leading regiment when it went into battle near Kearnstown, Va., on the 23d of March, 1862.

Specification 4th: In this that Brig. Gen'l. R.B. Garnett did so neglect to post his regiments properly in the battle near Kearnstown, Va., on the 23d of March, 1862, that they became mixed together when they should have been kept separate.

Specification 6th: In this that Brig. Gen'l. R. B. Garnett did during the battle near Kearnstown, Va., on the 23d of March, 1862, gave the order to fall back, when he should have encouraged his command to hold its position.

Specification 7th: In this that Brig. Gen'l. R. B. Garnett did during the battle near Kearnstown, Va., on the 23d of March, 1862, send an order by Major F. B. Jones, directing Col. W. H. Harman to retreat, notwithstanding his regiment had not yet been brought into action, and should have continued to advance on the enemy, as it was doing at the time, in obedience to orders from Major Gen'l. T. J. Jackson.

Signed, T. J. Jackson, Maj. Gen'l. P.A.C.S.

Witnesses:

Major F. B. Jones, 2d Regt. Va. Vols.

1st Lieut. A. S. Pendleton, AAG Maj. Gen'l. T. J. Jackson, P.A.C.S.

Col. John Echols, 27th Regt. Va. Vols.

Lieut. Col. J. R. Jones, 33d Regt. Va. Vols.[26]

No reply was forwarded, so a second copy was dispatched to Johnston's headquarters, and finally a third went to the Adjutant General's office.[27]

Embittered and angry about his arrest, Garnett was outspokenly hostile toward Gen. Jackson. In a letter to his father (who was then staying in Lexington, Virginia, at the home of his daughter, the wife of Professor Williamson of the Virginia Military Institute), he told of his arrest. Sandie Pendleton's mother, also of Lexington, in writing to her husband, General Pendleton, General Lee's Chief of Artillery, "Gen. Garnett, I am afraid, has the tendency of his family to free speech. He uses no measured language about Jackson's 'Winchester Folly', and says nothing but the timidity of the enemy prevented a total defeat."[28] (On April 8th, Gen. Johnston published a General Order expressing "his own sense of the admirable conduct of Major-General Jackson and his division, by which they fully earned the high reward bestowed by Congress: resolved by the Congress of the Confederate States of America That the thanks of Congress are due, and they are hereby tendered, to Maj. Gen. T. J. Jackson and the officers and men under his command for their gallant and meritorious service in the successful engagement with a greatly superior force of the enemy, near Kearnstown, Frederick County, Virginia, on the 23d day of March, 1862).[29] "Now, of course, we are all conjecturing what it can be for, and as

Bob Nelson* says he heard that Gen. Garnett behaved well on the battlefield, we suppose he may have said something 'improper' of the general to the men, though we have not heard one word of the sort, but his letter is an imprudent one, used as it has been, for the poor father read it today on the streets."[30] She adds that "William Garnett says 'Jackson is mad' (and) it seems to be a real comfort to him to think so."[31] Thus in Jackson's home town, Garnett's bitter letter and the public airing it received, both on the streets of Lexington and in the homes of people with whom Jackson was intimately associated, helped carry on this one-sided "crusade."

While in Harrisonburg, Garnett also wrote to his kinsman, R. M. T. Hunter, who had been Speaker of the House in the U.S. Congress, and was now a powerful figure in the C.S. Senate, having served as Secretary of State for the Confederacy, and who was also an associate of R. E. Lee's.[32] In answer to a letter from Hunter, Garnett states he is "very grateful...for your proffered aid."[33] He urgently requests "as speedy a trial as possible for by no other means could I be fully vindicated." He later states "the gravest fault lay at the door of the *Comdg. General,*" thus conveying his bias against Jackson to another quarter, a relative highly placed in the government. Garnett also enclosed his report of the battle of Kearnstown, even though he himself admits "I not only have no right to make it public, but might perhaps be censured for putting it into private hands." He closes feeling "well satisfied that you will do all that could be desired of a relative and friend..." In a postscript to the letter, Garnett writes that his AAG, Capt. Wingate, learned from Hunter "that Genl. Jackson had made a written complaint against me to the War Dept. last Jany..." and begs him to "procure a copy of this communication, which would be of great service to me... Forward if you please to this place." Garnett labels this complaint as a malicious and underhanded covert attack.[34] Not only is this description extreme, but the fact that Garnett was in a position to have procured for him confidential unrelated War Department communications seems inconceivable; yet it is only the start of a repeating pattern.

In a letter of April 13th, Lt. Pendleton wrote his mother that he had no advance knowledge of the arrest, but the reason that Garnett was deprived of active command was the impossibility of holding a court-martial at a time when the army was constantly moving.[35] It is clear there was no vindictiveness involved in Garnett's detention, particularly since he was permitted freedom to investigate the capital's silence concerning the charges.[36] Sandie continued that the arrest was necessary and a "most sorrowful affair." Garnett's decision to order back the 5th before it had actually entered the fight had lost the day. He adds that it was "no want of gallantry, only a sad, sad blunder" on Garnett's part.[37]

Having waited some time in Harrisonburg, Garnett addressed a communication to Jackson's headquarters, requesting information of his trial. He was informed that the charges had been sent out, but nothing further was known. On his application, permission was then granted for him to go to Richmond and endeavor to have his case given attention.

On May 6th, Garnett, in Richmond, addressed a letter to General Samuel Cooper, Adjutant and Inspector General, waiving all consideration of rank in the composition of the court if one could be convened. If this was impractical, he requests that the charges be suspended, and he be reassigned to active duty.[38] As was apparent, a trial at this time, with Jackson and his command actively engaged in operations, was impossible. In fact, this was the reason he was sent away from the army. It seems more likely that Garnett believed that the charge would be suspended, than to think a court would be convened. On the same day, in a letter to R. E. Lee, Cooper acknowledged that "it would be no easy matter to bring this case to trial at this particular time," and left it to Lee's discretion.[39] Cooper also enclosed a copy of the charge and a letter from Gen. Jackson, in which he expressed the opinion that "I have only to say that I have no desire to see the case pressed any further but that I regard Gen'l. Garnett as so incompetent a Brigade Commander that, instead of building up a Brigade, a good one, if turned over to him, would actually deteriorate under his command."[40] Lee's immediate response was "I see no possibility of bringing this case to trial as much as I feel for Gen'l. Garnett. If Gen'l. Jackson had closed his letter with the expression of his wish not to press the matter further, I should recommend it be dropped. But his closing remarks make it necessary as an act of justice to Gen'l. Garnett and the service to bring him to trial."[41] Lee mentally dismissed the charge of Neglect of Duty, yet it was under this charge that Garnett would be brought to trial. "As an act of justice," *Garnett* must be permitted to bring *Jackson* to trial; the defense must be against Jackson's opinions, and by subsequently proving them wrong, Garnett and "the service" would benefit. Thus, Lee appears to be

*Robert Nelson was an Episcopal clergyman, who occasionally acted as a supply minister of the Grace Episcopal Church in Lexington during the early days of the war.

sanctioning a direct attack against Jackson himself, rather than a defense of Garnett's behavior during battle.

By June 20th, Garnett was in Staunton drafting a sixteen page defense. It disclosed his cognizance of the contents of both Gen. Jackson's letter (mentioned above), and the sympathy held for him by President Davis and Gen. Lee, as revealed by quotations from private communiques within the text of his defense. Also quoted "on good authority" was a note from Gen. Jackson to Gen. Johnston written after Kearnstown regarding the battle.[42] Garnett's text was sent to Richmond on June 30th.

Five days earlier, on the 25th of June, the Secretary of War issued Special Orders No. 146, in which he suspends the arrest of Brig. Gen. Richard B. Garnett, and orders him to report for temporary duty with Major-General D. H. Hill.[43] It had previously been decided that Hill would be assigned to Gen. Jackson; now, not only had the major-general's charges been almost virtually disregarded, but Garnett would be reassigned to a command directly connected with Jackson. This reinstatement was ostensibly to fill a command position in Hill's Division during the forthcoming crisis around Richmond. At that time, before any fighting had taken place, it would have been more logical to keep Garnett unassigned until there was a specific need for his services. Commencement of these actions was June 26th, as Gen. Lee knew, having personally outlined them to his major-generals on June 23rd. Yet the Special Order was not issued until the 25th, which would provide Garnett with no time to become familiar with the situation and the men he was to command under extremely trying circumstances. This task would be next to impossible when Garnett's supposed whereabouts are taken into consi-

The Battle of Kearnstown

On March 23, 1862, the battle of Kearnstown was fought between General T. J. Jackson's Valley Army and the Federal forces under General James Shields. The audacity on the part of the Confederates to face an enemy far superior in numbers to their own led Shields to write in his Official Report: "Though the battle had been won, still I could not believe that Jackson would have hazarded a decisive engagement at such a distance from the main body without expecting reinforcements. So, to be prepared for such a contingency, I set to work during the night to bring together all the troops within my reach.... General Banks, hearing of our engagement on his way to Washington, halted at Harper's Ferry, and with remarkable promptitude and sagacity ordered back William's whole division...."[1] These 9000 men had to retrace their steps and along with every available man in Shield's rear, hurried by forced marches back to Winchester.

Jackson himself never considered Kearnstown a defeat. In his official report of March 29th, he wrote, "Though Winchester was not recovered, yet the more important object for the present, that of calling back troops that were leaving the Valley, and thus preventing a junction of Banks' command with other forces, was accomplished, in addition to his heavy loss in killed and wounded. Under these circumstances I feel justified in saying that, though the field is in possession of the enemy, yet the most essential fruits of the battle are ours."[2] When Kearnstown would come up in conversation over the following months, he would refer to it with pride. In writing to Mrs. Jackson on April 11th, he told her, "Although I was repulsed in the attempt to recover Winchester, yet the enemy's loss appears to have been three times that of ours. In addition to this, the great object which required me to follow up the enemy, as he fell back from Strasburg, seems to have been accomplished very thoroughly. I am well satisfied with the result. Congress has passed a vote of thanks, and General Johnston has issued a very grarifying order upon the subject, one which will have a fine effect upon my command. The great object to be acquired by the battle demanded time to make known its accomplishments."[3]

He was right. The re-concentration of Federal forces in the Valley and Lincoln's maneuvering of troops for the defense of Washington led to the downfall of McClellan's move on Richmond, and gave the Confederate capital a reprieve. Jackson and his Valley Army had a right to be proud.

[1] Official Records, Vol. XII, pt. 1, p. 341, 342.

[2] O.R., Vol. XII, pt. 1, p. 383, 384.

[3] "Memoirs of Stonewall Jackson" by Mary Anna Jackson, Courier — Journal Job Printing Co., Louisville, Ky., 1895, p. 248, 249.

deration: the reinstatement was forwarded to Gordonsville.[44] To suppose Gen. Garnett could receive these orders and return to Richmond in time to act upon them would be unreasonable. Therefore, the reinstatement appears to be more of a gesture than a serious effort at providing the army with an officer.

Gen. Garnett arrived in Richmond on July 11th, having "learned from private sources a day or two since that an order had been issued," restoring him to duty.[45] Although still officially under arrest when he left Richmond, the War Dept. was not made aware of his location, sending several messengers to Gordonsville with the news of his reinstatement. Garnett only discovered the contents of the Special Orders through personal informants, thus appearing in the capital too late to participate in the active command he was so eager to return to. On July 12th, Garnett addressed a communication to Gen. Cooper, in which he "urgently request(s) that as soon as the exigencies of the service will admit of it, that an investigation may be ordered...with regard to all the allegations, both general and special, made against me by Major General T. J. Jackson."[46]

It was arranged that Gen. Garnett's court-martial be conducted while the army recovered near Richmond. It is difficult to believe that any tribunal sitting in judgement on Garnett would not have been swayed by the knowledge of his public reinstatement. The Sec. of War and the Commanding General of the Army officially declared their faith in Garnett's innocence by assigning him to active duty just before the Seven Days. After such a testimonial from Lee, it would have proved difficult to bring a verdict of guilty. Another impediment to reaching a just decision was the untimely deaths or unavailibility of so many witnesses: Cols. S. V. Fulkerson and J. W. Allen were killed at Gaines Mill on June 27th, and Col. A. J. Grigsby was wounded. Cols. Echols and Cummings had both been wounded at Kearnstown, and had not yet returned to duty; Col. Harman had retired from the service. Finally, Maj. Frank Jones, who was considered to be the most important witness against Garnett, had been mortally wounded. Despite these serious drawbacks, the trial was scheduled. But on the day it was to begin, Gen. Lee ordered Jackson and his army to Gordonsville to stave off the threat of the massing Union forces: again, the trial had to be postponed.

On July 28th, the order was given for a mobile court to follow the army and convene at any time practical.[47] The presence of the court at Gordonsville pressured Jackson into making time for this trial, despite the fact that all of his attention should have been directed to formulating military strategy. Even though he had expressed willingness to participate in the trial, as commanding general in this new field of operation, he should have been freed from this added and unnecessary responsibility. The trial might better have been postponed until winter, but the authorities in Richmond yielded to Garnett's promptings, and the court convened on August 6th.

In his opening testimony, Gen. Jackson touched upon the various specifications.[48] In reference to the 1st and 2nd, he stated: "Gen'l. Garnett, after having moved with 4 of his Regts. some distance, became separated from 3 of them, and as I could not find him, it became necessary to give orders directly to regimental commanders to move their commands in the required direction." Of Spec. 3 and 4, he states: "I desired to keep the Infy. in advance (of the artillery) — the only Regt. I could find of Gen'l. Garnett's brigade was Col. Echols' (27th Va. Vols.) — which I directed to move forward, and keep in advance. Giving the order to Col. Echols — as no other Regt. was in sight (but the) 21st Va., I directed its commander, Col. Patten to move forward and support Col. Echols. I did not know where to find Gen'l. Garnett. He was not with his leading Regt." In speaking of the 6th and 7th Spec., he stated: "The reserve was ordered up and about the time our Infy. began to fall back, through the open space referred to, Col. W. H. Harman, Comdg. the 5th Regt. Va. Vols., which had not yet been put into action, came up and reported to me with his Regt. — I directed him to move forward and to support the troops engaged — soon after I observed him filing his Regt. to the rear — I moved towards him for the purpose of remedying this retrograde movement, as far as practical, but before it could be well effected, this Regt. was in the woods to the rear of the open space. When I entered the woods I saw Gen'l. Garnett moving to the rear, accompanied by his troops — He did not manifest any concern, and was not using any efforts to rally his men — I rode up to him and asked him why he did not rally his men, or try to do it — He told me he had done so, till he was hoarse. He then endeavored to rally his men. Part of them made a stand.... here in the woods for a few minutes until the 42d Regt. came up — could the troops have held their position five minutes longer before falling back there is good reason to believe, that we would not have been forced from that part of the hill in rear of the open space, as it would have given time for the 5th and 42d Regts. to have gotten into position before they were attacked by the enemy,

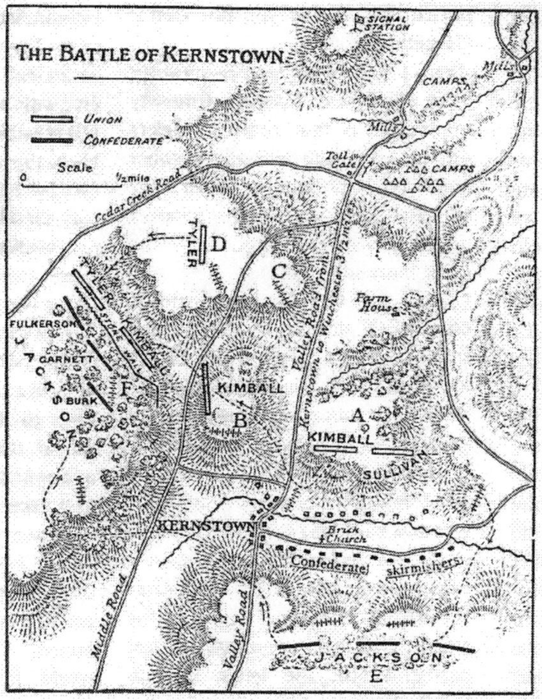

MAP OF THE BATTLE OF KERNSTOWN, VA., MARCH 23, 1862.

Based upon the maps in the "Official Records," Vol. XII., Part I., pp. 362-365. A represents the first position of Kimball's and Sullivan's brigades on the morning of March 23d. Sullivan remained to hold the Union left, while Kimball moved to the position at B, and finally to the main battle-field, F (evening of March 23d), where he joined Tyler, who had previously been in position first at C, and then at D, whence he advanced to oppose Stonewall Jackson in his flanking position at F, to which Jackson had marched by wood roads from his first position at E.

From "BATTLES AND LEADERS OF THE CIVIL WAR"

and possibly the last Regt. cut completely (off) which was coming up to take part in the engagement." Gen. Jackson further elaborates: "Gen'l. Garnett states in his official report that there was some confusion among the troops of his command. It was especially important that the diff. Reg. should have been kept separate, so as not to have become mixed with each other — when we were so much out numbered. Gen'l. Garnett further states in his report that he gave the order to fall back. It was important he should have held his position as long as possible, unless directed by myself to fall back — when I speak of the troops under Gen'l. Garnett's comd., I mean Infy., when he was, as he was, the 2nd officer in command on the field and I was not with him."

After this opening summary, Gen. Garnett commenced his questioning.[49]

Question by accused: What was your plan of the battle fought near Kearnstown, Va., on the 23d of March, 1862.

Answer by Gen. Jackson: First, to defeat the enemy by gaining heights on his right, which commanded his position, pressing on towards Winchester, then turning his right and getting in his rear.

Qs. by ac.: Did you communicate this plan to me before or during the action.

Ans.: I did not to my recollection.

Qs. by ac.: What was the character of the ground passed over by the troops previous to and during the action.

Ans.: The ground was what I called level but slightly undulating until we reached the hill upon which our troops fought — It was partly wooded and partly cleared. The hills also had some wood-land and some cleared land.

Qs. by ac.: Could you from the hills and woods, which obstructed the view, judge well of the strength and positions of the enemy before the battle commenced?

Ans.: From what I could see in front, I regarded his position in front too strong to be forced with my command, but believed it could be done by turning his right as I attempted. There were no troops of the enemy so far as I know on the hill where our force fought at the time we commenced the movement to occupy the hill. I saw the enemy send troops over there after we reached the hill. The enemy could have had troops concealed when the infantry engagement commenced without my knowing it from the position I occupied... When the order was given to occupy the hill, I deemed it necessary to bring on the engagement that evening, lest the enemy should bring up re-enforcements during the night and the time would not admit of delay, for the purpose of making much reconnaissance of the position. My information rec'd from Col. Ashby who had been in front of the enemy since the previous day was to the effect that the enemys Infy. force did not exceed 3 Regts. — so little force had been shown by the enemy, the previous day, that Col. Ashby, according to his report could with the aid of one regt. of Infy. had entered Winchester, that is the regiment in addition to his own command.

Qs. by ac.: Did you not say to me in the forenoon of the 23d of March, 1862, that you would take charge of the artillery belonging to the 1st Brigade, and did you not direct its movements during the day.

Ans.: I don't remember stating to Gen'l. Garnett that I would take charge of it, but I did give specific attention to it.

Qs. by ac.: When Col. Fulkerson first moved off to the left towards Winchester, did you or did you not direct me to send *one* regiment to support him?

Ans.: I am well satisfied that I did not.

Qs. by ac.: You state that you saw Gen'l. Garnett going to the rear accompanied by his troops — Had not a large number of his command passed you, before you saw Gen'l. Garnett.

Ans.: I do not remember seeing a great many pass through before. My impression is that some stragglers did pass through and the number may have been large — When I say he was accompanied by his troops, I mean by what appeared to be the main body of them.

Qs. by ac.: Did you not write to Gen'l. J. E. Johnston or one of his staff officers after the battle of Kearnstown, in substance, that you had been deceived with regard to the numbers of the enemy on that occasion — and had been obliged to retire before a superior force, but that you had done so in good order?

Ans.: I remember writing to Gen'l. Johnston, or one of his staff after the battle, that I had attacked the enemy and been repulsed. I feel pretty certain the letter was to Gen'l. Johnston himself, and in it I explained the reasons which induced me to make the attack — I do not remember saying anything about the order in which my troops fell back, but if I did, it expressed my opinion of the others — I was deceived as to the enemys strength and I am of the opinion that I so stated in my letter to Gen'l. Johnston (see pages 8 and 9 of this article for the specifics of this letter) — I think no blame was to be attached to Col. Ashby — as to the force he reported where he met the previous day was small and that the federal force had been leaving Winchester and had been moving towards (the) mts.

Qs. by ac.: Did any troops in the action besides those of Gen'l. Garnett fall back before being ordered by yourself.

Ans.: The mass of troops engaged fell back, and none of them (Infy.) by my order.... My foremost desire was to hold the position on the hill till after dark — Every other officer except Gen'l. Garnett who ordered his troops to fall back was, so far as I could ascertain, justified in doing so — From what I saw, the troops with Gen'l. Garnett were the first to give way. Had my other troops (not) fallen back after Gen'l. Garnett fell back, they must have been captured.

Qs. by ac.: State if you know.... why Gen'l. Garnett should be with his leading reg. rather than any other.

Ans.: I conceive it to be the duty of an officer to be with his leading troops, when they go into action — in order that he may post them as they go into action — and superintend them during the action — and watch the enemys movements and take every advantage that circumstances may enable him to do —

The testimony of Captain Pendleton began August 7th.[50] In reference to the first specification, he stated: "After moving into the wood I delivered the orders to Gen'l. Garnett to move on with his command to support Col. F. (Fulkerson), who was forming his Brigade on the left....on reaching the front, where Gen'l J. (Jackson) was, he directed me to find Gen'l. Garnett.... In searching for Gen'l. Garnett I spoke to Col. Allen of the 2d, Col. Ronald of the 4th and Lt. Col. Grigsby of the 27th Regts. — I thus know that these regiments were left there, but why I do not know." In reference to the 2nd Spec., he stated: "When in the woods where the 2d, 4th, 27th Regts. were — I was directed by Gen'l. Jackson, who was also there, to try to find Gen'l. Garnett, and direct him to move those three regts., as neither Lt. Junkin or Maj. Jones had been able to find him....I went towards the 33d Va. Vols. under Col. Cummings in search of Gen'l. Garnett. On reaching this regt. I did not see the Gen'l. and I immediately returned and informed Gen'l. Jackson I could not find him. I will state before going after the 33d Regt., I made diligent search and frequent inquiries as to Gen'l. G's (Garnett) whereabouts without gaining any definate information.... I will state in connection with Spec. 3d that the 1st regt. of Gen'l. G's (Garnett) Brigade that I saw become engaged was the 33d Va. and Gen'l. Garnett was not with that. That was in the beginning of the Infy. engagement, when the enemys skirmishers first began to appear in the wood beyond the open space on the ridge where the arty. was posted.... The 33d Regt. advanced as the enemy began to appear in the farther piece of woods. I will state I do not know where the other regts. of Gen'l. Garnett's Brigade were at this time as I did not see any of them advancing in connection with (the) 33d.

Qs. by ac.: Where was the most advance arty., and did it retire before or after the Infy.

Ans.: Two pieces of McL (McLaughlin) Battery and 4 of Waters'. They were on the right and in advance of the enemy Infy. while our Infy. did not move on to support them....I do not mean by their retiring they intended to leave the field, but to take up a new position, but on reaching this position found the Infy. retiring also, and had to move on to the rear—"

The testimony was interrupted and the trial adjourned as Gen. Jackson received information concerning the positioning of Union troops near Gordonsville which required his immediate attention. Within 48 hours, the Confederate forces were engaged in the battle of Cedar Run. There was no opportunity to reconvene the court, and no existing record indicates a verdict was ever rendered.

On September 5, 1862, pursuant to his earlier inclinations, Gen. Lee ordered Garnett's release from arrest, and instructed him to report to Gen. Longstreet for assignment to command.[51] Thus, a man who had 3 months experience as a brigadier, and who, during that time, had gained the reputation of pampering his brigade, disobeying his superior's orders, and who had been charged with Neglect of Duty, was rewarded with a new brigade.

For 120 years, since Garnett's letter was read aloud in the streets of Lexington, to contemporary historians, Garnett has invariably been portrayed as the innocent victim of Jackson's vindictiveness. Yet it was Garnett, not Jackson, who had relatives in high places to pursue his case; who had unrelated documents illegally sent to him; and who, without one word of defense from himself, had the sympathy and support of Gen. Lee and Pres. Davis. It is clear that Garnett, not Jackson, carried a grudge, and his attack has been relentlessly and senselessly pursued to this day. ★

Concurrent Events

☐ MEXICAN AFFAIRS

New Orleans papers contain details of the Mexican news of August 22, 1854:

Santa Anna had issued a decree enabling the religious orders to avail themselves of the civil tribunals to enforce the monks and clergy to an adherence of their vows.

☐ FINANCIAL AND COMMERCIAL

There is nothing new in the money market, and no failures are reported. The foreign exchange market closed dull, but firm.

June 30, 1861

☐ THE YEAR 2056

Scene — Parlor in the house of an elderly gent. in New York. Old gent. telegraphs to the waiter, who ascends in a baloon.

Old Gent.— John, fly over to South America and tell Mr. Johnson I will be happy to see him sup with me. Never mind your coat now; go.

John leaves, and at the end of five minutes returns.

John — Mr. Johnson says he will come; he has got to go to the North Pole for a moment, and then he will be here.

Old Gent.— Very well, John; now start the machine for setting the table, and telegraph to my wife's room and tell her that Mr. Johnson is coming, then brush up my baloon, for I have an engagement in London at 12 o'clock.

John flies off to execute his order, and the old gentleman runs over to the West Indies for a moment to get a fresh orange.

(From the Saturday Evening Post, New York, May 17, 1856.)

☐ NAPLES AND SICILY

The Neapolitans had evacuated Messina, Melazzo and Syracuse, and were being transported to Naples.

It was reported that Garibaldi was preparing for a descent on the mainland.

Revolutionary demonstrations had taken place at Naples.

August 6, 1860

General Garnett's Official Report

Head. Quarters 1st Brigade Va. Vols.
Camp Buchanan
Near Mount Jackson, Va.
March 30th 1862

Sir:

I have the honor to submit the following report of the operations of the 1st Brigade in an engagement with the Federal forces near Kearnstown, in the neighborhood of Winchester, on the 23*rd* of March, 1862. The Brigade comprised the following troops. The 2*nd*, 4*th*, 5*th*, 27*th* and 33*rd* Regt. Virginia Volunteers, commanded respectively by Col. Allen, Lieut. Col. Ronald & Cols. Harman, Echols, & Cummings, and three batteries: McLaughlin, Carpenter's and Waters.—

This command left Cedar Creek some 15 miles from Winchester by the Strasburg Turnpike, on the morning of the 23*rd* inst. — & after a march of 12 miles found Col. Ashby's Artillery engaged with that of the enemy's.

In accordance with instructions from the Major Genl. Commanding, the Infantry was placed in a woods a quarter of a mile west of the turnpike, and the Artillery in an intervening field.

After remaining a short time in this position, I was directed to place a regiment in the rear of a stone wall running perpendicular to the turnpike, and selected the 5*th* for this position. Soon after another Regt. was ordered to follow Col. Fulkerson's Brigade, which had been moved off in a westernly direction, and I detached the 2*nd* for this duty.

When the 2*nd* Regt. had proceeded some considerable distance, the remaining Regts. (4*th*, 27*th*, 33*rd*) were ordered to follow, after marching about half a mile through fields & open woods, the troops were temporarily halted near the edge of a body of timber. While in this position, the enemy opened upon them with their Artillery.

Here, Col. Fulkerson's Brigade, consisting of about 500 men, was formed in close column by division, and sent to the front across an open valley several hundred yards in width commanded perfectly by the enemy's Batteries, stationed on a wooded eminence opposite.—

I here rec'd an order thru Lieut. Junkin, aid to the Commanding Genl., to support Col. Fulkerson's command. The 33*rd* Regt. Col. Cummings was formed by divisions and follow the course taken by the command above mentioned, being some two hundred yards in rear of it.

Col. Fulkerson's troops bore to the left of the enemy's Batteries in the direction of a body of heavy timber — I deem it just to say here that Col. Fulkerson's command crossed this valley in the most perfect order, although subjected to a most terrible fire from the Federal Artillery.

When the 33*rd* Regt. had proceeded about half way across this valley, perceiving that the enemy had accurately obtained our range, & that they had been reinforced by several pieces of Artillery, and a large Infantry force making its appearance in front of their Batteries, and feeling confident that if the Regt. followed its present line of march, it would suffer severely. I ordered it to move obliquely to the left, so as to gain the protection of a ridge running parallel to the one occupied by the Federal Artillery.

After reaching the other edge of the slope, & finding that the Regiments of the Brigade had not come up owing, no doubt to a misapprehension of orders. (which were sometimes sent thru me, and at other times transmitted directly to the Regimental Commanders) I determined to take back 33*rd* Regt. to the main force. I sent my aid, Lieut. Williams, at the suggestion of Col. Cummings of the 33*rd*, to inquire after the command of Col. Fulkerson's of which I had not heard since its entrance into the heavy piece of timber, mentioned previously.

Lieut. Williams not returning, Col. Cummings & myself proceeded to the supposed positions of Col. F, whom we met in the rear of the woods spoken of. When I informed him of my position, and of my intention of returning with the 33*rd* to the main force of my Brigade, as nothing could be accomplished with our small force where we were. Col. F. replied that he was waiting orders, but eventually determined to withdraw his force in the same direction of mine.

Returning with the 33*rd* It was found that McLaughlin's Battery was moving along the ridge on the west side of which the 33*rd* & Col. F.'s troops were. Here these troops were halted. McLaughlin's at this point — unlimbering his guns and opening fire on the enemy's Batteries; which fire was briskly returned by them.

At this time, the 2*nd*, 4*th*, & 27*th* Regts. came up & also Carpenter's and Waters' Batteries. The 27*th* being ordered to advance with & support Carpenter's Battery, did so.

During the engagement of the Artillery, our troops were exposed to the shot & shell of the enemy, which were constantly passing over & bursting among them for over an hour.

An order was rec'd. betweem 4 & 5 o'clock P.M. for the infantry to move along the base of the ridge upon which the Artillery was moving & to support it. At this time I was informed by Capt. McLaughlin that a large body of the Federal Infantry had made its appearance on the right — of the turnpike. I proceeded to the crest of the hill, from which I could see two heavy columns moving in the direction of Kearnstown.

Just then Lieut. Col. Grigsby of the 27*th* reported to me that his Regt. was in advance, and separated from the rest of the Brigade by troops of other commands. I was proceeding in the direction of the 27*th* when dropping shots of musketry were heard in front, soon followed by general heavy discharges of small arms. The firing was so severe and continuous, I did not believe the Regts. (27*th* & 21*st*) could hold their positions. I accordingly commenced forming the troops in the rear in line to receive the enemy as he advanced, but in consequence of being applied to by Col. Grigsby & Major Paxton (27*th* Reg.) for reinforcement, the troops were ordered to advance, which they did with alacrity and enthusiasm.

The Regiments of the different Brigades had become somewhat intermixed, owing to the movements to support the Batteries, and it was impossible to form the different Regt. of the 1*st* Brigade regularly in line. The troops, as they came upon the fields of action entered into the engagement with utmost gallantry. Field Staff & Company officers cheering on their men, and mingling with them in the thickest of the fight — Non-commissioned officers & privates vied with each other how best they could sustain the well-earned reputation of the Brigade, fighting with the utmost determination & bravery. So heavy and destructive was the fire of all the troops engaged the enemy was driven back from his position on our left, and front, and his fire then at one time almost silenced, the Federals retreating to their reserves.

After sustaining our position for over an hour, in the face of a much superior force, which were being constantly reinforced: our ammunition in many instances being entirely exhausted: officers & men rapidly falling. A Cavalry demonstration also being made on our left, I gave the order for the troops to fall back.

While retiring, about a quarter of a mile from the scene of action, I met Col. Harman of the 5*th* Regt., who informed me he had been ordered to the front by the Commanding Genl., and now reported to me for orders. I directed him to form his Regt. in line below the crest of one of the spurs of the ridge running parallel with the late line of battle, which order he executed with the utmost promptness, highly creditable to the Regt. Knowing as they did that they were about to engage a vastly superior force, before which our left was retiring, I then proceeded to the rear to aid in rallying the troops. Upon my return I found the 5*th* hotly engaged with the enemy, gallantly supported on the right by a portion of Col. Burks Brigade (the 42 Va.) commanded by him in person. This small but chivalric band met & held in check the advancing columns of the enemy, covering the retreat of our troops on the left, & our Batteries on the right. Capt. McLaughlin's Battery in spite of the near approach of the enemy's infantry bravely retained its position sufficiently long to throw in several discharges of canister at a distance of only 150 yards, inflicting upon them heavy losses, & throwing them into confusion. Night now coming on, the troops last engaged being overcome by superior numbers, they were forced to withdraw. I accompanied the 5*th* Regt., falling back to our wagons beyond Newtown, on the Strasburg road.

I regret that I can say but little from personal observation with regard to the part taken by the Artillery of the 1*st* Brigade during the engagement having been entirely under the direction of the Commanding Genl., but I feel confident from the reports (herewith transmitted) of Commanders of Batteries' that they did highly effective service, and that officers & men nobly discharged their duty.

I find it difficult to point to individual instances of gallantry, where so many were conspicuous. The field officers being familiar to me, and generally mounted more particularly attracted my attention, and it is with profound pleasure & satisfaction that I state they were found in the midst of their men in the most exposed positions encouraging them both by words & example. Col. Echols of the *27th* whilst engaged in the early part of the action was severely wounded & had to be borne from the field, in consequence of the loss of this estimable officer the command of the *27th* devolved on Lt. Col. Grigsby who led it nobly during the rest of the day. Lt. Col. Ronald of the *4th* was early disabled by being thrown from his horse, & Major Pendleton effectively commanded the Regiment during the battle.

That officers, Non-Commd. officers & men did their duty — the heavy list of casualties suffered by the Brigade gives ample evidence. The reported losses of the enemy, & their failure to even harass our retreat, proves conclusively that the chivalric bearing of the command produced & exaggerated impression of the strength of our forces, when it is remembered that these troops had made a forced march of 25 miles on the previous day, carrying their knapsacks, over a heavy road; had marched 12 or 14 miles on the *23rd* & then engaged an enemy far superior in point of equipment & numbers inflicting on them with the able assistance of the other troops a heavy loss, too much praise cannot be awarded the gallant men of this Brigade (scarcely 1400 strong) who have thus amply sustained their former reputation won on other fields.

I respectfully refer you to the names of the officers, non commd. officers & privates who are particularly mentioned, to the reports of the Regimental Commanders & Commanders of Batteries. Accompanying this report will be found tabular list of the Killed, Wounded & Missing of the action of the *23rd*.

I desire to express my thanks to Major Mercer, Brigade QrMaster, Capt. Wingate, A.A.G. Capt. Sexton, Brigade Commissary, Lieut. O.C. Williams — A.D.C. & Mr. Elliott Johnston of Md., my volunteer aid, for their gallant (   ) valuable services — Capt. Wingate, Lieut. Williams & Mr. Johnston rendered gallant & efficient service in the field. Both Capt. Wingate & Mr. Johnston were with me during most of the battle, Capt. W. — horse being wounded during the Infantry engagement, occasioned our separation during the latter part of the day. I have to express my deep regret at the loss of Lieut. Williams, an officer of rare courage & promise who was captured early in the day while conveying a message to Col. Fulkerson, the circumstances attending his capture have not yet transpired. Capt. Wingate had his horse wounded during the musketry firing & thus became separated from me. He rendered great service in bringing up reinforcement when the 21 & 27 were needed. Mr. Johnston accompanied me during the entire day, except when engaged carrying orders, or rallying & encouraging troops. His conduct during the battle was characterized by the most dauntless bravery. His inspiring & chivalrous bearing challenged the admiration of all who saw him. I most earnestly recommend him promotion in the Confederate Service. It is also with peculiar satisfaction that I acknowledge my obligation to Lieut. Boswell of the Engineers Corps., & attached to the staff of the Commanding Genl.. He reported to me for duty early on the *23rd* & rendered me valuable assistance. He was particularly conspicuous while the Infantry was engaged, (and bravely rendered service whenever it was required). (He was frequently exposed to the severe fire of the enemy).

I omitted to mention in the proper place the soldierlike bearing of the *33rd* Regt. which I accompanied whilst crossing the Valley under a severe and trying fire from the enemy's batteries. Its steadiness on that trying occasion is deserving of the highest commendation.

In conclusion, I claim the heartfelt gratitude of the Country for the brave men have fallen, and the survivors who bravely provided their lives for our Cause.

It is eminently due to the Medical Staff of the *1st* Brigade to acknowledge the humane & fearless manner with which they discharged their duties on the Battlefield; ministering to our wounded wherever found, and after under the murderous fire of the enemy. ★

The original of Brig. Gen'l R. B. Garnett's official report is in the R. B. Garnett papers, Museum of the Confederacy, Richmond.

Concurrent Events

☐ FROM PANAMA

By the Panama Railroad Co's steamer *Guatemala*, we have recent news from Central America. An attempt at a revolution has been set on foot in Guatemala, but was discovered in time to prevent any serious results. It appears considerable distress exists among the laboring classes at present, and a party of them set on foot a plot to destroy the city by fire, probably for the purpose of pillage, and with this object they attempted to explode the powder magazine in the Government House, which contained some 60,000 pounds of powder. Fortunately, the fire was discovered just in time to prevent an explosion. It has not transpired who the leader of the plot was, but several hundred carpenters have been arrested as being engaged in it.

June 2, 1862

☐ The coffee crop of Costa Rica this year now appears to be 106,000 quintals, being about 30,000 over that of last year.

June 2, 1862

The Revolution in Printing Offices

(From the Cincinnati Columbian)

The recent Printer's strike in this city has revived in a practical way the attention of the public mind the importance of opening new avenues of employment for women. The "strikers" themselves opened wide the door to one avenue which is now being rapidly occupied. The movement of last winter in behalf of laboring women, prepared the public mind in a very considerable degree for hearty sympathy with the present determination of several of our leading publishers and Book and Job Printers to afford women an opportunity of earning a pecuniary independence by typesetting, and thereby reducing the oppressive competition among the employments which have heretofore been the reliance of working women.

A few months ago a movement in New York on the part of a typographical Union opened several printing offices to women, and one daily paper, the *Day Book,* is now set up by them, and the office has been converted into a school for teaching the Art preservative of arts, and the girls who graduate from it readily find work enough to do.

A few weeks ago, a Typographical Union in Pittsburgh, became obstropolous, and men were discharged and women put in their places, or at least in the places of such as seemed determined to have a share in the management of the business of the offices, because they were being paid large wages for work done in them.

Last week the Typographical Union of Cincinnati became dictatorial. Publishers resolved to be independent of the union and its members, and now women are setting type at cases where the strikers were makaing from $12 to $25 per week.

In Boston and New Haven women have been employed as composers in the offices of weekly papers for some time past, and their employers express themselves well satisfied. They render entire satisfaction also in New York and Pittsburgh, and wherever else they have been engaged....

But it is not a new thing for women to set type, nor is it true that merely as an outgrowth of any "women's rights movement" of the present day have women become Publishers and Editors of Newspapers. The women's rights agitation gave them aid and comfort no doubt, but they had ancient precedent for their assumption of Editorial control.

We are completing our arrangements to avail ourselves of the labor of female printers, and we are quite confident that typesetting is to become one of the regular employments of women, in all parts of our country, and must soon be encouraged in the great cities of Europe to the very necessary help of the most oppressed and most needful of all working people, the feeble seamstress. ★

(1853)

Projectiles

Bullets: The object of small-arms is to attain animate objects; their projectiles are, therefore, made of lead, and are generally known as *bullets.* They are both round and oblong; but in consequence of the great improvements that have been made of late, in adapting the principle of the rifle to small arms, the oblong ball is now very generally used in all military services, the round bullet being chiefly retained for use in case shot.

OBLONG

Round bullets are denominated by the number contained in a pound; and this method of denomination is often used to express the calibre of small arms; as, for instance, the calibre of the old musket was 17 to the pound, and the rifle was 32. In 1856, these two calibres were replaced by one of 24 to the pound, the new rifle-musket.

The oblong bullet is denominated by its diameter and weight; the new rifle-musket ball has a diameter of 0.58 in., and weighs 540 grains.

The *oblong bullet* at present used in the United States' service is composed of a cylinder surmounted by a conoid — the conoid being formed by the arcs of three circles. The cylinder has three grooves cut in it, in a direction perpendicular to its axis, to hold the grease necessary for lubricating the bore of the piece in leading, and to guide the bullet in its flight, after the manner of the feathers of an arrow.

A conical cavity is formed in the bottom, in which the gas of the charge expands, and forces the sides of the bullet into the grooves or rifles of the gun. From these grooves it receives a rotary motion around its long axis, which prevents it from turning over in its flight.

PERCUSSION

Percussion bullets are made by placing a small quantity of percussion powder, enclosed in a copper envelope, in the point of an ordinary rifle-musket bullet. The impact of the bullet against a substance no harder than wood is found to ignite the percussion charge, and produce an effective explosion.

These projectiles can be used to blow up caissons and boxes containing ammunition, at very long distances. ★

From "A Course of Instruction in Ordnance and Gunnery" by Captain J.G. Benton, D. Van Nostrand, New York, 1861.

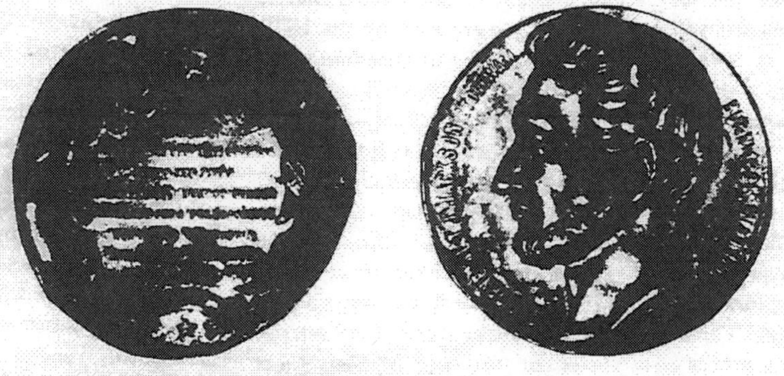

The Jackson Medallion

By
S. L. KOTAR

On October 13, 1862, the Congress of the Confederate States of America approved an Act, authorizing the President "to bestow medals with proper devices upon such officers of the armies of the Confederate States as shall be conspicuous for courage and good conduct on the field of battle; and also to confer a badge of distinction upon one private or noncommissioned officer of each company after every signal victory it shall have assisted to achieve." The act was published November 22, 1862, by the Adjutant and Inspector General, and a roll of honor was prepared, but no medals were ever officially issued by the government. The medallions that were issued during the war, such as the Davis Guard Medal, were manufuctured privately. The New Market Cross of Honor and the Southern Cross of Honor were distributed long after the war.

Thus it was that when the occasion arose to pay tribute to one of the South's great soldiers, Lieutenant-General T. J. Jackson, the Confederate government commissioned Frederick Volck to make a bronze equestrian statue of him. Volck, a refugee from the 1848 political turmoil of Europe, had come to the attention of the authorities in Richmond by displaying a bust he had secretly made of President Jefferson Davis. This highly acclaimed piece of sculpture was selected to be the model for John Archer's engraving of the Davis ten cent stamp.

Volck, who had previously made the death mask of Jackson, found it impossible to carry out his work in Richmond during the war, and left in August, 1863, for Nuremburg, Bavaria, a city renowned for its bronze castings. Using the measurements he had made, and an old uniform coat, boots and other clothing of the General's, he created a clay statue. He was advanced $28,000 in Confederate money (which was converted into approximately $5,000 in gold in England) for his expenses.

While Volck was in Europe, it is possible that he was contacted by the Marquis de Lafayette to sculpt Stonewall's likeness for a Jackson medal he wished to prepare as a tribute. The exact circumstances surrounding this coin are unclear, but information obtained from the Front Royal Museum in Virginia suggests that Lafayette, a descendant of the Revolutionary War hero, ordered 3,000 medallions made at his own expense to express his admiration of General Jackson and his Stonewall Brigade. The coin was engraved by Armand August Caqué, whose name appears under the truncation of the neck. The obverse of the coin shows a profile of the General, and the lettering "Lieut. General T. J. Jackson, Stonewall, Born 1821, Died 1863." (The birth year is incorrect; it should read 1824.) The reverse bears a list of battles: "Kernstown, Front Royal, Middletown, Winchester, Strasburg, Harrisonburg, Port Republic, Mechanicsville, Cold Harbour, White Oak Swamp, Malvern Hill, Cedar Mountain and Manassas." These are shown encircled by a sheaf of flowers on the left and corn on the right, bound by a ribbon, which bears nine battles: "Bull Run, Sudley, Harper's Ferry, Shephardstown, Fredericksburg, Chantilly, Martinsburg, Antietam and The Wilderness." Above "Bull Run" is a downward pointing 5-pointed star; beneath the listed battles, arms are pictured; below this is a shield, bearing the motto "Deo Vindici."

It was intended that every member of the Brigade be given a medallion, and they were shipped through the Federal blockade, in the charge of Colonel C. A. Lamar, of Georgia. They were stored in a warehouse in Savannah, and subsequently forgotten, until the building was razed years later.

A second accounting of these coins relates that they were cast in France in 1863 by order of the Confederate Congress. They were shipped to the CSA, but only got as far as Bermuda, where they were held up. In 1890, they were re-discovered and brought to the United States, where they were sold by the UCV for $4.50 a piece to raise funds.

In 1906, the mystery surfaced with the appearance of the coins. The one person it was hoped could clear up the speculation was Mrs. Mary Anna Jackson. Dr. Hoge wrote to her about this, and she replied on November 19th: "It is with regret that I am not able to substantiate the story of the veteran about the medalion of Gen. Jackson. Certain it is that *I* never sent to France, or anywhere else, and had any medalions of him made. I imagine I know the medalion that you speak of, as I have had several sent to me, with a miserable picture of Gen. J. on one side, and a list of the battles he fought on the other. I could never endorse them, because I see so little resemblance to the original, and the story of my having anything to do with it is all a fabrication. Isn't it strange how people will resort to such misrepresentations and frauds to make a little money, and to see the old *soldiers* duped this way is lamentable! I am sorry too that *you* have been deceived."

Mrs. Jackson's laconic description of the engraving as "miserable" and bearing "so little resemblance to the original" is certainly justified. The profile is bland, almost to the point of being featureless, and depicts the General as having only a partial beard with no mustache. If anything, it more closely resembles the carte-de-viste of the 1851 photograph of Jackson that was reprinted by Anthony during the war, and widely circulated in England and Europe, captioned "Brig. Gen. 'Stonewall' Jackson." There is no reason why Volck would have had to resort to using this image of the General to fashion his sculpture for the medal. If nothing else, he was aware that Jackson wore a full beard, and the absence of this on the medal raises the possibility that he was not the artist responsible for the image. It may be that his work was altered or specifically made to resemble the face Europeans had come to identify with the great leader.

Whatever the explanation, the Jackson medallion has become a part of Confederate history, and is a rare and highly prized part of any Jackson collection. ★

Concurrent Events

☐ The venerable Dr. Lyman Beecher, in a recent sermon at Boston, said—

"A preacher's ideas should stand out like rabbit's ears, so that the people can grab hold of them, hold on to them, and get something for their pains."

April 23, 1853

☐ The frequent use of asparagus is strongly recommended in affections of the chest and lungs; in fact, asparagus is one of the most wholesome, as well as agreeable vegetables we possess.

April 23, 1853

THE BUGABOO CORNER

A Debt Repaid

By
J. E. GESSLER

Earlier than now, before the War had reached this bitter stage — in point of fact, just prior to Manassas, a company of us and a company of them resided, temporarily, on opposite sides of a little rill, called B——Run. The Yankees were as brand-new on the job as we, and I must admit that being diametrically opposed politically had certainly altered everyone's appearance, to the point that we all wondered what those Federals looked like. They undoubtedly wondered the same about us, for it was that, once we had seceded, we all grew long tails, developed purple spots, and sprouted horns. At least, such was the opinion given of us in a little half-penny sheet, which establishment was situated so near the Virginia border, that its proprietors could not have helped but seen us, not only before the fatal hour of division, but several times thereafter. And if such a description was printed in the newspapers, it had to be a fact — that was that.

Since we had altered so considerably (albeit none of us ever jumped at the sight of each other), we naturally deduced the Yankees had experienced some sort of morphological amendment. The burning curiosity of our men reached a boiling point; it was then that a few of us spilt over the kettle's edge, and, at the risk of our lives, ranks and souls, dared Fate: we set a date to view those mysterious Northern Brothers of ours. The original communication was achieved through the device of a missive entrusted to a hollow of a certain tree, by which the enemy was known to pass while on picket duty. Though this was on our side of the water, unfamiliar tracks were often discovered by it, the general concensus being that, due to the heat of our Southern sun, the Yankees had been constrained to take cool refuge beneath its broad and leafy branches. Slight depressions in the earth convinced us that some fatigued, sunburnt sentry had there rested, if not actually committed the unpardonable sin of sleeping while on duty. A small white rag was wedged into the bark next to this hole as a signal to the tree's next visitant. After the simple letter was abandoned, we waited for the reply.

It was not long in coming. The next day, at sunset, we saw the flag gone, and, upon examining the hollow, found our answer. The terms had been agreed upon. We were to meet that night in the middle of the stream, approximately 30 yards down water from the tree. We could trade, talk, and share a drink. And so, we prepared to meet the bluebirds for the first time.

At moonrise, we set out. Discipline in the camp was rather loose, but considering the green recruits then, laxness was the rule, not the exception. There were four of us, and there were to be four of them. We packed our haversacks with some items to be traded — items that now would bring tears to our eyes, so needful are we of any staples, not to mention luxuries.

A light caught our eyes as we made our way toward the water — but it was only the moon shining silver off the stream, moving and flickering with the waverings of the current. After this first small apprehension, we let out our breaths, and plodded softly down the sandy mud of the little bank towards our rendezvous. The air seemed preternaturally still, the water a smothered gurgle as it tumbled gently over small stones in the creek bed. We heard each other's breathing; in the quiet, it sounded labored, fearful; yet not a one of us was afraid — so all had sworn before departing on this nocturnal journey. So onward we pressed until in sight of the designated spot — a large rock marked it on our side. We stopped, looked — but no movement was to be espied, betraying our trading partners' position. Our decision was made in the silence, in whispers: I would shout across, as the Yankees might be fearful of our "devilish" intentions — we may not have convinced them as yet of our sincerity. Sprague, a corporal and generally our leader by right of rank, was peculiarly white that night, a vague look about his pale eyes, as if preoccupied with some business of the

mind that allowed for no acceptance of the world as was presented to those blue orbs. Before I hailed the enemy's side of the run, I dared to ask where his thoughts were. But after he answered, I felt quite sorry I had. It seems the creek where we stood had been the scene of an early cavalry skirmish, just a mere exchange of bullets as passed for a greeting between the South and the North; but Sprague had been acquainted with the only casualty of the brief encounter. Not until that night had he recalled the place and not until we had struck out did he remember the promise Isaacs (the name of the fatally wounded cavalryman) had made to him: to greet him, and make good his word concerning a bargain that was struck between them, which had once brought Isaacs good fortune. The still night did not still our nerves, which had already begun to quaver; but we had made a bargain with the Unionists, and weren't about to back down. I called in a quavering voice, which I felt certain would be identified as the accents of a coward; but an equally unsteady voice answered me in the queer dialect of what I took to be a New Yorker.

"Yank!" I thus addressed him. "Yank! We brought what was promised. Choose a leader, and wade to the center of this stream with your goods. Carry no weapons! And perhaps we'll mark this meeting with a sip of something good." The terms were agreed to, and I was chosen to continue my leadership, having proven my abilities to the satisfaction of those who preferred to stay on shore. Sprague particularly seemed appeased, eager to stay hidden with his companions, and I must confess I understood why. So I stripped and bundled our goods up in a blanket, and, when I saw the Yankee on the mud doing the same, and that he had no more of a difference in natural appearance than any one man has from another, I felt reassured, and the calm that descends on those who command filled me with courage. I made bold to be the first to step in the rill, the Yankee, skin white in the moonlight, only a moment after me. I noticed his eager stare — checking for my devil's accoutrements — but I suppose he saw none of the promised monstrosity of person on or about me. If I had to say truly, I would state the soldier was a trifle disappointed, albeit simultaneously relieved. The way to the center was not far, only a matter of a few yards; but as we felt the cool stream climb to our knees, from the distance came the sound of horse's hooves thumping off the bank and splashing into the creek. We both stopped dead in our tracks, so to speak; I, thinking that an enemy officer had discovered our intentions, and was going to kill us all; my counterpart obviously believed the same, for he shot me a suspicious, frightened look just as I glared an accusation at him. But we both could have saved our mutual mistrust, for, as the rider approached, we could see he belonged to no army of this world. His form gleamed, a glow of bluish-white surmounted his head and shoulders, and although he wore what passed for a rather fine Confederate uniform, it was attended by the same eerie illumination which characterized his vividly wan face. All was suddenly deathly silent, for the ghostly rider reined in his steed. The intruder's features were now clearly visible. He had a youngish face, long straight nose, collar-length hair, over which he wore a simple forage cap, a contrast to his heavily decorated coat. But the fine uniform was soiled and slightly awry, the shining boots scratched, his horse's trappings sprayed with mud, the martingale tangled with bits of long grass and dead twigs. That beast gazed long at us two living soldiers, standing stock still, sinking in the mud, our eyes wide, our mouths dilated, but silent; our bravery turned to mush. We'd have taken flight, if it were not for the fact that we were anchored by horror. At this junction, the animal's livid master turned his face slightly toward our side of the creek, inclined his head barely an inch — at which I heard a piercing scream, muffled by the hands of my friends, I later learned from Sprague. And I then knew the rider to be the same Isaacs, the deceased cavalryman.

about whom our corporal had moments before spoken. His shriek acted as a signal for the ghost; at its height, the phantom turned his head, and that of his horse's, toward the Yankee, and began to move solemnly forward. But the Federal, seeing he was the spectre's target, moaned loudly, dropped his bundle in the creek, and fled, not stopping once to recover his uniform from the bank. A loud crackling of branches testified to the fact that his friends fled with him, all uttering exclamations of horror, notwithstanding their fate when discovered by their own pickets. I was on the point of flying, when the grim vision drew his sabre from its scabbard. Fearing to move lest I be impaled on the still shining steel, I waited to receive whatever fate should be meted out to me by the cavalryman's spirit. Instead of running me through, however, the tip of the weapon was aimed at the bundle, now bobbing in the weak current. The ghost of Isaacs retrieved the Yankee goods, and, with a slow yet firm gesture, the heavy blanket was delivered to my trembling arms, to rest atop our own wares. Isaacs turned to me, and, to my amazement, smiled; a different light glistened in those eyes, and the lips curved in a boyish look of mischief. Had his complexion not been so frightfully pale, I'd have taken him for a prankster of terrific nerve. Then he spoke, the sound of hollow wind punctuating his words:

"This is for Sprague. Take it to him." And here he once again inclined his head in the corporal's direction. "And with it deliver my undying thanks. I have kept my promise." With that, he bowed from the waist, turned his spectral steed, and rode off, sabre still in hand, humming a solemn melody, but in the most pleasant of tones. ★

The Battle Behind Kearnstown

FOOTNOTES

[1] "One Hundred Years at V.M.I." by William Couper, Garnett and Massie, Richmond 1939, Vol. I, p. 157.

[2] "Stonewall's Man, Sandie Pendleton" by W. G. Bean, Uni. of North Carolina Press, Chapel Hill, 1959, p. 60. See also, "I Rode With Stonewall" by Henry Kyd Douglas, Uni. of NC Press, Chapel Hill, 1940, p. 37.

[3] "Lee's Lieutenants" by Douglas Southall Freeman, Charles Scribner's Sons, New York, 1942, Vol. I, p. 314.

[4] "Mighty Stonewall" by Frank E. Vandiver, McGraw-Hill Book Co., New York, 1957, p. 333.

[5] "The Stonewall Brigade" by James I. Robertson, Jr., Louisiana State University Press, Baton Rouge, 1963, p. 55.

[6] "Lee's Lieutenants" Vol. II, p. 147.

[7] "The Stonewall Brigade" p. 54. See also "Lee's Lieutenants" Vol. I, p. 317.

[8] "Biographical Register of the Officers and Graduates of the U.S. Military Academy" by George W. Cullum, D. Van Nostrand, New York, 1868, Vol. II, p. 25.

[9] IBID, p. 25.

[10] IBID, p. 25.

[11] "Confederate Military History" by Jedediah Hotchkiss, The Blue & Grey Press, Vol. III, p. 597.

[12] "General Officers of the Confederate Army" by Marcus J. Wright, p. 66.

[13] Official Records, LI, pt. 2, p. 399.

[14] O.R., Vol. V, p. 981.

[15] "The Stonewall Brigade" p. 55.

[16] "Stonewall Jackson and the American Civil War" by G. F. R. Henderson, Longmans, Green and Co., New York, 1911, Vol. I, p. 190.

[17] From the R. B. Garnett MSS, Confederate Museum, Richmond.

[18] IBID.

[19] IBID. See also O.R., Vol. XII, pt. 1, p. 409.

[20] O.R., Vol. XII, pt. 1, p. 335. p. 336.

[21] "Life and Campaigns of Lieut.-Gen. Thomas J. Jackson" by R. L. Dabney, Blelock & Co., New York, 1866, p. 320.

[22] IBID, p. 321.

[23] O.R., Vol. XII, pt. 1, p. 379.

[24] Garnett MSS.

[25] "Stonewall's Man", letter of A. E. Pendleton to W. N. Pendleton, April 6, 1862, p. 59.

[26] Garnett MSS.

[27] IBID.

[28] "Stonewall's Man", letter of A. E. Pendleton to W. N. Pendleton, April 6, 1862, p. 59.

[29] O.R., Vol. XII, pt. 1, p. 384.

[30] "Stonewall's Man", p. 59.

[31] IBID, letter of A. E. Pendleton to W. N. Pendleton, April 13, 1862, p. 60.

[32] "The Wartime Papers of R. E. Lee", Clifford Dowdey, Editor, Little, Brown and Co., Boston, 1961, p. 21.

[33] Garnett MSS, letter to R. T. M. Hunter, April, 1862.

[34] IBID.

[35] "Stonewall's Man", p. 60.

[36] Garnett MSS, Garnett's defense, written June 20, 1862.

[37] "Stonewall's Man", p. 60.

[38] Garnett MSS, letter of May 6, 1862.

[39] IBID.

[40] Jackson's MS Letter Book, April 29, 1862.

[41] Garnett MSS, letter of Lee to Cooper, May 6, 1862.

[42] IBID, Garnett's defense, dated June 20, 1862.

[43] IBID.

[44] IBID, note on the bottom of Special Orders No. 146.

[45] IBID, letter of R. B. Garnett to Gen'l. S. Cooper, July 12, 1862.

[46] IBID, letter of R. B. Garnett to Gen'l. Samuel Cooper, July 12, 1862.

[47] IBID, letter of R. H. Chilton, A.A.G., to R. B. Garnett.

[48] IBID, "testimony of Major-Gen'l. T. J. Jackson, P.A.C.S."

[49] IBID.

[50] IBID, testimony of 1st Lt. A. S. Pendleton.

[51] O.R., Vol. XIX, pt. 2, p. 595.

☐ THE DEATH PENALTY

A bill had passed one branch of the Legislature of Wisconsin for the abolition of the death penalty in that State, and it is said there is every probability that it will become a law. A similar bill was before the Ohio Legislature recently, and was lost by but a single vote.

(From the Saturday Visiter, Pittsburgh, Pennsylvania, April 2, 1853.)

☐ JENNY LIND

The Courier learns directly, and from the very best authority, that the family difficulties which were recently said to have determined this distinguished artist to return to this country, do not exist in the least degree: and that there is no present prospect of our hearing her again from any cause whatever.

(April 2, 1853)

NOSTRUMS AND NOTIONS

"Conceive it or Not!"

By
ROBERT KAMER, M.D. AND
SAMUEL JACKSON, M.D.

When the word "generation" is used today, the usual connotation has to do with people of the same age group sharing their thoughts and culture. There is the '60's *generation* of long-haired rock'n'rollers, the '80's *generation* of computer advocates. There is *generation* of energy, the *generation* of an idea, and the *generation* gap. In the 1800's, however, *Generation* had a meaning synonymous with its Greek root word: to beget. Generation was the term used for procreation, and as much encompassed the whole phenomenon of sexual function.

The most common and perhaps the only agreement medical authorities had on generation, or reproduction in the 19th Century was that it was a mystery, which started with the Old Testament adage, "increase and multiply." In 1855, it was stated there were "no less than *two hundred and sixty-three* modifications"[1] of the basic theory of propagation. In "The Hydropathic Encyclopedia," the author declared: "It is enough for all practical purposes, to understand that sexual association is necessary for the continuance of our species;" the female supplies the ovum or germ of the future being, while the male communicates the vivifying (to give life to; animate) principle. More (or less?) specifically, "Hoopers Medical Dictionary" teaches us, "The part of the male, in the act of reproduction, is to deposit the semen (The seed or prolific liquor of animals to be admitted into the female vagina, and there, by its aura, to penetrate and impregnate the ovulum) in the vagina, at a greater or less distance from the orifice of the uterus. The function which the female discharges is much more obscure; some feel, at this moment, very strong voluptuous sensations; others appear entirely insensible; while others, again, experience a sensation which is very painful...."[2] At the present, however, it is purely conjectural about the general signs of fecundation (see Generation):[3] "at the instance of conception, the woman feels, it is said, a universal tremor, continued for some time, accompanied by a voluptuous sensation; the features are discomposed, the eyes lose their brilliancy, the pupils are dilated, the visage pale, &c.!"[4]

Numerous attempts have been made to determine precisely how generation takes place. The philosopher Spallanzani, fortunately, has provided us with four basic axioms:

1) Three grains of semen, dissolved in two pounds of water, are sufficient to give to it the fecundating virtue.

2) the spermatic animalcula are not necessary to fecundation.

3) that the *aura seminalis*, or seminal vapour, has no fecundating property. (On account of the "difficulty of conceiving" the passage of the semen to the ovarium, it was believed by some authorities that the *vapour* of the sperm, rather than the sperm itself, is responsible for fertilization).

4) a bitch can be impregnated by mechanical injection of semen.

Perhaps the entire "generation" of speculation was best summarized by Dr. Hooper, when he wrote, "Now, this phenomenon is one of those on which our senses, and even our mind, have no hold: it is one of those impenetrable mysteries of which we are, and, perhaps, shall ever remain ignorant."[5] ★

[1] The Hydropathic Encyclopedia by R. T. Trall, MD, Fowlers and Wells, Pub., New York, 1855, p. 443.
[2] LEXICON MEDICUM; or Medical Dictionary by Robert Hooper, MD, F.L.S., Harper & Brothers, New York, 1842, p. 386.
[3] IBID, p. 355.
[4] IBID, p. 387.
[5] IBID, p. 386.

Concurrent Events

☐ An inch and a half is the width of the cravat worn in Paris, as last reported. Some young ge lemen turn down the shirt collars. Any change that gives room and freedom to the breathing and vocal organs is commendable, even if it is an imported French fashion.

April 23, 1853

HALLET, DAVIS & CO.,

Warerooms, 409 Washington St., Boston, Mass.

We have received *Thirty-two Premiums* within the last twelve years for SUPERIOR PIANOS. We have recently perfected an entirely New Scale, which combines the quality of the Grand

Piano, which we shall be pleased to show to all who will call at our Rooms. We can also exhibit a Piano made in 1519, thereby showing the march of improvement.

A Full Warrantee will be given with each Piano of our make, which will relieve the purchaser from *all responsibility* as to its giving perfect satisfaction.

☞ Illustrated Catalogues will be forwarded to any address (containing Cuts with full description of our Pianos), by mail, *free of charge*.

WALKO & BARNUM,

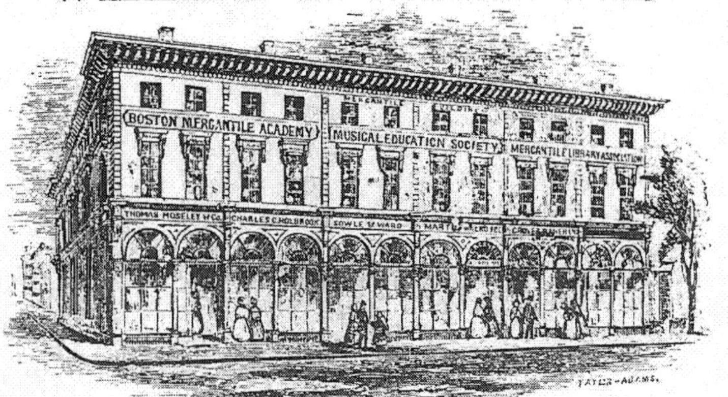

MANUFACTURERS OF FURS,

Children's Hats and Caps, and Ladies' Riding Hats.

W. & B. would call the attention of the public to their large assortment of SIBERIAN and RUSSIAN SABLES.

16 SUMMER STREET.

☞ ONE PRICE. N. B. CASH PAID FOR SKINS.

Watches, Jewelry, and Silver Ware.

THE SUBSCRIBERS HAVE ALWAYS IN STORE A FULL ASSORTMENT OF

FINE WATCHES

FROM THE CELEBRATED MANUFACTORIES OF

CHAS. FRODSHAM,
F. B. ADAMS & SONS,
E. FAVRE BRANDT,
ALFRED LAVELLETTE.

RICH JEWELRY.

BROOCHES, BRACELETS,
EARRINGS, AND RINGS,
OF
DIAMOND, PEARL, ETRUSCAN, MOSAIC,
CORAL, STONE, CAMEO,
AND FINE GOLD.

SILVER WARE,

OF STERLING SILVER.

TEA-SETS, PITCHERS, WAITERS, GOBLETS, VASES, CUPS, FORKS, SPOONS, PIE KNIVES, FISH KNIVES, AND OTHER FANCY SILVER FOR PRESENTS.

SILVER-PLATED WARE, OF ENGLISH AND AMERICAN PLATE, AS HANDSOME AS SILVER,
AT ONE-TENTH OF THE COST.

All of which they will be happy to show to their customers and to sell at reasonable prices.

SAMUEL T. CROSBY & CO.,

Jewellers and Silversmiths,

69 Washington Street, ... Three doors South of Court Street,

BOSTON.

SAMUEL T. CROSBY.
FRANK HUNNEWELL.

THE KEPI

JUNE-JULY 1983 $3.00

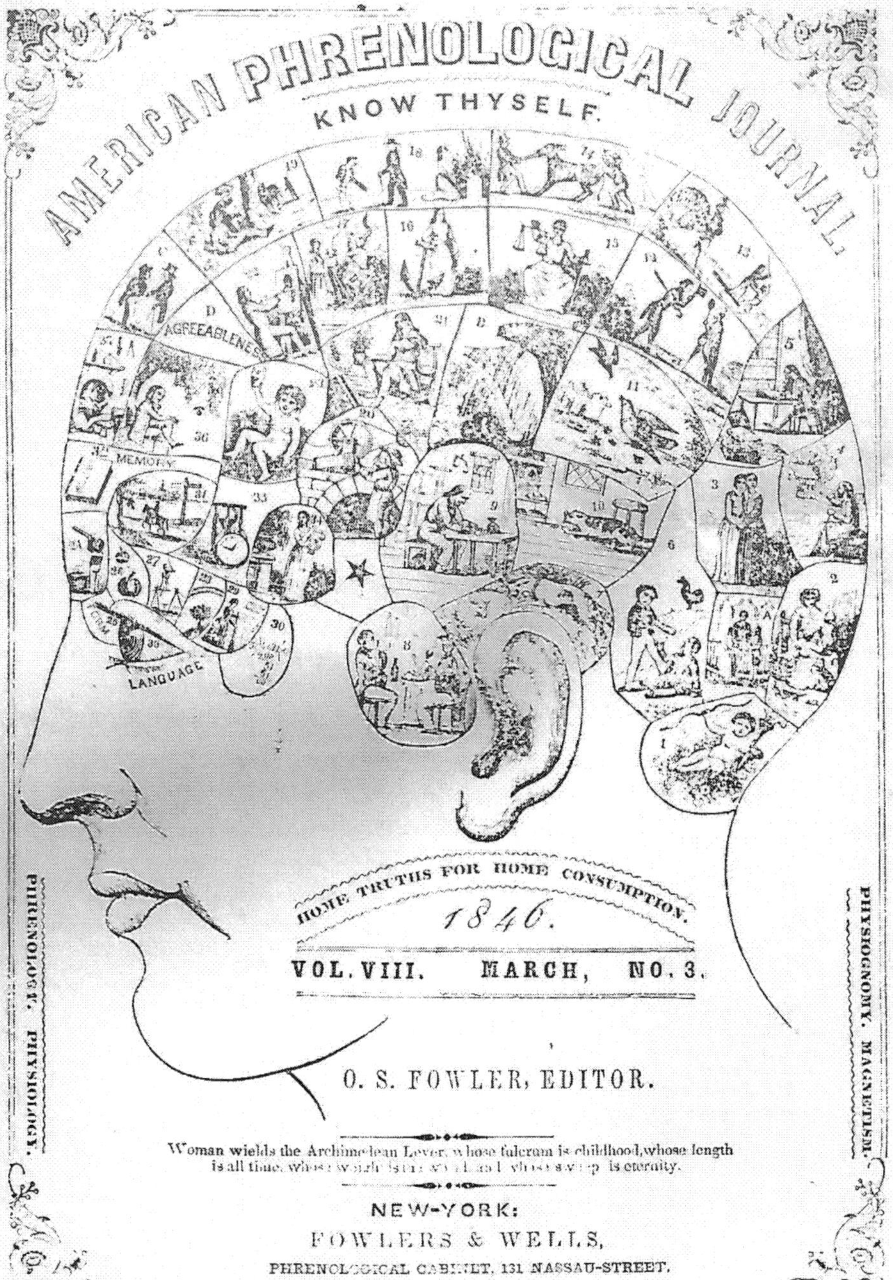

AMERICAN PHRENOLOGICAL JOURNAL.

KNOW THYSELF.

HOME TRUTHS FOR HOME CONSUMPTION.

1846.

VOL. VIII. MARCH, NO. 3.

PHRENOLOGY. PHYSIOLOGY.

PHYSIOGNOMY. MAGNETISM.

O. S. FOWLER, EDITOR.

Woman wields the Archimedean Lever, whose fulcrum is childhood, whose length is all time, [illegible] is eternity.

NEW-YORK:

FOWLERS & WELLS,

PHRENOLOGICAL CABINET, 131 NASSAU-STREET.

Terms, $1 a Year, invariably in advance. Ten Cts. a Number.

[illegible]

THE KEPI

JUNE-JULY 1983 VOLUME I, NUMBER 2

THE ARTICLES

Front Cover
From "the American Phrenological Journal," March, 1846

Back Cover
From "History of West Point," 1863.

Phrenological illustrations from the texts cited in the footnotes. West Point illustrations from the texts cited.

THE IRREGULARS

Editors-in-General S. L. Kotar and J. E. Gessler. THE KEPI, 12115 Magnolia Boulevard, P. O. Box 107, North Hollywood, California 91607.
Return postage must accompany all inquiries and submissions. THE KEPI reserves the right to edit, and copyright all letters.

GENERALLY SPEAKING

In this issue of THE KEPI, we have chosen to include topics which appear to vary greatly from one another, and yet are all interwoven into the culture of early Victorian America. The three main articles also reveal a common trend, which was taking hold of the world — Science. Technological developments in ordnance and gunnery, particularily in the new rifling made our Civil War unique — it became (without knowing it) the first modern war. Advancements were gripping other fields as well. Reapers, plows and new techniques in agriculture, such as the science of crop rotation, which is discussed in THE GENTLEMAN AGRONOMIST, freed men from the fields and allowed them to fill the ranks of both armies. The art of medicine was crawling slowly away from its superstitutions and ignorances and looking more to science for help in curing disease. In the cover story on Phrenology, adherents were predicting that character could be read with scientific certainty. Even religion was not beyond this new movement, involving itself in the scientific assessment of man's morality. In the military schools of West Point and the Virginia Military Institute, the scientific fields were stressed, to the exclusion of almost everything else. In the years just prior to the war, VMI actually offered scientific courses for the pursuit of agriculture. Part I of "Military Education in the United States: Boys into Cadets," deals with the USMA's system of cadet appointments and examinations. Part II, in the upcoming August issue, will deal with some comparisons and contrasts with West Point and VMI, as well as the direction that state school was heading in before war interrupted.

Also in this issue, we are featuring one of the Irregular Columns, TRACTS. Dealing with the consecration of a Jewish synagogue in 1860, it serves as a reminder of the many and varied cultures that made up our great nation and contributed to it.

LEXICON and WHIMSEY round out this issue, offering a touch of background and humor to the Civil War years.

As any "new concept," THE KEPI is growing and experimenting in its efforts to provide readers with topics and points of view not readily found elsewhere. We appreciate your comments and contributions, and we hope that with the enthusiasm and support of the Civil War community, we can make this magazine a useful tool in the study and appreciation of those turbulent years.

S.L. Kotar J.E. Gessler

Editors-in-General

Head-Quarters of THE KEPI
1st June, 1983

GENERAL ORDERS,
No. 1.
by Command of the Editors-in-General

Coming in the August-September issue of THE KEPI. Part II of "Boys into Cadets: Military Education in the United States: The Virginia Military Academy." Discussed is how VMI, as a Southern state school, differed and compared to the national USMA at West Point, and how the French influence was turning VMI away from the military, just prior to the War.

CAMPFIRE will contain an authentic, unpublished narrative of a man's struggle to retrieve his brother's body after the Battle of Perryville, Kentucky, in 1863.

NOSTRUMS AND NOTIONS deals with the serious and humorous aspects of hydrotherapy, with emphasis on the newfangled invention called "the shower."

TRACTS illustrates a type of religious pamphlet distributed to soldiers during the War; "Eloqution!" 19th-Century style is discussed and the Articles of War are continued.

UNCLASSIFIED

RATES FOR KEPI ADVERTISING: 50¢ a word one time; 40¢ a word 3 times
$100 full page ad; $55 half page ad; $30 quarter page ad. 20 words or more for 3 consecutive issues or quarter page ad or larger, logo run free.

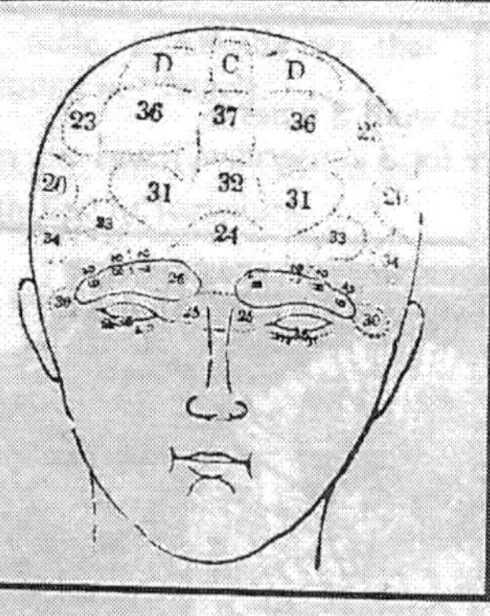

THE PHRENOLOGICAL DEVELOPMENT

By
Robert Kamer, M.D. and
Samuel Jackson, M.D.

Phrenology, as defined by one of its greatest American advocates, O.S. Fowler, "Points out those connexions (sic) and Relations which exist between *the conditions and developments of the BRAIN, and the manifestations of the MIND,* discovering each from an observation of the other."[1] In simpler language, Phrenology claims to be able to determine, by examining the bumps on the head, the moral and physiological developments of any creature.

The science of Phrenology was taken seriously by many people in the 19th Century. Its origins can be found in Europe, where it was first developed by Dr. Gall, a physician of Vienna, who observed that people with full and prominent eyes had the ability to memorize easily. This led him to look at other portions of the head and face to see if there were any other similarities between physical characteristics and mental ability. He soon discovered that by comparing the heads of people who had strong characteristics, he could find a corresponding similarity of structure on one part of their skull. For example, he concluded that people who are known to be extremely violent all have a bump on their heads in exactly the same place, while all people who are renowned for their generosity have a protuberance in exactly the same place, over a different part of the skull. Thus, by grouping together people with the same temperaments and examining their skulls, he was able to isolate specific areas on the head which would physically point to any emotion. If the bump, or *organ,*[2] as it was labled, was "large," so was the corresponding trait; if the organ was "small," this indicated a lack of development in that area. Specifically, a man whose organ for "Language" was large, expressed ideas very well and was a good orator. A man whose "Cheerfulness" was small was likely to be depressed and look at the dark side of things.

Concluding that certain manifestations of the mind (emotions) are always accompanied by particular manifestations of the brain (bumps), Drs. Gall and Spurzheim traveled throughout Europe, examining heads of inmates of prisons, hospital patients and citizens known to have extremes of character. By comparing and charting their results, they "confirmed" the existance of approximately 30 Phrenological organs.[3]

PHRENOLOGY IN AMERICA

While still in its infant stages, Phrenology came to America. O.S. and L.N. Fowler, two brothers, became intrigued with its possibilities, and studied what appeared to be one of the greatest discoveries of their age. Soon convinced that this new science was the key to man's mind, they took up the crusade, and carried it all over the country. Under the banners "Knowledge is Power,"[4] and "Self Knowledge is the greatest knowledge,"[5] they began publishing books and giving lectures, turning their enthusiasm into a lucrative business. In 1836, calling themselves "Practical Phrenologists," they printed "Phrenology, Proved, Illustrated and Applied," the first American work on the subject,[6] laying claim to many important improvements in the science of Phrenology," and giving the reader many examples of the practical application of its principles. For 50 cents a piece (and a minimum of 300 people per lecture), the interested 19th Century denizen could witness for himself, with almost circus-like wonder the astonishing performances of the Phrenologists. Men and women, young and old, famous and infamous were brought on stage to be examined. There, the "scientific mystic" would observe the *general size* of the *whole head,* then the *relative size* of its *several parts.* He would ask questions to determine the *temperament,* health, habits, and education of his subject. Once ascertained, he applied the balls of his fingers to the head. The first joint of the second finger was placed upon the *middle* of the organ (or bump) to be examined, and the first and third fingers upon the *sides* of the organ. Protuberances and depressions were thus "sized up." When every organ was thus measured, and the complete history of the subject revealed, the Phrenologist would give a characterization of that person. A study of President Andrew Jackson's head revealed that he possessed an extraordinary development of "Firmness," "Self-Esteem," "Benevolence," "Combativeness," and "Adhesiveness" (friendship, love of company, social, warm-hearted feelings. He had large "Veneration" and "Hope," and smaller "Ideality" (imagination, fancy, love of poetry).[7] Martin Van Buren's skull revealed that "Cautiousness" was his largest organ, "Secretiveness" almost equally large, "Ac quisitivensss" (love of property, desire to amass wealth) and "Destructiveness" large.[8] Henry A. Wise (Governor of Virginia) had a character, as revealed by Phrenologists, of towering ambition, accompanied with all the intellectual and propelling powers required to sustain himself in his aspirations after greatness. His *temperament* was of the most favorable kind, and his "Perceptive Faculties" developed to a very unusual degree.[9]

The most casual observations of these readings give the impression that they could as easily have been deduced by a simple study of personality, coupled with an acknowledgement of the individual's current status. Andrew Jackson's history and his political power certainly influenced the outcome of his phrenological interpretation. Martin Van Buren, less popular and less renowned, received a less than satisfactory rating. And one certainly must wonder if Henry A. Wise's skull would have told a different story after 1861.

BIGGER IS BETTER...

Phrenologists would argue that although they were aware of the personality of these famous men before applying their science to them, it was indeed the measurements of the head that determined their answers and not the reverse. Phrenology, which claimed to be "a new and complete system of intellectual and moral philosophy and professes to develop and illustrate the fundamental principles of human nature"[10] rested its support in these propositions:

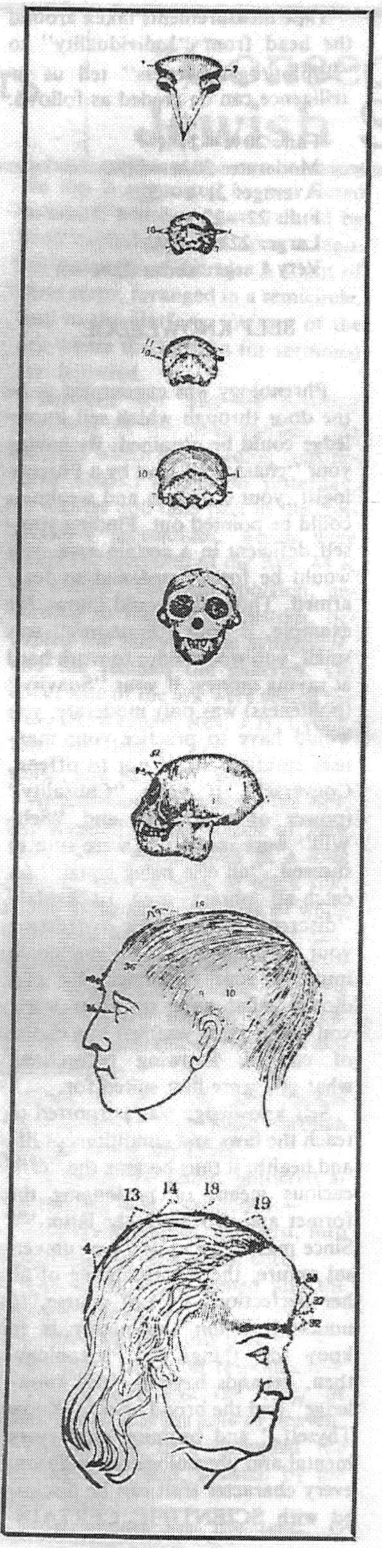

1) *The Brain is the organ of the mind,*[11] for "we have no knowledge of the operations of the mind, except through the medium of its physical organ, the body."[12] (Inasmuch as all parts of the body have one and only one function, it follows that the brain is devoted to the performance of mental operations. Since it has been proven by anatomical demonstration that the senses have their origin in the brain, the *remaining* mental operations are performed by the *remaining* portions of the brain. There being a definite correlation between the *size* of the brain and the *qualities* of mind, it follows that the *larger* the brain, the *greater* the strength of it. For example, the frog has a very small brain with correspondingly weak mental powers. The dog possesses a cerebral development far superior to the frog, and its intellect is proportionately greater. Human idiots possess brains vastly inferior to men of ordinary talents, and they in turn are inferior to a Benjamin Franklin.)[13]

2) *The Brain consists of as many different Portions or organs, as the Mind does of Faculties.*

Phrenologists stated that throughout nature "no single organ is known to perform more than one class of functions." As hearing, seeing and sensation have only one *exclusive* function, and are connected to only *one* area of the brain, it follows analogically that all the other functions have their own area of the brain from which they operate exclusively. Simply stated, you use your eyes for seeing; no other part of the body can see. You use your ears for hearing; no other part of the body can hear. The portion of the brain set aside for vision does not hear, and the organ for hearing does not see. These senses being separate and distinct, it follows that *all* senses of mental and moral deduction must also have their own separate and distinct area of the brain. Phrenologists mapped these areas, and determined the relative size of each. And, as size is the measure of power ("a principle too familiar to require proof"), the larger the organ, the stronger the emotion that corresponds to it, and *vice versa.* Since the shape of the brain may generally be ascertained by the form of the skull, and since the *exercise* of mind causes portions of the brain to expand (much the same as the exercise of a blacksmith's arm causes muscle to form), the measurement of the head as a whole and the various organs reveal character.[14]

No. 57. LARGE.

To facilitate the interpretation of the phrenological conditions, the following Table was developed:

VERY LARGE organic conditions (or bumps) are sovereign Kings over character and conduct and direct and sway their possessor.

LARGE organic conditions have a powerful and almost controlling influence over feelings and conduct.

FULL organs play subordinate parts, exerting more real than apparent influence.

AVERAGE ones have considerable, yet limited influence, but it is mainly in *combination* with large ones that they affect character.

MODERATE faculties are below par, exert limited influence and leave character defective in these respects.

SMALL organs are so deficient as to be easily perceived and leave their possessor weak and faulty in these points.

VERY SMALL ones render the individual almost idiotic in these functions.[15]

Tape measurements taken around the head from "Individuality" to "Philoprogenitiveness" tell us intelligence can be graded as follows:

Fair: 20¼—21¼
Moderate: 20¾—21¼
Average: 21¼—22
Full: 22—22¾
Large: 22¾—23¾
Very Large: above 23¾

SELF KNOWLEDGE

Phrenology was expounded to be the door through which self knowledge could be obtained. By having your "character" read by a Phrenologist, your strengths and weakness could be pointed out. Finding yourself deficient in a certain area, you would be forewarned and so forearmed. Thus you would know, for example, if your "Economy" was small, you would have to work hard at saving money; if your "Suavity" (politeness) was only moderate, you would have to practice your manners carefully so as not to offend. Conversely, if your "Causality" (power of reasoning) and "Self-Will" were large, you were sure to succeed, "all else being equal." (A catch-all phrase used to explain "discrepancies.") By correcting your personality flaws, you could improve your character. By predicting what your strengths were, you could guide yourself in a choice of careers, knowing beforehand what you were best suited for.

Self knowledge was purported to teach the laws and conditions of life and health; it thus became the "efficacious means of prolonging the former and increasing the latter."[16] Since man is the epitome of universal nature, the representative of all her perfections — "of course, to understand *him* thoroughly is to know *all* things."[17] Phrenology, then, expands beyond "self knowledge" and the broad axiom "Know Thyself," and promises that every mental and physiological condition, every character trait can be disclosed with SCIENTIFIC CERTAINTY, leaving nothing dark or doubtful in the SCIENCE OF THE MIND.[18] With their "new sun," there would never arise the need for psychiatry or analysis — everything was to become calculated and predictable.

"PROOFS"

Armed with Fowler's "Self Instructor," the serious student (for Phrenology must be *studied,* rather than merely read),[19] could tell, by a careful look, all about his neighbors, his partners, the Minister or the barber. If any of these people were characterized by a large

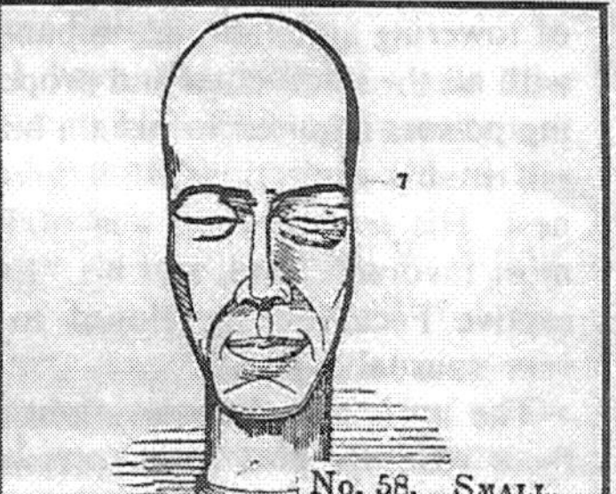

No. 58. SMALL.

mouth, it was A FACT that they had a corresponding quality of mind, while a small one indicated a lesser amount of mentality. A man who stuttered had serious flaws in his mental processes, and reasoned disjointedly. Hair color, skin tone, the way a child laughed were all immediate insights into the innermost recesses of the soul. How was this possible? Because GOD had caused the inherent character of every being to push out through every organ in the body.

This, according to Phrenology, is particularly evident in the heads of the various races of mankind, and serves as a further "proof" of this science.

By applying these new "scientific principles" to the skulls of American Indians, Phrenologists "discovered" the common features of an extreme development of "Destructiveness," "Secretiveness" and "Cautiousness."[20] Phrenology, being a science of *observation* and *correlation,* merely required the 19th

Century man to read in his newspaper about the Indians' "cruel, blood-thirsty, and revengeful disposition," to believe this was true. If reports said all Indians were bloodthirsty, and this was confirmed by their physiological organs, then it had to be true: thus, Phrenology was "proved."

The African race furnished another "striking instance" between character and phrenological development. It was determined they possessed, in general, large and very large "Adhesiveness" (propensity to associate, desire to love and to be loved), "Philoprogenitiveness" (love of offspring), "Hope," "Veneration," "Marvellousness" and "Tune." The size of their heads was generally "small" or "moderate." Phrenology summed up the race (although admitting that individual exceptions were frequently met with) as predisposed to lead a life of ease and idleness, inspired with song, but having few thoughts, little depth of intellect, feeble judgement, fondness for dress, vanity and a belief in ghosts![21] Unfortunately, "science" was called upon to substantiate the grounds for man's prejudices and provide a rationale for his "Institutions."

Phrenologists also delighted in pointing out the differences between the sexes. According to their measurements, the female organs of "Adhesiveness" and "Philoprogenitiveness" were so strongly developed as to elongate, and even deform the head, affording a "sure sign" by which to distinguish skulls. The "timidity," "trepidaton" and "anxiety" of the sex was proverbial, in accordance with which their organs of "Cautiousness" were larger than the male, while "Combativeness" and "Destructiveness" were much smaller.[22]

No. 89. LARGE.
FORM, SIZE, AND COLOR.

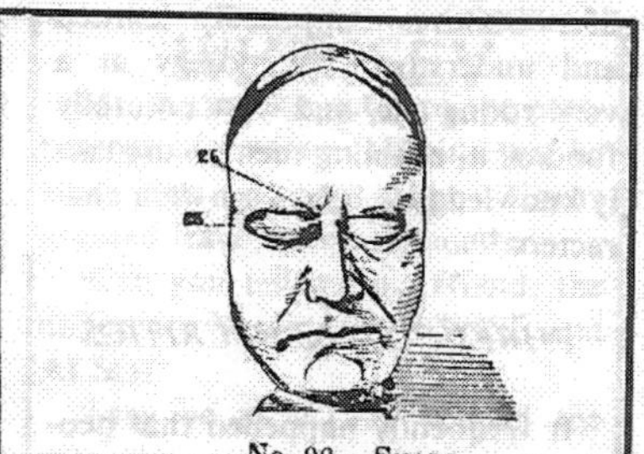

No. 90. SMALL.
FORM, SIZE, AND COLOR.

TEMPERAMENTS

For a proper Phrenological assessment to be given, the *temperament* (quality, organization, health, habits and activity) of the individual needed to be determined. The four broad temperaments were classified as:

1) *LYMPHALICK or PHLEGMATICK* — in which secreting glands were the most active portion of the system, producing soft, abundant flesh, languor of the pulse and mental functions. This temperament was often found among Pennsylvania Germans and Negros.

2) *SANGUINE* — in which the arterial system (particularly the blood) was most active. Those people had strong pulse, blue eyes, light hair, and strong animal passions.

3) *BILIOUS* — in which the muscular system predominated. That resulted in an athletic form, dark hair and eyes, and a considerable force of mind and character.

4) *NERVOUS* — in which the brain and the nervous system gave rise to the highest degree of excitability and activity of physical and mental powers. Clearness and rapidity of thought, delicate skin, fine, light, thin hair characterized this group, as well as a power and endurance of mind and body.[23]

"PRACTICAL PHRENOLOGY"

The Fowler brothers predicted that Phrenologists would be set up in cities to be consulted by parents to state what education and choice of occupations were adapted to their children.[24] As "proof" of the value of such a service, an example was cited: a Methodist clergyman brought his son, Frank, to O.S. Fowler, and said that he had examined the lad last summer, and it was stated that he had unusual arithmetical powers, of which they were not yet conscious of. Upon hearing the declaration, however, the Reverend tested his boy, and a year later the truth of the prediction was confirmed: Phrenology revealed an important truth his own observations had failed to detect.[25]

Carrying this one step further, the Practical Phrenologists championed their art as being invaluable in the choice of business partners, customers, friends, servants, and even spouses.[26] It was purported that a good way to improve many an odd and otherwise lost minute was by studying the Phrenological bust (which could be purchased from Fowler's shops) kept standing on your mantel. Thus, in two or three years time, you could acquire any amount of knowledge useful in daily

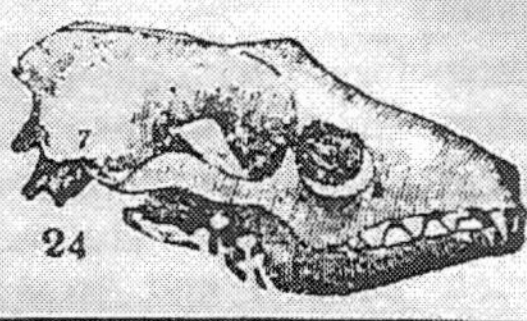

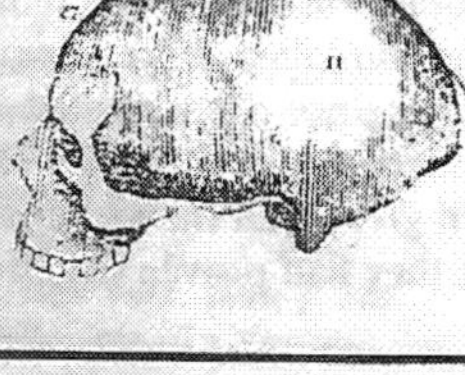

life. Children supposedly learned and understood Phrenology at a very young age, and were naturally fond of it, enabling them to use early knowledge to help form their characters.[27]

PHRENOLOGICAL CRITICS

It frequently happened that people disagreed with what the Phrenologist said about their character. The "scientist" explained that one basic reason for this was that "man does not know himself, because all previous systems of mental philosophy are obscure and contradictory."[28] It also happened that even a man's friends often "misinterpreted" his abilities, as in the case of a clergyman, ascribed at a public examination to have small "Language." The audience dissented, declaring him to be one of the most rapid speakers in the country. The Phrenologist felt himself vindicated, however, when, upon attending the clergyman's next sermon, found his manner of speaking rapid, yet his style by no means copious or flowing.[29]

The second reason for dissention with Phrenological readings was that man lived in a highly artificial state of society, which concealed and perverted nature.[30] The form of "perversion" was not clearly stated.

Scientists and religious men were among those who found fault with Phrenology, and their criticisms often appeared in newspapers. These written challenges resulted in prompt, and sometimes emotional responses.

Critics charged that since specific nerves leading to the organs of seeing, hearing and smelling had been identified, none for the emotions had been found. The Fowlers replied that this was not a material point, since Phrenologists *never* relied upon dissection and therefore would not discuss the anatomy of the brain.[31] (This, in reality, *was* the crux of the issue, and since there was no anatomical defense, Phrenologists simply chose to ignore it.)

Anti-phrenologists charged that this new "science" would lead to Fatalism, since it indicated that the CREATOR made men unequal. He gave some men very large organs for vice or murder, and under these circumstances, such an individual would be *compelled* to commit wrongful acts, and in fact had no free will at all. Other men, having been born with large benevolence, were therefore *obliged,* by their physical organization, to be kind and generous.

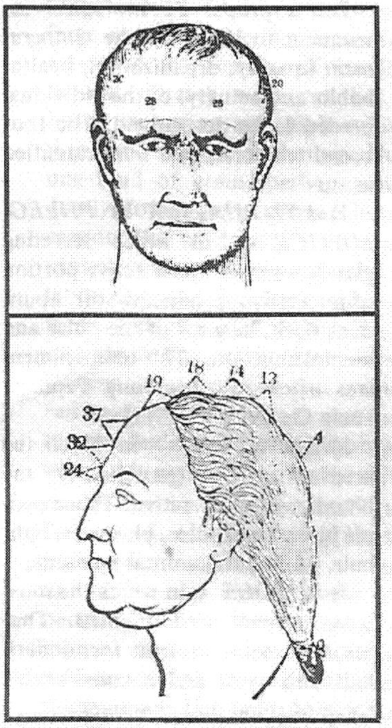

Phrenologists rebutted with the statement it was self-evident that God either did or did not influence Human behavior. Therefore, as far as Phrenology was concerned, the argument bore equally against the Deity having *any* hand in conduct. Further, if a man had a large organ for destruction, it was because he was a murderer, and not the reverse: instead of *compelling* a man to kill, the bump was merely the indication that he *had* killed. Phrenologists added that since there existed organs for "Veneration" and "Hope" (which gives man the ability to worship), they had proven God does exist.[32]

When asked how Phrenology could explain Regeneration (a sudden conversion from faithlessness to belief), when the size of the organs do not suddenly change, advocates answered that the only change is one of *direction* — the original nature and character of the subject does not change — much like the steamboat going upstream does not change its mechanism for going downstream, only its direction.[33]

It was also charged that Phrenologists were profiting from their science by charging $5.00 a piece for courses, requiring fees for consultations and selling Phrenological apparatus and books, which generally sold for 25% higher than other books. To this, the indignant reply was that they had a right to charge like any doctor or lawyer, and, in fact, the very charging of money imparted a certain dignity to the profession.[34]

SHORTCUT TO KNOWLEDGE

Why did Phrenology attract so much attention and gain so many adherents? Primarily because it was promoted as the scientific and certain road to all knowledge. "Know Thyself" and "Knowledge is Power" are very appealing slogans, and in a time when mental function and disfunction were very little understood, it seemed like the perfect answer. On the surface, it was deceptively clear and logical. If bigger was better, than it stood to reason that a man with a large head *ought* to be smarter than one with a small head; "everyone" had seen a red-haired man with a fiery temper, so it seemed clear that red hair was a characteristic of temper; the concept that beautiful people had beautiful souls, so deformed people must be correspondingly deformed inside, was undoubtedly an invitingly simple theory.

Phrenology also promised to be a shortcut to the metaphysical, a realm which had always been difficult to understand. It promised to promote morality and "true religion" by teaching "natural theology" and "natural religion," thus doing away with bigotry. After all, it was claimed, if there was a "scientific" explanation, then it had to be true. If one race was "naturally superior," that was the way God wanted it. Phrenology also appealed to people's affinity to classify and catalogue; by having the head "read," that person would know exactly where he stood in comparison to others. Additionally, it must have been very flattering to have Positive Proof that you were a good person, or better yet, proof that you were a good, *moral* person. The measurements of the head could be worn like a medal as tangible evidence of superiority. It provided ways of life, diet and exercise that held out the alluring possibilities of a long, happy life.

A knowledge of Phrenology offered the adherent an almost unlimited command over the minds and feelings of his fellow men. For all these benefits, then, it must have indeed seemed like a "new sun" rising.

"OR WHOLLY FALSE..."

The one hypothesis that turned out to be correct was that "Phrenology is either wholly true or wholly false."[35] Phrenology did not withstand the advance in science and the birth of psychology. Its "logical and observable" laws, as well as its claim of deducing from the universal to the particular were replaced by more accurate interpretations. Physical form could not be equated to mental or moral development; individual emotions did not have separate "organs" in the brain, and size did not correspond to strength.

At its most harmless stage, Phrenology might be equated to astrology: the vague, "apply to all situations" type of prediction found in newspapers today are similar to the reading of heads in the 1800's. For 50 cents, any citizen could be amused by attending a "publick lecture" and hear how the mayor had a propensity for "Secretiveness" or the town drunk a predisposition for vice.

At its most dangerous, this pseudo-science claimed to be able to tell a good man from a bad, a moral from an immoral one. It is hard to imagine how many lives may have been ruined by having a bump in the wrong place, or how many children were pushed into occupations they were not suited for, based simply on their exterior cranial development, or how entire races were discriminated against because a UNIVERSAL LAW said they were inferior, or aggressive, or timid and weak-minded.

Phrenology was a "science" of taking it for granted. Its arguments were filled with "moreovers," "of courses," similarities, rhetoric and semantics. An argument in support of the LAW, "all structure corresponds with character," was of a bullet bouncing off the tough skin of a crocodile. This meant that because its skin was strong, the reptile's sinews were also strong. But when referring to humans, Phrenologists included *morality,* as well. In applying the test to man (for, according to Phrenology, if the science itself was true, LAWS must apply equally to all living creatures), there could be no person with strong sinews *or* strong moral fiber, because bullets do not bounce off skin. It is, perhaps, a classic case of deduction gone awry; of attempting to reach a truthful conclusion from a false premise: it may have worked *logically,* but it did not provide a *factual* conclusion.

Edgar Allan Poe, in "The Purloined Letter"(1845), summed up the dangers of faulty reasoning (which applies to Phrenology) in the words of his detective, Dupin: "What is true of *relation* — of form and quantity — is often grossly false in regard to morals, for example. In this latter science it is very usually *un*true that the aggregated parts are equal to the whole. In chemistry also the axiom fails. In the consideration of motive it fails; for two motives, each of a given value, have

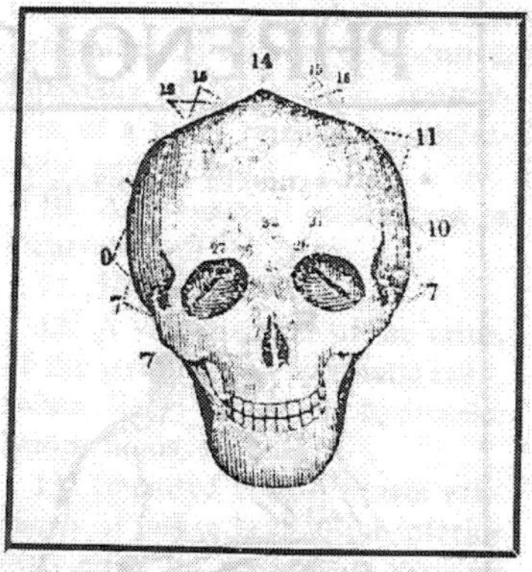

not, necessarily, a value when united, equal to the sum of their values apart. There are numerous other mathematical truths which are only truths within the limits of *relations.*"

The sincere student of human nature could hardly be blamed for investigating Phrenologists' claims — to discover Truth though a mere application of fingertips was too tempting an avenue to ignore. Far less frustrating and time-consuming than trial-and-error, and easier than a lifetime of hard, often unrewarding study, Phrenology must have seemed as though it might be able to live up to its promises. Particularly satisfying to anyone is the hope that Humankind may be piercing the veil of hidden truth — on the very threshold of one of God's ultimate horizons. To comprehend the infinite has always been the irresistible goal of humans. Phrenology was one of the now-discarded stepping stones toward that never-ending compulsion to *know.* ★

PHRENOLOGICAL TYPES

A WELL-BALANCED ORGANIZATION.

WASHINGTON

George Washington is a prime example of a well balanced temperament, combining all the conditions of power, activity and susceptibility, allowing neither icy coldness nor passions burning heat to rule his well governed feelings. He had sound common sense, far-seeing sagacity and brilliance.

SHARP AND ANGULAR, OR EXCITABLE.

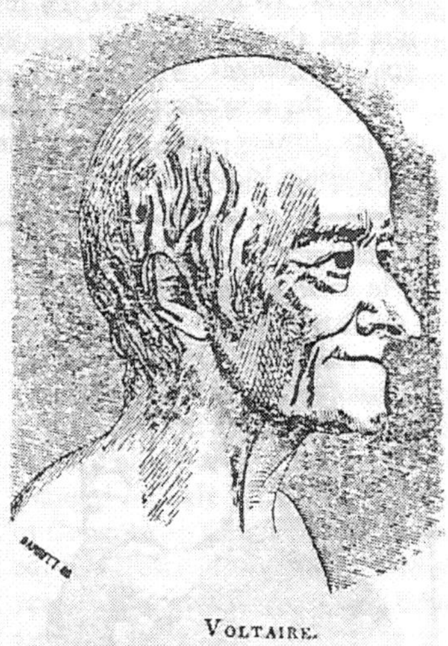

VOLTAIRE.

Some men resemble closely members of the animal Kingdom and have corresponding characteristics. Daniel Webster, the "Lion of the North" takes hold of great undertakings which he pursues with indomitable energy, roars so that a nation quakes, demolishes his victim and is an intellectual King among men. He bears a strong resemblance in his stout form, square face and mouth, large nose and open countenance, to the King of beasts.

THE LION FACE.

DANIEL WEBSTER

Francois-Marie Arouet de Voltaire is a prime example of the "Sharp and Angular" face. Since structure corresponds with character, Phrenology reveals he had ardent desires, intense feelings, keen susceptibilities, quick temperament and a great deal of passion, or what we call SOUL. This temperament prevails in brilliant writers who are too refined and sensitive for the mass of mankind.

FOOTNOTES

1 "Practical Phrenology" by O.S. Fowler, Fowlers and Wells, New York, 1853, p. 5.
2 IBID, p. 5.
3 "Phrenology Proved, Illustrated and Applied" by O.S. and L.N. Fowler, Fowlers and Wells, New York, 1836, p. 25.
4 "The Illustrated Self-Instructor in Phrenology and Physiology" by O.S. and L.N. Fowler, Fowlers and Wells, New York, 1855, p. 1.
5 IBID, p. 1.
6 "Phrenology Proved, Illustrated and Applied," p.iii
7 IBID, p. 287.
8 IBID, p. 287.
9 IBID, p. 286.
10 IBID, p. 7.
11 IBID, p. 7.
12 IBID, p. 7.
13 "Practical Phrenology," p. 7-9.
14 IBID, p. 7-9.
15 "Self-Instructor," p.vii, viii.
16 IBID, p. 9, 10.
17 IBID, p. 10.
18 IBID, p. 11.
19 IBID, p. vii.
20 "Phrenology Proved," p. 29.
21 IBID, p. 31, 32.
22 IBID, p. 32, 33.
23 IBID, p. 39.
24 IBID, p. 421.
25 IBID, p. 274.
26 IBID, p. 428.
27 IBID, p. 430.
28 IBID, p. 415.
29 IBID, p. 264.
30 IBID, p. 415.
31 IBID, p. 354.
32 IBID, p. 380-406.
33 IBID, p. 412, 413.
34 IBID, p. 421.
35 "Practical Phrenology," p. 5.

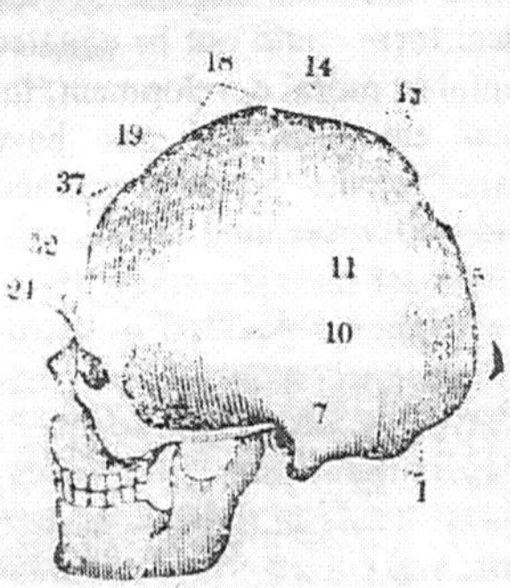

Poetic License

THE VOLUNTEER'S BURIAL

By Park Benjamin

'Tis eve; one brightly beaming star
Shines from the eastern heaven afar,
To light the footsteps of the brave,
Slow marching to a comrade's grave.

The northern wind has sunk to sleep;
The sweet South breathes, as, low and deep,
The martial clang is heard, the tread
Of those who bear the silent dead.

And whose the form, all stark and cold,
Thus ready for the loosened mould,
And stretched upon so rude a bier?
Thine, soldier, thine! the Volunteer.

Poor Volunteer! the shot, the blow,
Or swift disease hath laid him low;
And few his early loss deplore—
His battle fought, his journey o'er.

Alas! no wife's fond arms caressed,
His cheek no tender mother pressed,
No pitying soul was by his side,
As lonely in his tent he died.

He died—the Volunteer—at noon;
At evening came the small platoon
That soon will leave him to his rest,
With sods upon his manly breast.

Hark to their fire! his only knell—
More solemn than the passing bell;
For, ah! it tells a spirit flown,
Unshriven, to the dark unknown.

His deeds and fate shall fade away,
Forgotten since his dying day,
And never on the roll of Fame
Shall be inscribed his humble name.

Alas! like him, how many more
Lie cold upon Potomac's shore!
How many green unnoted graves
Are bordered by those placid waves!

Sleep, soldier, sleep! from sorrow free,
And sin and strife. 'Tis well with thee.
'Tis well: though not a single tear
Laments the buried Volunteer!

—Evening Post.

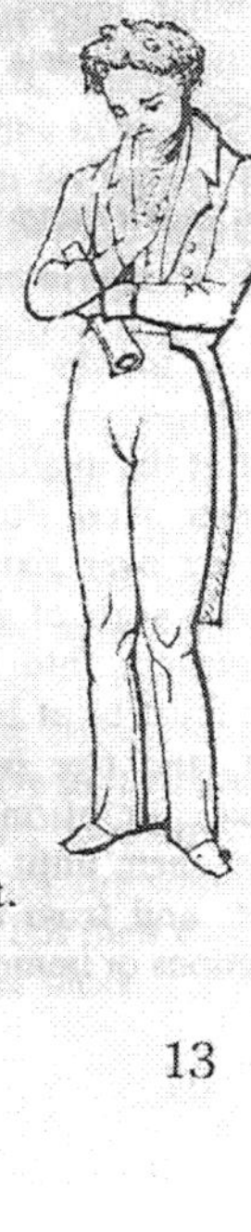

"I FIGHTS MIT SIGEL!"

By Grant P. Robinson

I met him again, he was trudging along,
His knapsack with chicken was swelling:
He'd "Blenkered" these dainties, and thought it no wrong
From some secessionist's dwelling.
"What regiment's yours? and under whose flag
Do you fight? said I, touching his shoulder;
Turning slowly around he smilingly said,
For the thought made him stronger and bolder:
"I fights mit Sigel!"

The next time I saw him his knapsack was gone,
His cap and canteen were missing,
Shell, shrapnel, and grape, and the swift rifle-ball
Around him and o'er him were hissing.
How are you, my friend, and where have you been,
And for what and for whom are you fighting?
He said, as a shell from the enemy's gun
Sent his arm and his musket "kiting:"
"I fights mit Sigel!"

And once more I saw him and knelt by his side,
His life-blood was rapidly flowing;
I whispered of home, wife, children and friends,
The bright land to which he was going;
And have you no word for the dear ones at home,
The "wee one," the father or mother?
"Yaw! yaw! said he, "tell them! oh! tell them I fights"—
Poor fellow! he thought of no other—
"I fights mit Sigel!"

We scraped out a grave, and he dreamlessly sleeps
On the banks of the Shenandoah River;
His home or his kindred alike are unknown,
His reward in the hands of the Giver.
We placed a rough board at the head of his grave,
"And we left him alone in his glory,"
But on it we marked ere we turned from the spot,
The little we knew of his story—
"I fights mit Sigel!"

—Moore's Rebellion Record

Boys into Cadets:
Military Education in the United States
PART I
"Cadet Life at West Point"

By
S. L. Kotar and J. E. Gessler

The two principle military institutions in this country prior to and during the Civil War were the United States Military Academy at West Point, and the Virginia Military Institute located at Lexington, Virginia. Both schools sent officers into the ranks of that conflict, and these men contributed greatly to the strategies that shaped those bloody years. But before the War Between the States called them to serve their respective causes, these young men were educated under a framework of Army rules and regulations. Why they entered West Point, their cadet appointments and the first weeks of their life at the Military Academy are examined in Part I.

In the mid 1800's it was generally known that West Point provided officers for the army, but specific information on the Academy and its workings was scarce. In "Cadet Life at West Point," J.B. Kinsman, after his father expressed the wish of seeing his son become a cadet, writes: "I knew not a single law relative to the Military Academy. I only knew that a certain number of young men were there trained for military duty, under the direction of the general government; that they were appointed by some high authority at Washington; that they were supposed, by all our neighbors at least, to graduate there with unlimited mathematical and physical attainments; and — as the effect, probably, of eating raw beef, and drinking each day a ration of rum and gunpowder, stirred with a ramrod — to hold the little matter of human life in the most supreme contempt."[1]

J.B. Kinsman hailed from the East. In the West, young Hiram Ulysses (Ulysses S.) Grant was informed by his father during Christmas vacation in 1838 that he had applied for the appointment to West Point. Grant later wrote: "I really had no objection to goint to West Point, except that I had a very exalted idea of the acquirements necessary to get through. I did not believe I possessed them, and could not bear the idea of failing."[2]

Other boys knew even less regarding the Military Academy. After reporting and "seeing the movements and learning the duties which I had to perform I came to the conclusion that I never could consent to live the life."[3] Similar ignorance caused many conditional cadets to drop out within a year.

"Perform with facility and accuracy..."

Upon application to the War Department, a prospective cadet would be informed of the qualifications needed to enter West Point. "Candidates must be over sixteen and under twenty-one years of age, at the time of entrance into the Military Academy; must be at least five feet in height, and free from any deformity, disease, or infirmity, which would render them unfit for the military service, and from any disorder of an infectious or immoral character. They must be able to read and write well, and perform with facility and accuracy the various operations of the four ground rules of arithmetic, of reduction, of simple and compound proportion, and of vulgar and decimal fractions...the candidate must write in a fair and legible hand, and without any material mistakes in spelling, such sentences as shall be dictated by the examiners; and he must answer promptly, and without errors, all their questions in the above-mentioned rules of arithmetic; failing in any of these particulars, he will be rejected."[4] (In 1812, the qualifications were: fourteen to twenty-one years of age, shall be well versed in reading, writing, and arithmetic, and that he shall sign articles, with the consent of his parent of guardian...)[5]

"The opportunities and the prestige"

Aside from a scientific education, entrance into West Point was an attractive proposition for boys in a financially embarressed position. This was an inducement for R.E. Lee to apply, as stated in William Fitzhugh's letter of recommendation. "His own age (eighteen I believe) and the situation of his mother require that he should lose no time in selecting the employment to which his future life is to be devoted."[6]

Another inducement to apply to the USMA, as in the case of U.S.

Grant, was a father's desire to prevent his son from following the life of a "down-the-river trader."[7] When a letter from the U.S. Senator from Ohio arrived, Jesse Grant (Ulysses' father) mentioned the appointment for the first time. " 'Ulysses, I believe you are going to receive the appointment.' 'What appointment?' I inquired. 'To West Point; I have applied for it.' 'But I won't go,' I said. He said he thought I would, *and I thought so too, if he did.*' "[8]

For Thomas Jonathan Jackson, the opportunity to receive a free education was a necessity. As with many other boys, the attraction was not the life of a soldier, but rather the opportunities and the prestige afforded by the Military Academy. The strict physical and mental regime he hoped would improve his health, and was a challenge he felt equal to meet. Acknowledging his education was incomplete, Jackson nevertheless felt he could "make it up in study. I know I have the energy and I think I have the intellect."[9]

James Ewell Brown Stuart was studying at Emory and Henry College when his appointment came through. He had expected to become a teacher; his application to West Point was a sudden decision. Perhaps he felt the security and the challenge of an army life offered more opportunity than a career as a pedagogue. For Dabney Maury, an appointment saved him from an undistinguished career as a lawyer. Maury summed up his law instructor's attitude towards him in these words: "I fear he (Judge Lomax) realized from the first that I would not prove a bright and shining light in my adopted profession..."[10]

J.B. Kinsman's inspiration to ascribe to the life of a soldier came from his father. The elder Kinsman had been so impressed by the arrival of an army officer at their home in pursuit of a deserter, that he "unaccountably fixed his determination" that his son, if possible, go to West Point and become a soldier. "I had never fully appreciated my military propensities until this morning."[11]

The opportunity of a free education in exchange for military service to the United States was a great incentive to the fathers who wanted to elevate their sons' positions in life, and for the young men themselves, who felt their own futures appeared rather bleak.

APPOINTMENT

The next step in their quest for a West Point cadetship was to obtain the much sought after nomination from a member of Congress. In 1843, it was established that "By provision of law, each Congressional and Territorial district, and the District of Columbia, is entitled to have one cadet at the Military Academy, and no more. The district appointments are made on the nomination of the member of Congress representing the district at the date of the appointment. The law requires that the individual selected shall be an actual resident...Also, appointments 'at large,' not to exceed ten, are annually made." Application must be made by letter to the Secretary of War at any time by the candidate himself, his parent, guardian or by any friend. His name is then placed on the register with no preference given on account of priority. "The appointments will be made annually in the month of February or March, on the applications made within the preceding year. The claims of all the candidates on the register will be considered and acted upon...as a general rule, no person will be appointed who has had a brother educated at the institution."[12]

The appointment of a young man to the Military Academy was an important part of the patronage of the office, and usually reserved to reward the fathers of those who had helped to elect the Congressman. Upon investigating charges that the appointment system led to the extension of the Congressman's personal interest, and that they tended to favor the wealthy and powerful, the authorities instituted an official inquiry into the pecuniary circumstances and occupation of the cadets' fathers. From 1842-63, it was determined that only one-eighteenth of those admitted could have received more than a common education if left to their own resources, and a still smaller number of officers in the army had any means of support beyond their pay.[13]

THE ACADEMY. (Looking Southeast.)
Erected, 1815; Destroyed by Fire, Feb. 19th, 1838.

Prior to 1843, when the regulations requiring an applicant to live in the district he was appointed from were changed, Congressmen could give these vacancies to their peers, thus allowing one such representative to accumulate several appointments. When in 1838 Jesse Grant wrote to the Honorable Thomas Morris, U.S. Senator from Ohio, regarding an appointment for his son, he found that it had been given away to a Senator from Pennsylvania. This led him to finally write to the Honorable Thomas L. Hamer, an estranged friend, who immediately filled out the proper forms, taking this opportunity to renew the friendship. The appointment was approved shortly thereafter.[14]

Providence and determination took a hand in T.J. Jackson's appointment to West Point. When a vacancy opened up in Lewis County, Virginia, four candidates appeared for the preliminary examination in Weston. Gibson Butcher won the nomination, and left for the Academy in May of 1842. Unable to bear the rigid military discipline, he returned home. He immediately advised young Jackson of the vacancy, who quickly gathered letters of recommendation and traveled to Washington, D.C., to present himself in person to Samuel L. Hays of the House of Representatives. Jackson presented a letter of resignation from G.J. Butcher, and petitions addressed to John C. Spencer, the Secretary of War, from important men of Lewis County. Congressman Hays forwarded these letters, as well as his own recommendation to the Secretary. The conditional appointment was approved, and Jackson left for New York.[15]

Nominations to West Point were far easier to obtain for sons of influential families, such as the Maurys or Lees. Maury was the son of a navy captain and member of a distinguished family which included Commodore Matthew Fontaine Maury. Lee, son of "Lighthorse" Harry Lee of Revolutionary War fame was endorsed by five Senators, three representatives, an older brother, Charles Carter Lee, his half-brother, Henry Lee IV, his teacher, W.B. Leary and the family advisor, William Fitzhugh, the latter writing directly to the Secretary of War. Robert presented this letter in person to Calhoun, then, having "brought to bear all the influence his family could exert,"[16] he waited. He was appointed March 11, 1824, to be admitted July 1, 1825.

"School of the Soldier"

After being instructed to bring "the articles marked thus," and if able, the sum of $60.95, to be deposited with the treasurer,[17] the cadet was directed to report between the 1st and the 20th of June.[18] Upon his arrival at the Military Academy, the "plebe" was directed to his quarters in the barracks and checked in by upperclassmen. Here they were given the rudiments of military etiquette "receiving a severe rebuke at every superfluous word, as well as for every unmilitary movement."[19] Shortly afterward, the plebes were sent to the Commissary's store. They were issued a military blanket, pillow, looking-glass, bucket, washbowl, cocoanut dipper, soap, broom, candles, candlestick, bottle of ink, twelve sheets of letter-paper, an arithmetic, slate and two slate pencils.[20]

The new "late importations" received their first instruction in the Manual of Military House-keeping. This consisted of keeping their room in order and the consequences for non-compliance. The first opportunity for the new cadets to view their class came at the signal for repairing to the parade-ground for dinner roll-call. "There were now about sixty of us, and a motley looking crowd it was. The pale-faced taciturn youth of the city, the weather-beaten heroes of the plow, the gaping New Englander, the scowling

THE ACADEMIC BUILDING.

THE OBSERVATORY AND LIBRARY. (Looking Southeast.)

Southerner, the tobacco-chewing Westerner, all, from Maine to Texas, were represented.''[21] After a march in close order to the Mess Hall, which resulted in a most unmilitary spectacle and an awkward meal, cadets were then ordered to the barber to have their hair cropped, followed by instruction in the School of the Soldier.

Academic study consisted of recitations in arithmetic, grammar, geography and other subjects the cadets would be tested in for the entrance exams. An hour in length, these classes were taught by third year cadets superintended by the professors.

"Rigid Examination"

The fourth class was called to the hospital, where they were ordered to strip and present themselves before the surgeon and assistant surgeon stationed at West Point, and another medical officer specifically ordered there for this duty. This "rigid examination" was used to disqualify those whose health would impair his efficiency as a soldier, under the following guidelines:

1. Feeble constitution and muscular tenuity; unsound health from whatever cause; indications of former disease; glandular swellings, or other symptoms of scrofula.
2. Chronic cutaneous affections, especially of the scalp, or any disorder of an infectious character.
3. Severe injuries of the bones of the head; convulsions.
4. Impaired vision from whatever cause; inflammatory affections of the eyelids; immobility or irregularity of the iris; fistula lachrymalis, and c., and c.
5. Deafness; copious discharge from the ears.
6. Loss of many teeth, or teeth generally unsound.
7. Impediment of speech.
8. Want of due capacity of the chest, and any other indication of a liability to a pulmonic disease.
9. Impaired or inadequate efficiency of one or both of the superior extremities on account of fractures, especially of the clavicle, contraction of a joint, extenuation, deformity, and c., and c.
10. An unnatural excurvature or incurvature of the spine.
11. Hernia.
12. A varicose state of the veins, of the scrotum and spermatic cord, (when large) sarcocele, hydrosele, hemorrhoids, fistulas.
13. Impaired or inadequate efficiency of one or both of the inferior extremities on account of varicose veins, fractures, malformation, (flat feet, and c.,) lameness, contraction, unequal length, bunions, overlying or supernumerary toes, and c., and c.
14. Ulcers, or unsound cicatrices of ulcers likely to break out afresh.

(1862)[22]

Height was determined by standing on a platform, questions were asked concerning hereditary consumption and eyesight was tested 'by some colors, coarse paint, etc., at the opposite side of a twelve by fourteen room. In short, a very few minutes sufficed to show whether or not I was physically qualified for my profession.''[23]

The academic examination was conducted in the library. The board was composed of the superintendent of the Military Academy, a high officer of the engineering corps, the Commandant of Cadets and the professors. Cadets were asked to write on the blackboard adding and conversion of decimals and fractions: were asked to read from a book, and submit samples of spelling and penmanship by writing dictated sentences, such as, "Remember now thy Creator in the days of thy youth."[24] The standards were not meant to be strict. "Sam French, a New Jersey Quaker boy who entered with Sam Grant, found the examination so trivial that, in later years, he had difficulty in recalling it: "I only remember Capt. W.W.S. Bliss (principal assistant professor in mathematics) asking us some questions in a polite manner, and then dismissing us."[25]

THE MESS-HALL. (Looking Southwest.)
Erected, 1815; Demolished, 1852.

THE ENCAMPMENT.

SUMMER ENCAMPMENT

The next morning, the Cadet Corps was ordered into summer encampment with the newcomers doing most of the heavy work. Tents were pitched, each with three or four occupants to simulate life in the field. Inspection was performed twice daily and the more advanced members of the fourth class were put on guard as sentinels. Muskets were issued as soon as the plebes became more proficient in marching.

Shortly after moving into camp, orders were published announcing the successful applicants. Those who were dismissed left immediately; new uniforms were issued as rapidly as possible for those who would officially become the fourth class on July 1st. In 1802, Congress had enacted a law that the pay of cadets be set at $16/month and two rations. In 1845, the pay was set at $24/month and no rations. Twelve years later, this amount was increased to $30/month. The term of service began July 1st. Cadets were required to serve four years at the Academy (although at various times a fifth class was attempted) and four years of army service. This had been changed on July 5, 1838 by an Act of Congress — formerly cadets served five years, including the four years of cadetship which had been established by Congress by the act of April 29, 1812.[26]

Wearing the cadet uniform made the new plebe feel more at home. R.E. Lee was issued the "regulation gray uniform, four pair of white trousers, a blue fatigue jacket and trousers, two silk stocks, and a cap (black leather, bell crowned, seven inches high, with a polished leather visor, diamond-shaped yellow plate, an eight inch black plume for dress parade, and, for less formal use, a leather cockade...the eagle and the yellow scales that could be fastened in front or under the chin.)"[27] U.S. Grant was measured for a uniform coat of sheep's gray (swallow-tailed and with a high quilted collar) for $10.88, a gigtop cap for $2.44, and white pantaloons at $10.50 a dozen.[28]

The summer encampment continued until the end of August, during which time the new cadets were drilled, walked post and became familiar with the peculiar peer system by which the third class demonstrated their superiority in the form of hazing.

Tri-weekly dances were held from 8 P.M. to 9:50 P.M. in the Academic building. Instruction in dance was allowed during the encampment, but members of the fourth class "were not expected to attend the dancing-parties, unless it were to accompany some relative who was visiting the Point...(but were permitted) under certain restrictions, to bathe for an hour in the Hudson."[29]

At the end of August, the furloughed class returned, and those plebes of the fourth class ("Seps")[30] who had been excused for the sum-

THE NORTH BARRACKS. (Looking Northeast.)
Erected, in 1817; Demolished, in 1851.

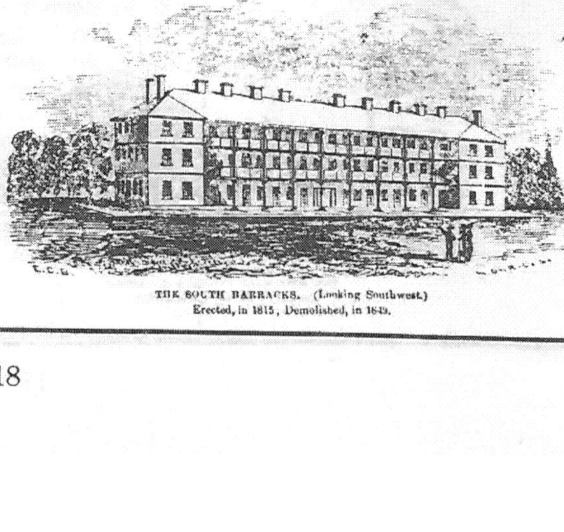

THE SOUTH BARRACKS. (Looking Southwest.)
Erected, in 1815, Demolished, in 1849.

mer, reported. They were examined, as had the earlier arrivals, and upon successful completion, became part of the fourth class. The tents were then struck and the Corps marched into winter quarters to prepare for the advent of the academic session.

THE WEST POINT HOTEL. (Looking Northeast)

Over the next four years these men would compete amongst themselves in academic standing and conduct. Examinations in January and June would weed out many of their number until graduation, when only a fraction of their number were left. (In the class of 1846, one hundred and sixty-four entered;[31] only fifty-nine graduated.)[32] These boys "of whom few had received either social or educational advantages of a very high order"[33] before they had come to West Point were now leaving as brevet second lieutenants in the United States Army.

THE CHAPEL. (North Front)

FOOTNOTES

1 "Cadet Life at West Point" by J. Burnham Kinsman, T.O.H.P. Burnham, Boston, 1862, p. 25.

2 "Personal Memoirs of U.S. Grant" by Ulysses S. Grant, Press of J.J. Little & Co., Nos. 10 to 20 Astor Place, New York, 1885, Vol. I, p. 32.

3 "Stonewall Jackson" by Lenoir Chambers, William Morrow & Co., New York, 1959, p. 50.

4 "United States Military Academy at West Point, Admission by Competitive Examination, Extract From Report of Visitors, for 1863," F.C. Brownell, New York, 1863, p. 559, 560.

5 "History of West Point" by Captain Edward C. Boynton, D. Van Nostrand, New York, 1863, p. 344.

6 "R.E. Lee" by Douglas S. Freeman, Charles Scribner's Sons, New York, 1934, Vol. I, p. 39.

7 "Captain Sam Grant" by Lloyd Lewis, Little, Brown and Company, Boston, 1950, p. 55.

8 "Personal Memoirs of U.S. Grant," Vol. I, p. 32.

9 "Family and Early Life of Stonewall Jackson" by Ray Bird Cook, Charleston Printing Co., Charleston, 1948, p. 84, 85.

10 "Recollections of a Virginian" by Dabney H. Maury, Charles Scriner's Sons, 1894, p. 19.

11 "Cadet Life at West Point," p. 24-26.

12 "United States Military Academy at West Point, Admission by Competitive Examination," p. 559.

13 "History of West Point," p. 230.

14 "Captain Sam Grant," p. 56, 57.

15 "Family and Early Life of Stonewall Jackson," p. 84-90, 92.

16 "R.E. Lee," Vol. I, p. 39-44.

17 "Cadet Life at West Point," p. 42,43.

18 "History of West Point," p. 267.

19 "Cadet Life at West Point," p. 53, 54.

20 IBID, p. 57.

21 IBID, p. 57.

22 "United States Military Academy at West Point, Admission by Competitive Examination," p. 560.

23 "Cadet Life at West Point," p. 100, 101.

24 IBID, p. 105.

25 "Captain Sam Grant," p. 68.

26 "The Centennial of the United States Military Academy at West Point, New York, 1802-1902," Government Printing office, Washington, 1904, Vol. II, p. 97.

27 "R.E. Lee," p. 51, 52.

28 "Captain Sam Grant," p. 68.

29 "Cadet Life at West Point," p. 143.

30 IBID, p. 148.

31 "Recollections of a Virginian," p. 22.

32 "Biographical Register of the Officers and Graduates of the U.S. Military Academy at West Point, New York" by Bvt. Major-General George W. Cullum, D. Van Nostrand, New York, 1868, p. 180.

33 "Recollections of a Virginian," p. 22.

WHIMSEY

Mr. J. once objected to the competency of a witness alleging that he was non-compos. The Court granted leave to test the matter.

'Can you tell me my friend, the difference between LIKEWISE and ALSO?'

'May be az I can,' replied the witness.

'Go on sir, let us hear.'

'Well you see Col. P. is a lawyer.'

'Very well,' said the counsel.

'And you is a lawyer ALSO.'

'Very well.'

'Col. P. is LIKEWISE a gentleman.'

'Very well.'

'But you is not likewise.'

The lawyer was dumb.

(Boston Cultivator, 1845)

"Bones," said Ginger, "which had you rather ride in, a stage coach or a steamboat?"

"Why I'd rather ride in a stage coach bekase if it upsets THAR YOU is; but if de steamboat blows up WHAR is you?"

(Boston Cultivator, 1845)

PRICES CURRENT

Feathers are DOWN — cheese is LIVELY and holds its own — dealers in grindstones complain of HARD TIMES — holders of whiskey are somewhat STAGGERED by recent transactions — the corn-stock market is a GRAIN lower — sad-irons are FLAT, and the holders have burnt themselves — cutlery is DULL generally except in one article which it is NEEDLESS to mention. The coal market looks black and heavy — it has BIT-TWO-MEN to our certain knowledge. Newspapers come in well — the smut is found not to injure the crop, Patent medicines DRUG the market, and Brandreth is considered below PARR. Boots and shoes are a PEG LOWER — Parasols OPENED rich last week, and are consequently a SHADE above PA without any MISS-take.

(Norfolk Democrat, 1845)

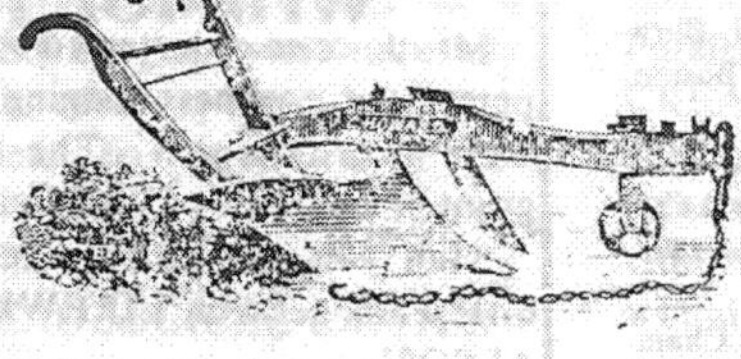

THE GENTLEMAN AGRONOMIST

Rotation of Crops

It is indeed a matter of astonishment, that at this day after so great improvements have been made in agriculture, and so much light thrown upon the subject by scientific men of almost every country, any should be found, who will not give assent to the doctrine of rotation; or who will not admit that a judicious system of alternation of Crops should be adopted by every one who would cultivate the earth with success. Yet it is no less true than strange, that there are thousands who still cling to the old way of cropping their fields continually with the same kind of grain or plant, until they hardly make a return of the seed they have received. To such men, farming will prove to be a sinking business. These too, are the very men whom we always hear finding fault with their crops, and almost charging the Wise Dispenser of all things, as being less favorable with them than with some of their more wise and prudent neighbors. They do not once think, that they have violated the laws of nature, and placed every obstacle in the way of her performing her kind offices. The truth is, they have tired out the land. It has become exhausted of the specific food which the plant requires, and utterly refuses to produce. Such farmers may sow, but they cannot reap; and if they persist in this ruinous course, they must sooner or later feel the evil effects of an empty purse. No man ought to expect a return for his labor, unless he give back to the soil, in some form, a part at least, of what he takes from it. It is to obviate in a great measure the evils consequent upon the practice of taking all, and returning nothing to the land, that is proposed to be effected by the rotation system. We do not say, that none of those who follow the old and hackneyed path, have raised large crops, or been in a degree successful in the business of farming; on the contrary, we know that by heavy manuring and high cultivation, land may be made to produce one kind of grain, perhaps abundantly; for a number of years in succession. But we do say, that the soil cannot without great expense of labor and manure, produce two or three crops of a kind without a perceptible falling off in the product. Neither is it contended that manure can be dispensed within the rotation system. But one manifest advantage which this system has over the other modes of cultivation is, it enable the farmer so to economize in the management and use of his manure, as that he may receive a two fold benefit from it. For instance, his manure may be applied in an unfermented state to roots and other hoed crops, and they receive their supply of nourishment from it, and at the same time it remains unimpaired for the use of the small grains that succeed. We will now proceed to state a few general principles on which this system is founded. And they are principles that are established by scientific investigations and experiments, and should be familiar to everyone engaged in this most laudable pursuit.

And 1st. Soils, however fertile, or highly cultivated, will lose their productiveness, if continually cropped with the same kind of plant.

2d. The degree in which a plant impoverishes the soil, depends much on the amount of food it returns to it, in the decomposition of its stalks and roots that remain.

3d. One plant draws its nourishment from a depth in the ground, another from the surface.

4th. Some plants receive nearly all of their food from the earth, while others are fed almost wholly from the atmosphere.

5th. The cultivation of the small grains, renders the land foul; the hoed crops tend to free it from weeds.

6th. Those plants that are permitted to ripen their seeds, are great exhausters of the soil; while those that do not mature their seeds, exhaust it comparatively little. If these principles are founded in truth, they will serve us a guide in arranging our different crops of grain, roots and grasses into a regular system of rotation. In making choice of crops to alternate with, care should be had to select those that are best adapted to the soil. In this, the judgement of the farmer will enable him to decide; as he is best acquainted with the nature of the land that he cultivates. It will also be an easy matter for him to determine which kind of plants should, and which should not succeed each other.

In conclusion, we would say that a system, whose operations conform so completely to nature's laws, and the effects of which are so well calculated to improve the soil, and the condition of the farmer, commends itself to the consideration of all, and it is to be hoped that this subject will receive the attention which its importance demands. ★

(Albany Cultivator.)
Boston Cultivator,
December 13, 1845.

TRACTS "Consecration of a Jewish Synagogue"

Interesting And Imposing Ceremonies — Addresses by Rev. Dr. Raphall And Rev. Messrs. Isaacs And Bondi.

The consecration services of the new synagogue of the congregation B'nai Israel, situated at the corner of Forsyth and Stanton streets, took place yesterday, which, according to the Jewish mode of reckoning time, is called the 17th day of Abb., 5620. The building was formerly occupied as a place of worship by the Baptist demonination, and has been remodelled for the purpose to which it is now devoted.

In the centre of the synagogue is located the *el-menor* or reading desk, where the *chazan* (corresponding to the minister in Christian churches) officiates. It is an elevated platform, seven or eight feet square, surrounded by railings, with steps leading to it on either side. Near the front of it is the inclined desk, covered with a mantle of red and gilt satin, upon which are laid the holy books used in divine services.

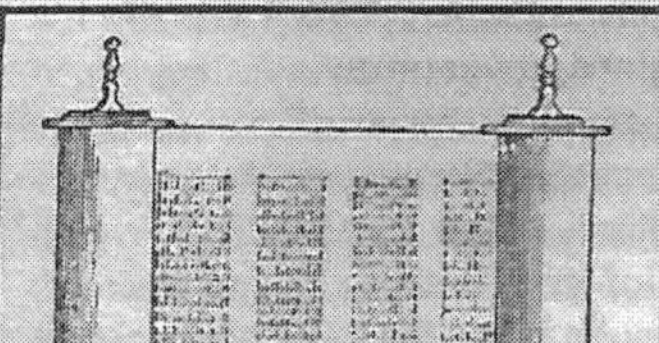

Form of a Synagogue Roll of the Pentateuch.

In that portion of the building where in other churches the chancel and altar are located is the ark. Here, in a semicircular recess, the *Hichol* or sacred scrolls of the law of Moses are kept. The whole device of the arc is in imitation of the portico of a Grecian temple. An ornamented architrave projecting over the top is supported by two fluted columns, which are surmounted by Ionic capitals and rest upon octagonal pedestals of marble. A flight of three steps, arranged in a semicircle, lead to the platform in front of the ark where the lectures (or sermons) are delivered.

Between five and six hundred members of the congregation (the number which the building comfortably holds,) assembled yesterday, to witness and take part in the consecrative ceremonies, which were under the direction of the Rev. M.R. De Leeuw. The male portion, according to the Jewish custom, occupied the body of the synagogue, and the females the galleries. On either side of the ark were seated the Rev. Dr. Raphall, Rev. J.H. Lyons, Rev. Mr. Isaacs, Rev. E.L. Goldsmith, and Rev. Mr. Bondi.

At half-past two o'clock the Rev. Dr. Raphall, Chief Rabbi of New York, who was dressed in a long flowing gown or toga of black silk, assumed the *talette,* a cloak of white satin with purple and black stripes, and commenced the services, which were conducted in the Hebrew tongue. Going outside the synagogue, he knocked at the door and exclaimed:—

Open the gates of righteousness for me that I may enter through them to praise the Lord.

To which the choir, gathered at the reading desk, responded:—

This is the gate of the Lord, into which the righteous shall enter.

The doors were opened and the *chazan* chanted:—

How awful is this place!

To which the choir responded:—

This is none but the house of God, and this is the gate of heaven!

The *chazan* then ascended the reading desk, and, assisted by the choir, sang:—

How goodly are thy tents, O Jacob! thy tabernacles, O Israel! And in the greatness of thy benevolence, will I enter thine house; in reverence of thee, will I bow down towards the temple of thine holiness. O Lord! I have loved the habitation of thine house and the dwelling place of thy glory. I, therefore, will worship, bow down, and bend the knee before the Lord, my Maker. And I will offer my prayer unto thee, O Lord, in an acceptable time: in thine abundant mercy, O God, answer me in the truth of thy salvation.

At the conclusion of which a procession of little girls to the number of thirty, between the ages of five and ten years, dressed in white with blue sashes and spangles, with baskets of flowers in their hands and wreaths of green leaves upon their heads, entered the synagogue, and marching to the front took up a position on either side of the space between the ark and reading desk. The officers and honorary officers of the congregation, covered with the *talette,* then brought the sacred scrolls of the Holy Law to the door, and thence proceeded to the ark while the *chazen* and choir sang:—

Blessed is he who cometh in the name of the Lord!

We bless you from the house of the Lord.

A low rumbling noise was then heard, and the heavy curtains in front of the ark and the circular doors glided back, disclosing the recess in which the scrolls were to be deposited. These scrolls, five in number, with two rollers, upon one of which the parchment folds as it comes off the other, contain copies of the five books of Moses, and were beautifully encased in velvet, satin and silk, with gold and silver ornamentation.

The seven circuits around the synagogue were then commenced, the children strewing the bearers of the scrolls with flowers each time they passed them, until the floor beneath is covered with the offerings from their little baskets.

The *chazen* and choir chaunted hymns, and as the procession arrives before the unveiled ark each time the bearers were relieved of their willing burthen by new aspirants to this highly esteemed honor.

At the termination of the seventh circuit the sacred scrolls were carried up the steps and placed in the ark. The lighting of the perpetual lamp, a duty devolving upon the oldest member in the congregation, then took place. The lamp, which was after the fashion of those used by the ancients, with the taper at one extremity, was suspended in front of the ark, upon which its never dying light is intended to shine.

The *chazen* than chaunted:—

Blessed are thou, O Lord, our God, King of the Universe, who hath preserved us alive, sustained us and brought us to enjoy this season.

To which a response was made by the congregation.

After which the Rev. Dr. Raphall, donning the *talette* as before, advanced to the platform in front of the ark, which latter he touched with the hem of his garment, and reverently kissed it. Then turning to the people, he read from his text the words, "How beautiful are thy tents, oh Jacob," and addressed them in English...

The *chazan* and choir then chaunted a prayer for the government, in which the federal, State and city officials were remembered...

After a chaunt by the *chazan* and choir as before, Rev. Mr. Bondi, late of the Norfolk street congregation, followed with an address in German. He felt happy, he said, in aiding in the consecration of a synagogue. The Jews had been free in Holland, France and Belgium, but in America religious liberty was guaranteed to the fullest extent. There were no limits here to the practice of any religion, and the Jews had good reason to be proud of their new synagogue. The Jews **might belong to different nations, but the covenant kept them together in one solid bond of opinion and feeling.**

After the various addresses came the consecration prayer, and the exercises closed with the singing of the hallelujah. ★

(THE NEW YORK HERALD,
Morning Edition — Monday,
August 6, 1860)

SIGNALS

Dear Editors:

I have read KEPI and found it to be truly outstanding. The Garnett material was well presented and of interest. Your publication is neither dry or dull and even contains a little humor. Humor, you must realize, is normally banned from historical publications for fear they may become interesting. Enclosed is my check for a year's subscription and I will be looking forward to future issues. I predict KEPI will become a highly successful publication.

Thomas J. Evans,
Vienna, Virginia

Dear Editors:

I received my copy of THE KEPI and just wanted you to know I think you did a terrific job! Most folks have no idea of the effort that goes into ANY publication, especially one as nice as yours! Congratulations!! I read almost every article and approve of your intent to delve into "contemporary feelings" and not sit back and judge a century later as is so often done.

Bill Bellinger,
Atlanta, Georgia

Dear Editors:

I am *very* impressed with your magazine and would like to subscribe to THE KEPI. I wish you much success with your "new concept in Civil War reporting".

Marylou K. Gillman,
Prospect, Conn.

Dear Editors:

I have just received your First Issue of THE KEPI and I really love it. I would like to subscribe to the magazine.

Darrell W. Stridh,
Whittier, Calif.

(We would like to thank everyone who has written us with kind words and encouragement. Your support is greatly appreciated. E's-in'G.)

Dear Editors:

I have received, approved and enjoyed the copy of THE KEPI. I took it to our chapter meeting and let the members see it, and also explained how our museum happened to receive the excellent coverage about the "Stonewall" Jackson medallion. I feel this publication will be of great interest to our visitors. Thanks again for THE KEPI, and I do wish you great success for the publication.

Mrs. Dewey Wood,
Warren Rifles Conf. Museum
Front Royal, Virginia

(Mrs. Wood supplied us with invaluable information on the Jackson Medallion and is secretary of the excellent museum in Front Royal, Va., which has several of the medallions, as well as the shell tribute Mrs. Jackson made for her husband. The museum is open from April 15—Nov. 1 and is a very worthwhile trip. E's-in-G.)

LEXICON

A TABLE OF SIMPLE AND DIPHTONGAL VOWELS REFERRED TO BY THE FIGURES OVER THE LETTERS IN THIS DICTIONARY

ENGLISH SOUNDS	FRENCH SOUNDS
1. a The long slender English *a*, as in fate, paper, etc.	*é* in *fée, épée.*
2. a the long Italian *a*, as in fa, fa-ther, pa-pa, mam-ma.	*a* in *fable, rable.*
3. a. The broad German *a*, as in fall, wall, wa-ter.	*â* in *âge, Châlcus.*
4. a. The short sound of the Italian *a*, as in fat, mat, mar-ry.	*a* in *fat, matin.*
1. e. The long *e*, as in me, here, me-tre, me-dium.	*i* in *mitre, epitre.*
2. e. The short *e*, as in met, let, get.	*e* in *mette, nette.*
1. i. The long diphtongal *i*, as in line, ti-tle.	*aï* in *laïque, naïf.*
2. i. The short simple *i*, as in pin, tit-tle.	*i* in *inn′e, tittr′e.*
1. o. The long open *o*, as in no, note, no -tice.	*o* in *globe, lobe.*
2. o. The long close *o*, as in move, prove.	*ou* in *mouvoir, pouvoir.*
3. o. The long broad *o*, as in nor, for, or; like the broad a.	*o* in *or, for, encor.*
4. o. The short broad *o*, as in not, hot, got.	*o* in *hotte, cotte.*
1. u. The long diphthongal *u*, as in tube, cube, cu-pid.	*iou* in *Cioutat, chiourme.*
2. u. The short simple *u*, as in tub, cup, sup.	*eu* in *neuf, veuf.*
3. u. The middle or obtuse *u*, as in bull, full, pull.	*ou* in *boule, foute, poule.*
oi. The long broad o, and the short i, as in oil.	*oï* in *cycloide, heroique.*
ou. The long broad o, and the middle obtuse u, as in thou, pound.	*aoû* in *Aoûte.*
Th. The acute or sharp *th*, as in *th*ink, *th*in.	
TH. The grave or flat TH, as in THIS, THAT.	

When *G* is printed in the Roman character, it has its hard sound in *get, gone, &c. as go, give, geese, &c.; when it has its soft sound it is spelled in the notation by the consonant J*, as giant, ginger, *ji-ant, jin-ger.* The same may be observed of *S:* the Roman character denotes its hard sound in *sin, sun, &c.; its soft sound is spelled by z* as *rose, raise, &c.; roze, raze, &c.*

AID-DE-CAMP, ade-de-kang′. s. An officer who attends the General that has the chief command of the army, to carry his orders to the inferiour officers. *Ash.*

This word, like most other military terms from the French, is universally adopted; but the polite pronunciation of the nasal vowel in the last syllable is not to be attained by a mere Englishman.

ALGEBRA, al′je-bra. s. 84. A peculiar kind of arithmetic.

To APPROPINQUE, ap-pro-pink′. v. a. To draw near to. *Not in use.*

To BESLUBBER, be-slub′bur. v. a. To daub, to smear.

BIRCHEN, bur′tshn. a. 103, 405. Made of birch. An Englishman may blush at this cluster of consonants for a syllable; and yet this is unquestionably the exact pronunciation of the word; and that our language is full of these syllables without vowels.

BLINKARD, blink′urd. s. 98. One that has bad eyes; something twinkling.

COLT'S-TOOTH, kolts-tooth′. s. An imperfect tooth in young horses; a love of youthful pleasure.

COLURES, ko-lurz′. s. Two great circles supposed to pass through the poles of the world.

COMMANDER, kom-man′dur. s. He that has the supreme authority, a chief; a paving beetle, or a very great wooden mallet.

GRIG, grig. s. A small eel; a merry creature.

GRIP, grip. s. A small ditch.

To GRIPE, gripe. v. a. To hold with the fingers closed; to catch eagerly; to seize; to close, to clutch; to pinch, to press, to squeeze.

GRIZZLY, griz′zle. a. Somewhat gray.

HUMORIST, yumur-ist. s. One who conducts himself by his own fancy, one who gratifies his own humor.

This word is often, though improperly, used for a jocular person.

To LAMM, lam. v. a. To beat soundly with a cudgel. *A low word.*

LANDLOPER, land′ lo-our. s. 98. A landman; a term of reproach used by seamen, of those who pass their lives on shore. This word is improved by seamen into the more intelligible word *Land-lubber.*

LANGUOR, lang′gwur. s. 166, 344. A faintness which may arise from want or decay of spirits.

LANSQUENET, lan′sken-net. s. A common foot soldier; a game at cards.

This word, as a game at cards, is altered by the vulgar into *Lambskinnet.* This is something, at least, which they understand; and this very intelligibility confirms them in the corruption.

MALE, mále. a. Of the sex that begets young not female.

To MULCT, mŭlkt. v. a. To punish with fine or forfeiture.

MUM, mŭm. s. Ale brewed with wheat.

MUMPS, mŭmps. s. Sullenness, silent anger; a disease.

MUSKCAT, mŭsk'kăt. s. The animal from which musk is got.

To POZE, pòze. v. a. To puzzle.

PRATINGLY, prà'tĭng-lè. ad. 410. With tittle-tattle, with loquacity.

To SCREAK, skrèke. v. n. 227. To make a shrill or harsh noise.

To SCONCE, skŏnse. v. a. To mulct, or fine.

SHEKEL, shĕk'k'l. s. 102. An ancient Jewish coin in value about two shillings and six pence sterling.

SOWINS, sòù'ĭnz. s. 323. Flummery made of oatmeal, somewhat soured.

STULTILOQUENCE, stŭl-tĭl'lò-kwĕnse. s. 518. Foolish talk.

SUPERVACANEOUSNESS, sù-pĕr-vă-kà'nè-ŭ-nĕs. s. Needlessness.

VAMPYRE, văm'pire. s. Vampyres were imaginary beings; supposed to be the souls of guilty persons, who tormented the living by sucking their blood when asleep. The belief of these beings was very common about a century ago in Poland and some parts of Germany.

To WHIZ, hwĭz. v. a. To make a loud hissing noise.

Articles of War

An Act For Establishing Rules And Articles For The Government Of The Armies Of The United States.*

SECTION 1. *Be it enacted, by the Senate and House of Representatives of the United States of America, in Congress assembled,* That, from and after the passing of this act, the following shall be the rules and articles by which the armies of the United States shall be governed:

ARTICLE 1. Every officer now in the army of the United States shall, in six months from the passing of this act, and every officer who shall hereafter be appointed shall, before he enters on the duties of his office, subscribe these rules and regulations.

ART. 2. It is earnestly recommended to all officers and soldiers diligently to attend divine service; and all officers who shall behave indecently or irreverently at any place of divine worship shall, if commissioned officers, be brought before a general court-martial, there to be publicly and severely reprimanded by the president; if non-commissioned officers or soldiers, every person so offending shall, for his first offense, forfeit one-sixth of a dollar, to be deducted out of his next pay; for the second offense, he shall not only forfeit a like sum, but be confined twenty-four hours; and for every like offense, shall suffer and pay in like manner; which money, so forfeited, shall be applied, by the captain or senior officer of the troop to which the offender belongs.

ART. 3 Any non-commissioned officer or soldier who shall use any profane oath or execration, shall incur the penalties expressed in the foregoing article; and a commissioned officer shall forfeit and pay, for each and every such offense, on dollar, to be applied as in the preceeding article.

ART. 4 Every chaplain commissioned in the army or armies of the United States, who shall absent himself from the duties assigned him (excepting in cases of sickness or leave of absence), shall, on conviction thereof before a court-martial, be fined not exceeding one month's pay, besides the loss of his pay during his absence; or be discharged, as the said court-martial shall judge proper.

ART. 5. Any officer or soldier who shall use contemptuous or disrespectful words against the President of the United States, against the Vice-President thereof, against

*These rules and articles, with the exceptions indicated by the notes annexed to articles 10, 20, 65, and 87 remain unaltered and in force at present.

the Congress of the United States, or against the Chief Magistrate or Legislature of any of the United States, in which he may be quartered, if a commissioned officer, shall be cashiered, or otherwise punished, as a court-martial shall direct; if a non-commissioned officer or soldier, he shall suffer such punishment as shall be inflicted on him by the sentence of a court-martial.

ART. 6. Any officer or soldier who shall behave himself with contempt or disrespect toward his commanding officer, shall be punished, according to the nature of his offense, by the judgement of a court-martial.

ART. 7. Any officer or soldier who shall begin, excite, cause, or join in, any mutiny or sedition, in any troop or company in the service of the United States, or in any party, post, detachment, or guard, shall suffer death, or such other punishment as by a court-martial shall be inflicted.

ART. 8. Any officer, non-commissioned officer, or soldier, who, being present at any mutiny or sedition, does not use his utmost endeavor to suppress the same, or, coming to the knowledge of any intended mutiny, does not, without delay, give information thereof to his commanding officer, shall be punished by the sentence of a court-martial with death, or otherwise, according to the nature of his offense.

ART. 9. Any officer or soldier who shall strike his superior officer, or draw or lift up any weapon, or offer any violence against him, being in the execution of his office, on any pretense whatsoever, or shall disobey any lawful command of his superior officer, shall suffer death, or such other punishment as shall, according to the nature of his offense, be inflicted upon him by the sentence of a court-martial.

ART. 10. Every non-commissioned officer or soldier, who shall enlist himself in the service of the United States, shall, at the time of his so enlisting, or within six days afterward, have the Articles for the government of the armies of the United States read to him, and shall, by the officer who enlisted him, or by the commanding officer of the troop or company into which he was enlisted, be taken before the next justice of the peace, or chief magistrate of any city or town corporate, not being an officer of the army,* or where recourse cannot be had to the civil magistrate, before the judge advocate, and in his presence shall take the following oath or affirmation: "I, A. B., do solemnly swear, or affirm (as the case may be), that I will bear true allegiance to the United States of America, and that I will serve them honestly and faithfully against all their enemies or opposers whatsoever; and observe and obey the orders of the President of the United States, and the orders of the officers appointed over me, according to the Rules and Articles for the government of the armies of the United States." Which justice, magistrate, or judge advocate is to give to the officer a certificate, signifying that the man enlisted did take the said oath or affirmation.

ART. 11. After a non-commissioned officer or soldier shall have been duly enlisted and sworn, he shall not be dismissed the service without a discharge in writing; and no discharge granted to him shall be sufficient which is not signed by a field officer of the regiment to which he belongs, or commanding officer, where no field officer of the regiment is present; and no discharge shall be given to a non-commissioned officer or soldier before his term of service has expired, but by order of the President, the Secretary of War, the commanding officer of a department, or the sentence of a general court-martial; nor shall a commissioned officer be discharged the service but by order of the President of the United States, or by sentence of a general court-martial.

ART. 12. Every colonel, or other officer commanding a regiment, troop, or company, and actually quartered with it, may give furloughs to non-commissioned officers or soldiers, in such numbers, and for so long a time, as he shall judge to be most consistent with the good of the service; and a captain, or other inferior officer, commanding a troop or company, or in any garrison, fort, or barrack of the United States (his field officer being absent), may give furloughs to non-commissioned officers or soldiers, for a time not exceeding twenty days in six months, but not to more than two persons to be absent at the same time, excepting some extraordinary occasion should require it.

ART. 13. At every muster, the commanding officer of each regiment, troop, or company, there present, shall give to the commissary of musters, or other officer who musters the said regiment, troop, or company, certificates signed by himself, signifying how long such officers, as shall not appear at the said muster, have been absent, and the reason of their absence. In like manner, the commanding officer of every troop or company shall give certificates, signifying the reasons of the absence of the non-commissioned officers and private soldiers; which reasons and time of absence shall be inserted in the muster-rolls, be remitted by the commissary of musters, or other officer mustering, to the Department of War, as speedily as the distance of the place will admit.

ART. 14. Every officer who shall be convicted before a general court-martial of having signed a false certificate relating to the absence of either officer or private soldier, or relative to his or their pay, shall be cashiered.

ART. 15. Every officer who shall knowingly make a false muster of man or horse, and every officer or commissary of musters who shall willingly sign, direct, or allow the

*By Sect. II of Chap. 38, August 3, 1861, the oath of enlistment may be administered by any commissioned officer of the army.

signing of muster-rolls wherein such false muster is contained, shall, upon proof made thereof, by two witnesses, before a general court-martial, be cashiered, and shall be thereby utterly disabled to have or hold any office or employment in the service of the United States.

ART. 16. Any commissary of musters, or other officer, who shall be convicted of having taken money, or other thing, by way of gratification, on mustering any regiment, troop, or company, or on signing muster-rolls, shall be displaced from his office, and shall be thereby utterly disabled to have or hold any office or employment in the service of the United States.

ART. 17. Any officer who shall presume to muster a person as a soldier who is not a soldier, shall be deemed guilty of having made a false muster, and shall suffer accordingly.

ART. 18. Every officer who shall knowingly make a false return to the Department of War, or to any of his superior officers, authorized to call for such returns, of the state of the regiment, troop, or company, or garrison, under his command; or of the arms, ammunition, clothing or other stores thereunto belonging, shall, on conviction thereof before a court-martial, be cashiered.

ART. 19. The commanding officer of every regiment, troop, or independent company, or garrison, of the United States, shall, in the beginning of every month, remit, through the proper channels, to the Department of War, an exact return of the regiment, troop, independent company, or garrison, under his command, specifying the names of the officers then absent from their posts, with the reasons for and the time of their absence. And any officer who shall be convicted of having, through neglect or design, omitted sending such returns, shall be punished, according to the nature of his crime, by the judgment of a general court-martial.

ART. 20. All officers and soldiers who have received pay, or have been duly enlisted in the service of the United States, and shall be convicted of having deserted the same, shall suffer death, or such other punishment as, by sentence of a court-martial, shall be inflicted.*

ART. 21. Any non-commissioned officer or soldier who shall, without leave from his commanding officer, absent himself from his troop, company, or detachment, shall, upon being convicted thereof, be punished according to the nature of his offense, at the discretion of a court-martial.

ART. 22. No non-commissioned officer or soldier shall enlist himself in any other regiment, troop, or company, without a regular discharge from the regiment, troop, or company in which he last served, on the penalty of being reputed a deserter, and suffering accordingly. And in case any officer shall knowingly receive and entertain such non-commissioned officer of soldier, or shall not, after his being discovered to be a deserter, immediately confine him, and give notice thereof to the corps in which he last served, the said officer shall, by a court-martial, be cashiered.

ART. 23. Any officer or soldier who shall be convicted of having advised or persuaded any other officer or soldier to desert the service of the United States, shall suffer death, or such other punishment as shall be inflicted upon him by the sentence of a court-martial.** ★

From the "Revised Regulations For The Army Of The United States, 1861, Philadelphia: J.B. Lippincott & Co., Publishers, pp. 499-503.

Concurrent Events

The steamship Europa, from Liverpool, August 18th, via Halifax, arrived in Boston yesterday forenoon, Sept. 2, 1854. The news she brings:

The object of the re-assembling of the Vienna Conference was stated to be to declare that the propositions of England and France as to the basis for a treaty of peace are in conformity with the protocol of Vienna and to ascertain the most fitting means to ensure the existance of Turkey and to connect it with the general balance of power in Europe.

☐ MEXICAN AFFAIRS

New Orleans papers contain details of the Mexican news of August 22, 1854:

Santa Anna had issued a decree enabling the religious orders to avail themselves of the civil tribunals to enforce the monks and clergy to an adherence of their vows.

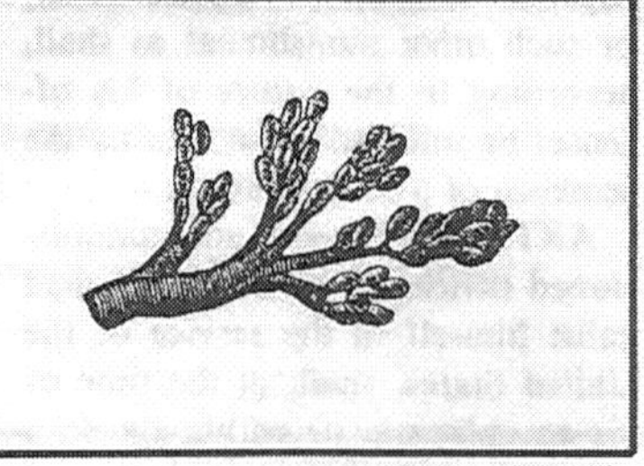

*No officer or soldier in the army of the United States shall be subject to the punishment of death, for desertion in time of peace.—*Act 29th May,* 1830.
**See above.

THE NEW CADET BARRACKS.

THE KEPI

12115 Magnolia Blvd.
P. O. Box 107
No. Hollywood, CA 91607

BULK MAIL

THE KEPI

AUGUST-SEPTEMBER 1983 $3.00

THE KEPI

12115 Magnolia Blvd.
P. O. Box 107
No. Hollywood, CA 91607

BULK MAIL

THE KEPI

AUGUST-SEPTEMBER 1983 VOLUME I, NUMBER 3

THE ARTICLES

THE IRREGULARS

Editors-in-General S. L. Kotar and J. E. Gessler. THE KEPI, 12115 Magnolia Boulevard, P. O. Box 107, North Hollywood, California 91607.
Return postage must accompany all inquiries and submissions. THE KEPI reserves the right to edit, and copyright all letters.

Generally Speaking

In this issue of THE KEPI we are concluding our series on military education with an article on the Virginia Military Institute. VMI, unlike the National School at West Point, was a state school that depended upon support from influential men to move appropriations through the Virginia Legislature, and also their donations to support expansion into non-military pursuits. The policies at VMI were influenced by these wealthy men, whose private goals helped direct the Institute toward the creation of an agricultural school. This article follows the course of the Virginia Institute from its inception, through its reasons for instituting the military form of education, to its eventual digression into courses for the civilian graduate. Different from the usual material presented on this famous school, THE KEPI offers its readers an alternate view of the policies and principles that influenced education in the United States.

We are also introducing in this issue two new Regular Columns: CAMPFIRE; ECHOES OF WAR and UNOFFICIAL REPORTS. The former features an article written by Thomas J. Evans, a record of a first-hand account of how a Union soldier's body was taken home after he was killed in the Battle of Perryville, Kentucky. Mr. Evans is a resident of Fairfax County, Virginia, where his Civil War articles appear in several local historical publications. He is a regular contributing writer for "North-South Trader," and is currently editor of the John S. Mosby Society's periodical, "The Partisan Review." Mr. Evans' work will be highlighting ECHOES OF WAR over the coming months, with more unpublished, first-hand accounts of the war. The Editors are pleased to present these excellent insights in THE KEPI, adding the histories of the unsung soldier to the drama of the Civil War.

The second column being introduced is UNOFFICIAL REPORTS, or the 1860's newspaper correspondents' versions of battles and leaders. Exactly as written and presented to the anxious folks back home, UNOFFICIAL REPORTS represents a continuing policy of reporting the mid-19th Century as it was actually viewed by those who lived it.

Also featured in this issue is TRACTS, presenting a discussion of a typical soldier's religious pamphlet; "ELOCUTION!", a study of learning the art of public speaking; and "SOAK IT!", a humorous look at the "newfangled" concept of taking a shower.

S.L. Kotar *J.E. Gessler*
Editors-in-General

Head-Quarters of THE KEPI
1st August, 1983

GENERAL ORDERS,
No. 2
by Command of the Editors-in-General

Coming in the October-November issue of THE KEPI is an article dealing with the controversial delay of General R.S. Ewell in ordering an attack on the right flank of the Union army in the Wilderness, on the 6th of May, 1864. Included is a letter written by Campbell Brown, Ewell's assistant adjutant-general, disputing Gen'l. Early's claim to any credit for the evening attack.

ECHOES OF WAR continues with another unpublished narrative of how two religious sects, the Covenantors and Quakers ran, the underground railroad prior to the Civil War.

MEDICAL MANUAL features an article on mesmerism and its advocates' rationale on how minds were actually magnetized when in a trance.

TRACTS contains an 1862 report from a missionary describing Sorcery and Sacrifices among the Hindoos, and the regular columns, CONCURRENT EVENTS and WHIMSEY also appear, and the Articles of War continue.

UNCLASSIFIEDS

RATES FOR THE KEPI ADVERTISING: 50¢ a word 1 time; 40¢ a word 3 times; $100 full page ad; $55 ½ page ad; $30 ¼ page ad. Logo run free; all typesetting at no charge.

Phil Barber

Antique newspapers and paper Americana bought and sold

GENUINE CIVIL WAR NEWSPAPERS: **New York Times** 1863 (front page headlines) $4.50; **Harper's Weekly** (lavishly illustrated) $8.00. Choice condition, free lists, satisfaction guaranteed.
PHIL BARBER Box 8694-K Boston, MA 02114

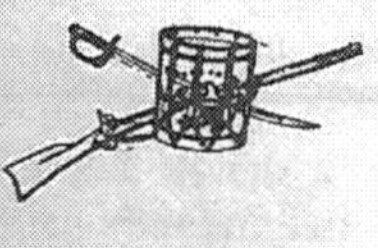

12,000 Civil War Books, Paper, & Images
from 75 leading dealers, authors, publishers:
at the
3rd Great American Civil War Book, Paper, & Image Fair
SAT., OCT. 22, 1983, 9 AM-5 PM at the Elks Lodge
8421 Arlington Blvd. (U.S. 50), Fairfax, VA (D.C. suburb)
Details: CWBF, 10453 Medina, Richmond, VA 23235, 804-272-5558

Leading Specialists In The American Civil War and Related Subjects

ESSENTIAL PRESS
BOOKSELLERS & PUBLISHERS
WHOLESALE DISTRIBUTOR

OFFICIAL ATLAS — FINALLY REPRINTED (9-1-83), $29.95. NEW LIST $1.00. ESSENTIAL PRESS, 10453K MEDINA, RICHMOND, VA 23235.

ZULLO & VAN SICKLE BOOKS
12137 Darnestown Road
Gaithersburg, MD 20878

OFFERING Civil War Books, rare editions, personalized service and extensive title search. Over 7000 volumes in stock. Free catalogues, published monthly. Inquiries welcome. Phone 1-301-963-7878.

Investment quality
ANTIQUE FIREARMS • UNIFORMS
EDGED WEAPONS
ACCOUTREMENTS

and hundreds of other items related to the American military from all periods between 1775-1917
Emphasis on the Civil War and Indian War periods.
Offered for sale in our
UNIQUE CATALOGS

The largest and best of their type in the country.
Issued bi-monthly, totalling 180 pages of fascinating, concise, photo-illustrated listings every year.

ANNUAL SUBSCRIPTION RATE $6.00
FREE RENEWAL TO ACTIVE BUYERS

The source for a wide range of Americana.

DALE C. ANDERSON CO.
DEPT A 32 TAYLOR ST • HAMPTON, NH 03842

Bellinger's Military Antiques

Bimonthly catalogs of guns, swords, belt plates, books, leather goods, knives, documents, personal items, images, accoutrements, etc. of the Indian War period and earlier with emphasis on the Civil War. Detailed descriptions and fair prices. $6 for six issues. Overseas subscription $10 (airmail $13.50).
Bill Bellinger - Box 76371- K
Atlanta, Georgia 30328

WANTED . . . WANTED
CIVIL WAR ITEMS
BOOKS, DOCUMENTS, TIN TYPES, LETTERS, ENVELOPES, REUNION BADGES, DIARIES, FLAGS, CONFEDERATE MONEY & BANK NOTES.
ALSO HAVE CIVIL WAR ITEMS FOR SALE

by John L. Heflin, Jr.

5708 BRENTWOOD TRACE
BRENTWOOD, TN 37027
PHONE 1-615-373-2917

THE CIVIL WAR ROUND TABLE ASSOCIATES

P. O. Box 7388
Little Rock, Arkansas 72217

Campfire; Echoes of War

"Carrying Isaiah Home"

By
Thomas J. Evans

[This interview was given to Tom Evans by James W. Lusk, a resident of Southern Indiana. Mr. Lusk, now long deceased, left no relatives, and never married. He was the last of the Lusk family in Indiana, which makes the death of his Granduncle, the subject of this narrative, even more tragic.]

"Of course I will be glad to tell you all about my Granduncle being killed in the Battle of Perryville, Kentucky, for I have heard it a thousand times from my Grandfather. My Granduncle's name was Isaiah Lusk and he was a cousin of Lieutenant Samuel Reid who the Grand Army of the Republic Camp in town was named after. Sam Reid was killed in the battle at Richmond, Kentucky, and his body never was positively identified. His folks never gave up hope he would come home and everytime they heard of a veteran with mental problems, "shell shocked" they got to calling it in later wars, why they would go and see if it was him. They must have spent a fortune after the war hunting around and never did find out what became of him.

Well, Isaiah Lusk took a notion to go to the army and help save the Union. Like a lot of the boys, he was afraid the war would be over before he had a chance to get in the fighting. They took to "beating up" a company over in town, and he knew all the boys joining it, and nothing would do but he had to go and join up too and go with them to teach those Rebels a lesson. Company D, 38th Regiment Indiana Volunteer Infantry they became. They went down to Elizabethtown, Kentucky and trained some so when Confederate General Bragg set out to invade the state they were ready for him. The Rebels fought right along up through the state and the main Union Army met them there at Perryville. They got into a real bloody battle on October 8, 1862.

Day after the battle a fellow rode out from town to tell Grandpa his brother Isaiah was killed in the fighting. That regiment had thirty boys killed that day, about a hundred and twenty five wounded, and they lost some prisoners too. Their flag was so shot up it was taken back to Louisville and put on display for folks to see. Now there wasn't anything for Grandpa to do but to go on down there and carry Isaiah home for a decent burying. It was a good hundred miles to Perryville so he hitched two horses to a light spring wagon and had Grandma pack him a bunch of food and stuff and a jug of water. He threw in some blankets, a shovel, and some feed for the horses and took off for town. Folks there gave him some letters in case he saw any of the boys, and asked him to bring back any news about the local boys who were wounded.

Grandpa made thirty miles to Louisville that first day and stopped over to buy a coffin but couldn't find one of the right size and had to give it up. He drove all the next day hearing news about the battle and how the Confederate Army was retreating. He keep meeting wagons loaded with wounded soldiers and every schoolhouse and church down there was being turned into a hospital. Well, it was getting late and he was tired and was talking to a Kentuckian about where to find a place to stay when this fellow up and says, "Why you will come home and take supper with me and stay at my house." So he did and the fellow wouldn't take a dime for it and said he was just sorry to hear about Isaiah being killed and wished he could do more.

Now it was late on the third day since Grandpa left home before he got to Perryville. He hunted around and found a field hospital and went over to see what was going on, and he had some real luck when he found several Indiana boys from the 38th Regiment. Some of Isaiah's friends had been detailed there as nurses and to help around, and they wouldn't have it any way but what they would go with Grandpa the next morning and help him dig Isaiah up. They knew just where he was and some of them had helped bury him. Grandpa allowed as how he would appreciate the help and passed out the mail he had brought. Then he asked around and wrote down who was wounded and how they were recovering.

Now Grandpa knew he would be another three days getting back home so he reckoned as how he had better see about getting Isaiah embalmed. There were several embalming surgeons had tents or rented outbuildings around there and were doing a good business. All they had to do was lay some planks or a door across some saw horses or barrels for a table and they were ready to work. Why, they would take a dead officer that hadn't been shot up too bad and when they got done stand him up in his open coffin and use him for a sign advertising their business.

Grandpa went over and stood around and watched one at work. They brought in an officer who had just died, stripped him down and washed him, and took his uniform and had it washed where it was bloody. Then the embalming surgeon pumped out the officer's blood with a little pump he had, and then he turned around and pumped that fluid they use back in. The embalming surgeon then massaged the face until that officer's eyes closed up good and tight then laid two flattened lead bullets on his eyelids to hold them shut.

The Embalming Surgeon says, "I use silver quarters to close eyes back home but don't dare to do it here for they would be gone and in somebody's pocket the minute my back was turned." Says, "Can't use copper pennies either because they sometimes cause the eyelids to turn green."

Well now, Grandpa told him he wanted to hire him but he wouldn't have none of it when he found out Isaiah was a private. Claimed he couldn't make any money off of privates and wouldn't sell any of his coffins either. Said he didn't have enough as it was with all the wounded officers dying. Grandpa went and talked to another one and didn't do any better and finally had to give it up. He didn't know what else to do so he went on back and told the Indiana boys what had happened. They said for him not to worry for they knew how to deal with embalming surgeons and

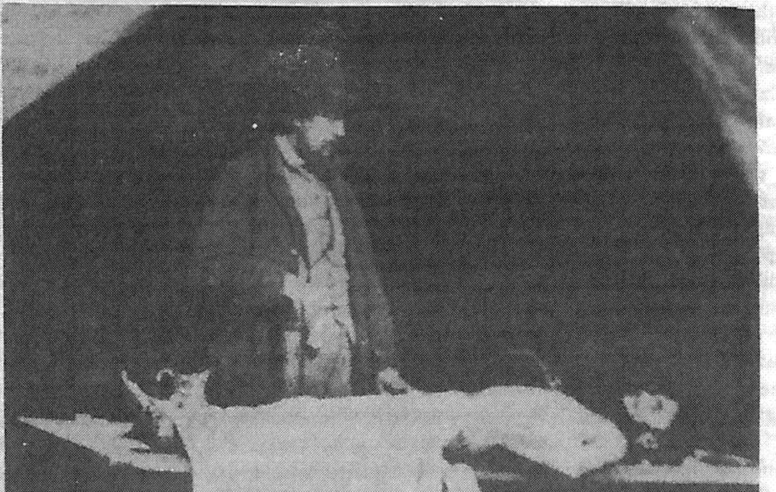

Working from an open tent, an Embalming Surgeon prepares a dead soldier for shipment home to relatives.

Very little was needed by an embalming surgeon to open his business. This barn near the battlfield served as a temporary office.

would go and take care of it in the morning.

Grandpa sat around a fire with the boys and they got to talking. The boys had some sad stories to tell about the battle. A sergeant of one company was laughing and joking with the boys and the next instant one of those cast iron shot from a cannon, grape shot I believe they call it, went plumb through his stomach. They dragged him over and propped him up against a tree and saw they couldn't do nothing for him he was bleeding so bad. He wanted a smoke so they got his pipe out of his knapsack and fixed it for him. He sat there under that tree calmly smoking until he bled to death.

There was another fellow there who saw this solid iron cannon ball rolling along the ground and put out his foot thinking he would stop it. The boys hollered at him not to do it but he went ahead. It took off his foot at the ankle as clean as if it had been amputated by a surgeon. They carried him to the hospital and took good care of him. He was coming along good when he up and took gangrene poisoning and died in no time.

Then there was this boy who had done good in the fighting. Soldier next to him had been hit and he had to brush that fellow's blood off of his uniform. It didn't seem to bother him and he went right on fighting. He had the bad luck to have to stand picket guard the night following the battle. Coming back to camp in the dark he tripped and fell in a little ravine. Well, a hospital wagon had chosen that ditch to dump a load of amputated limbs, and he fell in there with all of those arms, legs, feet, and hands and raised a terrible ruckus. They thought the Rebels were attacking the way he carried on so. Some boys went and pulled him out and something had gone wrong with his mind. He wouldn't say nothing to anybody and the surgeons didn't know what to do for him. He got discharged and ended up in a Soldier's Home I heard. They said he acted just like a little child but wouldn't say a word.

And there was a lot of boys lost an arm or a leg. Those big old soft lead bullets they used would just shatter a bone when they hit it and the surgeons couldn't do nothing but saw it off. After the war the Government got busy and made them artificial limbs but they weren't as good as they make today. Used willow wood for making them because it was so light and tough. Where those big hospitals were located, like around Washington, D.C., they say there is hardly any willow trees left along the creeks today they cleaned them out so.

There was some strange things happened, too. This one boy had a premonition he was going to get killed in the battle. Just came over him clear as day he said. He wrote a letter home saying goodbye to his family the night before the fighting, and he called his two closest friends together and gave one his watch and the other his ring to send home when he was killed. His friends tried to jolly him out of it and told him there wasn't much chance of any of them getting hurt. He wouldn't have none of it and wouldn't even eat. Said it would be a waste of food. Well now, the battle started and as they were marching along his friends heard a loud plop. They looked at him and he was plumb white in the face. A bullet had clean buried itself in the stock of his rifle and never even so much as scratched him. He said he guessed he was alright now and it wasn't his time to go after all. He was in the thick of the fighting and came out without a wound. The two boys he had given his watch and ring to hold was both killed. He found their bodies on the battlefield with their pockets turned out and everything they had on them gone. He never did get his things back.

Now Grandpa was wore out and had to give up talking and go to bed under the wagon. He was so worried he hardly slept at all that night. Come morning the boys had him a cup of hot coffee fixed and a good breakfast, and then they walked over to the hospital tents to see who had died during the night and quite a few had. The wounded boys left there were mostly the ones hurt too bad to send to a better hospital.

Soon up rides a boy on a mule passing out handbills and says, "My employer is an embalming surgeon, best in the business. He is new here and wants your trade. Give these here handbills to your officer friends. You can see his price is right."

The boys winked and made signs for Grandpa to keep quite and somebody asks, "When will this fellow be getting here?"

Says the boy, "Should be on the Springfield Pike by now so he will be here by midmorning. You won't have any trouble recognizing him. He is a big man driving two black horses. His coffins are following in some wagons later in the day."

So the boys all crowded around and took those handbills and promised they would pass them out to their sick and wounded officers and would give that embalming surgeon any business that came their way. Grandpa was a little surprised they were so glad to see that boy and that they asked him so many questions about that embalming surgeon. Well, Grandpa decided he had better go on and dig up Isaiah. The boys said to wait a few minutes as they had to get a couple of things and then some of them would go and help him. Soon up comes two privates who knew Isaiah, "Slim" and "Shorty." Those boys all had nicknames for each other and hardly ever called one another by the right name. They brought along one wild looking fellow Grandpa had not seen before who had real long hair falling down over his face, a long nose, and always had a big grin on his face. Grandpa said he made him think of a possum looking out of a brush pile. Of course this fellow's nickname was "Possum." Slim said he would drive the horses since he knew where to go, so he and Grandpa got up on the wagon seat and Shorty and Possum climbed in the back and sat down. Grandpa noticed Slim had a captain's coat and sword rolled up in a bundle and Shorty had a jacket with great big sergeant's stripes on it stuck in a bag with what appeared to be several pistols. They said the owners didn't need them right then and they had borrowed them. Possum was in his full privates uniform with all his equipment and had his rifle with a bayonet on it. They drove over the battlefield and saw folks still picking up dead Confederates and burying them. Fellow named Squire Bottom took and gave a hill top on his farm for a place to bury those Rebel boys and they were burying them there.

They rode on to the place where Isaiah was killed. He had been laying behing a log shooting at the Confederates when he shouted, "Boys, I just got me a Johnny Reb! Whoo—! Then he dropped his head down and was quiet. While he had his mouth open yelling a bullet went in it and came out the back of his neck killing him instantly. Never even broke his teeth! He took such a grip on his rifle they had trouble getting it away from him later they said.

Next to the road they had buried a bunch of the Union Army boys and those that were known had their names written in pencil on little flat boards stuck in the ground. The boys went right to where Isaiah was and had him up in no time since he wasn't buried very deep. He was just wrapped in his army blanket. They loaded him in the wagon and Slim said they would drive down the road aways and make a deal with an embalming surgeon.

They drove down the road a short piece and parked the wagon in the woods behind a curve in the road. Possum winks, and gets out with his rifle, and goes up the road and acts like he is on guard duty. Soon along comes this wagon pulled by two black horses and they recognize it is that embalming surgeon the boy on the mule was working for.

Possum steps out in the middle of the road and says, "Halt! Give the password!"

"I don't know any password," says the Embalming Surgeon. Says, "Let me by, you fool, or it will go hard with you!.

Possum looks kind of mean, steps back, cocks his gun and points it at the Embalming Surgeon and says, "I am the one that knows the password and has this here gun and here you are calling me the fool. I will blow a hole in you a rabbit can run through just to show you I mean business!"

"Wait now," says the Embalming Surgeon, "Don't get too hasty there! I am new at this army business."

Says Possum, "For all I know you may be that Rebel spy who is going around saying he is an embalming surgeon. I can get a medal if I shoot him and maybe a promotion too. What kind of business did you say you were in?"

"Now don't get excited here," says the Embalming Surgeon, "I am not a Rebel, just a poor man trying to make a living helping others in their time of need. I do follow the embalming trade, however, and keep my charge reasonable to all."

"Sergeant of the Guard," shouts

These dead soldiers standing in their coffins served as advertisements for the work of the Embalming Surgeon.

Possum, "I may have caught that Rebel spy! Best fetch the Captain up here quick!"

Now out of the woods hustles Slim and Shorty dressed in those borrowed uniforms holding big revolvers in their hands and looking mighty mean.

"Private, who is this man?" says Captain Slim acting all official like.

Says the Embalming Surgeon, "I am an embalmer new to this area and just want to do a little business." Says, "I tried to tell your man that, and I want you to watch him for he may shoot me yet!"

The three boys then back up and huddle in a little group and whisper to each other. Every so often they stop and look at the Embalming Surgeon real stern and he begins to get kind of scared as each minute passes.

Leaving the group Sergeant Shorty walks back and says to the Embalming Surgeon, "Just how much do you charge for your work?"

"Why I charge a hundred and fifty dollars for a general, seventy five for a colonel, and fifty dollars for doing a major. I don't like to deal any lower than that in rank but I will do a captain for forty dollars if I can work him in," says the Embalming Surgeon.

Asks Sergeant Shorty, "And just how much do you charge the folks back home?"

"That isn't a fair question," says he. Says, "I got overhead out here and the bodies are full of holes. I may only charge ten dollars to folks back there but I don't have to work in the open under these terrible conditions."

Sergeant Shorty walks back to the group and they start whispering and suddenly Possum says real loud, "I say let me shoot him!"

Captain Slim tells Possum to be quiet, then he walks back to the Embalming Surgeon and says, "Well, you may just be who you say you are since you have an honest face. We will have a test and see. The General's nephew who was killed in the battle is in a wagon just ahead and you can work on him. If you do a good job, we will pay you what you get at home and I will write a note to the General and he will see you get all the business you can handle. Also I will tell you the password to keep fellows like this private here from wanting to shoot you. On the other hand, if you can't do the job, why you are the Reb we are looking for and I am going to shoot you on the spot."

"Boys, I am ready to take your test," says the Embalming Surgeon.

They all went over to the wagon where Grandpa had been listening to the whole affair and by now had figured out why everyone had been so interested in talking to that boy on the mule. The boys told Grandpa to take a walk in the woods and they would call him when the Embalming Surgeon was finished. He walked aways and took a seat under a tree and read his pocket testament while he waited. Shorty called pretty soon and when he returned Isaiah was wrapped up in his blanket again.

"That body sure ain't as fresh as it might be, in fact, that boy is pretty far gone. You better put a wet towel full of soda water on his face or he will turn black soon," says the Embalming Surgeon. Says, "I can let you have a coffin in a couple of hours if the General can wait."

"Afraid he doesn't have the time," says Captain Slim, "but you have passed the test. Now the password is "Hang Jeff Davis" and you see any sentry or officer you just holler it right out. And here is a note to introduce you to the General, General Orders that is, and be sure and give him my regards when you find him. And here is ten dollars in greenbacks for your fine services. Anything happens to me I will have the boys here send for you."

The relieved Embalming Surgeon hastily departed in anticipation of the promised business, and the boys shook hands with Grandpa and wished him a safe journey home. He thanked them and then they changed out of those borrowed uniforms and disappeared into the woods. They were country boys, you know, and went back through those woods to camp in no time.

Grandpa headed on out for home stopping at a house along the road to inquire about getting some soda. They told him he had best see a Granny lady at the next house as she knew all about such things. She turned out to be a little old lady about eighty years old, but nothing would do but for her to make up some soda water and give Grandpa one of her best towels to use. She was as spry as could be and climbed right up in the wagon and bathed Isaiah's face and wrapped it in a wet towel. She said she had just got home from the church nearby where she was nursing the wounded boys, Rebels and Yanks alike. She wanted Grandpa to stay while she cooked up a bite to eat but he said he had to go on. She made him take some bacon sandwiches made with cold biscuits, and he said all his life they were the best he ever ate. Of course she wouldn't take any money and said to tell his mother how sorry she was about her boy.

Grandpa finally got back to Louisville and still couldn't find a coffin because of the demand for them. He gave up and telegraphed back home to the folks to start building one. You could buy coffin handles, hinges, and such hardware in most stores in those days and people could make their own. Grandpa had been sleeping under the wagon mostly since he left home, so he figured he would just push on and stop when he was too tired to go no more. Before he could get out of the city a shoe came loose on one of the horses, and a colored blacksmith — a freeman, opened up his shop at night and fixed it. When he found out what Grandpa was doing traveling so late, why he wouldn't charge nothing for the job.

Grandpa finally got home and had been gone most six days and was plumb wore out along with being upset about Isaiah being killed since they were very close to each other. They had the coffin ready and they put Isaiah in it right here on this porch. They went right on out to the cemetery where a grave had been dug by the neighbors, and had a quick little ceremony and buried him. My Great Grandmother, Isaiah's mother, took it real hard and took to her bed for a spell. The family got Isaiah a white gravestone and had the date he was killed carved on it and if you go back to the old Covenanter Cemetery you will see it there. They never did get him a government stone although he was entitled to one, and just to look you could never tell he was ever a soldier or that he gave his life for his country.

Come the end of the war Isaiah's comrades would stop by and see Grandpa. Slim came out and he and Grandpa rode out to the cemetery. Slim said he was glad Isaiah was buried there with family members where it was so quite and peaceful and not off somewhere in an unmarked grave like so many of the boys on both sides ended up.

Shorty never made it through the war. A little piece of cannon ball wounded him on top of the head. He and all the boys with head injuries were left to last for treatment at a field hospital because there wasn't much the surgeons could do for them in those days. They took a little piece of iron about the size of a dime out of his head wound and a piece of cloth from his cap that the fragment had carried in there. He never regained consciousness and died right there at the hospital.

Possum came by too. He was a kind of simple minded fellow but everybody liked that boy. Perryville kind of harmed him too in a way. They didn't bury those dead Rebels as soon as they should have. Hogs got to some of them first and Possum came along and saw what was going on. Vowed he would never eat hog meat again. The boys said after that the smell of pork cooking would make him sick to his stomach. He married well after the war and bought a big farm down in the river bottoms. For meat he had him some fish traps, raised chickens, and run a beef cow or two. Wouldn't have a hog on his place. His poor wife used to go and stay with her parents once in awhile just to get a bite of bacon.

So that is how my Granduncle Isaiah was killed during the Civil War and how my Grandfather drove a wagon down to Perryville, Kentucky and carried him back home wrapped in his army blanket.

Interview with James William Lusk at his Walnut Ridge farm, Washington County, Indiana during October, 1951. The Military Service Records, National Archives, Washington, D.C. indicate Private Isaiah R. Lusk was killed in action at Perryville, Kentucky. "Report Of The Adjutant General Of The State Of Indiana," Adjutant General's Office, Indianapolis, 1866, Vol. V, (1861-1865), p. 205 lists Isaiah R. Lus*h*, (name misspelled), as a private in Co. D, 38th Regiment Indiana Volunteer Infantry, with his date of muster as September 18, 1861. The organization of the 38th Regiment is shown in "Official Army Register Of The Volunteer Force Of The United States Army," Adjutant General's Office, Washington, 1865, Part VI, p. 92. Casualties of the 38th Regiment are contained in "Centennial History Of Washing-

PERRYVILLE AND THE SOUTHERN INVASION OF KENTUCKY

In the summer of 1862, General Bragg, newly appointed commander of the Confederate forces massed at Tupelo, Mississippi, needed a plan of attack that would recover territories lost after the retreat from Shiloh and Corinth earlier that year. Union troops controlled nearly all of Tennessee and Kentucky, as well as the waterways from St. Louis to Vicksburg. The Federals, under the command of G.W. Morgan, occupied the Cumberland Gap, and the main army, commanded by General Don Carlos Buell, controlled the area around Nashville.

Moving his forces to Chattanooga to save it from the advancing enemy, as well as to protect Georgia, Bragg found the way now open for him to move further into Tennessee and perhaps to Kentucky.

Forrest and John H. Morgan were sent ahead to harrass the Federal lines of communication while the Confederates made their move. On August 14th, Kirby Smith started for Rogers's Gap, to confront Morgan at Cumberland Gap. Discovering he was too strongly provisioned, Smith moved into Kentucky and marched for Lexington. En route, the Federals resisted his advance at Richmond, resulting in a battle outside the town. The Confederates won a decisive victory, killing 206, wounding 844 and capturing 4,303. In light of these activities in his rear, G.W. Morgan evacuated Cumberland Gap and moved through eastern Kentucky to Greenup.

Meanwhile, Bragg had left Chattanooga and was heading north for Louisville, with Buell in pursuit. The Federal garrison at Munfordville surrendered before Bragg's show of arms, resulting in 15 killed, 57 wounded and 4,076 prisoners. The Confederates were placed in strong positions to await the arrival of Buell, but the understanding that the Federal commander had been reinforced precipitated a general pullback to Bardstown. With Kirby Smith's troops still at Lexington representing Confederate occupation, Richard Hawes was installed as the Provisional Governor of Kentucky. The expected recruits from Kentucky did not come, however, and the gray army, reduced by the arduous campaign, found itself outnumbered as Buell marched from Louisville to confront them on October 1st. Sill and Dumont marched to Frankfort to threaten Smith, while the rest of the Union army followed Bragg.

Bivouacking near Perryville on the 7th of October, General Hardee formed his men for battle on the morrow. The country having been in a severe drought for several months, water was at a premium, and the Confederates stubbornly defended the few pools left in Doctor's Ford. This water was fought over on the 8th, and the Confederate offense was successful until late in the day, when their line was threatened. Darkness brought to the Confederates the realization that the Federals had brought superior numbers against them, and that a renewal of the conflict the next day was too dangerous. Hoping for an entire concentration of Confederate forces, a movement was made toward Harrodsburg. When it was evident Buell was not going to follow up the Battle of Perryville, Bragg then determined to retreat to Knoxville via the Cumberland Gap.

The Confederate invasion of Kentucky was thus concluded, and eventually both Buell and Bragg were deprived of their respective commands for not having defeated the other.

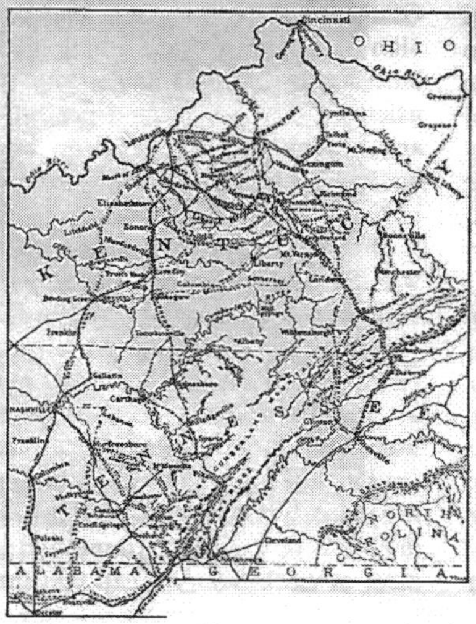

BRAGG'S INVASION OF KENTUCKY.

BRAGG —x—x— SMITH —+—+—
BUELL'S ARMY ....
RAILROADS

ton County Indiana," by Warder W. Stevens, B.F. Bowen & Co., Indianapolis, 1916, p. 308. Death of Isaiah Lusk at Perryville is also mentioned, IBID, p. 952. Information concerning the donation of land at Perryville by Squire Henry P. Bottom for the establishment of the Confederate Cemetery is contained in "Battles And Leaders Of The Civil War — Bragg's Invasion Of Kentucky," by General Joseph Wheeler, CSA, Castle Books, N.Y., 1956, (Reprint), Vol. III, (Illustration), p. 23. Route of the Springfield Pike through the battlefield is shown on the map, IBID, p. 24. The small gravestone of Isaiah Lusk may be seen in the Lusk family section of the Covenanter Cemetery, Walnut Ridge, Washington County, Indiana. Date of death is shown as October 8, 1862, date of the Battle of Perryville, Kentucky.

Boys into Cadets:
Military Education in the United States
PART II
"Military Discipline for a Civilian Life"

S. L. Kotar and J. E. Gessler

The Virginia Military Institute, located in Lexington, Virginia, in the upper Shenandoah Valley, was known as the "West Point of the South,"[1] prior to and during the Civil War. But unlike its Federal counterpart that served the entire country, VMI was, until 1859, exclusively a state school. Established in 1839 by the approval of the Virginia Legislature, VMI traced its origins back only a few short years to the citizens of Lexington. Members of the Franklin Debating Society, voicing the sentiments of the community, wished to rid Rockbridge County of the ill-disciplined troops that guarded the Lexington arsenal. Desirous of keeping the arsenal but not the soldiers, a plan was devised to create a military school, with cadets to watch over the old muskets and accoutrements. In exchange for this guard duty, the young men enrolled at the Virginia Military Institute would be given an education at state expense. The martial aspect of the school was in keeping with its duty to protect the arsenal and served the additional purpose of maintaining discipline, for there was fear in the state that no one could trust arms to "frolicsome boys."[2] Further developing this thought in an editorial against the establishment of VMI, it was argued, "Out of ten cadets appointed from this Congressional District to West Point, not one remained to graduate, not from the want of talent, but from the restive spirit of insubordination, which so shamefully characterizes our youth." (Buchanan Advocate, January 20, 1836.) Superintendent Smith would write in later years that he had to rely heavily on discipline for many bad subjects were sent to VMI to be "reformed."[3]

Principal Professor

The principal professor chosen for the infant institution was Francis Henney Smith, a Virginian and alumnus of West Point. Chosen over Joseph Reid Anderson because, as rumor had it, the Board of Visitors liked his wife better,[4] Smith had little actual military background. After graduating from the United States Military Academy in 1833, he served three years in the Army before resigning[5] as a second lieutenant to teach mathematics at Hampton-Sydney College, a seminary in eastern Virginia. As principal instructor, Smith was to teach mathematics at VMI, as well as instruct the new cadets in Infantry Tactics and military police, the science and practice of artillery and civil and military engineering. "Old Specks' " shortcomings in the latter categories were soon revealed, and Cadet Sergeant Sumpter, a West Point dropout, took over the actual drilling.[6]

The other instructor, John Thomas Lewis Preston, a member of the Franklin Debating Society, and an early advocate of the Institute, was a graduate of Washington College (situated next to VMI at Lexington). He had no military background, and was to teach Modern Languages.

State and Irregular Cadets

Like West Point, which it was modeled after, the entrance requirements were low.[7] Applying to the Board of Visitors, who were appointed yearly by the governor, an applicant was required to furnish written testimonials of his good moral character, and demonstrate he was well instructed in reading, writing and the four ground rules of arithmetic, including vulgar fractions.[8] If accepted, the cadet was required to attend the Institute for three years. This was changed to four years in 1845, because cadets were so lacking in education they needed a first year for basic subjects to be taught. (A fifth class was added twice over the following years to further aid the unprepared cadet, but in neither instance did it last more than a trial period.)

Those boys who could pay for their education were allowed to enter, and became "pay cadets," or irregular cadets, as opposed to "state cadets." Upon graduation, the pay cadets were free to pursue any course they chose, but, by an Act passed in March 1842, those who had been educated by the state were required to teach in one of the

1 2 3 4 5

THE VIRGINIA MILITARY INSTITUTE IN 1857. 1—Honorable Samuel McD. Moore's residence; 2—Porter's Lodge; 3—Superintendent's quarters and V.M.I. headquarters; 4—Colonel Gilham's quarters; 5—Colonel Williamson's quarters; 6—Cadet Barrack; 7—Laundry; 8—Hospital; 9—Mess Hall. (From a print made by E. Sachse and Company, of Baltimore, and published by C. Bohn, of Washington and Richmond. Entered according to the Act of Congress in the year 1858 by Casimir Bohn in the Clerk's Office of the District Court of the District of Columbia. The superintendent's quarters, erected in 1847, was razed

6 7 8 9

in 1860, when another house was built back on the line of the quarters of Colonels Gilham and Williamson, thus permitting enlargement of the parade ground westward. At the extreme left is the residence of Honorable Samuel McDowell Moore, acquired by the Institute in 1908. The porter's lodge, the only V.M.I. building shown on the print which was not destroyed by the Federal army on June 12, 1864, was enlarged in 1878 and converted into quarters for a professor. With the single exception noted, the Institute, seen from this direction, had not changed in May-June, 1864.)

schools of the Commonwealth for two years. In neither case was involvement in the state militia required. It was hoped that these graduates would enter the volunteer state forces and help train Virginia's citizen-soldiers, and this argument was used to promote the often unpopular appropriations VMI placed before the legislature. Some did join, but it was not until the threat of war brought attention to the armed forces that "the proverbial inefficiency of the militia"[9] noticeably improved.

New Professors

In 1841, Thomas H. Williamson, a West Point dropout, and roommate of Colonel Smith, was employed as a professor of Tactics and Drawing, and served as Commandant of Cadets. By 1845, he was instructing cadets in Engineering and Science of War, Infantry Tactics and Drawing. In 1846, William Gilham was appointed Professor of Physical Sciences, instructor of tactics and Commandant of Cadets. A West Point graduate, Gilham served a year in Florida, then spent three years as an assistant in the Philosophical Department at the USMA. After a brief period with the Army of Occupation in Mexico, he was given a leave of absence and reported to VMI.[10]

The Mexican War brought national attention to West Point, as its graduates proved the worth of a military education by serving with distinction in every branch. VMI was represented by 26 former cadets, most serving with the volunteer forces. Still a young school, the Virginia Military Institute graduated 13 men the year American forces won victories at Buena Vista and Vera Cruz.[11]

VMI Expands

In 1847, the Board of Visitors expanded the scope of VMI by approving a course in architecture. Captain Williamson was selected to teach it, and as no suitable text was available, he wrote one.[12]

In January, 1851, Prof. Gilham submitted "a rather elaborate" report to the governor and the Virginia Legislature on soil analysis and manures.[13] He also suggested that the same general chemistry course as was taught at the USMA be used at VMI, as well as including instruction in mineralogy, geology and a course in agricultural chemistry. Rationalizing the need for such a course, he stated that the greater portion of the educated community was interested or engaged in agricultural pursuits. He had ascertained that 4/5's of the cadets' fathers were farmers, so most VMI graduates would inherit land. This being the case, he argued, why teach them courses fit only for professors, engineers or draftsmen? (At the same time, he requested the Professor of Chemistry at VMI be named the "State Agricultural Chemist;" this latter part was defeated "by hostile interests.")[14]

Professor J. T. L. Preston

In 1851, Brevet-Major Thomas Jonathan Jackson was appointed Professor of Natural and Experimental Philosophy and Artillery Tactics. Jackson, a West Point graduate in 1846, had served gallantly throughout the Mexican War in the artillery, being promoted to first lieutenant and breveted captain, then major, for meritorious conduct in the battles of Contreras, Churubusco and Chapultepec.[15]

Following the war, he served on court-martial duty, then went to Florida, where he served briefly before being appointed to a professorship. Taking a leave of absence, he reported to VMI and assumed command there, Major Gilham being away on a two month vacation.

Gratuitous Rank

The castellated walls and military attire of the cadets must have reminded Jackson of the USMA, but there were distinct differences between the two institutions. Unlike West Point, where the superintendent was chosen from the Corps of Engineers, and all professors belonged to the Army, VMI was a state institution with no authority over any Federal or state forces. The staff had always been granted a gratuitous military rank: Smith, as Principal Professor, was a 'Major" as of November 21, 1839; Preston was made a "Captain" by the Board of Visitors on April 30, 1840. By July of 1841, Smith was elevated to "Colonel," and later the professors were elevated to "Major." This rank granted no authority outside the Institute, and it was recommended that all VMI officers be commissioned by the state, and that this "be confined entirely to this establishment (VMI) as a separate military post independent of any organization with the residue of the

Professor William Gilham

militia."[16] In 1847, the question again arose concerning the VMI officers commissioned by the state, but nothing was done.[17] Finally in 1851, the superintendent and tactical officers were commissioned by Virginia, but this conferred no rank for them in the militia. As late as 1860, when the Virginia Legislature recognized the VMI officers as part of the military establishment, their commissions still conferred no rank in the active militia. The situation was equally unclear as pertaining to cadets.

On February 18, 1845, an Act was passed, requiring the State Adjutant General to include the roster of the Corps of Cadets in his annual return of the militia of the state, but there was never any intention of involving VMI with the volunteers. In 1859, the Corps of Cadets was called out at the last minute to present themselves at the execution of John Brown at Charlestown, but there had been much concern about the advisability of such a move, as parents of cadets might object, because the young men were not bound to do anything but guard the arsenal.[18] As late as May 1, 1862, the Board of Visitors "were unanimous in their disapprobation of the cadets being in any wise subjected to the risk of battle unless in the immediate defense of the Institution **at Lexington...**and positively forbidding any portion of the corps being sent to join General Jackson."[19]

Boy-Power

Another difference between the Federal and state military schools was that VMI had no horses. In consequence, there was no instruction in cavalry tactics and when drilling cadets in the rudiments of artillery drill, it was necessary to have the heavy pieces moved by ropes and boy-power.

Discipline and Mutiny

Discipline and the cadets themselves were somewhat different from those at the national Academy. Three months before Major Jackson reported, the first class had "mutinied,"[20] and disobeyed the direct orders of acting Superintendent Preston. They compounded the seriousness of their offence by taking the other classes with them, as they abandoned their duties and went to watch a trial at the courthouse. Upon Smith's return, his decision was to dismiss the entire first class and it was only after an appeal to the Board of Visitors that they were reinstated. Hazing, or "quizzing," new cadets at the beginning of summer was also a problem at VMI, and in 1847, this practice, as well as the cadets depredations, pillaging and stealing, became so serious that Col. Smith had to be recalled from vacation to restore order. The subject was taken up again in 1852, and in 1856, when eight cadets were dismissed for maltreating a new plebe.

Discipline at VMI further suffered: "Characteristic of the [18] 50's was the use of arms by cadets against each other. These offences were frequent and usually met with dismissal. Stabbing with a bayonet, use of a knife, carrying concealed weapons after a threat, were among some of the methods employed."[21]

In January of 1852, a second cadet mutiny occurred, this time involving a Cadet Captain. A private refused to obey the orders given by the captain, threats were exchanged, and the entire first class blacklisted the officer.

A third mutiny occurred in 1858, when two cadets were expelled for hazing plebes. Cadets refused to attend dress parade "in disobedience to an express order" from acting Superintendent Preston, unless their classmate was reinstated. Preston dismissed them all, stating they had "an inadequate sense of evil."[22]

Changes in Curriculum

In 1855, the Virginia State Agricultural Society submitted a proposal to VMI, stating it would defray the expenses of a Professor of Practical Agricultural Chemistry, if VMI would establish such a course. Prompted by Prof. Gilham's involvements in agriculture, the Board of Visitors received the idea favorably, but pointed out VMI owned no farm, and the matter was delayed due to limited barrack space. Interest was maintained, however, and later in the year, the cadets were taken on a field trip to Petersburg to attend the Agricultural fair.

In 1856, Col. Smith reported to the Board of Visitors that a course in physical geography and one on Dew's "Lectures on History" had been introduced. The latter provided a foundation of "the divine institution of slavery."[23]

The annual report of the superintendent to the Board of Visitors in 1857 contained a suggestion that a Chair of Military Strategy be created, but it was not acted upon. In May of 1858, he stressed the need to teach cavalry tactics at VMI because of the recent reorganization of the militia, but, as with previous requests for horses, the decision was not favorable, the terrain prohibiting it.

A General Scientific School

In 1858, the Board could not make enough appointments for pay cadets, so they decided to admit cadets from other Southern states.

Also in 1858, Prof. Smith went to Europe, a trip largely financed by Colonel Philip St. George Cocke, a leading enthusiast in agriculture. Upon his return, Smith submitted a report on "Scientific Education in Europe" to the Virginia Legislature, then prepared a lengthy proposal for the Board of Visitors at VMI, outlining his suggestions. This paper of June 24, 1859, suggested that VMI be organized as a **General Scientific School** with 3 special schools of application:

1. Agricultural
2. Engineering
3. Fine Arts

This was the system used at the Ecole Polytechnique that he had visited in France. Further elaborating on his plans for VMI, Smith suggested that the first two years all students attend classed under the present system, then, after reaching this point, candidates for regular graduation would complete the current syllabus, while those who desired to enter the three Special Schools would attend those classes. He recommended the Board approve at once departments of Scientific Agriculture, Human Physiology, Anatomy and Veterinary Medicine, Fine Arts, History, Political Economy and English Studies, and approve the erection of halls for an agricultural museum also to be used for Models and Drawings.[24] The Board was receptive to this report, especially with the receipt of two donations, amounting to $30,000. They quickly established a Chair of Agriculture, appointing Gilham professor, an additional Chair of Engineering, a Chair of Fine Arts, a Chair of Physiology Veterinary Medicine and Biology and a Chair of Military History, Strategy, Modern History and Political Economy. Thus, five months before John Brown's raid on Harper's Ferry, and less than two years before the opening of the Civil War, the Virginia Military Institute, the "West Point of the South," was radically altering its curriculum in favor of non-military pursuits.

War Approaches

In November of that year, Smith was suggested as the officer to be in charge of John Brown's execution, but because the state had not confirmed his rank in the militia, this was impossible.[25] The Corps of Cadets was called out by the governor to present themselves at Charlestown, where the artillery detachment was commanded by Major T.J. Jackson.

A month afterward, Gilham was appointed to write an instruction manual for the militia. (His text, written over the following months, proved to be overly long, and was used only briefly during the war, before being discarded.)[26]

In 1860, the legislature approved the VMI appropriation bill, making their total annuity$15,000 and $20,000 for building funds **for the agricultural pursuits.** The only significant change that pertained to the military pursuits at VMI came from the Board of Visitors, which decreed that when entering the Institute, a cadet also entered the service of Virginia, under the military command of those appointed to govern it.

In September, 1860, the first "special student" entered the VMI School of Agriculture, and while Virginia readied herself for war, the VMI Board of Visitors made preparations to send one of their graduates to Europe, to prepare him for the Chair of Chemistry. They also intended to send Gilham to Europe, so he could devote himself to the study of agricultural education and "other such matters," but war intervened.

It was only in January of 1861 that VMI acknowledged the fact that a portion of the Corps of Cadets might be called into military service, and began to stress military instruction. The boys themselves, taking their cue from Smith,[27] were supporters of secession, and were involved in several altercations in Lexington concerning this. The Virginia Ordinance of Secession brought both Institute and town together, and by April 21, 1861, when Major Jackson marched the young soldiers to war, they were given enthusiastic support.

West Point of the South

The drilling of recruits by the VMI cadets in the early months of the war, perhaps more than any single factor, helped the Confederacy send organized troops into the field.[28] VMI graduates were elected into officerships, and the quiet Professor of Natural and Experimental Philosophy became the South's greatest hero.

But although Col. Cocke wrote in 1849: "there are not a few who now look to the Institute as a West Point for the South in case of disunion," VMI was distinctly a state institution, probably best described by those who originated it: the object of the school was to prepare young men for the varied work of civil life; the military feature, thought essential to discipline, is not primarily in its scheme of education. What it had done since 1839 was educate Virginia's young men, give them the rudiments of a military background and send them off to teach other youths throughout the Commonwealth.

VMI was never meant to be the West Point of the South. Nationalistic pride and need cast it in that light, and the war ended too soon for it ever to be fulfilled.

Principal Professor Francis Henney Smith

TRACTS

RELIGIOUS SERVICES IN COLONEL GROESBECK'S BRIGADE, GENERAL POPE'S DIVISION

"PEACE BE STILL"

By Rev. Allen Kelley

Religion was a very strong influence during the anguished years of 1861-65. In the prayers of men were pleas for victory, mercy, help. Reverence to the Almighty for battles won were common in official reports; songs of devotion sprang from breasts spared to fight another day. Perhaps in no other period did Americans cling so desperately to their beliefs. The Bible was used as a justification by both the slave owner and the abolitionist to "prove" their philosophies; the clamor for army chaplins was loud and constant and whole brigades turned out to hear a sermon; regeneration, or sudden conversion to faith, baptisms, and revivals were common amongst soldiers looking for salvation. The need was always there for words of comfort and religious instruction. When there were not enough ministers of the gospel to supply the need, tracts, or small Christian pamphlets were found in abundance. They were, in effect, short written sermons that passed from hand to hand, or were read aloud, as men gathered around their campfires to share a moment's meditation. In this way, the illiterate shared with the educated, the private sat by the general, to hear the words of men sing the praises of God.

The American Tract Society in New York was largely responsible for publishing these little dissertations on faith. "Peace, Be Still," a 64 page paper-bound book with a dark green cover, stamped in gold with a guard at attention before a tent, may be a design typical of the period. The patriotic camp scene made it easy for the soldier to identify with, and the small size, approximately 4½x3", was convenient for him to carry in his pocket. Written by the Reverend J.C. Ryle, B.A., Rector of Helmingham, Suffolk, the thoughts presented on Mark 4:37-40 were written in the first person, to lend the lesson an air of familiarity, as a teacher to his pupil. The Biblical quotation, from whence the title was taken was presented first: "And there arose a great storm of wind, and the waves beat into the ship, so that it was now full. And he was in the hinder part of the ship, asleep on a pillow: and they awake him, and say unto him, Master, carest thou not that we perish? And he arose, and rebuked the wind, and said unto the sea, Peace, be still. And the wind ceased, and there was a great calm. And he said unto them, Why are ye so fearful? How is it that ye have no faith?"

The following pages discuss five basic lessons to be learned from the passage: "Learn first of all that *following Christ will not prevent your having earthly sorrows and troubles.*" This was demonstrated by the fact that the disciples were anxious and afraid of pain and death when the ship was threatened by destruction. "Learn, in the second place, that the *Lord Jesus Christ is truly and really man,*" as evinced by the fact he was "asleep on a pillow." "Learn in the third place, that *there may be much weakness and infirmity even in a true Christian,*" as demonstrated by the conduct of the disciples. They were impatient, distrustful and filled with unbelief, very much like men of every age. "The plain truth is, that there is no literal and absolute perfection among true Christians so long as they are in the body." "Learn, in the fourth place, *the power of the Lord Jesus Christ,*" the striking example of which is his rebuke of the wind and sea, which resulted in a great calm.

"Learn, in the last place, *how tenderly and patiently the Lord Jesus deals with weak believers.*" Using the disciples, examples are sited, revealing their lack of faith and pettiness, as they argued who among them should be the greatest (Mark 9:34), or when they wished to call down fire from heaven upon a village that had not received them. (Luke 9:54). The Lord, however, did not dismiss them, but led them forward step by step.

The readers of the tract are urged to deal gently with beginners and backsliders, and in summary concluded: "remember these five lessons, and you will do well;" study the gospel and let your example be a guide for others.

Geared toward the Protestant faiths ("Listen not for a moment to the wretched argument of the Roman-Catholic, when he tells you that the Virgin Mary and the saints are more sympathizing than Christ."), the tracts served the citizen-soldier, and perhaps made him feel a bit closer to home and heaven, when he read their words and shared their thoughts.

"Elocution!"

By J. E. Gessler

Elocution, according to Walker's Dictionary (1825), means, "The power of fluent speech; eloquence, flow of language; the power of expression or diction." One hundred and forty years ago, before television, radio and other mass communication media, oratory was more than just a luxury. If a person wished to be thought of as well-educated "polished" and gregarious, he would have to speak well. Debating societies, popular organizations of the time, served the dual purpose of discussing important issues, as well as entertaining townspeople for a night. It was generally expected that anyone wishing or possessing a high social standing would join such groups and take an active part in town affairs.

Another inducement to study elocution was politics: public speaking was a necessity, particularly for long-term stumping. One had to shout over large audiences long before electronic amplifiers were invented. Lecturers of any sort would need to speak for many hours each day; ministers were certainly not exceptions to this practice. Lawyers counted upon a successful presentation to win their cases, depending heavily on word usage and inflection. All such occupations relied heavily on the ability to speak clearly and at length: an orator had to catch and hold his audience's attention in order to sway public opinion in his favor. Skill in elocution was a treasured power.

C.P. Bronson, author of "Elocution; or, Mental and Vocal Philosophy: Involving the Principles of Reading and Speaking; and Designed for the Development and Cultivation of Both Body and Mind, in Accordance with the Nature, Uses, and Destiny of Man," had at least twenty-four editions of his book printed by 1845. In an advertisement for the book, its stated design was "to explain and illustrate all the principles of READING and SPEAKING, and show how the voice may be *cultivated* and *strengthened*, and the *breath* economized..." It was claimed that one would be enabled "to *Read*, *Speak*, and *Sing*, and blow on *Wind Instruments*, for HOURS in SUCCESSION, without HOARSENESS or injurious EXHAUSTION...," and also to "tend to the *Preservation* and *Care* of diseases of the *Throat and Lungs*, *Dyspepsia*, and other complaints incident to Public Speaking and a Sedentary Life." In addition to numerous "Oratorical and Poetical Portraits," are forty-four mouths illustrating the position to produce the sounds of our language, and engravings "exhibiting all the Passions...with appropriate examples," as well as "An immense number of Questions, and Subjects suggested for *Lyceums*, *Debating Societies*, and *Social Parties*..."[1]

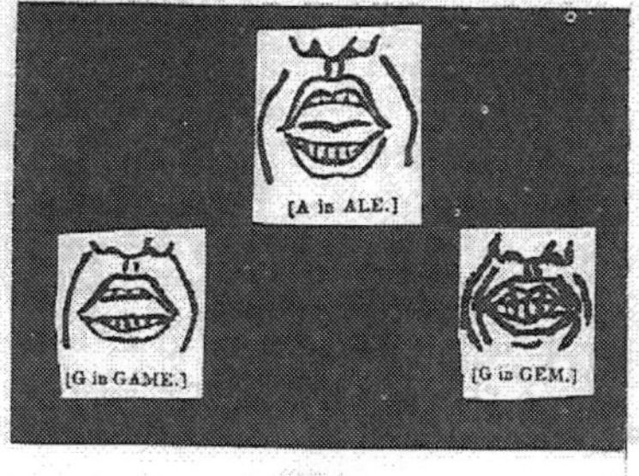

In explaining the fundamental principles of elocution, it was believed that proper exercise of the vital organs was necessary for good health. "To secure a perfectly healthy distribution of the vital fluids throughout the body, and a free and powerful activity of the mind, there must be a full and synchronous action in the *brain*, the *lungs*, and the viscera of the abdomen; the soul operating naturally, on the dorsal and abdominal muscles, and thus setting in motion the whole body."[2] The usual testimonials for such a system of instruction came mainly from ministers and professors.[3]

"Physiological Introduction" and "Oratorical and Poetical Gestures" preceed the actual instruction of elocution. The former consists of representations of the nervous, muscular and osseous systems, clothed and unclothed human figures, postures, and engravings which showed how air entered the lungs. Each was accompanied by an explanation of how the body worked, with an emphasis on the importance of Nature, the power of God, and the glory of "MAN...a finite representation of the Infinite Deity!" According to Bronson, "the Soul is the architect, and the body its workmanship..."[4]

In discussing the nerves of respiration, elocutionists traced the nerves of the vocal organs, of the heart and lungs, and of the main body. Claiming the roots of these nerves "are in the cerebellum...a receptacle of life," they concluded that "intensity of thought, carking cares, &c., impede respiration, and infringe on the laws of health...inducing dyspepsia, and even consumption..."[5] They warned women against "tight lacing" of corsets, as being nothing less than a form of slow suicide, and harmful to their offspring. Also included were "attitudes" (postures), with a warning to beware of too much stiffness and too much laxity of the muscles.[6] The actions of the diaphragm and of the heart were illustrated to the student,

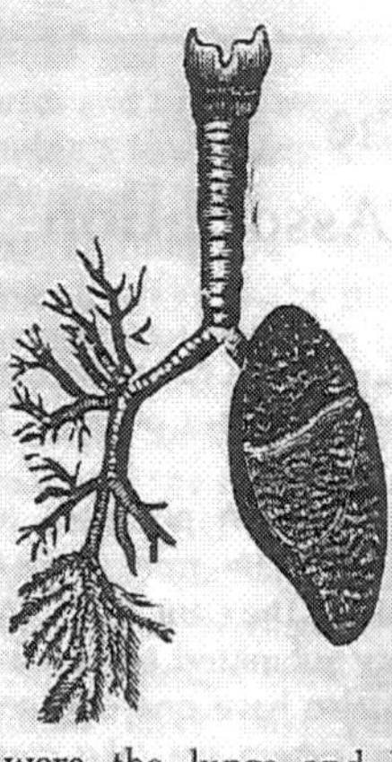

as were the lungs and bronchial tubes. Included was the remedy for bronchitis: "...free use of cold soft water over the whole body, and bandages wet with the same, placed about the chest and neck, to be removed every few hours, as they become dry."[7] This "cure" employed the methods of hydrotherapy; and, in practice, the two vastly different subjects employed many of the same philosophies.

ORATORICAL AND POETICAL GESTURES.

The "Oratorical and Poetical Gestures" were illustrations "designed for studies; they involve every variety of Thought and Feeling, and their modes of manifestation; some are to be imitated, others avoided, because of their awkwardness: judge ye."[8] These were early representations of what we now call "body language." The "passions" were taught by engravings depicting such emotions as Tranquility & c., Hatred, Aversion, Revenge, and Horror. Once mastered, the future orator entered into the actual instruction of elocution, with the reminder that "This system unfolds the true Philosophy of MIND and VOICE, in accordance with the *nature of Man*, and the *structure of Language*."[9] The elements were presented, to be repeated until the

TRANQUILLITY, &c. REVENGE. ANGER, RAGE, FURY. MIRTH, JOLLY LAUGHTER.

learner became comfortable in pronouncing them correctly. Then combinations were given, which increased in difficulty, to cultivate the voice and ear, while practice perfected the mind and body to the laws that should govern them. Students were taught mouth position by first practicing the vowels, then moving on to the consonants. To prepare the student for participation in social functions, lessons emphasizing technique were provided. (Italics indicate emphasis.)

Change of the seat of accent: On her *en*-trance, she was en-*tranced* at being es-*cort*-ed by a grand *es*-cort: I es-*say* to make an *es*-say to ex-*ile* the *ex*-iles: ex-*port* the *ex*-ports, without ex-*tract*-ing the *ex*-tracts for the ex-*tract*-ers: the *ab*-ject fel-lows ab-*ject* the gifts, and the *ab*-sent minded ab-*sent* themselves from the party: he ab-*stracts* the *ab*-stracts and at-*trib*-utes the *at*-tri-butes to others: I lay the *ac*-cent on the ac-*cent*-ed vowel, and af-*fix* the *af*-fix to the final syllable, and make *aug*-ment in the right place and aug-*ment* the word in *Au*-gust, and thus make the idea au-*gust*.[10]

Topics of debate were used to stimulate the mind and voice: 1. Were any beings ever *created* angels? 2. Is it *right* ever to do *wrong*? 3. Why was a revelation *necessary*? 4. May we not protect our *person* and *character* from assault? 5. Does *civilization* increase *happiness*? 6. Which excites more curiosity, the works of *nature*, or the works of *art*? 7. Ought a *witness* to be questioned with regard to his *religious* opinions, or *belief*? 8. Was the general *bankrupt* law a *benefit* to the country? 9. Why are we disposed to *laugh*, even when our best *friend* falls down? 10. Which is the greatest, *faith*, *hope*, or *charity*? 11. Should *controversy* interrupt our *friendship* and *esteem* for each other? 12. Have *christians* any right to *persecute* each other for their *opinions*?[11]

Exercises in speech also served to instruct the student in the proper mode of living: "INDUCING DISEASE. There is no doubt, that the seed of a large number of diseases are sown in childhood and youth; and especially in our progress in obtaining what is called, an EDUCATION. The bad habits of position in and out of school, and our unhealthy mode of living, contribute very essentially to the promotion of various diseases, particularly dyspepsia, liver and lung complaints, and headaches. Hence, we cannot be too watchful against sitting in a crooked position, nor too prudent eating, drinking, and sleeping, as well as in our clothing, and our lodging apartments. Let us put forth every effort in the performance of our duties, be they physical, intellectual, or moral."[12]

Humor was also included in the instruction of elocution: "Anecdote. *Use of Falsehood.* A jury, which was directed by the Judge, to bring in a certain prisoner guilty, on his own confession and plea, returned a verdict of "Not Guilty;" and offered, as a reason, that they knew the fellow to be so great a liar, they did not believe him."[13]

These exercises were designed not only to teach elocution, but the author's conception of morals, as well. It was suggested to: "Let the pupils memorize any of the proverbs, laconics, maxims, or questions, and recite them on occasions like the following: 'When they first assemble in the school-room; or, meet together in a social circle: let them also carry on a kind of conversation, or dialogue with them..."[14] It is easy to imagine the effect of Bronson's moralizing in the form of lessons that were perhaps repeated daily, then memorized. Bronson also quoted the favorite sayings of the Phrenologists, another popular "science" of the day: "Know Thyself" and "The proper study of Mankind — is Man."[15]

The art of public speaking, 19th Century fashion, was taught as an incorporated part of the culture it belonged to. Similar in this respect to education, religion, medicine and even science, learning elocution required "swallowing" the philosophies of the instructor. Then, on their own, these pupils went out to the podium, city halls and stages of the country to expound their "cultivation of both body and mind."

FOOTNOTES

¹ "Elocution; or, Mental and Vocal Philosophy: Involving the Principles of Reading and Speaking; and Designed for the Development and Cultivation of Both Body and Mind, in Accordance with the Nature, Uses, and Destiny of Man" by C.P. Bronson, A.M., M.D., Louisville: Morton & Griswold, 1845, p. 3 (preceeding title page).

² IBID, p. 2.
³ IBID, p. 3.
⁴ IBID, p. 5.
⁵ IBID, p. vi.
⁶ IBID, p. ix, x.
⁷ IBID, p. xi.
⁸ IBID, p. xvi.
⁹ IBID, p. 17.
¹⁰ IBID, p. 72.
¹¹ IBID, p. 162.
¹² IBID, p. 127.
¹³ IBID, p. 120.
¹⁴ IBID, p. 154.
¹⁵ IBID, p. 5.

A History of the John Pelham Historical Association

The John Pelham Historical Association was founded by Margaret Vogtsberger of Hampton, Virginia, on March 17, 1982 – the 119th anniversary of the Battle of Kelly's Ford, where John Pelham received his mortal wound. Miss Vogtsberger is a member of the Williamsburg Civil War Round Table and initiated the project of erecting a monument to Pelham where he fell at Kelly's Ford.

The purposes of the John Pelham Historical Association are: (1) tc study the life and career of Major John Pelham, the cavalry campaigns of the Army of Northern Virginia, and other aspects of the Civil War; (2) to establish and maintain an archives—consisting of letters, documents, maps, photographs, and relics—relating to Major John Pelham; (3) to maintain, preserve, and erect, when appropriate, the ten monuments and markers to Major John Pelham; and (4) to promote and stimulate interest in the life and career of Major John Pelham.

Our organization publishes a bimonthly newsletter, from about 8-10 pages, called "The Cannoneer." Most articles are submitted by our members. We also have one convention each year, where we take care of business and tour Civil War sites.

Our organization celebrated its first anniversary on March 17, 1983. In less than a year we've gained 75 members from all over the country. In 1982 we contributed $50 to replace a post of a state marker about Pelham which was hit by a car near Anniston, Alabama.

Our membership is open to anyone who has an interest. Dues are currently $10 a year. If interested in joining, please write to: Ms. Jennifer Young, Secretary, John Pelham Historical Association, 4021 Hallman Street, Fairfax, VA 22030.

FOOTNOTES

¹ "One Hundred Years at the Virginia Military Institute" by William Couper, Garrett and Massie, Richmond, 1939, Vol. 1, p. 76.
² "Buchanan Advocate," January 20, 1836.
³ "History of the Virginia Military Institute" by Francis Henney Smith, p. 242-244.
⁴ "The Virginia Military Institute" by Jennings C. Wise, J.P. Bell Co., Lynchburg, Va., 1915, p. 39.
⁵ "Biographical Register of the Officers and Graduates of the United States Military Academy" by George W. Cullum, D. Van Nostrand, 1868, p. 425.
⁶ "One Hundred Years," p. 72.
⁷ IBID, p. 46,47.
⁸ Articles of the V.M.I., Article 11, 1839.
⁹ Letter of William Richardson to F.H. Smith, April 27, 1841.
¹⁰ Cullum, p. 594.
¹¹ "One Hundred Years," p. 164.
¹² IBID, p. 165.
¹³ IBID, p. 240.
¹⁴ IBID, p. 241.
¹⁵ Cullum, Vol. II, p. 150.
¹⁶ "One Hundred Years," p. 159.
¹⁷ IBID, p. 165.
¹⁸ IBID, Vol. II, p. 7; letter of William H. Richardson to Gov. Wise, Nov. 17, 1859.
¹⁹ IBID, p. 149.
²⁰ IBID, Vol. I, p. 237-239.
²¹ IBID, p. 257, 258.
²² IBID, p. 338.
²³ IBID, p. 316.
²⁴ IBID, p. 340,341.
²⁵ IBID, Vol. II, p. 7; see also Ch. 34, Sec. 22, p. 176 of the Virginia Code
²⁶ IBID, p. 26.
²⁷ IBID, p. 78.
²⁸ "Lee's Lieutenants," by Douglas Southall Freeman, Charles Scribner's Sons, New York, 1942, p. xxiii.

Unofficial Reports

"On Sunday, the 16th of March, Captain Morgan, with 40 of his men, suddenly appeared at Gallatin, Tenn., 28 miles the other side of Nashville. After catching all the Union men in the place, and confining them in a guard-house, Captain Morgan, dressed in a Union uniform, proceeded to the telegraph office at the railroad depot, a short distance from the town. Entering the office, the following conversation took place between Captain Morgan and the telegraph operator, a blustering fellow:

Captain Morgan — 'Good-day, sir. What news have you?'

Operator — 'Nothing, sir, except it is reported that d——d rebel, Captain John Morgan, is this side of the Cumberland, with some of the Cavalry. I wish I could get sight of the d——d rascal, I'd make a hole through him larger than he would find pleasant.'

While thus speaking, the operator drew a fine Navy revolver and flourished it as if to satisfy his visitor how desperately he would use the instrument in case he should meet with the rebel captain.

'So you know who I am?' quietly remarked Captain Morgan, continuing the conversation.

'I have not that pleasure' replied the operator.

'Well, I am Captain Morgan,' responded the gentleman. At these words, the operator's cheeks blanched, his knees shook, the revolver dropped from his hands, and he sunk to the floor. He literally 'wilted.'

After the frightened individual had recovered himself sufficiently, Captain Morgan required him to telegraph some messages to Louisville, among others one to Prentice, of the *Journal*, politely offering to act as his escort on his proposed visit to Nashville. Then taking the operator with him as a prisoner, Captain Morgan with his men awaited the arrival of the train from Bowling Green for Nashville.

In due time, the train came thundering in. Captain Morgan at once seized it, and taking five Union officers, who were passengers, and the engineer of the train prisoners, he burned to cinders all of the cars, with their contents, and then filling the locomotive with turpentine, shut down all the valves and started it toward Nashville. Before it had run 800 yards, the accumulation of steam caused it to explode, shivering it into a thousand atoms. Captain Morgan then started southward with his prisoners, and made his way safely to the Confederate camp."

(Atlanta *Confederacy*, 1862.)

SIGNALS

Dear Editors:

Congratulations on your magazine THE KEPI. It certainly is a welcome addition for the Civil War enthusiast. Your unique perspective regarding the format should prove to be quite educational and informative.

Tom LaPorte,
Bristol, Conn.

Dear Editors:

I enjoyed the first issue of THE KEPI. I look forward to my next. Please find enclosed my check for the first years subscription.

Robert Newsome,
Middletown, Kentucky

Dear Editors:

A Civil War magazine that covers the entire period, not only the military aspect, is much needed today. We can not truly understand the war unless we understand the times it was fought.

Mike Cavanaugh,
Editor, CWBX
Philadelphia, Pa.

Dear Editors:

The concept of THE KEPI is a refreshing one; providing a background of events and attitudes is an overlooked aspect of Civil War histories and one that I think is absolutely necessary to get an understanding of why things happened as they did.

Phil Barber,
Boston, Mass.

Dear Editors:

Enclosed is our check for a subscription to THE KEPI. It's a real professional publication. I have a great interest in 19th Century life so I particularly enjoyed your first issue. Good luck on your new endeavor.

Willard Carpenter,
Rensselaer, N.Y.

NOSTRUMS AND NOTIONS

"SOAK IT!"

by
Robert Kamer, M.D. and
Samuel Jackson, M.D.

In the mid 1800's, a "panacea — a cure everything"[1] for disease was sweeping the world. Everyone, from housewife to royalty was singing the praises of the Water-Cure, or Hydrotherapy. No illness was beyond its curative powers; no doctor need ever be called, for "Nature is the true physician;"[2] no medicines ("alias poisons")[3] were ever required. "If those who are predisposed to dyspepsia, or consumption, or rheumatism...will but *bathe* daily, they may laugh at all disease... 'that flesh is heir to.' "[4] Pure water was the proof of the "law of things, that every disorder and evil had its antidote at hand... whenever any disease prevails may be found its cure-all — a provision beautifully in keeping with that benevolent tendency of every law of nature."[5]

Water, it was believed, constituted 3/4's of the whole mass of blood, 7/8's of the substance of the brain and more than 9/10's of the various colorless fluids and secretions;[6] it was the only medium through which waste or extraneous ingredients could be expelled. Being the only "detergent" and the only material capable of circulating in all the tissues of the body,[7] it was reasoned that by soaking in a tub or wet sheet, pure liquid would enter the body and all foul, polluted material which caused disease would be removed. This principle was called "Endosmose and Exosmose," and was explained by the following example. A glass tube was filled with salt water, closed at the end with a piece of bladder, then im-

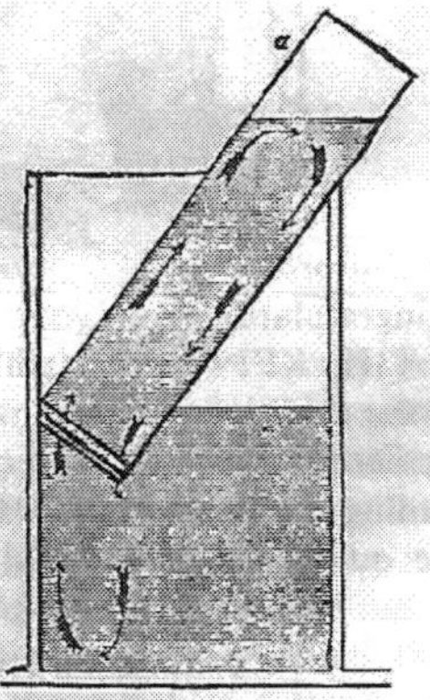

ENDOSMOSE AND EXOSMOSE.

mersed in a beaker of water. The clear water drew out the salt, and filled the tube until the salt content in both tube and bladder became equal.[8] In case this experiment failed to convince a person of the "purifying efficacy" of water, it was suggested to take any man who was not accustomed to daily bathing, and

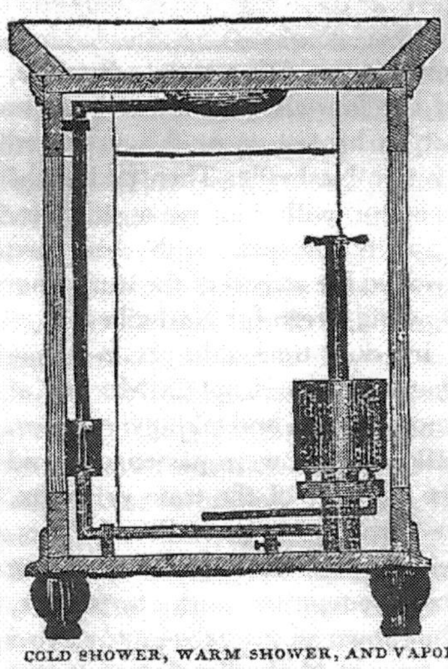

COLD SHOWER, WARM SHOWER, AND VAPOR-BATH COMBINED.

who lived off pigs, chickens, wine and cigars, and "pack" him in a wet sheet. After an hour or two of soaking, "the intolerable stench will convince all persons who may be present that his blood and secretions were exceedingly befouled, and that a process of depuration is going on rapidly!"[9]

There were many different ways to take the Water-Cure; or, in the language of the day, "The hydropathic appliances embrace all the usual methods of vapor, warm, tepid, cool and cold bathing, besides a variety of processes which have their origin in the development of Water-Cure as a system."[10] There was the wet sheet (the process of packing an invalid in a wet sheet, then wrapping many flannel blankets around that), the douche (a stream of water directed to the back of the neck or joints), the sitz bath (a low tub for soaking the lower organs of the pelvis and abdomen), the shallow, half, foot and head baths, the plunge bath, the vapor bath, ear and eye baths, which might be summed up by the familiar axiom, "Soak it!"

Perhaps the most easily misused and dangerous[11] of all the hydropathic baths was the SHOWER. Although the favorite of allopathic physicians, the use of the shower was sadly mismanaged. The shower, or sprinkles of water upon the body ("whether it should ever be taken on the head is a controverted proposition.")[12] was regarded as the very worst bath for an invalid to commence with. It was recommended that patients work up to a shower slowly, and that only those who were active and had a pretty well balanced circulation could profit by it.[13] The shower was contra-indicated in the extremely nervous, those subject to headache, palpatations of the heart, determination of the lungs or severely constipated bowels. It was managed so injudiciously in home practice as to work more mischief than benefit.[14]

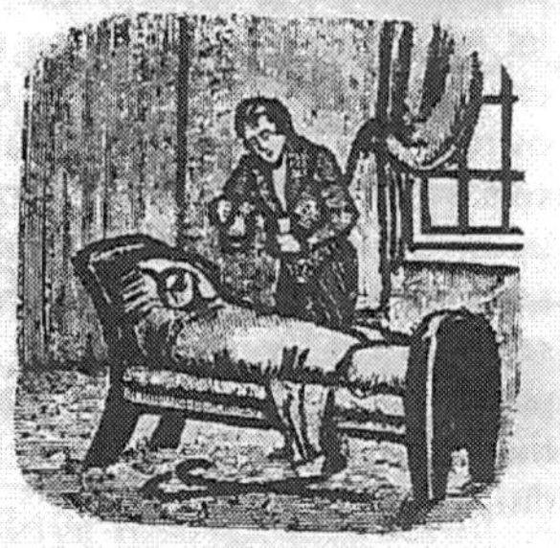

THE WET-SHEET PACKING.

Aside from all its drawbacks, the shower did offer two advantages, the first being a method of torture. 19th Century penologists discovered that the terrors of cold water often brought prompt obedience from felons, and it was subsequently written into the law books. Prisoners were said to dread it worse than barbarous flagellation.[15] (The comment from the reviewing physician was that the shower was more humane, but liable to do injury to those susceptible to head affections!)

Convenience was the second advantage offered by the shower. It could be taken indoors, year round, thus relieving men from bathing in the river, and risking a chill or worse in bad weather or winter, when bathing was not practiced at all, resulting in not more than a dozen baths per year.[16]

Various tubs, douche pails and devices were assembled to facilitate the soaking and pouring that hydrotherapists prescribed. One invention, recommended in 1845 by the American Phrenological Journal, was called "Bates's Sliding Shower Bath." Nothing less than a *"public blessing,"*[17] this simple apparatus rendered bathing convenient *"at all times,* and particularly so *on rising,"*[18] when it was the most agreeable and beneficial. If employed before dressing (for if we do not take our bath till after we are dressed, we will likely be too busy to find time), when the body is moist and warm from sleep, the *reaction* will be greatest. Bates's shower bath, which consisted of a sliding top which could be lowered, changed and raised in a quarter of a minute, could be kept in the bedroom; when not in use, it became an ornament. An attachment for taking the *vapor* was also available, allowing families in 1845 to enjoy that which was employed by the Romans. The possession of this shower (at a cost of $10 to $25) would enable its user to "dismiss the doctor,"[19] and enjoy "the greatest personal comforts and family blessings that can well be obtained."[20]

THE HEAD-BATH.

Hydrotherapy and showering had good points and bad, comforts and tortures. But if nothing else, they at least made the world smell a little nicer.

Bates's Shower Bath

Taking A Shower

Doctor.— Well, how did your wife manage her shower bath, deacon? Deacon.— She had real good luck. Madame Mood told her how she managed. She said she had a large, oiled silk cap, with a cape to it, like a fireman's that came all over her shoulders, and— Doctor.— She's a fool for her pains — that's not the way. Deacon.— So my wife tho't. Doctor.— Your wife did nothing of the sort, I hope. Deacon.— Oh, no, doctor, she used an umbrilly. Doctor.— What, used an umbrella, what the mischief good did the shower bath do her? Deacon.— She said she felt better. Her clothing did not get wet a mite. She got under the umbrilly for half an hour, till all the water had trickled off, and said 'twas cool and delightful, and just like a leetle shower bath in summer. Then she took off her things; and rubbed herself for half an hour after. (SCIENTIFIC AMERICAN, New York, September 9, 1848.)

FOOTNOTES

1 "American Phrenological Journal," August 1845, p. 281.
2 "The Hydropathic Encyclopedia" by R.T. Trall, M.D., Fowlers and Wells, New York, 1855, Vol. II, p. 6.
3 IBID, p. 3.
4 "Phrenological Journal," p. 281.
5 IBID, p. 273.
6 "Hydropathic Encyclopedia," p. 3.
7 IBID, p. 3.
8 IBID, p. 7.
9 IBID, p. 23.
10 IBID, p. 22.
11 IBID, p. 33.
12 IBID, p. 34.
13 IBID, p. 34.
14 IBID, p. 34.
15 IBID, p. 34.
16 "Hydropathic Encyclopedia," p. 282.
17 IBID, p. 282.
18 IBID, p. 282.
19 IBID, p. 283.
20 IBID, p. 283.

Articles of War

An Act For Establishing Rules And Articles For The Government Of The Armies Of The United States.

ART. 24. No officer or soldier shall use any reproachful or provoking speeches or gestures to another, upon pain, if an officer, of being put in arrest; if a soldier, confined, and of asking pardon of the party offended in the presence of his commanding officer.

ART. 25. No officer or soldier shall send a challenge to another officer or soldier, to fight a duel, or accept a challenge if sent, upon pain, if a commissioned officer, of being cashiered; if a non-commissioned officer or soldier, of suffering corporeal punishment, at the discretion of a court-martial.

ART. 26. If any commissioned or non-commissioned officer commanding a guard shall knowingly or willingly suffer any person whatsoever to go forth to fight a duel, he shall be punished as a challenger; and all seconds, promoters, and carriers of challenges, in order to duels, shall be deemed principals, and be punished accordingly. And it shall be the duty of every officer commanding an army, regiment, company, post, or detachment, who is knowing to a challenge being given or accepted by any officer, non-commissioned officer, or soldier, under his command, or has reason to believe the same to be the case, immediately to arrest and bring to trial such offenders.

ART. 27. All officers, of what condition soever, have power to part and quell all quarrels, frays, and disorders, though the persons concerned should belong to another regiment, troop, or company; and either to order officers into arrest, or non-commissioned officers or soldiers into confinement, until their proper superior officers shall be acquainted therewith; and whosoever shall refuse to obey such officer (though of an inferior rank), or shall draw his sword upon him, shall be punished at the discretion of a general court-martial.

ART. 28. Any officer or soldier who shall upbraid another for refusing a challenge, shall himself be punished as a challenger; and all officers and soldiers are hereby discharged from any disgrace or opinion of disadvantage which might arise from their having refused to accept of challenges, as they will only have acted in obedience to the laws, and done their duty as good soldiers who subject themselves to discipline.

ART. 29. No sutler shall be permitted to sell any kind of liquors or victuals, or to keep their houses or shops open for the entertainment of soldiers, after nine at night, or before the beating of the reveille, or upon Sundays, during divine service or sermon, on the penalty of being dismissed from all future sutling.

ART. 30. All officers commanding in the field, forts, barracks, or garrisons of the United States, are hereby required to see that the persons permitted to suttle shall supply the soldiers with good and wholesome provisions, or other articles, at a reasonable price, as they shall be answerable for their neglect.

ART. 31. No officer commanding in any of the garrisons, forts, or barracks of the United States, shall exact exorbitant prices for houses or stalls, let out to sutlers, or connive at the like exactions in others; nor by his own authority, and for his private advantage, lay any duty or imposition upon, or be interested in, the sale of any victuals, liquors, or other necessaries of life brought into the garrison, fort, or barracks, for the use of the soldiers, on the penalty of being discharged from the service.

ART. 32. Every officer commanding in quarters, garrisons, or on the march, shall keep order, and, to the utmost of his power, redress all abuses or disorders which may be committed by any officer or soldier under his command; if, upon complaint made to him of officers or soldiers beating or otherwise ill-treating any person, or disturbing fairs or markets, or of committing any kind of riots, to the disquieting of the citizens of the United States, he, the said commander, who shall refuse or omit to see justice done to the offender or offenders, and reparation made to the party or parties injured, as far as part of the offender's pay shall enable him or them, shall, upon proof thereof, be cashiered, or otherwise punished, as a general court-martial shall direct.

ART. 33. When any commissioned officer or soldier shall be accused of a capital crime, or of having used violence, or committed any offense against the persons or property of any citizen of any of the United States, such as is punishable by the known laws of the land, the commanding officer and officers of every regiment, troop, or company, to which the person or persons so accused shall belong, are hereby required, upon application duly made by, or in behalf of, the party of parties injured, to use their utmost endeavors to deliver over such accused person or persons to the civil magistrate, and likewise to be aiding and assisting to the officers of justice in apprehending and securing the person or persons so accused, in order to bring him or them to trial. If any commanding officer or officers shall wilfully neglect, or shall refuse, upon the application aforesaid, to deliver over such accused person or persons to the civil magistrates, or to be aiding and assisting to the officers of justice in apprehending such person or persons, the officer or officers so offending shall be cashiered.

ART. 34. If any officer shall think himself wronged by his Colonel, or the commanding officer of the regiment, and shall, upon due application being made to him, be refused redress, he may complain to the General commanding in the State or Territory where such regiment shall be stationed, in order to obtain justice; who is hereby required to examine into said complaint, and take proper measures for redressing the wrong complained of, and transmit, as soon as possible, to the Department of War, a true state of such complaint, with the proceedings had thereon.

ART. 35. If any inferior officer or soldier shall think himself wronged by his Captain or other officer, he is to complain thereof to the commanding officer of the regiment, who is hereby required to summon a regimental court-martial, for the doing justice to the complainant; from which regimental court-martial either party may, if he thinks himself still aggrieved, appeal to a general court-martial. But if, upon a second hearing, the appeal shall appear vexatious and groundless, the person so appealing shall be punished at the discretion of the said court-martial.

ART. 36. Any commissioned officer, store-keeper, or commissary, who shall be convicted at a general court-martial of having sold, without a proper order for that purpose, embezzled, misapplied, or wilfully, or through neglect, suffered any of the provisions, forage, arms, clothing, ammunition, or other military stores belonging to the United States to be spoiled or damaged, shall, at his own expense, make good the loss or damage, and shall, moreover, forfeit all his pay, and be dismissed from the service.

ART. 37. Any non-commissioned officer or soldier who shall be convicted at a regimental court-martial of having sold, or designedly, or through neglect, wasted the ammunition delivered out to him, to be employed in the service of the United States, shall be punished at the discretion of such court.

ART. 38. Every non-commissioned officer or soldier who shall be convicted before a court-martial of having sold, lost, or spoiled, through neglect, his horse, arms, clothes, or accoutrements, shall undergo such weekly stoppages (not exceeding the half of his pay) as such court-martial shall judge sufficient, for repairing the loss or damage; and shall suffer confinement, or such other corporeal punishment as his crime shall deserve.

ART. 39. Every officer who shall be convicted before a court-martial of having embezzled or misapplied any money with which he may have been intrusted, for the payment of the men under his command, or for enlisting men into the service, or for other purposes, if a commissioned officer, shall be cashiered, and compelled to refund the money; if a non-commissioned officer, shall be reduced to the ranks, be put under stoppages until the money be made good, and suffer such corporeal punishment as such court-martial shall direct.

ART. 40. Every captain of a troop or company is charged with the arms, accoutrements, ammunition, clothing, or other warlike stores belonging to the troop or company under his command, which he is to be accountable for to his Colonel in case of their being lost, spoiled, or damaged, not by unavoidable accidents, or an actual service.

CONCURRENT EVENTS

Re-Population of Virginia — The re-population of Virginia in the rear of our victorious armies, by settlers from the free States, has already begun. Buyers of lands at and in the vicinity of Manassas have appeared, but they experience a difficulty in purchases. The real owners are chiefly rebels, and are fugitives from their possessions. Of course, purchases will not be made of any but the owners, and they must be loyal, to insure future protection to the transactions. The result will be, in the absence of loyal owners, that strangers will take possession in the manner of the squatters of the West, and leave to the future the settlement of title, which will doubtless be confirmed to the new holders in process of time. In this way, and in various others ways, the deserted wastes of Eastern Virginia will be re-occupied by a people who will make them blossom as a rose, and who will afford protection and renumerative employment to the colored laboring classes whom the war will have emancipated.
(Leslie's Illustrated Newspaper Supplement, April 26, 1862.)

FROM KANSAS — Col. Sumner has returned to Fort Leavenworth, leaving only a small number of troops at Lawrence. An attempt recently being made to arrest Mr. Spear, of the Kansas Tribune, on a charge of resisting Jones' authority as sheriff, Mrs. Spear dashed a dish of hot water in the face of the Deputy Sheriff, who beat a hasty retreat, to the infinite amusement of his posse of U.S. Soldiers. The soldiers are said to be generally Free State men, — and say that they are afraid they would shoot either too high or too low if called upon to fire at their liberty-loving citizens. The evidence taken by the Congressional Committee, so far is said to be very full and explicit as to the great amount of fraud and outrage at the election. (*Saturday Evening Post*, May 17, 1856.)

THE KEPI

12115 Magnolia Blvd.
P. O. Box 107
No. Hollywood, CA 91607

BULK MAIL

THE KEPI

OCTOBER-NOVEMBER 1983 $3.00

"A New Concept in Civil War Reporting"

THROWING UP BREASTWORKS IN THE WILDERNESS. FROM A SKETCH MADE AT THE TIME.

From "BATTLES AND LEADERS OF THE CIVIL WAR"

CONFEDERATE LINE WAITING ORDERS IN THE WILDERNESS.

MICHIGAN CIVIL WAR COLLECTORS SHOW

SATURDAY OCTOBER 15, 1983

SHOW HOURS — 9:00 a.m. - 4:00 p.m.

EXHIBITOR SET-UP — Friday Oct. 14 7:00 p.m. - 10:00 p.m. & Saturday Oct. 15 7:00 a.m. - 9:00 a.m.

135 tables available for Civil War Memoribilia and Relics to Buy, Sell and Trade. Guns, Bayonets, Paper Items, Swords, Buttons, Images, Books and Other Related Items.

☆ *NO FLEA MARKET MATERIAL WILL BE PERMITTED* ☆

(Security Guards will be provided for exhibits Friday night)

DISPLAY AWARDS WILL BE GIVEN

MICHIGAN INN

16400 J. L. HUDSON DR. — SOUTHFIELD (DETROIT), MICHIGAN 48075 — 313-559-6500

(John Lodge Freeway - U.S. 10 - Service Drive Between 8 & 9 Mile Roads)

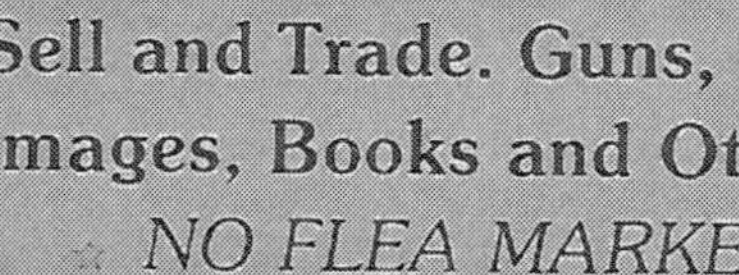

Plenty of Free Parking, Food available on premisis. A block of rooms for overnight accomodations has been set aside for show guests. Reservations for rooms can be made by contacting the Hotel at the number listed above.

For tables and additional information contact:

Dave Parks
18387 Ash Creek Dr.
Mt. Clemens, Michigan 48044
313-286-6462

THE KEPI

"A New Concept in Civil War Reporting"

OCTOBER-NOVEMBER 1983 | VOLUME I, NUMBER 4

THE KEPI is published bi-monthly for $3.00 per issue or $18.00 per year. Canadian single copy, $3.50, subscription, $21.00 per year. United Kingdom single copy, $4.50, subscription, $28.00 (air mail) per year. Correspondence to the editors may be directed to: THE KEPI, 12115 Magnolia Blvd., North Hollywood, CA, 91607. Unless expressly stated, THE KEPI neither endorses nor is responsible for the content of advertisements appearing in its pages. THE KEPI requests authors to query with specific topics before submitting articles and assumes no responsibility for unsolicited manuscripts not accompanied by return postage. Work not fully documented and footnoted where applicable will not be considered. Please allow 4-6 weeks for delivery of first issue.

ILLUSTRATIONS

Cover taken from "The Official Military Atlas of the Civil War."

"The Burning of the Wilderness," p. 7, from "Battles and Leaders of the Civil War."

"Sketch of the Battle of the Wilderness," p. 3, from "The Official Military Atlas of the Civil War."

"Strange Effect of the Draft," p. 18, from "Harper's Weekly," Saturday, August 30, 1862.

"The Galvanic Process," p. 22, from "The Diseases of Woman," by Frederick Hollick, M.D., T.W. Strong, New York, New York, 1856.

"The Process of Magnetizing," p. 22, from "Lectures on the Philosophy of Mesmerism," by Rev. John Bovee Dods, Fowlers and Wells, New York, New York, 1850.

THE REGULARS

THE IRREGULARS

GENERALLY SPEAKING

This issue of THE KEPI features an article by Samuel J. Martin, entitled, "Gordon's Flank Attack." While researching his biography of Lieutenant-General Richard S. Ewell, Mr. Martin discovered a letter written by Campbell Brown, Ewell's assistant adjutant-general, during the Wilderness Campaign. This document (reprinted in this issue), written by Brown to dispute Jubal Early's claim for the success of John Gordon's attack on the Federal right, May 5, 1864, has added a new dimension to the modern historians' interpretation of this critical day. Brown asserts that Ewell was delayed from investigating Gordon's proposed attack because he was summoned to General R.E. Lee's headquarters on the Confederate right. The military background and the new speculations this information has given rise to are developed in the feature article, and the ensuing one, titled "Repercussions." Samuel J. Martin's discovery also proves to the modern researcher that the study of the War is alive, and old interpretations are subject to inspection and change. While all new information must be carefully analyzed, its discovery brings those tumultuous years closer to our scrutiny and understanding, and with it, it holds the promise of more new discoveries, as documents, filed in archives and private collections, are brought to light. All of the questions concerning "America's bloodiest war" will never be answered, but with the constant quest for truth, Civil War scholars will find the field alive with challenge.

ECHOES OF WAR features another unpublished narrative by Thomas J. Evans, describing the underground railroad, and the two very different religious sects that ran it. Very little has been written about the people who risked their lives to help escaping slaves, or of the methods they used. We are pleased to add "The Underground Railroad in Washington County, Indiana" to this limited store of knowledge. Complementing this is a petition from the Society of Friends, which outlines their peaceful principles, as presented to the Convention of North Carolina.

MEDICAL MANUAL investigates the 19th Century belief that electricity, as a component of blood, was "the spark of life" that actually became magnetized in the brain. TRACTS contains a missionary's account of life among the sorcerers of India and the 1862 Christian's prayer for their salvation.

The regular columns, CONCURRENT EVENTS and WHIMSEY, represent 19th Century life and humor, and, in a new feature, MILITARY REVIEW, books pertaining to soldiers and battles of the Civil War are examined by contemporary scholars.

S.L. Kotar *J.E. Gessler*

Editors-in-General

Head-Quarters of THE KEPI
1st October, 1983

GENERAL ORDERS,
No. 3
by Command of the Editors-in-General

Coming in the December-January issue of THE KEPI is the original, unpublished prison diary of Private H.H. Wiseman. This Confederate soldier of the 1st Tennessee Heavy Artillery was captured at Fort Morgan, Alabama, in August, 1864, and sent to Elmira Prison, New York. Few letters and diaries written by Confederate prisoners are extant, making Private Wiseman's observations a valuable addition to the knowledge of the Civil War.

Also featured is an article dealing with the "reconstruction" of the Confederate soldier, from President Lincoln's first pardon offer in December, 1863, through the final U.S. legislation in 1898, to President Jimmy Carter's restoration of citizenship to Jefferson Davis in 1978.

ECHOES OF WAR continues with a narrative of how a Union soldier, Thomas Evans, of the 54th Pennsylvania Volunteers, won the Medal of Honor for capturing a rebel flag at the Battle of Piedmont, Virginia, in June, 1864.

The regular columns will also appear, and the Articles of War continue.

Gordon's Flank Attack

by
Samuel J. Martin

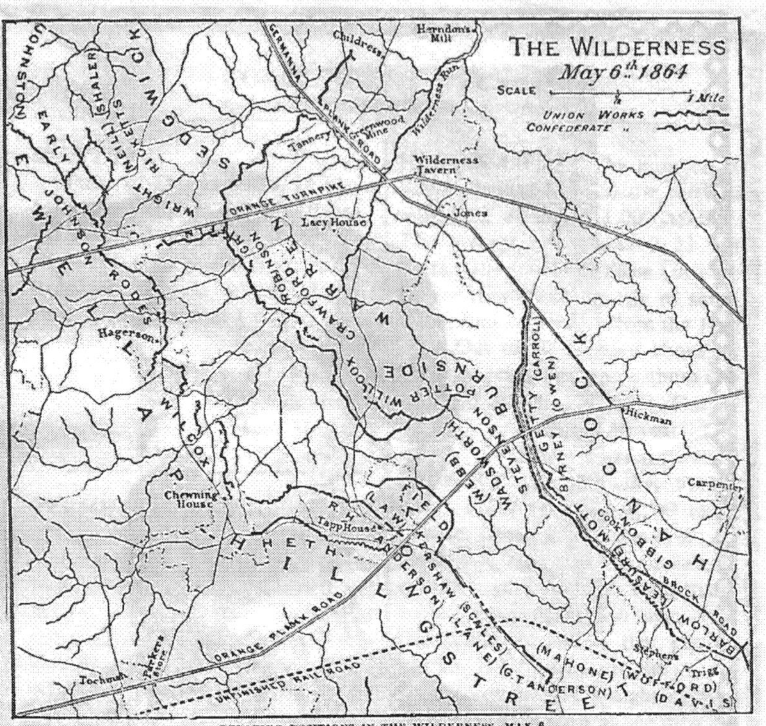

RELATIVE POSITIONS IN THE WILDERNESS, MAY 6.

For the most part the troops are indicated by divisions, and when a name indicates a brigade it is inclosed in parentheses. It should be noted that Griffin's line, before connecting with Wright, extended a short distance parallel with the Orange turnpike. Wadsworth, early in the morning, advanced south from near the Lacy house to the Orange Plank road, and formed across that road as indicated. Leasure's brigade, of Stevenson's division, coming into the line in Gibbon's first position, advanced north across Hancock's front to the Plank road. On the morning of the second day Webb, of Gibbon, fought on, and north of, the Plank road, while Gibbon's other two brigades, Owen and Carroll, were supporting Getty on, and north of, the road. Gibbon had general charge of Hancock's left, and Birney of Hancock's right.—EDITORS.

Lieutenant-General Richard S. Ewell, commander of the Second Corps of the Army of Northern Virginia, anxiously awaited the dawn of May 6, 1864. His men stood poised in the Wilderness — the jungle of scrub oak and brambles west of Fredericksburg — ready to attack. His orders from Robert E. Lee, twice repeated the prior evening, gave the General two options: he could either assault the Federal right to sever their supply line to the Rapidan River, or move south to reinforce A.P. Hill's Third Corps.[1] Ewell had chosen to take the offense.[2]

Ewell's line, straddling the Orange Turnpike, faced east. He had Jubal Early's and Edward Johnston's Divisions north of the road, while Robert Rodes' Division was posted south of the narrow dirt trail. West of the Brock Road, which ran southeast through the Wilderness, the enemy had two corps fronting Ewell: General John Sedgewick, commanding the Federal VI Corps was to his left, and General G.K. Warren, commanding the V Corps on his right.[3]

Two and one-half miles south of Ewell, A.P. Hill's troops were formed across the Plank Road, which was parallel to the Orange Turnpike. They were opposed by General Winfield Scott Hancock's II Corps.[4] Robert E. Lee, accompanying Hill, was directing the Southern strategies, while Ulysses S. Grant was in charge of the Army of the Potomac's plan of battle.

The Confederate cannons boomed at 4:30 A.M. Breaking from behind their entrenchments, Early's men stormed toward the Federal right.[5] They met and scattered a thin wall of skirmishers as they raced ahead.[6] The Yankees, having planned a 5:00 A.M. attack on Ewell, were formed when Early's troops moved out.[7] They rushed forward to greet the Rebel challenge. The two foes clashed in the dense brush between the lines.[8]

For half an hour, the adversaries struggled. Seeing that he could not smash through the Union's superior numbers, Early finally withdrew to his breastworks.[9] Once behind the lattice of logs and debris they had raised the previous night, Early's regiments turned to defend their bulwark. They saw the enemy coming — not only directly toward them, but also up and down Ewell's line.[10] Above the Orange Turnpike, Sedgewick threw charge after charge upon Early's and Johnson's Divisions. Each was repelled with an equal vigor.[11] To the south, along Rodes' front, Warren was not so daring. While his skirmishers crept forward, the bulk of his force never reached the Confederate entrenchements.[12] They showed their presence, but not their nerve.

Frustrated by being pinned in position (and prevented from carrying out his orders for the day), Ewell waited for a break in the action. It came at 6:30 A.M. [13] As a stillness settled over the field, a messenger galloped in from the right. The General expected word that the Yankees had given up their assault. Instead, he was told that Rodes had spotted a Union force in route to the gap between his and A.P. Hill's locations.[14] Ramseur's Brigade, who had rejoined the corps that morning from picket duty along the Rapidan River, made up Ewell's only reserve.[15] The General rushed the

JOHN BROWN GORDON

North Carolinians to the opening and they got there just in time to thwart the Federal's thrust.[16]

About 7:30 A.M., the Yankees renewed their attack on Ewell's left.[17] Although delivered indecisively, the blow made sense in light of a message from Cowles' cavalry. They alerted the General to Union troops approaching from the north.[18] It seemed that Grant was trying to hold the Rebels in place as he set up a surprise assault on their left flank. Fortunately, the General had just been reinforced. Robert D. Johnston's Brigade arrived after a grueling 66-mile march from Hanover Junction.[19] To rebut the menace to his left, Ewell sent them on to Early with instructions that he was to personally put them into position.[20]

Despairing that he would never be able to carry out Lee's wishes for the day, Ewell informed the commanding general of the situation.[21] He then steeled himself for the battle to re-open on his flank.

An excited John Gordon, one of Early's brigadiers, rode up to corps headquarters about 9:00 A.M.[22] After dismounting, the ex-lawyer ran up to Ewell and blurted an amazing story: the Federal right was in the air. If the General would only give the order, Gordon would swoop down on the enemy and rout them from their position.[23] Worried over an expected assault on his left, Ewell was not convinced of the opportunity. He referred to Cowles' report as he demanded proof of Gordon's claim. Dismissing the threat to the north as just a feint,[24] the Georgian went on to explain that he had scouted Sedgwick's right with three separate parties.[25] On the last trip, he had accompanied his men and seen for himself that the way was clear for an attack.[26] At that moment, Early came back from his assignment of posting R.D. Johnston's Brigade.[27] Gordon re-told his tale, but Early rejected the idea. He countered by noting that Burnside's Corps had crossed the river that morning and was reinforcing the Federal right. He further argued that since Ewell had no reserves, the plan was too risky. If the assault failed, it would not only expose the General's corps to counterattack, but would also jeopardize Lee's whole army.[28] Gordon answered defiantly. He would assume responsibility if disaster resulted from the attack.[29] When Early stubbornly refused to budge from his position, Ewell was forced to settle the confrontation.

In his memoirs, Early wrote that "General Ewell concurred with me in this opinion..."[30] Gordon stated that "...Ewell...agreed...to the impolicy of making the morning attack..."[31] In his official report, Ewell recalled that "...Early thought the attempt unsafe. This necessitated a personal examination..."[32]

While all participants in the affair agreed that the assault was postponed, none commented in detail on what followed. Most historians have deduced from the meager evidence that Ewell wasted the rest of the day in vacillation. Shelby Foote concluded that Ewell was "...unable to bring himself to unleash Gordon..."[33] Freeman admitted "It is unknown when or to what extent [the General] made reconnaissance of his own." But he accused Ewell of allowing "...the fateful afternoon to pass without an offensive blow."[34] And Clifford Dowdey reported that Ewell "...did nothing..."[35]

What really happened? A recently discovered memo, written by Campbell Brown, Ewell's assistant adjutant general, sheds a revealing light. Objecting to Early's memoirs, in which he took credit for the eventually successful attack, Brown notes that Ewell was **not** dissuaded by Early's arguments that morning. The General supported Gordon's proposal to move at once against the Union's right flank.[36] Early, however, managed to delay the assault by insisting that **he**—not Ewell — examine the ground in question.[37] This was proper military procedure, and the General, a stickler for army etiquette, acceded.

Brown also disclosed that just as this decision was made, the general was summoned south to see Robert E. Lee.[38] This kept Early from beginning his investigation. As the senior division commander, Early was responsible for the corps during Ewell's absence. He was obliged to stay at headquarters until the General returned.

Ewell's conference with Lee on the morning of May 6, 1864 — never before documented — adds a highly interesting input to the events of that day. And while Brown did not reveal what took place at the meeting, circumstantial evidence presents a clear view of the probable discussion. To set the stage, A.P. Hill, too, had been attacked about 5:00 A.M.[39] But unlike Ewell, whose lines held fast, Hill was routed.[40] A disaster was avoided only by the timely appearance of Longstreet's Corps. Arriving at 6:30 A.M.[41] from a two-day march from Mechanicsville, Longstreet's men tore into the advancing enemy; and threw them back.[42] By 8:30 A.M., both adversaries were entrenched in their original positions.[43] Hancock renewed his assault at 9:00 A.M., but this time was unable to gain ground.[44] And while the bitter skirmish continued, Lee and Longstreet plotted to take the offense.[45]

About 10:00 A.M., Major General M. L. Smith, Lee's Chief Engineer, joined the strategy session.[46] He had exciting news. Just back from exploring the enemy's position, he reported finding an abandoned railroad cut that led to Grant's left flank.[47] No one was guarding the approach. To exploit the opportunity, Longstreet began to organize an assault force. It was probably at this moment that Ewell arrived on the scene. Lee undoubtedly repeated his request that the General bring his corps south to support the coming attack. Lee was anticipating a massive blow, and needed every man he could spare. But since Ewell's men stayed in place that day, Lee did not give that order. Ewell must have cited the fact that

THROUGH THE WILDERNESS.

THE BURNING WOODS, MAY 6—RESCUING THE WOUNDED. FROM A SKETCH MADE AT THE TIME.

[Before the dark of the night had completely settled, fire swept across the bloody Confederate right, spread by a brisk wind. Two hundred men that could not be brought to safety, perished in this merciless aftermath.]

the Federal's right flank was also in the air, and proposed an independent assault by his troops. While Lee evidently agreed to the Second Corps being on their own, he did not direct Ewell to act in concert with Longstreet. Both corps attacking at the same time would have been the best strategy.

When Ewell arrived back at his headquarters. Early became free to leave on his scouting mission. He was gone for an interminable length of time. And while Ewell waited in vain for word from the irascible Early, he saw the opportunity of the morning slipping away. The General heard Lonstreet's battle, expected to occupy the enemy on the right, begin at 11:30 A.M.[48] It died away in less than an hour. As a stillness settled upon the field, Federal horsemen were reported to be approaching from the north (where Early was supposedly poking the bushes, hoping he would not find blue-coats). Ewell passed this news on to Lee, and was told to call on Fitz Lee's cavalry if he needed support.[49] The musketry to the south flared anew at 2:00 P.M., but quickly abated.[50] A half an hour later, a courier from Ramseur arrived at headquarters. He brought word that the Yankees were moving left and might be massing there.[51] Ewell no doubt groaned with dismay. If the enemy concentrated on his flank, his chance of turning theirs was gone. He asked Gordon to check out the rumor.[52] About 4:15 P.M., the fighting resumed to the south.[53] It was still raging when Early finally appeared. Gordon was with him.[54] Ewell must have been shocked to hear Early announce that he had indeed found the enemy's right flank to be unprotected.[55] Burnside, who had been in reserve there, had moved south to share Hancock's affair with Longstreet and Hill.[56] Early recommended attack.[57] But before the General could act, Lee arrived.[58] Lee, too, had suffered a frustrating day. After a dazzling initial success, the Rebel blow on the right had floundered, when Longstreet was mistakenly shot and wounded by his own men.[59] It took four hours, interrupted by a feeble assault by Burnside at 2:00 P.M.[60] before a necessary re-organization had been completed. By that time, Grant had restored his line. Lee's attack at 4:15 P.M. had been repelled.[61]

The situation on the left was presented to Lee. As Ewell explained that Early had wasted the afternoon confirming what Gordon had discovered before 9:00 A.M., a cold silence greeted his comments. Early was to blame, but Lee held Ewell accountable. He limited his words to a terse order to advance.[62]

Gordon had a different recollection of the meeting. Claiming that neither Ewell nor Early said anything about the flanking opportunity, he credited himself as the one to first tell Lee that the Federals' left was exposed.[63] The Georgian said that Early "...vigorously protested...that the movement was too hazardous..."[64] But when Gordon went into detail on his findings, Lee sided with him.[65] He even asserted that Lee rebuked both of the senior officers by ignoring the chain of command to personally order Gordon to attack.[66] Considering that Ewell had already met with Lee in the morning and certainly discussed the situation on the left, Gordon's affirmation must be judged as fantasy, written 38 years after the incident. Edward Steere, an autho-

JUBAL ANDERSON EARLY

rity on the Battle of the Wilderness, agreed. Even without Campbell Brown's evidence, he felt "Gordon's version is open to serious question."[67]

The attack was set up to hit the Union from three sides. Gordon would strike from the north. When the enemy in the entrenchments turned to face this threat, Pegram's Brigade would rush forward from the west to smash the abandoned front. And R.D. Johnston's regiments were to follow Gordon, but veer left into the Yankee rear, and cut off all avenues for escape.[68]

The initial Confederate burst at 6:00 P.M. brought all that Gordon had promised.[69] His blow scattered both Federal brigades on the flank.[70] But as the sun fell below the western horizon, the attack began to stall. Peagram's troops bore right, and instead of blasting the men the Georgian had disrupted, they ran into a solid front, further down the Yankee's line. Colonel John S. Hoffman, directing the Confederate charge, was compelled to withdraw.[71] And when a portion of Gordon's force strayed to the east, joining Johnston in his rush to the rear, too few were left on course.[72] They met a hurriedly formed defense, set up at a right angle to the Union breastworks, and were repelled.[73] Unable to reform their troops in the dark in the depths of the Federal encampment, Gordon and Johnston decided to take their brigades back to the safety of the Rebel line. They returned with 600 prisoners, including the Federal Generals Shafer and Seymour.[74]

Gordon rued to his dying day that if he had been allowed to attack in the morning, he would have won the war with one blow.[75] Early argued that the Confederates were lucky to escape that night. He wrote that only the darkness saved Gordon from the enemy discovering "...the disorder on our side...and availing himself of our condition."[76] Ewell stayed out of the controversy. To avoid disparaging Early (a typical, thoughtful trait of the General), he limited his remarks in his official report to just the vague truth.[77]

When Brown wrote his memo, his purpose was to damn Early. He finished with the comment, "Early arrogates almost the whole credit to himself. Oh, Jubal!"[78] Had he known that Gordon would later distort the facts to favor his own reputation (at the expense of Ewell's), Campbell probably would have added these words, "Oh, John!"

FOOTNOTES

1 "War of the Rebellion: A compilation of the Official Records of the Union and Confederate Armies," U.S. Government Printing Office, Washington, D.C., 1880-1901, Vol. XXXVI, Part 2, pp. 952, 953.

2 "The Wilderness Campaign" by Edward Steere, The Stackpole Co., Harrisburg, Pa., 1960, p. 292.

3 IBID, p. 292.

4 IBID, p. 292.

5 IBID, p. 319.

6 IBID, p. 319.

7 IBID, p. 319.

8 IBID, p. 319.

10 IBID, p. 320.

11 "The Wilderness Campaign," p. 320.

12 IBID, p. 321.

13 IBID, p. 322.

14 O.R., Vol. XXXVI, Part 1, p. 1081.

15 IBID, p. 1081.

16 IBID, p. 1081.

17 "The Wilderness Campaign," p. 319.

18 O.R., Vol. XXXVI, Part 2, p. 962.

19 O.R., Vol. XXXVI, Part 1, p. 1071.

20 "War Memoirs" by Jubal Anderson Early, Indiana University Press, Bloomington, Indiana, 1960, p. 348.

21 O.R., Vol. XXXVI, Part 2, p. 961.

22 "Lee's Lieutenants" by Douglas Southall Freeman, Charles Scribner's Sons, New York, New York, 1944, Vol. III, p. 369.

23 O.R., Vol. XXXVI, Part 1, p. 1077.

24 O.R., Vol. XXXVI, Part 2, p. 962.

25 "Reminiscences of the Civil War" by John B. Gordon, Civil War Times Illustrated, Gettysburg, Pennsylvania, 1974, pp. 243, 244.

26 IBID, p. 244.

27 "Lee's Lieutenants," Vol. III, p. 369.

28 "War Memoirs," p. 348.

29 "Reminiscences of the Civil War," p. 256.

30 "War Memoirs," p. 348.

31 "Reminiscences of the Civil War," p. 260.

32 O.R., Vol. XXXVI, Part 1, p. 1071.

33 "The Civil War" by Shelby Foote, Random House, New York, New York, 1974, Vol. III, p. 182.

34 "Lee's Lieutenants," Vol. III, p. 370.

35 "Lee's Last Campaign" by Clifford Dowdey, Bonanza Books, New York, New York, 1960, p. 170.

36 Campbell Brown, Ewell-Brown Papers, Manuscript Unit, Tennessee State Library and Archives, Nashville, Tennessee.

37 Ewell-Brown Papers

38 IBID.

39 "The Wilderness Campaign," p. 328.

40 IBID, p. 332.

41 IBID, p. 340.

42 IBID, p. 342.

43 IBID, p. 356.

44 IBID, p. 391.

45 "Lee's Lieutenants," Vol. III, p. 361.

46 "The Wilderness Campaign," p. 393.

47 IBID, p. 393.

48 IBID, p. 396.

49 O.R., Vol. XXXVI, Part 2, p. 961.

50 "The Wilderness Campaign," p. 416.

51 O.R., Vol. XXXVI, Part 2, p. 961.

52 IBID, p. 961.

53 "The Wilderness Campaign," p. 423.

54 "Reminiscences of the Civil War," p. 258.

55 "War Memoirs," p. 348.

58 "Reminiscences of the Civil War," p. 258.

59 "The Wilderness Campaign," p. 405.

60 IBID, p. 416.

61 IBID, p. 427.

62 "Reminiscences of the Civil War," p. 258.

63 IBID, p. 258.

64 IBID, p. 258.

65 IBID, p. 258.

66 IBID, p. 258.

67 "The Wilderness Campaign," p. 435.

68 "The Wilderness Campaign," p. 440.

69 IBID, p. 440.

70 "Reminiscences of the Civil War," p. 249.

71 "The Wilderness Campaign," p. 442.

72 IBID, p. 443.

73 IBID, p. 442.

74 O.R., Vol. XXXVI, Part 1, p. 1071.

75 "Reminiscences of the Civil War, p. 261.

76 "War Memoirs," p. 350.

77 O.R., Vol. XXXVI, Part 1, p. 1071.

78 Ewell-Brown Papers.

[The following is from a letter of Campbell Brown's; the reports are copied by him, and all notations are made in his hand.]

In consequence of what I cannot but consider a willfully incorrect statement of Genl. Early's in his "Memoirs," I here copy a letter from Genl. Jno. B. Gordon endorsed on a report of Major Cowles, which is also copied. They refer to Gordon's brilliant attack on the right flank of the 6th Corps, for which Early claims the credit — while all he deserves is that of having prevented its being made six hours earlier and consequently far more completely there being no change in the enemy's position of which any one but he was ever aware except that a small scouting party sent out by there on the opposite side of an impassable swamp that day half a mile to our extreme left, had returned to the main body.

(Copy) "Out Post on Germanna Road"
May 6th 1864
8 o'clock AM

Brig. Genl. Gordon—

Genl. The enemy (Infantry) are moving upon my pickets to the left of Mrs. Willis's house on the Germanna Road. A Reg't. have passed & still coming — A force should be sent in this direction at once — A Reg't. for skirmishing until more could be brought up if necessary. This move, if continued, flanks you about two miles to your left.

Respectfully,
W.H.H. Cowles
Maj. Comd'g. 1st N.C. Cav.

Endorsed is the following note
in Genl. Gordon's own handwriting

Genl. Early or Ewell—

This must be a feint — I don't think it can be intended for a serious movement. This cavalry could keep us informed; till the movement here could be made. I still think it best to make the movement I spoke of, & it will check any move to our left. If made, I should like to order at once. I could feel of them (sic) very soon & then look after this flanking party.

J.B. Gordon, B.G.

Now I know of my own personal knowledge that Genl. Ewell was in favor of this but begged out of it by Early's strong personal appeals *until* he could go to examine the ground himself, when he at once ordered an attack. But the delay thus occasioned was considerable, as Genl. Early found Genl. Ewell just summoned to Genl. Lee & he had to delay the investigation until his return. Yet Early arrogates almost the whole credit to himself. Oh Jubal!

[Reprinted by permission of the Tennessee State Library and Archives.]

WHIMSEY

Slightly So. — A man once relating a story of being on a locomotive that struck a cow standing on the track, and threw her several yards into the field, where she lit squarely upon her feet, with her feet, with her head towards the train, and strange to say, she wasn't hurt a mite.

''But didn't she look scared?'' inquired a listener.

''Well, I don't know whether she was scared or not, but she looked a good deal discouraged.''

(Yankee Notions, New York, February, 1859.)

"No Mistake at all Sir!"

A sailor having purchased some medicines of a celebrated doctor, demanded the price.

'Why,' says the doctor, 'I cannot think of charging you less than seven-and-sixpence.'

'Well, I'll tell you what,' replied the sailor, 'take off the odds, and I'll pay you the even.'

'Well,' returned the doctor, 'we won't quarrel about trifles.'

The sailor laid down sixpence, and was in the act of walking off, when the doctor reminded him of his mistake.

'No mistake at all, sir,' said the sailor; 'six is even and seven is odd, all the world over; so I wish you a good day.'

'Get you gone,' said the doctor; 'I've made fourpence out of you yet.'

(Boston Cultivator, Boston, Massachusetts, December 13, 1845.)

RICHARD STODDERT EWELL

REPURCUSSIONS

Campbell Brown's assertion that General Ewell was summoned to Lee's headquarters on the morning of May 6, 1864, thus delaying his investigation of Gordon's report has far more serious and wide-ranging implications than he realized. If it is true, it necessitates an entire reconstruction of events, and places blame for the costly delay not on Ewell or Early, but at the door of the commanding general himself.

The events of May 6, 1864, have always presented frustrating inconsistencies: none of the three men involved had the same recollection of the events. General Gordon, who first discovered the unprotected Federal flank, reported that upon confirmation of the fact, he rode to Corps headquarters, where he found both Ewell and Early. Upon his arrival, he explained the situation and asked for permission to attack. When Early attempted to dissuade Ewell, Gordon went so far as to assume all responsibility in the event of failure.[1] When Ewell deferred a decision, Gordon was forced to wait. According to him, it was not until late afternoon when R.E. Lee appeared at Ewell's headquarters, that he had a chance to personally describe the situation to him. Early vigorously protested, but Gordon's description of the possibilities convinced the commanding general enough to order the attack, and rebuke both Ewell and Early.[2] The attack was made, but it was already too late in the day for complete success. In his official report, Gordon stated that "...had the movement been made at an earlier hour and properly supported, it would have resulted in a decided disaster to the whole right wing of General Grant's army, if not in its entire disorganization."[3]

Jubal Early, on the other hand, when writing his memoirs, stated that he was not present at Ewell's headquarters when Gordon arrived, having been in the field posting Johnston's Brigade. On his arrival there, "...the subject was mentioned to me by General Ewell, and I stated to him the danger and risk of making the attack under the circumstances..."[4] Stating firmly that Ewell concurred with him, he mentioned neither Ewell's morning visit to Lee, nor Lee's afternoon visit to Ewell. He made no reference at all to having personally investigated Gordon's claim, or of General Ewell having done so, or the reason why it was not confirmed. His account simply states that in the afternoon, when the Federal threats had been removed "...at my suggestion, General Ewell ordered the movement which Gordon had proposed."[5]

General Ewell, in his official report, states that he got word from Gordon "through General Early in person."[6] He stated he could not decide whether or not to allow Gordon to proceed, without personal examination, which he resolved to make. He continued by saying that "order unavoidable causes" delayed his investigation, and that he made a personal examination "as soon as other duties permitted." Then Ewell wrote: "I ordered the attack and placed Robert D. Johnston's brigade of Rodes' division (that morning arrived from Hanover Junction) to support Gordon."[7]

In an attempt to fit together these three versions, Campbell Brown's letter takes on great significance. In the first report he quotes, that of Major Cowles, dated May 6, 1864, 8

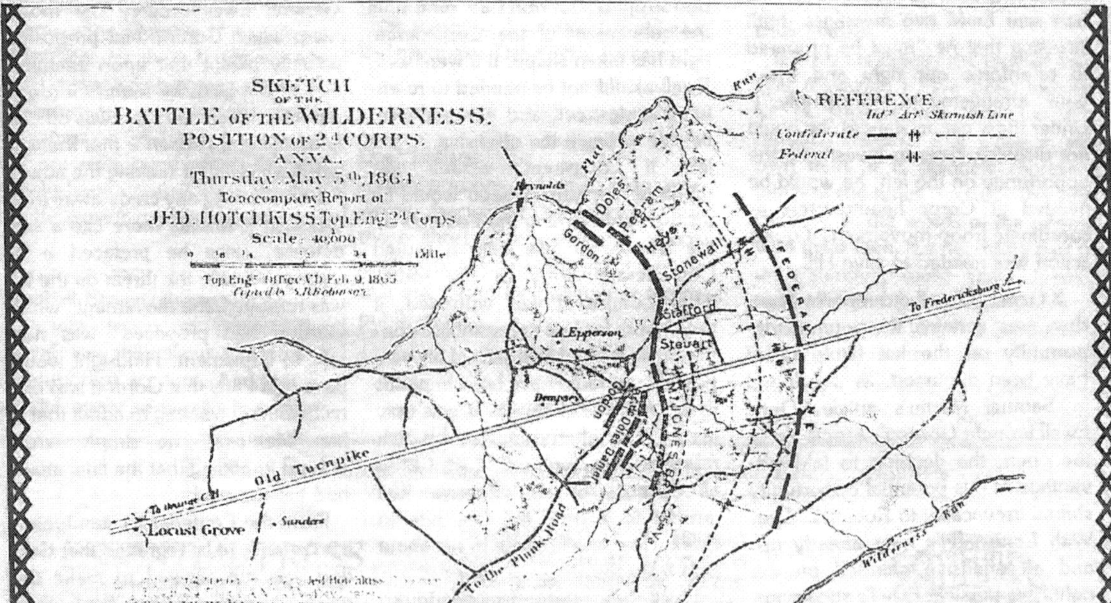

o'clock AM, the cavalryman mentions the movement of Union troops that flank Ewell's men by two miles to the left. On it, John Gordon endorses to **either Gen.'l Early or Ewell** that this must be a feint and he "still think[s] it best to make the movement I spoke of..." Clearly, then, by the time he received Cowles' report, Gordon had **already** spoken to both his superiors about the proposed move, and was urging one to come to a rapid decision. Presumably he had returned to his own command, and there received the message from the cavalry. Forwarding it to Corps headquarters, where he believed Ewell and/or Early to be, his endorsement may have been read by Early (who, as agreed upon by all sources, disapproved of the idea), but may not have been seen by Ewell. The Lieutenant-General may, in fact, have been planning to personally investigate Gordon's assertion, when he was suddenly called to Lee. Already aware of the severe pressures being put on the Confederate right, it would have been logical for him to assume he was being summoned to receive orders that would necessitate his reinforcing Hill. (On May 5th, Lee had sent Ewell **two** messages, both stressing that he "must be prepared to re-enforce our right and make your arrangements accordingly.")[8] Under these circumstances, he could not dispatch Early to investigate the opportunity on the left; he would be needed at Corps headquarters to coordinate troop movements if quick action was needed to save Hill.

If General Ewell did report to Lee, then most certainly the potential opportunity on the left flank would have been discussed, as developed in Samuel Martin's article. Once Ewell brought Gordon's report out in the open, the decision to take advantage of this potential opportunity shifted irrevocably to Robert E. Lee. With Longstreet's men already up, and an offensive planned on the right, it is possible Ewell's suggestion sounded too risky, especially in light

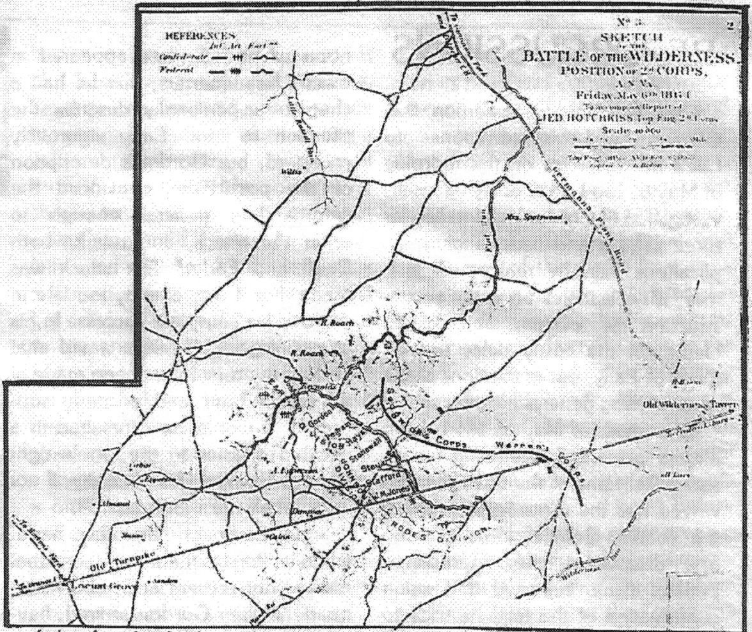

of the fact he had not had time to substantiate it. In all probability, as he had done so often with General Jackson in the past, Lee's orders to Ewell would have been to investigate the position of the Union forces, attack only if success was certain, and be prepared at any moment to reinforce Longstreet and Hill. Or he may simply have said, investigate Gordon's report, but don't act on it until the movement of the Confederate right has taken shape. If it went well, Ewell would not be needed to re-enforce Longstreet, and it would then be safe to begin the offensive on the left. If Longstreet's assault was repulsed, the urgent need would be on the right: turning the Federals on the left would little help a routed Confederate Army to the south. After Longstreet was wounded, it took hours for Lee to reestablish control of his army. Uncertain of his own position, it would have been impolitic to order Ewell to attack. It was only in the late afternoon, "in his first relatively free moment,"[9] and after all chances of an offensive had ground to a halt, did Lee ride to Ewell. Now was the time to see about help from the left.

Ewell then returned to headquarters with either vague or specific instructions not to risk attack. The movement of his troops to the south had been delayed, but so had the chance for swift action.

It is probable that upon his return, both Ewell and Early investigated Gordon's claim. Ewell mentions in his official report that **he** did make a personal reconnaissance. Early's statement that "at my suggestion, General Ewell ordered the movement which Gordon had proposed" actually meant that upon examination of the field, he found Gordon's report to be correct, and was officially notifying his superior that the way was now clear for making the attack. Far from taking any credit away from Gordon, it sounds more like a self-defense, since he prefaced it by stating that after the threat on the left was removed, the movement "which Gordon had proposed" was now safe to implement. Hindsight would have told Early that Gordon was correct; without wanting to admit that in his "Memoirs," he simply wrote around it, adding that the final attack had his approval.

From the Confederate standpoint, it is certainly to be regretted that Gordon was not allowed to make the move against the Federal flank when it was first suggested. But, as Douglas Southall Freeman pointed out,[10] Lee did not blame either John Gordon or Jubal Early for the delay, as both were promoted to positions of responsibility shortly thereafter. He is not known to have censured Ewell, either, although if his lieutenant were to blame, his inability to make a choice and his delay in confirming Gordon's opportunity was a serious defect in his command ability. On the other hand, if General Ewell was in favor of permitting Gordon to take the offensive, and was only delayed by a summons from his commander, then it is **Lee** who must bear the burden of failure, and Ewell, who was so sarcastically labled as "perplexed or weary or hypnotized,"[11] must be exonerated from a large portion of the blame.

For his own part, Ewell wrote little in his own defense. But perhaps what he did not write is more significant. If Ewell was called away by his commander, explained the situation on the left flank, but came away with equivocal orders from Lee, to include this in his official report and place the blame squarely where it belonged would not serve the cause he loved. Rather, it would only hinder it by causing instability and perhaps a lack of confidence in the Commanding General of the Army of Northern Virginia. It was already felt by some that "nothing seemed as satisfactory as upon the opening of the campaign of 1863. The troops were not in the same spirits...",[12] and it might have seemed more prudent to write an ambiguous account of the delay.

While it may be argued that Campbell Brown, the son of Ewell's wife, was trying to protect the general, it is still difficult to ignore a contemporary account. And if R.E. Lee did arrive at Ewell's headquarters that fateful afternoon, and stare silently at Gordon's two superiors, perhaps his reticence was not rebuke, but a silent acknowledgement of his own mistake.

FOOTNOTES

1 "Reminiscences of the Civil War" by John B. Gordon, Civil War Times Illustrated, Gettysburg, Pa., 1974, pp. 255-56.

2 IBID, p. 258.

3 "War of the Rebellion: A Compilation of the Official Records of the Union and Confederate Armies," U.S. Government Printing Office, Washington, D.C., 1880-1901, Vol. XXXVI, Part 1, p. 1078.

4 "War Memoirs" by Jubal Anderson Early, Indiana University Press, Bloomington, Indiana, 1969, p. 348.

5 IBID, p. 348.

6 O.R., Vol. XXXVI, Part 1, p. 1071.

7 IBID, Vol. XXXVI, Part 1, p. 1071.

8 O.R., Vol. XXXVI, Part 2, pp. 952, 953.

9 "Lee's Last Campaign" by Clifford Dowdey, Bonanza Books, New York, New York, 1960, p. 170.

10 "Lee's Lieutenants" by Douglas Southall Freeman, Charles Scribner's Sons, New York, New York, 1944, Vol. III, pp. 443, 444.

11 IBID, p. 370.

12 "I Rode With Stonewall" by Henry Kyd Douglas, The University of North Carolina Press, Chapel Hill, North Carolina, 1940, p. 274.

SIGNALS

Dear Editors:

The first two issues of THE KEPI have arrived and I was delighted to receive them. I enjoyed both, and see the possibilities for your magazine. I think you're on the right track.

Vernon H. Crow
Fresno, California

Dear Editors:

I am very pleased with your first two issues. It is definitely a new concept.

Nancy Melton
Manager, George Washington
Office Museum
Winchester, Virginia

Dear Editors:

THE KEPI is exactly the type of Civil War magazine I have been looking for. It supplies me with valuable research material that I use in my own field. It is uniquely representative of the times. To study the 19th Century, **in the words of the people who lived then,** is a necessary part of understanding those years. I have read all three issues cover-to-cover, and have given them a permanent place in my library.

Glenn Scott
Joplin, Missouri

Dear Editors:

The issue you kindly sent along convinces me that THE KEPI has limitless potential. I will do all that I can to help make THE KEPI a success.

Bill Howard
Delmar, New York

Dear Editors:

The portions of the magazine which fascinated me most were those short historical asides which dealt with everyday life and times of that era. Good luck with your future publications.

Ronald K. Pelton
Social Science Dept.
Los Angeles City College

Tracts

EDITED BY REV. ALLEN KELLEY

"Sorcery and Sacrifices among the Hindoos"

In the missionary field occupied by the American Madura mission are a class of sorcerers, called in Tamil Kordangkies. They profess to have the spirit of some god come upon them; and thus, when questioned, they give answers showing what should be done to remove disease, or where stolen property may be found, etc. They also have twelve lotus seeds, representing the twelve signs of the zodiac, with which they perform divination. They are much sought after, and I very often hear in the villages the sound of the little drum, which they use constantly while invoking the divine power, and then continue it when giving the responses. Yesterday, on the afternoon of the Sabbath, in a village near Melur, I heard the sorcerer's drum; and I know four or five such persons living near me. A few days ago I visited the house of a man who had the day before buried a son of about twelve years. I asked him if he consulted the Kordangkies, and what they said about his son. He spread out the fingers of both his hands, indicating that he had consulted many, and I had reason to believe that he had at least five. All, he said, assured him that his boy would recover. And now here were the father and mother sitting, sighing, and mourning in their loneliness and sorrow, trusting not in that sympathizing, comforting Saviour who wept at the grave of Lazarus. If things occur as the sorcerer predicts, he gets much credit; and when they do not, he says, "It was the writing in the head," or fate, and the unreflecting people, in their ignorance, accept the excuse, and very likely seek him again the next time a person is ll. I have sometimes tried to undeceive a crowd of the deluded people by writing in Tamul a word or name on a tract, and putting down a sum of money, having told the sorcerer he should have it if he would tell what was written. Of course he always fails, and yet he seems very little ashamed. "My stomach wants. I must get a living," he replies; and strange to say, the people are usually satisfied with this foolish excuxe, and go on as before consulting these deceivers. How sad is their slavery to ignorance, deception, and sin!

The favor of the gods or evil spirits is often sought by sacrifices. Let me tell you of one I saw some months since at a place fourteen miles from Melur. I went out on the Sabbath afternoon with a catechist to a village, where we found the people, particularly women, engaged in boiling rice as an offering of thanksgiving to a goddess, Paravathi Amman, for recent rain. A temporary temple, like a booth, had been built, and the idol within was highly decorated with flowers and tinsel. Near the entrance, seated on straw, was a Brahmin, the great man and lord of the village. He and the company listened to an exposition of the ten commandments and the way of life for the most part quietly, and without much of a spirit of disputation, until it began to grow dark. They then politely gave me leave to withdraw, but as I intimated a desire to stop, linger and observe their mode of worship and sacrifice, they made no serious objection. I then retired to a little distance and sat down, the Brahmin kindly sending me a portion of his straw as a seat. Women came with their dishes of rice, presented them to the idol, and gave a portion to the priest. Persons were busy, as is usual on such occasions, in giving the worshippers sacred ashes, made of burnt cow-dung, with which they rub their foreheads, necks, bare arms, and breasts. Lights were placed before the idol, and in the entrance of the temple, and after a time a lamb was brought, water was twice poured over it, and then its head was cut off. Afterwards four or five full-grown sheep and numerous fowls were thus sacrificed, the heads only going to the priest, while the offerers took home the bodies of the animals to feast upon. Meanwhile drums were beaten, and the women of the crowd kept up a strange noise by the trilling of the tongue. Those who had sacrificed were constantly hurrying away in the darkness with the bleeding, quivering victims in their hands. Altogether it was a wild and strange scene, the like of which I have not before witnessed in a life of nearly thirteen years among the Hindoos. I have before seen great numbers of goats on their way to be sacrificed at the famous temple of Alagur, twelve miles from Melur, and also more than once the headless trunks of goats and sheep borne on a pole between two men, but never had I seen an actual sacrifice, reminding me so forcibly of the everywhere felt want in the human soul of some mode of expiating sin, and the truth of revelation, that "almost all things are by the law purified with blood; and without shedding of blood is no remission." What Christian heart can refrain from the desire and prayer that these poor idolators may soon be led by the Spirit to accept of Him who "now once in the end of the world hath appeared, to **put away sin by the** SACRIFICE of HIMSELF."

Thank God that you know of the Lord Jesus Christ, the one only sacrifice for sin, and see that your hope and trust are in **him and in him alone.** And also, I pray you, do all in your power by praying and giving for the salvation of these perishing Hindoo's, whose trust is in deceivers, vain oblations, and lying vanities.

T.S.BURNELL

(American Messenger, September, 1862.)

Echoes of War

"The Underground Railroad in Washington County, Indiana"

By

Thomas J. Evans

[This interview was given to Tom Evans by James William Lusk at his Walnut Ridge farm, in Washington County, Indiana, in December, 1951. Mr. Lusk had known many Civil War veterans when he was a boy, and had lost a granduncle, Isaiah Lusk, killed in the Battle of Perryville, Kentucky, on October 8, 1862.]

"Now the Civil War started hereabouts thirty years before Fort Sumter because this county was the location of the first stations on two 'underground railroads' used by runaway slaves escaping to Canada. The reason they called it that was one slave owner said if once a runaway slave got into the hands of an abolitionist, no information could be found about him, and those abolitionists must have an underground road to spirit them away.

Now Indiana was never a slave state, however around here we are only thirty miles from the border with Kentucky. There were plenty of slaves in Kentucky even though President Lincoln, who was born there, saved the state for the Union during the war. The slave catchers from down there all wanted to try and capture any runaways before they made the first underground station here in a free state. After this first station those runaways just disappeared and didn't show up again until they appeared in Canada.

So we had slave catchers, real mean fellows, riding in here and the law was on their side in those days. Then on the other side of the law sneaking around here were the 'Blackbirders.' Those rascals would kidnap any free black person they could find and take them South and sell them. Between them, there were a lot of undesirable characters riding about this county for at least thirty years before the war.

Now there were two different underground railroad lines had a station in this county. The Quakers ran one and moved runaways from one Quaker neighborhood to another throughout the free states. They were quiet folks, good neighbors, and they didn't believe in violence. They had to outsmart the slave catchers, and they did a good job of it, but they got beat up and threatened because the rough element knew their religion wouldn't let them fight back.

The other station was run by the Covenantors, a religion my grandfolks belonged to. I took you down the ridge and showed you that station when they were tearing the old log house down. You saw that house had a full basement built out of limestone rocks with an extra wall across the back end leaving a secret little room. You could enter the basement through the outside door and to just go in there and look around you would never guess there was a room hidden behind that end wall. You could only get in that secret room through a trap door in the floor above. There was a rug over that trap door and a bed pulled over the rug. There were hundreds of runaway slaves in that hiding place over the years.

You know the Covenanters were run out of South Carolina because they were bitterly opposed to slavery. You couldn't be in that religion and own slaves. They weren't against violence and didn't turn the other cheek like the Quakers when it came to slavery. They were of Scotch-Irish descent and they were always ready to fight for any principle whatever. They didn't take much stirring up. All you had to do was shout, 'Blackbirding', and it was like waving a red flag in front of a bull. They would charge right out and take on any number of blackbirders, numbers made no difference to those boys when they were riled up. They carried double-barreled shotguns with the barrels cut off short and the stock sawed down so they could use them easily while riding horseback. They would load each barrel with a double charge of buckshot and when they fired, the only safe place was directly behind that gun. Blackbirders and slave chasers moved around mostly at night, so one of those 'blackbird guns' as they called them beat a pistol of rifle for any shooting in the dark. There used to be two of those guns in the attic until I sold them to an antique gun dealer for parts. One belonged to my Grandfather and the other one to my Granduncle I told you about who was killed in the Battle of Perryville, Kentucky during the war.

It's no secret who were the conductors of the local stations. James Thompson ran the Quaker one, and he took a lot of abuse and was beat up pretty bad once or twice. Isaiah Reid ran the Covenanter station, but he didn't have much trouble since anyone riding up to his house and threatening him would have stopped a load of buckshot and everybody knew it. If the slave catchers got to watching the Quaker station, why Isaiah Reid would send a few boys around to bring that runaway to his station, and then he would send him North using his own route through the Covenanter settlements. Isaiah influenced the vote pretty much in this area so the sheriff was not about to bother him, and if he had trouble with the law at any time I don't remember hearing of it.

The blackbirders were the worst lot going. There was a black couple by name of Rouse had a little farm out from town. The man was out rounding up the cows for milking one evening when a couple of blackbirders came through and stole his wife. Martha was her name, Martha Rouse. The whole county was up in arms because they were good folks, always helping out, and they chased that bunch clean to Louisville and got the law on them. Because she was black, the court wouldn't let that woman testify in her own behalf that she was a free black. Slaves had no rights in court then. The local folks got together and raised some money to buy her back, but those blackbirders refused to sell her, and she was sent South and sold and put to work on a plantation.

Blackbirders were always coming in here. They took a couple of boys, little fellows, from around the Knobs area, but they caught them before they could get across the state line and brought those kids back home. You had to watch out for any strangers you saw hanging around.

There was an old man lived down the ridge here, Covenanter he was, heard there was blackbirding going on in the county. He was nigh being eighty, but he saddled up a mule and took right out into the night. His women folks had a fit carrying on and wanted him to wait for help,

STRANGE EFFECT OF THE DRAFT.

"One of the most singular Phenomena of the day is the remarkable increase of QUAKERISM in the Sixth Ward in this City."

(Forthcoming Municipal Statistics.)

but he wouldn't have none of it. He knew that mule was safer to ride at night than his horse and he didn't want to waste time. He made for a piece of woods along the main road, and soon along comes this team and wagon with two blackbirders sitting up on the seat.

'Boys, that is a free man you have got in there and I aim to have him,' says the old man.

'You are an old fool,' says one blackbirder, and they both pull out pistols and start shooting.

Well, the old man cut loose with both barrels of his blackbird gun and killed those two fellows, both horses in that team, and durned near killed the black man that was tied up inside the wagon. The old man soon had help when a bunch of boys heard the shooting and rode over to take care of things. They pulled those horses over a hill and left them for the buzzards, and towed the wagon away to be taken apart. The two blackbirders were no problem for those boys were used to disposing of such rascals where nobody would find them. They carried them to an isolated church cemetery where nobody lived closeby, dug them a grave, buried them, stuck up a flat piece of field stone for a marker, and even threw on a few flowers when they was finished. Folks that saw the new graves would figure someone from down in the river bottoms had buried folks there (they didn't keep any records then), and anybody was welcome to use a cemetery without charge. Come Resurrection Day, the good folks in the cemeteries around here will sure be surprised when all the blackbirders and slave catchers that have been resting there start rising up around them. There are probably a hundred or so scattered around the county I would guess.

Now they had to ride in town and get a doctor to work on the black the old man had shot by accident. Since he was a freeman, they told the doctor he was shot in a hunting accident, and that was the end of it. If he had been a slave escaping, it would have been harder to do something for him. I believe Jimmy Thompson, the Quaker, had to use a sharp pocket knife to cut the lead pellets out of a runaway that had been shot by someone trying to collect a reward. He did a good job of it, and that slave lived and sent him word down from Canada he wasn't even scarred up to amount to anything.

So that is why I say there was a war going on some thirty years before the Civil War ever got started. There was thousands of runaway slaves passed through this county using the two underground railroad stations here, and there was shooting and fighting aplenty to move them through. You could get killed just as dead here as in any battle fought during the war. But there was a lot of folks around that believed in slavery even though Indiana wasn't a slave state. They believed slaves was property, pure and simple. Around this county, the war wasn't as much fought about state's rights as it was slavery. So now you know how things were in those days prior to the Civil War."[1]

FOOTNOTES

[1] Interview with James William Lusk at his Walnut Ridge farm, Washington County, Indiana, during December, 1951. Verification Sources:

"Centennial History of Washington County Indiana" by Warder W. Stevens, B.F. Bower & Co., Indianapolis, 1916.

p. 269. James L. Thompson identified as head of the Quaker underground railroad station in the county.

p. 277. Describes removal of shot from wounded runaway slave Tom Redding by James Thompson using a pen knife.

p. 282. Isaiah Reid identified as head of the Covenanter underground railroad station in the county.

pp. 282,283. Tells of kidnapping of Martha Rouse.

p. 441. Mentions two black boys stolen from county in a blackbirding incident.

REBEL AUTHORITIES
Memorial.

The following memorial was laid before the State Convention of North-Carolina by the yearly meeting of Friends, on the subject of bearing arms:

At a stated meeting for sufferings, representing North-Carolina yearly meetings of Friends, held at Deep River, on the fourteenth of fourth month, 1862, the subject of our present sufferings, on account of our conscientious scruples against bearing arms, claiming the considerate deliberation of the meeting, and believing it right to embrace our privilege to petition those in authority, we therefore adopt the following:

To the Convention of North-Carolina, in Convention Assembled:

Your petitioners respectfully show that it is one of our fundamental religious principles to bear a faithful testimony against all wars and fightings, and that in consequence we cannot aid in carrying on any carnal war.

This is no new principle of our Society, but one which was adopted at its rise, as the doctrine taught by our Saviour and followed by his disciples for more than two hundred years, and has ever been and is now held as one of our fundamental and vital principles, and one that we cannot yield or compromise in any degree whatever.

We would further show that the

whole number of our members in the confederate States is less than ten thousand, while in the United States the number probably exceeds two hundred thousand, and who bear the same testimony against all wars and fightings; and that in every nation and clime where our Society exists, it is at this day, as heretofore, maintaining this precious principle of peace, and that we everywhere in this respect speak the same language and mind the same thing.

We may further show that, according to the best information we can obtain, until the present time, Friends of North-Carolina have not been called on to aid in the battlefield or military camp; but now our peaceful principles are in a measure disregarded, *and many of our members are drafted to take part in the conflicting armies, while we understand our brethren in the United States are not.*

We have enlisted under the banner of the Captain of our soul's salvation, Jesus Christ, the Prince of Peace; therefore, in obedience to his express command, we cannot fight in or aid directly or indirectly, in any carnal wars. But your petitioners would represent to you that we believe it to be our moral and religious duty to submit to the government under which we live, and to the laws and powers that be, or suffer patiently their penalties.

We love our homes and our country much, but at the same time we love our religious principles more; therefore, your petitioners would most respectfully ask that you grant us the enjoyment of this important religious principle.

We own no God but the God of love, peace, mercy, and judgement, whose blessings we invoke, and whose wisdom we implore to be with you in your *legislative* deliberations.

Signed on behalf and by direction of the meeting.

Nathan F. Spencer,
Clerk.

(From "Moore's Rebellion Record," Supplement, pp. 752, 753.)

WHIMSEY

On a Republican Stage

Mr. Charles Kean, when once playing Richard, at New Orleans, observed, as he was seated on the throne, and the curtain was rising, that his noble peers wore their hats or caps in his presence. With his truncheon to his lips he contrived a stage whisper that said, "Take off your hats; you are in the presence of the king." "And what of that?" roared high-reaching Buckingham, looking round at the audience, and smacking his own cap tighter on his circumspect head; "what of that? I guess we know nothing of kings in this country." The New Orleaners were in raptures, and the king sat corrected. — **Habits and Men.**

(The Country Gentleman, Albany, New York, May 3, 1855.)

Love, the toothache, smoke, a cough, and a tight boot, are things, which cannot possibly be kept secret very long.

It has been humanely given to Members of Congress to admire their own speeches, or else they never could speechify so much as they do!

(The Country Gentleman, Albany, New York, April 26, 1855.)

A READY WITNESS. — During the course of a trial at the recent assizes, a learned counsel handed a document to a witness — a quiet going north countryman, under cross-examination — which the latter appeared to peruse with great attention. After he had finished, and the paper had been handed back to the learned gentleman, the following colloquy ensued:

Counsel — "Have you read that document?"

Witness — "Yes."

Counsel — "Have you ever seen it before?"

Witness — "Yes, I think I have."

Counsel (evidently thinking the moment had arrived to crush his victim) — "Now, sir, on your oath, tell me when you **last** saw that paper."

Witness — "Why, I should think it cannot be much more than half a minute since!"

The cool and ready way in which the witness gave the answer, set the court in a perfect roar of laughter, in which the judge heartily joined.

(Yankee Notions, New York, February, 1859.)

MEDICAL MANUAL

"Humbug and Collusion: The Science of Mesmerism"

by
Robert Kamer, M.D. and
Samuel Jackson, M.D.

What has come to be called hypnotism in the 20th Century was regarded as the science of Animal Magnetism in the 19th. Also known as Mesmerism, it was proclaimed to be the science of the mind with god-like powers.[1]

Advocates in the 1850's readily agreed that skeptics had, until then, a valid reason for questioning the wonderful phenomena produced by magnetism. Practioners had held audiences and patients alike in mystery as to the explanations of how Mesmerism worked, "and perhaps pronounced them inscrutable to the 'human intellect.' "[2] But as the science grew more acceptable and popular, being endorsed by such public figures as Pierpont, the Fowlers, Gilbert, Neal and Wayland,[3] rationales were needed to quiet the most vocal critics, whose cry was "Humbug and Collusion!"[4] Dismissing the unbelievers charge that there were more mysteries in mesmerism than in any other science, by the reply, "there is no science in the universe, but what has some incomprehensibilities,"[5] the advocates promised miraculous uses, and cures, if the correct techniques be used. By having oneself magnetized, nervous disorders, seizures and insanity could be eliminated, and it was a safeguard against fever and ague. Magnetized patients could be operated on, even have an arm amputated, without the slightest pain. And in the highest degree of Mesmeric trance, the subject actually attained the state of clairvoyance. "Local Magnetism," or the magnetizing of some part of the body, was useful in case of accident or injury, where only a portion of the body needed to be anesthetized. The use of it, then, was one of "Heaven's unchanging truths."[6]

Shocking Charges

"The Why and the Wherefore" of mesmerism, according to the 19th Century mind, was *electricity*. It was their belief that the air was composed of about 1/5 oxygen, 4/5 nitrogen, *plus* hydrogen and electricity. (Oxygen and electricity were the principles of flame and of animal life, while nitrogen extinguished both.)[7] Following this line of thought, it was felt that this air was drawn into the lungs; oxygen and electricity were transmitted to the blood, while the nitrogen was exhaled. The iron in the blood was instantly rendered magnetic, or positive and cherry red in color, under the influence of electricity, and oxydized by the oxygen. The friction of the blood going through the veins caused the electro-magnetic power to escape from the circulating system, for which it had a strong affinity. This electro-magnetic power was transported to the brain by the nerves, where it was secreted as *nervo-vital fluid*, or animal galvanism.[8]

The blood, after having given off its electrical power, appeared dark in the veins, and was negatively charged. When it returned to the lungs (on the principle that negative and positive rush together) it was recharged, and made positive again. From this, it was perceived that every organ and muscle was polar, by the negative-positive principle, and that was how life was sustained.[9]

Phrenological Poles

Phrenologists, adapting this philosophy into their own "science," believed that every organ and emotion had a *magnetic pole* in the face. Hence, when an organ was diseased, that portion of the countenance would appear drawn, so to express the feeling or sentiment of that organ. Based upon observation, it was determined the pole of self-esteem was between the mouth and nose — thus the action that produced the curl of the upper lip, expressing scorn, contempt and pride. The pole for firmness was near the edge of the upper lip, its actions producing a compression, as in the common expression, "Stiff upper lip." The pole of the liver was in the cheek, resulting in a flush when disease was present.[10]

Nervo-Vital Fluid

It was believed (at least by the galvinists) that the brain was filled with electrically produced nervo-vital fluid; there was no air, no blood and no oxygen. Thus, when one was in good health, he was considered to be "fully charged." When

This plate represents a lady undergoing the Galvanic process, for an affection of the lungs.

he came in contact with an individual whose brain was wanting its due measure of nervo-vital fluid, "the former gently and unchangeably holds his mind upon the latter and by the action of the WILL, the fluid will pass from one brain to the other, until the equilibrium between the fluids in the two brains is attained...This is MAGNETISM."[11] Over a period of time, this person will yield to the magnetic power of the mesmerizer and sweetly slumber in its inexpressible quietude.[12] This enables the person performing the magnetizing to exert "not only a PHYSICAL, but a MENTAL, and MORAL power."[13] It was admitted, however, that no one could be mesmerized against his will.

That Nervous Feeling

Electricity was considered to be the *only* substance that came in contact with the mind, so it was believed that all sensations were transmitted to the brain by this medium. In explaining the phenomenon of hearing, 19th Century galvinists taught that air was too "gross"[14] a substance to pass the barriers of the ear and enter the brain. Only electricity, "which is the only correspondent or mediator between mind and matter, laying its brilliant hand upon both parties," could bring them into communication and produce sound.[15]

Sensation or feeling was believed to start in the nerves, but was only *felt* in the mind. If a finger was burned, it was only through the agency of the electro-magnetic fluid that sensation was conveyed to the mind. The mind feels it, and it is only by habit that it is associated with the finger. The postulate that only the mind feels was "proven" by the example of an amputee; if the correspondent nerve at the end of the stump was touched, the patient would yet associate the feeling with the end of the finger.

Spiritualism

Spiritualism was another name for Mesmerism, animal magnetism, mental electricity and galvanism. In appealing to Christians, proponents argued that electricity was the sole means through which God came in contact this universe. This was evident "because it is electricity, as it exists in the human system, through which our spirits come in contact with matter."[16] Applying this as a proof of spiritualism, they stated that since it is acknowledged that the Infinite Spirit can produce infinite results by the energies of his Will, so the finite spirit (mankind) can produce a finite result by its WILL. This will stirs the nervo-vital fluid in the voluntary nerves, causing them to vibrate. The galvanic vibration causes the nerves to contract and the muscles raise the arm.

The PROOF for such a statement was given by way of an experiment: let a continuous shock be applied to a dead body and the muscles will contract and contort. "The reason for this is galvanism. And what is galvanism but electricity in a charged form; so that instead of giving the system a sudden shock like electricity, it produces a singular vibrating sensation upon the nerves, causing them to contract."[17]

Mesmeric Trances

There are five degrees of Mesmeric trance:

1. The whole body of the subject can be attracted by the conjoint action of mental and physical energies of the magnetizer.
2. The body of the subject can be attracted by mental *or* physical energies.
3. The subject can neither hear nor answer any person but the magnetizer.
4. The subject can taste, smell and see what the magnetizer tastes, smells and sees.
5. Clairvoyance.[18]

Clairvoyance is the state of having the nervous system so charged with nervo-vital fluid as to be rendered Positive and in an exact equilibruim with external electricity. The great quantity of super-charged nervo-vital fluid passes out through the pores of the skull, rendering it transparent. Uniting with external electricity, or Galvanic light, in the air, the whole universe is thrown upon the mind, allowing the clairvoyant to see everything.

The only problem with attaining this "second sight," is if the subject be charged with too much or too little electricity, or the weather be damp and unfavorable to electricity.

The process of Magnetizing

Shooting Star

The 19th Century mind was fascinated by electricity, and faced with the need to rationalize complex problems with seemingly incomprehensible solutions, the new field of electricity appeared to be a perect answer. Why shouldn't it be the spark of life," or the medium

God chose to contact finite man through? Following this logic, it seemed like a ready explanation for the mysticism that surrounded mesmerism. Working without the system of scientific investigation, microscopes, X-Rays or even basic blood analysis, research into galvanism was one of observation and deduction. Woven around Christian doctrine and a few scientific "laws," advocates pressed forward, proclaimed mesmerism as the "soverign cure for all human woes."[19] Similar in this respect to the prolific patent-medicine dispenser, who claimed to have a tonic for every ache, pain and chill, the field and philosophy of mesmerism would eventually be all but abandoned. It came and went, a shooting star in the quest for the whys and wherefores of the human mind.

FOOTNOTES

[1] "Lectures on the Philosophy of Mesmerism" by Rev. John Bovee Dods, Fowlers and Wells, New York, New York, 1850, lecture III, p. 30.
[2] IBID, lecture I, p. 11.
[3] IBID, p. 10.
[4] IBID, p. 9.
[5] IBID, p. 8.
[6] IBID, lecture IV, p. 58.
[7] IBID, lecture II, p. 20.
[8] IBID, p. 20.
[9] IBID, p. 22.
[10] "Fowlers' Practical Phrenology" by O.S. Fowler, Fowlers and Wells, New York, New York, 1853, p. 61.
[11] "Philosophy of Mesmerism," lecture I, p. 16.
[12] "The Diseases of Woman" by Frederick Hollick, M.D., T.W. Strong, New York, New York, 1856, p. 254.
[13] "Philosophy of Mesmerism," lecture II, p. 19.
[14] IBID, lecture IV, p. 55.
[15] IBID, p. 56.
[16] IBID, lecture III, p. 34.
[17] IBID, lecture II, p. 22.
[18] IBID, lecture V, p. 60.
[19] IBID, lecture VI, p. 72; see also "Diseases," p. 254.

CONCURRENT EVENTS

SPORTING

Match of Quoits will take place for a silver cup, at the East Brooklyn Shades, Myrtle Avenue, between Walworth and Spencer Streets, East Brooklyn. Play on Monday, July 1, 1861. Commences at 1 o'clock. Admission free.

J.J. Fletcher,
Proprietor.
June 30, 1861

PERSONAL INTELLIGENCE

It is hard to believe that Mrs. Cunningham, of Burdell and bogus baby notoriety, has got married; yet such is actually the fact. Women of any class must be very scarce in San Francisco, and, as for the gentleman who has replaced the unfortunate Burdell, there is no fear of his having excited either the envy or jealousy of any living thing.

(New York Herald, August 16, 1860.)

A SLAVER CONDEMNED

On Tuesday last, the case of the schooner Ann, which had been pending for several weeks in the circuit court of the United States, was concluded. It will be recalled that this vessel was seized by the Marshall, on suspicion of having been built and fitted out expressly for the slave trade. The Court decreed the vessel to be condemned as a slaver. J.R. Nelson, Esq. appeared on the part of the prosecution, and Reverdy Johnson, Esq. for the defense. — [Baltimore Sun.]

[The vessel is condemned, but where are the men that built, owned or navigated it on its nefarious errand? We want THE MEN.]

(The Emancipator, April 2, 1840.)

HUMAN LIFE — The number of human beings now on the earth is estimated at about 1,000,000,000. The average age of life is 33 3/8 years. One-fourth of those born die before they reach the age of seven years, and one-half before the 17th year. Out of 100 persons about six reach 60 years, while only about one in 1,000 attains 100 years. Out of 500 only one reaches 80 years. Out of the 1,000,000,000 living persons, 333,000,000 die annually, 91,000 daily, 3,730 every hour, 60 every minute, consequently one every second. The loss, however, is balanced by the gain in births. Women are generally stronger than men until their 50th year, after that period, less so. Both births and deaths are more frequent in the night than in the day. One-fourth of the men are capable of bearing arms. There are now in the world 3,064 known languages, and in which upwards of 1,100 religions or creeds are preached. The more civilized a country is, the more full of vigor, life and health are the people, and hence the idea that education enfeebles and degenerates the human frame is far from the truth.

The total population of Chicago is 137,030, an increase of 27,768 in two years.

(1863)

The President contributed $650 to the sum collected by the efforts of Mrs. Smith, to provide a Christmas dinner for 30,000 sick and wounded soldiers in Washington.

(1862)

MILITARY REVIEW

Custer Victorious: The Civil War Battles of General George Armstrong Custer by Gregory J.W. Urwin. Illustrated, 308 pp., index. Associated University Presses, Inc., East Brunswick, N.J. 1983.

Custer Victorious by Gregory J.W. Urwin is much more than just another biography of the flamboyant cavalryman George Custer. Mr. Urwin's richly illustrated, tightly researched and finely written book is a much needed and important addition to Custer scholarship.

In the forward to his new volume, Mr. Urwin admits that a huge mass of material has already been published on Custer; much of the work centering on Custer's defeat at Little Big Horn in 1876. Mr. Urwin argues that too much scholarly attention has been paid to but a single incident in Custer's military career and that Custer's brilliant (and well documented) Civil War service has been neglected in favor of historical speculation over the boy general's last battle.

Mr. Urwin writes that Custer is "less a subject of American historians than a victim." Custer is portrayed in Urwin's text as the epitome of the tarnished American hero, a man of humble background whose meteoric rise to fame during the Civil War won him great public acclaim, but also many jealous enemies among the Union officer corps. Urwin argues that Custer's early death allowed envious fellow officers the opportunity to destroy Custer's military reputation and distort the historical record against the boy general. *Custer Victorious* attempts to smash the stereotyped image of Custer promoted by dissatisfied contemporaries and tries to chronicle the story of America's most controversial soldier at the height of his fame, the Civil War period.

Custer Victorious re-evaluates the cavalry leader and succeeds, perhaps for the first time, in compiling a full record of Custer's invaluable service to the Union during the Civil War. In a fast-paced narrative, Urwin relates the story of Custer and his devoted red-scarfed Michigan Brigade from Bull Run to Appomattox. Urwin's brilliant retelling of the Trevilian Station battle is especially moving.

Where Mr. Urwin's book lapses is in his general background history of the war. Urwin generally succeeds in building a contextual history around his biographical subject, but his essay comes into error on at least one major point. Urwin greatly simplifies and does great injustice to General Grant when he grossly misrepresents Grant's strategy and tactics in the 1864 spring campaign. Urwin somehow fails to realize that there was more behind Grant's formula for victory than sheer muscle. Describing Grant as a "big brown bear stunned by a crack on the noggin" (p. 167) or "more like a mad bull than a master of military chess" (p. 167) negates a great wealth of Grant scholarship, including General J.F.C. Fuller's standard *The Generalship of U.S. Grant*.

Urwin's impression of Grant might have been enhanced had he not relied so heavily upon the diary of Colonel Charles Wainwright. Urwin freely quotes many passages critical of Grant from the officer's diary, forgetting perhaps, that Wainwright seldom wasted a good word on any contemporary and his opinions should be used guardedly.

Still, Urwin is writing a book about Custer and perhaps general errors can be excused. Urwin's Custer scholarship is original and fresh. The author is undoubtedly enamored with his subject but it is to his great credit that the book remains one of the most objective Custer biographies to appear in recent times. Mr. Urwin's purpose is clearly to rescue Custer from the clutches of axe-grinding historians and hacks, but in only 290 pages he persuasively argues that perhaps it is only fitting that the public realizes what an inspiring and brilliant leader Custer was during the Civil War.

Urwin's Custer research seems to falter on only one minor point. On page 255, in describing Custer's role in the events leading up to the Appomattox surrender, Urwin writes that Custer passed under flag of truce to the Southern lines and was hastened to General Longstreet, whereupon Custer demanded immediate unconditional Confederate surrender. Although Longstreet did relate this incident in his memoirs, historians generally concur that Custer's bold declaration was delivered to General John B. Gordon and not Longstreet. Mr. Urwin did not list Gordon's memoir *Reminiscences of the Civil War* in his bibliography, but Gordon does mention the Custer meeting on page 439 of the 1903 Scribner's edition.

Such disputes however, do not detract from the historical value of *Custer Victorious*. Mr. Urwin has done a great service to Civil War students and all who are interested in the fascinating career of the North's boy general. *Custer Victorious* is a book that destroys old cliches, re-evaluates historical judgements and questions the methodology and ethics of the historical profession. The book is bound to cause some controversy, and will, no doubt raise the ire and stir the tired blood of many historians. *Custer Victorious* is an important new biography, a book that compells historians, students and the general public to take a second look at a much-maligned figure in American popular culture.

William F. Howard
Delmar, New York

Articles of War

An Act For Establishing Rules and Articles For The Government Of The Armies Of The United States.

ART. 41. All non-commissioned officers and soldiers who shall be found one mile from the camp without leave, in writing, from their commanding officer, shall suffer such punishment as shall be inflicted upon them by the sentence of a court-martial.

ART. 42. No officer or soldier shall lie out of his quarters, garrison, or camp without leave from his superior officer, upon penalty of being punished according to the nature of his offense, by the sentence of a court-martial.

ART. 43. Every non-commissioned officer and soldier shall retire to this quarters or tent at the beating of the retreat; in default of which he shall be punished according to the nature of his offense, by the sentence of a court-martial.

ART. 44. No officer, non-comissioned officer, or soldier shall fail in repairing, at the time fixed, to the place of parade, of exercise, or other rendez-vous without leave from his commanding officer, before he shall be regularly dismissed or relieved, on the penalty of being punished according to the nature of his offense, by the sentence of a court-martial.

ART. 45. Any commissioned officer who shall be found drunk on his guard, party, or other duty, shall be cashiered. Any non-commissioned officer of soldier so offending shall suffer such corporeal punishment as shall be inflicted by the sentence of a court-martial.

ART. 46. Any sentinel who shall be found sleeping upon his post, or shall leave it before he shall be regularly relieved, shall suffer death, or such other punishment as shall be inflicted by the sentence of a court-martial.

ART. 47. No soldier belonging to any regiment, troop, or company shall hire another to do his duty for him, or be excused from duty but in cases of sickness, disability, or leave of absence; and every such soldier found guilty of hiring his duty, as also the party so hired to do another's duty, shall be punished at the discretion of a regimental court-martial.

ART. 48. And every non-commissioned officer conniving at such hiring of duty aforesaid, shall be reduced; and every commissioned officer knowing and allowing such ill practices in the service, shall be punished by the judgement of a general court-martial.

ART. 49. Any officer belonging to the service of the United States, who, by discharging of firearms, drawing of swords, beating of drums, or by any other means whatsoever, shall occasion false alarms in camp, be ordered by the sentence of a general court-martial.

ART. 50. Any officer or soldier who shall, without urgent necessity, or without the leave of his superior officer, quit his guard, platoon, or division, shall be punished, according to the nature of his offense, by the sentence of a court-martial.

ART. 51. No officer or soldier shall do violence to any person who brings provisions or other necessaries to the camp, garrison, or quarters of the forces of the United States, employed in any parts out of the said States, upon pain of death, or such other punishment as a court-martial shall direct.

ART. 52. Any officer or soldier who shall misbehave himself before the enemy, run away, or shamefully abandon any fort, post, or guard which he or they may be commanded to defend, or speak words inducing others to do the like, or shall cast away his arms and ammunition, or who shall quit his post or colors to plunder and pillage, every such offender, being duly convicted thereof, shall suffer death, or such other punishment as shall be ordered by the sentence of a general court-martial.

ART. 53. Any person belonging to the armies of the United States who shall make known the watchword to any person who is not entitled to receive it according to the rules and discipline of war, or shall presume to give a parole or watchword different from what he received, shall suffer death, or such other punishment as shall be ordered by the sentence of a general court-martial.

ART. 54. All officers and soldiers are to behave themselves orderly in quarters and on their march; and whoever shall commit any waste or spoil, either in walks of trees, parks, warrens, fish-ponds, houses, or gardens, corn-fields, inclosures of meadows, or shall maliciously destroy any property whatsoever belonging to the inhabitants of the United States, unless by order of the then commander-in-chief of the armies of the said States, shall (besides such penalties as they are liable to by law) be punished according to the nature and degree of the offense, by the judgement of a regimental or general court-martial.

ART. 55. Whosoever, belonging to the armies of the United States in foreign parts, shall force a safeguard, shall suffer death.

ART. 56. Whosoever shall relieve the enemy with money, victuals, or ammunition, or shall knowingly harbor or protect an enemy, shall suffer death, or such other punishment as shall be ordered by the sentence of a court-martial.

ART. 57. Whosoever shall be convicted of holding correspondence with, or giving intelligence to, the enemy, either directly or indirectly, shall suffer death, or such other punishment as shall be ordered by the sentence of a court-martial.

From the "Revised Regulations For The Army Of The United States, 1861," Philadelphia: J.B. Lippincott & Co., Publishers, pp. 506-508.

THE KEPI
12115 Magnolia Boulevard
North Hollywood, California 91607

BULK MAIL

THE KEPI

DECEMBER 1983-JANUARY 1984 $3.50

"A New Concept in Civil War Reporting"

The Prison Diary of Private H. H. Wiseman
by William F. Howard

Capture the Flag
by Thomas J. Evans

Pardon From the Adversary That You Despise
by Everett K. Cooper

THE KEPI

"A New Concept in Civil War Reporting"

DECEMBER 1983-JANUARY 1984 VOLUME I, NUMBER 5

THE KEPI is published bi-monthly for $3.50 per issue, or $18.00 per year. Canadian single copy, $4.00; subscription, $21.00 per year. United Kingdom single copy, $4.50; subscription, $28.00 (Air Mail) per year. Correspondence to the editors may be directed to: **THE KEPI**, P.O. Box 4542, North Hollywood, CA. 91607. Unless expressly stated, **THE KEPI** neither endorses nor is responsible for the content of advertisements appearing in its pages. **THE KEPI** requests authors to query with specific topics before submitting articles, and assumes no responsibility for unsolicited manuscripts not accompanied by return postage. Work not fully documented and footnoted where applicable will not be considered. Please allow 6 weeks for delivery of first issue.

ILLUSTRATIONS

Cover taken from the collection of William F. Howard, depicts Confederate prisoners awaiting transfer to Union prisons.

Elmira Prison photographs from "Prisons and Hospitals," edited by Francis Trevelyan Miller, Castle Books, New York, 1957.

"Wounded Prisoners," and "They never fail..." from "The Soldier's Story," by Warren Lee Gross, Lee & Shepard Publishers, Boston, 1876. By courtesy of Lavinia Berdge.

Medal of Honor, from the collection of Thomas J. Evans.

Medical College of Georgia, from "The Literary Emporium," 1846.

Back Cover from "Harper's Weekly," Nov. 1, 1862.

THE REGULARS

THE IRREGULARS

GENERALLY SPEAKING

In this issue of THE KEPI, the editors are proud to announce our expanded format and new look. The support and encouragement of our readers has enabled us to make THE KEPI longer, thus providing more articles, illustrations and short features, which make TK distinctive. We will continue to present original footnoted works on the War, while including our regular features that deal with life in 19th Century America.

We would like to bring to your attention that TK has a new mailing address: THE KEPI, P.O. Box 4542, North Hollywood, California, 91607.

Our cover story is "The Diary of Private H.H. Wiseman." Wiseman was a young Confederate who was paroled at the fall of Vicksburg, and later rejoined his old regiment in the forts guarding Mobile Bay. Grant scholars have dismissed the idea that Rebel soldiers captured at Vicksburg ever re-enlisted, but here, in Private Wiseman's account, is an entire regiment that was dispersed and then reformed farther south. This original Confederate diary was found beneath the floorboards of a Long Island, New York home, and was donated to our author, William F. Howard received his B.A. from Manhattanville College in American Studies. A lifelong student of American history, he has authored numerous articles on the Civil War, and has recently completed work on his first full-length book, an edited collection of Civil War soldiers' letters.

"Capture the Flag," by Thomas J. Evans, concerns the story of Private Thomas Evans, 54th Reg. Penna. Vols., who won the Medal of Honor during the Civil War. Since the native Welshman could neither read nor write, his story was passed down by word of mouth, and is presented, for the first time, to KEPI readers.

"Pardon From the Adversary That You Despise" deals with the restoration of citizenship to the Southern people. Supplementing the article and Appendixes which give a fascinating insight into the words and actions of the men who conquered, and the men who despised." Author Everett K. Cooper has written numerous articles for The Confederate Philatelist, Civil War Times, Illustrated, Paper Money *and* North-South Trader. *He is an active Civil War Round Table member, and has served as president of the Harrisburg, Pennsylvania, and Houston, Texas, Round Tables.*

MEDICAL MANUAL contains an article on the status of the health care profession in 1860, and gives modern readers a look at the pride and the concerns of physicians on the eve of civil war.

MILITARY REVIEW evaluates a Kent State University Press republication of a 1964 work, retitled "Denmark Vesey's Revolt," by John Lofton. This groundbreaking work presents the life and times of a controversial black man in pre-Civil War South Carolina, and is a must for those studying slavery, and the causes and events leading to disunion.

Our regular/irregulars, CONCURRENT EVENTS, WHIMSEY and the Articles of War, complete this issue.

As we enter the holiday season and the new year, we would like to thank you for being part of our "new concept in Civil War reporting," and wish you the best for 1984.

S. L. Kotar *J. E. Gessler*
Editors-in-General

Head-Quarters of THE KEPI
1st December, 1983

GENERAL ORDERS,
No. 4
by Command of the Editors-in-General

Coming in the February-March issue of THE KEPI is an article on "The Fighting 69th" New York (Irish) Regiment. These men, whose motto was "Never Retreat From The Clash Of Spears," fought throughout the Civil War, from First Manassas to the pursuit of the Army of Northern Virginia in 1865. It ranked 6th in battle casualties of all Union Regiments, and its history is an example of determination and spirit.

Also featured is an article examining the costly Confederate indecision that delayed Gen'l. James Longstreet's transfer from Lee's army to the Western theatre in 1863. The resulting empty victory at Chickamauga and the blame for the lost opportunity, which reached to the top of the Confederate hierarchy, are explored.

The regular/irregular columns will also appear, and the Articles of War continue.

"THE PRISON DIARY OF PRIVATE H.H. WISEMAN"

Edited by
William F. Howard

Most of the estimated 215,000[1] Confederate soldiers that were captured and confined in Union prison camps during the Civil War come down through history as unknown soldiers. Few letters and diaries written by Confederate prisoners are extant. Illiteracy, a shortage of writing materials and harsh prison conditions all combined to deprive the imprisoned Southern soldier of his historical voice.

The diary of Confederate private, H.H. Wiseman, recently discovered by Mrs. John J. Gavin beneath the floor-boards of her Long Island, New York, home fills a small part of this void in historical scholarship, and offers a rare glimpse into the life of one young Southerner who fought and died in our nation's bloodiest war.

H.H. Wiseman was 28 years old when he joined the Confederate army at Covington, Tennessee on February 16, 1862.[2] He volunteered to serve twelve months as a private in Captain H.J. Maley's company, Colonel Andrew Jackson's 1st Tennessee Heavy Artillery. This regiment was organized May 10, 1862 at Fort Pillow with ten companies which had previously seen service with the Tennessee Artillery Corps.[3]

After embarking with the regiment on the steamer *Golden Age* for Vicksburg, Private Wiseman served in that city's defenses and was heavily engaged in the bombardments of June 28, 1862, and again from July 12-27, 1862.[4] When Vicksburg fell on July 4, 1863, and Colonel Jackson was compelled to surrender his hungry regiment, Wiseman was paroled in keeping with the terms of capitulation negotiated by Generals Grant and Pemberton.

As a paroled prisoner, Private Wiseman swore that he would "not take up arms again against the United States,"[5] but his Confederate loyalty was hardly intimidated. Wiseman, and surviving members of the 1st Tennessee (about 176 men), and remnants of the 1st Battalion Tennessee Light Artillery paroled at Port Hudson, rejoined the Confederate army at Marrietta, and in January, 1864, the two units were consolidated into one by order of Lieutenant-General Polk.[6]

With Lieutenant Colonel Robert Sterling commanding, the regiment moved to Fort Morgan, Alabama on April 3, 1864.[7] Fort Morgan was the most elaborate of the three fortifications guarding the approaches to Mobile Bay, a haven for Southern blockade runners and the appealing target of renewed Union interest in August, 1864.[8] The troops in Fort Morgan numbered only 600 men, not nearly enough to counter a coordinated Union attack, but Brigadier-General Richard L. Page, commanding the garrison, "determined to hold the fort to the last man."[9]

In the end, the Union war machine prevailed. After General Gordon Granger landed 1,500 Federals on nearby Dauphin Island, investing Fort Gaines on August 3, 1864, and Admiral David Farragut stormed past the Rebel defenses into Mobile Bay, capturing the Confederate ironclad, *Tennessee*, the fate of Fort Morgan was sealed forever.[10] In a siege that lasted until August 22, the Federals poured over 3,000 shells into the fort. Finally, with his garrison reduced to ruins and further resistance impossible, General Page surrendered Fort Morgan on August 23, 1864. The campaign for opening Mobile Bay ended in the rubble of Fort Morgan: for the second time in his military career, Private Wiseman was a prisoner.

This time, however, parole would be long in coming. Wiseman was taken to New Orleans, where the captured garrison embarked on the Union gunboat, *U.S.S. Benville*, for New York on September 18, 1864.[11] He arrived at Fort Columbus, New York Harbor, September 28, and in the evening was sent to Governor's Island. From Castle Williams, New York, Wiseman was transferred to Elmira Prison, arriving there December 5, 1864.[12]

Elmira Prison, in southwestern New York, existed for only a single year, but in that time 2,963 of the 12,123 Confederates imprisoned there perished from disease and other causes. It is remembered as being "among the worst of the Northern compounds," and compares well to the horrors of Andersonville.[13]

Wiseman endured the conditions at Elmira, writing daily entries in the diary he received from Sergeant C.D. Allaire of the 162nd New York Infantry and involving himself in camp routine.

The following are extracts from his original; spelling and grammar are maintained.

THE DIARY OF PRIVATE H.H. WISEMAN:

Fort Morgan, August 23, 1864

Surrendered to the Yankees today. Shiped on the U.S.G. Boat Benville for New Orleans. Arrived their on the night of the 24th and stayed on board until the 25th when we was taken to the prison in Costen Press No. 4 wheare we got verry little to eat. I am thankful to a kind Providence for verry excelent health.

Sept. 18

Left N.O. on the Steam Ship Cassandra for New York. Had quite a pleasant trip. The wind blew a prety stif gale one day & night. Several of the Boys was sea sick. It did not affect me. Anchored at N.Y. on the morning of the 28 and in the evening was sent to Govinors Island and put in Castle Williams.

Nov. 9th

I have been quite unwell since I have been heare. Cold & chills. The health of the Boys is verry good. Their is 300 of us & only four deaths since we left New Orleans. The election is over & Abe is President four years longer. The City of New York polled 90010000 votes 37000 majority for McClenen. Gold 249.

10th

Genl. Butler has the bay & rivers full of Gun Boats and armed vessels to surpress Riots in the City. Gold 256¼. My health is improving. Some of the Boys have the Small Pox. Two have died with it. The Prisenors heare are divided in Companys of 50 to each room or casemate which is 25 ft. square and we are verry much crowed. Their is scarcely room for us to sleepe. The floor is anything but Soft to a man with one Blanket. The Provo M raised row today with Sarjt. who has us in charg about our grub and says we have not been getting wat was allowed to us. Gold 244 1/8

Nov. 12th

The weather heare has been verry fine and continues soe. Their has been no ice heare yet. Our grub is some better today or rather some little larger peices. Their was noathing occurd today to brake the dull monotomy of a Prison life all the Talk Amoung the Boys is when we will be Exchanged and once more tread Dixie soil. Gold 244

Sunday Nov. 13th 1864

This has been quite a dull day, no news of importance. My health is improving I hope a kind Providence will continue his kindness to me protect me from all harme & diseas and permit me to see my friends again. Gold 242. The Weather has changed and is quite Cool.

Nov. 14th 1864

This morning is quite Cold & plenty of Ice the first I have saw. Gold 240.

Nov. 15th

Snowing today & quite cold. No news today. Gold 245.

Nov. 16th 1864

Quite pleasant today. Nothing of importance. No news of an exchange soon. Gold 238.

Nov. 17th

Weather pleasant. 413 Reble prisonors arrived last night from Ship Island two of their number died at sea. No paper today don't know the price of gold.

Nov. 18th

The lot of 400 Prisenors are leaving this morning for Harts Island. Gold 226.

Nov. 19th 1864

Today is quite pleasant. The last of the Prisenors left this morning. Gold 215.

Nov. 20th

This is quite a glumy & rainy day. I have a sevear cough & quite unwell. Gold 218.

Nov. 21st

Sevearl of the Yankee recruits brough last Evening have been carrying on quite a brisk trade with our Boys buying Rings. They have plenty of money & pay two & three dollars apiece. They gave several of the Boys money & tobacco. This is a glumy & fogey day. Gold 224.

Nov. 22nd 1864

This is a verry nice day. Cleare and warm. One of the Alabama Boys died today. The health of the Prisenors is verry good. I heard from Home today through Moriner in a letter to R. Goodman. Gold 222.

Nov. 23

Quite cold & cleare Today. Three months ago we were Captured. It seams six months. Gold 222 ¾.

Nov. 24th 1864

This is Thanksgiveing day among the Yankees. The Sargents in charge of the rooms had a verry nice dinner sent by Head Quarters. Turkey, cak, Wine, fruits & several other artickles. The dinner was divided amoung all the Prisenors. Col. Bomford[14] sent it. Gold 222.

November 25th 1864

The Yankees fired a salute today in honor of the avacuation of New York City by the British. Received a letter today from Uncle Billie was glad to heare from home. Gold 219.

Nov. 26th 1864

This is a beautiful day cleare & warm one almost allmost imagins that he was in Dixie. Gold 218.

Sunday Nov. 27th 1864

Several of the Alabamas have gone to the Hospital with the small pox. We can heare nothing. Sherman in Georgia & Hood in Tennessee. This is a beautiful Saboth. Oh how I wish I was with deare friends in Dixie. I hope God will continue his mercies to me & I will soon be Home Again. Gold 226 5/8.

Nov. 28th 1864

No news of importance. The Papers say their is a chance for an exchange soon which if soe will be welcom news to us.

Nov. 29th 1864

The news from the South is encouring. Hood is neare Nashville & it is hoped that he will soon redeam Tenn. Sherman it seames meets strong opisition wheare ever he

Elmira Prison, New York, 1864, Changing of the Guard.

goes. If the People of Georgia will only doe their duty his grand army will soon come to greif. Gold 232.

Nov. 30th

I had the pleasure taking Dick Munford by the hand today. He came over to bring me some clothing & money which was verry acceptable. Dick is in good health and looking finely. Gold 232½.

December 1st 1864

The news from the South is quite flatering. Sherman it seames that he has nearly ran his cours. Hood is close apon Nashville. Several of the Boys has been sent to the Hospital, some with coalds & fever and some with Small Pox. It is only variloid think as it is very light. Gold 226.

December 2

Their is no news of interest today their is a report that we will leave heare soon. Gold 228.

December 3, 1864

We are orderd to leave today some say for Harts Island and some say for Elmyra. The Papers report Hood to have got badly whiped by Thomas at Franklin. Gold 228 ⅓.

Sunday Decem. 4th, 1864

We have not left yet. I do not think we will leave today. My health is better than it has been for sometimes. Several of the Boys are getting sick with Pneumonia & Small Pox. We have just received Orders to leave for Elmyra.

December 5th 1864

We left Govenors Island yesterday evening. We left Jersey City 6 Ock on the cars which ran verry slow. We arrived at Elmyra late in the evening when we were marched some miles to the Prison. Some 8 or

10 of the Boys jumped off the cars & made their escape. Three of them from our Reg. R.G. Goodman, Dr. Clements and James Leear.

December 6th 1864

I do not like this place as well as Gov. Island. Their is some 8000 prisoners heare and very much crowed. The most of them are verry dirty & Lousey. The place is low & flat plenty of mud. The weather verry cold. I feare that it is verry sickley heare. I get as much grub as I want. My health is not verry good. I still have a verry seveare cough & coald. Amoung all the Prisenors heare I found no one that I know. Elmyra is a verry nice place and appeares quite a large Town.

December 7th 1864

The more I see of this place the worse I want to leave it. I am informed that neare 4000 of the Prisenors heare have made application to take the Oath. They are a low down set of men & will rob you of all you have. Knock you down at night if your caught out of your quarters.

December 8th 1864

This is a very coald morning. I feare we will suffer heare this Winter. Their is a report in camp that we will soon be exchanged. I hope it will prove true. It is rumored that Hood has Nashville. We had the Sisters of Charity to see us today distributing the Tracts & Testaments.

December 9th 1864

There is no news of importance in camp. It is snowing & quite coald. Two cases of Small Pox was sent to the Hospital. It is reported that we are exchanged & will soon be sent away. Oh that it is true.

December 10th 1864

Still snowing this morning it is several inches deep. The Yankees had us all out to muster us. They cept us in the snow until we were nearly froze. They have muster every ten days and roll call twice a day. Their is considerable sickness heare Phenumonia Coald and Small Pox.

Sunday Decem. 11th 1864

This morning is some warmer quit snowing. Clothing & Blankets are being ishued to the men that is most in neede of them. Nothing of importance transpierd today.

Decem. 12th 1864

Verry coald last night. Snowing today. Wrote Uncle Billie today. Their is some 250 of our Boys in the Guardhouse. You can find every Kind of a man heare you want. The Preacher, Doctor, Lawyer, all kind of Workman, Thieves & Rascals of the lowes order. Some of our men die every day & night.

Decem. 13th 1864

The Yanks had anothe of their Musters this morning. Cept us out for an hour. Some of the men nearly froze. The Papers state this morning that a general Exchange has commenced if soe I hope we will start for Dixie soon.

Decem. 14th 1864

Some warmer today. We do not get half wood enough, have to stay in bead in ceep warm.

Decem. 15th 1864

We suffer with coald their is some 250 in a room and only two stoves & not wood enough to burn half the day. Some of [the men] never get neare the fier. The room is crowed soe it is verry uncomfortable. Snowing this evening.

December 16th 1864

We have news that Sherman has been badly whiped neare Savanah. Nothing transpierd of importance in camp today.

December 17th 1864

We had quite as able ferment last night. Rained & froze as fast as it fell all night soe slick this morning that

Wounded prisoners pressed back by guards.

Elmira barracks, 1864.

it is allmost imposible to get about you can see the Boys falling all the time. We had thaw today no end to the mud. Some more Prisenors come in today. 150 of them mostly sick and wounded.

A lieutenant-commander or a major shall be exchanged for officers of equal rank, or for eight privates or common seamen.

Sunday December 18th 1864

This is quite a pleasant day, the first day the sun has shined since I have been heare. The sick have but pore attension heare. We have lost two good men of Company B Since we come heare. I hope God in his goodness will preserve me from all sickness.

December 19th 1864

We had rain last night and their is no end to the mud today. The Yankees have one of the Rebs promenading the camp with a barel shirt to the tune of the roughs march for forgeing an order on the Sutler.

Decem. 20th 1864

Mustered again this morning got soe coal that I have a seveare pain in my side. Three more of our Reg. died this morning. Joe Brown, Church & Lewelen.[15] I feare we will all be sick heare.

December 21st 1864

Snowing today no news of importance today.

Decem. 22nd 1864

This is the Coaldest day we have had a greate many of the men are getting sick. Jim Hughlet is quite sick. My health is some better. We suffer heare with Coald. It is impossible to get neare the stove.

December 23d 1864

This is a bitter coald day. The Yankee Papers say they have completly routed Hoods army.

December 24th 1864

Their is a rumor that we will be Exchanged soon. I trust it is soe for I fear that we will nearly all die if we stay heare tel Spring. This is Christmas eve and a dull and glumy one.

Sunday Decem. 25th 1864

Christmas

Chritmas day and a verry nice day for this country. I thought last Christmas was dull anough to me but nothing compared to a Christmas in a Yankee Prison. This is the third Christmas that I have spent from Home & Friends. I thank God that he has permited me to see this Christmas while many of my fellw Prisenors have died & gone to their long home. I am reading a verry good Book Powers of Religion on on the Mind by Lindley Murry. We have a library heare we get plenty Books to read. The Rebs have a verry good School which is largely attended. You can see grown men learning to write and some in their Abs.

December 26th 1864

Christmas is over & I am no nearer Dixie. This morning is quite warme & verry windy. All the men in the two wards was sent to the gard House.

Decem. 27th 1864

The weather is quite warme today and this is the mudyest place I ever was in. The Yanks say Sherman has Savanah.

Decem. 28th 1864

No news of Interest today dull and glumy. Their is a great deal of sickness in Camp. Several cases of Small Pox but in a light form.

Decem. 29th 1864

Snowed all night & snowing this morning. I have been quite unwell for several days. Their is a rumor that Hood has whiped Thomas at Columbia. I hope it is true.

December 30th 1864

Quite coald & snowing this morning. My cough is some better today. One of the Rebs who was on detail doing some work for the Yanks made his escape today.

December 31st 1864

Snowing today. Muster day again. It is rumored that we will soon be exchanged. This is the last day of the weeke the last day of the month and the last day of the year and I pray God that it may be the last of this Cruel War. I return thanks to a Kind and merciful God that he has permitted me to the end of the Year 1864. And Oh may his Kind and protecting care still remain with me during the Comeing Year and protect me from all harme and enable me to walk holy and upright before him and do his will. And Oh may he soon deliver me from this prison and soon meet with loved ones at Home.

Sunday Jany. 1st 1865

This is the first day of a new Yeare but quite Coald and glumy day. Oh may God in his mercy bring this war to and end this year and restore Peace & Happyness to our once Peaceful land. The papers of today says that arangement have been made for a general Exchange and that Wilmington, N.C. is the point of delivery. I pray God that it is soe and that we may soon be delivered from this Prison which to stay in in a liveing death And meet with loved & dear ones at Home.

January 2d 1865

No news in camp today. Pat Winn is quite sick. I have felt verry unwell today, had a chill.

January 3d 1865

This is a pleasant day for this Country. The Sun is shining once more. The rumor today is that we will certainly be Exchanged in a few days.

January 4th 1865

This is the coldest day we have had. It is rumored that 12000 sick & wounded prisenors are to be exchanged shortly.

January 5th 1865

Quite coald today. Pat Winn is quite sick with Coald. I am verry unwell today.

[No further entry until:]

February 10

Went to Hospital. I am 9th. The Federals deserve credit for the treatment of our sick. The Hospital is some 130 ft. long cealed well has 20 large windows three large coal stoves to each Room, a good clock, a Thermommest at each end of the Room, good cots & beds. The diet is verry good for the sick. I was well pleased with the treatment. I sighned the Parol yesterday. The order is that we leave today for Dixie.

Feby. 12th 1865

Jim Hughlet was sent for yesterday Evening to come to HdQts. and he was informed that they have just received orders from Washington City that if Hughlett & myself would take the Oath we could go to our Homes. They sent Hugh. to see me. I sent word to them that the Oath was something I never could take. That I wanted to go out of the servis honorable and did not want to disgrace my country, my wife & relatives. They said at HdQtrs. that the oath had been obtained through some friends of ours. I do not know who they are and exspect that they thought they was doing me a favor but I do not thank then for any such kindness. If it had been for anything els but taking the Oath I would have been under many obligations to them.

Private Wiseman was paroled February 9, 1865. But unfortunately he did not live to enjoy his freedom. Taken ill with chronic diarrhea, he was transferred to U.S. General Hospital, West's Buildings, Baltimore, Maryland, on February 15. Here, on February 16, 1865, Private H.H. Wiseman died.[16]

Elmira barracks and tents for overcrowding, 1864.

"They never fail who die
In a great cause. * * *
They but augment the deep and sweeping thoughts
Which overpower all others, and conduct
The world at last to Freedom." *Byron.*

FOOTNOTES

[1] James I. Robertson, *The Civil War,* (Washington: U.S. Government Printing Office, 1963), p. 46.

[2] War Records of Private H.H. Wiseman, 1st Tennessee Heavy Artillery. Military Records Section, The National Archives.

[3] Tennessee Civil War Centennial Commission, *Tennessians in the Civil War: A Military History...* (Nashville, Tennessee, Civil War Centennial Commission, 1964), p. 119.

[4] *Ibid,* p. 121.

[5] National Archives.

[6] *Tennessians in the Civil War,* p. 121.

[7] *Ibid,* p. 121.

[8] Mark M. Boatner III, *The Civil War Dictionary* (New York, David McKay Company Inc., 1959), pp. 298, 299.

[9] *Tennessians in the Civil War,* p. 122.

[10] War Department, *War of the Rebellion: The Official Records of the Union and Confederate Armies* (Washington, U.S. Government Printing Office, 1902), Vol. XXXIX, pt. 1, p. 402.

[11] Civil War Diary of Private H.H. Wiseman, the William F. Howard Collection.

[12] National Archives.

[13] *The Civil War,* p. 46.

[14] Colonel J.D. Bomford of the 16th Infantry commanded Governor's Island from September 6, 1864, to February 16, 1865. Michael J. Marcelle, "A Longing For Dixie: The Reflections of H.H. Wiseman, Confederate Prisoner of War," (Unpublished manuscript), 1978.

[15] James W. Brown, Co. B, died 12/20/64; buried at Elmira Prison cemetary, grave number 1074. Thomas A. Church, Co. A, died 12/23/64, grave number 1098. G.M. Lewallen, Co. A, died 12/19/64, grave number 1069, Marcelle, n.p.

[16] National Archives.

BIBLIOGRAPHY

Public Documents

Robertson, James I. *The Civil War.* Washington: U.S. Government Printing Office, 1963.

Tennessee Civil War Centennial Commission. *Tennessians in the Civil War: A Military History...* Nashville: Tennessee Civil War Centennial Commission, 1964. Part I.

U.S. War Department. *War of Rebellion: The Official Records of the Union and Confederate Armies.* Washington: U.S. Government Printing Office, 1902. (Series I) Vol. XXXIX, Part I.

Books

Boatner, Mark M. III. *The Civil War Dictionary.* New York: David McKay Company, Inc., 1959.

Foote, Shelby. *The Civil War: A Narrative.* New York: Random House, 1974. Vol. III.

Johnson, Rossiter. *Campfires and Battlefields: A Pictorial History of the Civil War.* New York: The Civil War Press, 1967.

Lindsley, John Berrien (ed.). *The Military Annals of Tennessee:* Confederate...Nashville: J.M. Lindsley & Co., Publishers, 1886.

Page, Richard L. "The Defense of Fort Morgan" in *Battles and Leaders of the Civil War.* Vol. IV, edited by Robert Underwood Johnson and Clarence Clough Buel. New York: The Century Co., 1887.

Articles and Periodicals

Huffstodt, James. "Campaigning For Mobile," *Civil War Times Illustrated,* XXI, No. 1 (March, 1982), 8-17.

Kaufhold, John A. "The Elmira Prison Camp," *North/South Trader,* V, No. 1 (November—December, 1977), 24-29.

Lord, Francis A. (ed.). "Diary of a Soldier — 1864," *North/South Trader,* IV, No. 1 (November—December, 1976), 12—26.

Other Sources

Letters from Mrs. John J. Gavin to the author, March 10, April 5, April 16, 1982.

Marcelle, Michael J. "A Longing for Dixie: The Reflections of H.H. Wiseman, Confederate Prisoner of War." (Unpublished manuscript), 1978.

The date was not well noted: since the birth was near midnight, shortly before or after, it might have been January 20 or 21, 1824. There was no birth certificate, and even the doctor who delivered the child could not remember with certainty. The boy himself was never sure, and to compound matters, West Point recorded his birth year as 1823. He was born in a state that "seceded from itself" during the Civil War and went for the Union, while its native son went with the rest of the Old Dominion for the Confederacy. He rose from the rank of major to lieutenant-general, and when he died in 1863, the whole world knew his name. His greatness has endured, and in the new year, on January 20 and 21, 1984, we celebrate Thomas Jonathan Jackson's birthday.

THE BARTER SYSTEM

In 1861, neither North nor South was prepared for the enormous number of prisoners that would fall into their hands. During the course of the war, it was estimated 400,000 were taken captive, and their food, clothing, shelter and confinement became a problem that proved persistent and difficult to solve.

When it became apparent that large volumes of prisoners needed attention, Confederate and Federal authorities bought old warehouses, employed the recently evacuated camps of instruction, and filled the civil jails and penitentiaries with the enemy. Young boys, old men and invalid soldiers were delegated to act as guards, but as the numbers swelled, it was necessary to pull troops from the field to watch the prisoners.

A custom of parole had been developed in 1861, which was a simple agreement between both sides that prisoners would be released under an oath not to take up arms until officially exchanged. This alleviated the need for the captor to guard and feed his captives, and freed the prisoner from a trip to the interior, and an indeterminate confinement.

A typical oath in 1861, prescribed to the men of the Third Infantry, commanded by Major C.C. Sibley, who had surrendered to the Confederates, ras as follows: To The Authorities Of The Confederate States Of America:

I give my word of honor as an officer and a gentleman that I will not bear arms nor exercise any of the functions of my office under my commission from the President of the United States, against the Confederate States of America, during the existence of the war between the said Confederate and United States, unless I shall be exchanged for another prisoner or prisoners of war, or unless I shall be released by the President of the Confederate States. In consideration of the above parole, it is understood that I am free to go and come wherever I may see fit, except that I shall not attempt to enter or depart from any fort, camp, or garrison of the Confederate States without the sanction of its commanding officer.

Paroled soldiers' names would go on a list, and they were then released to their own lines. They were not permitted to go back into the ranks or perform any military duties until exchanged. When the returned captives on one side were equalled by the captives on the other, the names of the soldiers on parole were checked off, and they were allowed to resume their places in the army.

In July of 1862, official terms were reached by Union and Confederate leaders, which established a cartel for exchanging prisoners.

Haxall's Landing, on James River, Va., July 22, 1862.

The undersigned, having been commissioned by the authorities they respectively represent to make arrangements for a general exchange of prisoners of war, have agreed to the following articles:

ARTICLE 1. It is hereby agreed and stipulated that all prisoners of war held by either party, including those taken on private armed vessels known as privateers, shall be discharged upon the conditions and terms following:

Prisoners to be exchanged man for man and officer for officer; privateers to be placed upon the footing of officers and men of the navy.

Men and officers of lower grades may be exchanged for officers of a higher grade, and men and officers of different services may be exchanged according to the following scale of equivalents:

A general commanding-in-chief or an admiral shall be exchanged for officer of equal rank, or for sixty privates or common seamen.

A flag-officer or major general shall be exchanged for officers of equal rank, or for forty privates or common seamen.

A commodore carrying a broad pennant or a brigadier-general shall be exchanged for officers of equal rank, or twenty privates or common seamen.

A captain in the navy or a colonel shall be exchanged for officers of equal rank, or for fifteen privates or common seamen.

A lieutenant-colonel or a commander in the navy shall be exchanged for officers of equal rank, or for ten privates or common seamen.

A lieutenant or a master in the navy or a captain in the army or marines shall be exchanged for officers of equal rank, or six privates or common seamen.

Masters' mates in the navy or lieutenants and ensigns in the army shall be exchanged for officers of equal rank, or four privates or common seamen.

A lieutenant-commander or a major shall be exchanged for officers of equal rank, or for eight privates or common seamen.

Midshipmen, warrant-officers in the navy, masters of merchant vessels, and commanders of privateers shall be exchanged for officers of equal rank, or three privates or common seamen.

Second captains, lieutenants, or mates of merchant or privateers, and all petty officers in the navy, and all non-commissioned officers in the army or marines shall be severally exchanged for persons of equal rank, or for two privates or common seamen, and private soldiers or common seamen shall be exchanged for each other, man for man.

ARTICLE 2. Local, State, civil, and militia rank held by persons not in actual military service will not be recognized, the basis of exchange being the grade actually held in the naval and military service of the respective parties.

ARTICLE 3. If citizens held by either party on charges of disloyalty or any alleged civil offense are exchanged, it shall only be for citizens. Captured sutlers, teamsters, and all civilians in the actual service of either party to be exchanged for persons in similar position.

ARTICLE 4. All prisoners of war to be discharged on parole in ten days after their capture, and the prisoners now held and those hereafter taken to be transported to the points mutually agreed upon at the expense of the capturing party. The surplus prisoners not exchanged shall not be permitted to take up arms again, nor to serve as military police or constabulary force in any fort, garrison, or field-work held by either of the respective parties, nor as guards of prisons, depots, or stores, nor to discharge any duty usually performed by soldiers, until exchanged under the provisions of this cartel. The exchange is not to be considered complete until the officer or soldier exchanged for has been actually restored to the lines to which he belongs.

ARTICLE 5. Each party, upon the discharge of prisoners of the other party, is authorized to discharge an equal number of their own officers or men from parole, furnishing at the same time to the other party a list of their prisoners discharged and of their own officers and men relieved from parole, thus enabling each party to relieve from parole such of their own officers and men as the party may choose. The lists thus mutually furnished will keep both parties advised of the true condition of the exchange of prisoners.

ARTICLE 6. The stipulations and provisions above mentioned to be of binding obligation during the continuance of the war, it matters not which party may have the surplus of prisoners, the great principles involved being, first, an equitable exchange of prisoners, man for man, officer for officer, or officers of higher grade exchanged for officers of lower grade or for privates, according to the scale of equivalents; second, that privateers and officers and men of different services may be exchanged according to the same scale of equivalents; third, that all prisoners, of whatever arm of service, are to be exchanged or paroled in ten days from the time of their capture, if it be practicable to transfer them to their own lines in that time; if not, as soon thereafter as practicable; fourth, that no officer, soldier, or employee, in the service of either party, is to be considered as exchanged and absolved from his parole until his equivalent has actually reached the lines of his friends; fifth, that the parole forbids the performance of field, garrison, police, or guard, or constabulary duty.

John A. Dix, **Major-General.** D.H. Hill, **Major-General, C.S. Army.**

This system of exchange was adhered to in the East whenever possible, but as the tide of war changed, Federal leaders felt the return of Southern prisoners to the South aided the Confederacy far more than the return of Union soldiers aided their cause. They argued that the sick and wounded Northern men were eventually discharged from the Army, while the majority of rebels returned to fight. Therefore, feeling they were justified, the parole system was virtually suspended on May 25, 1863. This rapidly increased the number of men in confinement, and prisons sprang up everywhere.

In May, 1864, some barracks on the Chemung River near Elmira, New York, were converted into a prison. Three months later, almost 10,000 Confederates were held there. The wooden structures were totally insufficient for the amount of prisoners, compelling more than half of them to live outside in tents. The freezing cold of an upper New York winter caused the sick rate to rise and conditions worsened as a result of a feud between the prison commandant and the surgeon.

On December 4, 1864, the Inspecting Officer reported that at Elmira Prison both meat and flour were bad, and that 1,166 prisoners had not even one blanket. The death rate that winter averaged 5% a month.

During and after the War Between the States, both sides criticized the other for their inhuman treatment of prisoners. Bitterness ran high, as death rates were calculated, and charges drawn. The triumphant Union forces arrested Captain Henry Wirz of Andersonville, charged him with murder, in violation of the laws and customs of war and convicted him. His execution on November 10, 1865 did not put an end to the terrible tragedy of prison deaths, it only accentuated it. For the truth was, shifting political policies and the fortunes of war had more to do with the treatment of prisoners than any commandant. As in any war, the good and the bad, the competent and the incompetent all contributed to the management of men's lives. Neither South nor North was innocent from blame, and the study of men's diaries written while in prison is done to better understand, not to blame.

Echoes of War

"Capture the Flag"

by
Thomas J. Evans

[This story was told to Tom Evans by his father, who'd had it repeated many times over by his father. It concerns a former Welshman named Thomas Evans, who had won the Medal of Honor during the Civil War. The author's father was named after this hero, and he, in turn, passed the name on to his son. The following is a narrative of this event, as told by Thomas Evans to the author's grandfather.]

"Since you are interested I will tell you the story about my capturing a Confederate flag down in Virginia during the war. Now, Northern Virginia is the prettiest country on God's earth, particularly in springtime, and it is hard to believe so much of the war could take place there. I was in towns down there like Berryville and Leesburg where a man could be content to live out his days if he wasn't a Welsh Yankee like me.

Well, in June of 1864, my old Regiment, the 54th Pennsylvania Infantry was marching south from Cedar Creek, Virginia led by General David Hunter. He was a native Virginian who had stayed with the Union Army. He was not well liked by the Rebels so we figured we were going to have to fight almost every day. I was starting to think an old man like me shouldn't have joined this regiment with all the young boys in it because they never seemed to get tired marching. We had been pretty well chewed up at New Market with General Sigel a couple of weeks before. The Rebels even used some military school cadets there and they had fought like tigers. We were down then, but not out, and now we were marching back spoiling for a fight.

We skirmished right along down the Shenandoah Valley, as pretty farm land as you ever saw, however we noticed the residents were right unfriendly. On the 5th of June, a Sunday, we were shooting away at each other near a little place named Piedmont. I don't know much about the tactics of the battle that followed, for we privates left such matters to the generals to worry about. Company D that I was in was down behind a ridge in some woods, and the Rebel artillery was making it right hot all around us. Case shot and pieces of shell were dropping all over and every now and then a tree limb would fall down among us boys. About noon the Captain ordered us to put bayonets on our rifles and then he moved the company up to the top of that ridge, where we lay down and fired a volley at the Confederates located in some woods on our front. We had surprised the Rebels on their flank and they were right mixed up. We had just loaded up when all of a sudden the Captain ordered us to charge them. We jumped up with a big yell and took out across an open field for all we were worth hoping those Rebels wouldn't start shooting before we could get in among them. When it came to shooting you couldn't beat those boys.

In front of me a Confederate officer in a gray uniform was waving his sword and shouting, 'New Market! New Market! Remember New Market!'

'Boys, shoot that officer before he starts a rally!' shouted one of our officers.

Several of us cut loose at him and he spun around, dropped his sword, and fell to the ground. The Rebel soldiers that had been collecting around him disappeared. Suddenly the biggest Rebel I ever saw was right in front of me waving a flag.

'Rally, Boys, Rally!' he was shouting.

Well, I was glad my bayonet was on my rifle instead of at my side, but I had never used it before except as a candle holder or to roast a little meat. Now here I was with my rifle unloaded and about to attack this giant who was about the size of a full grown grizzly bear. I hauled off and hit that flag staff a right smart whack with the bayonet and it must have really stung that Johnny's hands. He sort of lowered the flag and I didn't know what else to do but grab the cloth with my left hand and pull.

'Let go Reb!' I shouted.

'Let go yourself!' says he.

We pulled back and forth and went around in a circle, him about to win since he was using two hands to my one. I finally managed to raise my rifle and says, 'Drop that there flag or I will pin you to a tree!'

Well now, he saw it wasn't any use and gave up and I sent him to the rear because the minnie balls were flying around right thick yet and there wasn't any use of us both getting killed. I handed that flag to the Lieutenant and got off another shot or two before the Confederates started to retreat. We lit right in after them and drove them a piece, and that was all for that day.

As soon as things got quieted down I got to thinking about that Confederate officer and thought his sword would make a nice relic of the battle. We hadn't been winning many of the battles up to now and I would like to have that sword to remember this day. I wandered on back towards where he had been shot and noticed a group of boys standing around a dead Confederate general. General 'Grumble' Jones they called him, and if every boy who later claimed to have shot him had put a bullet in him, why, he would have been so heavy it would have taken a four horse team to have moved his body off of the battlefield.

Now, I am not one who claims to have shot the General, and as a matter of fact, I do not recall of having seen him at all during the charge. I beat it on over to that Confederate officer and had the surprise of my life. He was laying there dead and standing over him with tears running down his cheeks was our Colonel Jakey Campbell. Would you believe that out of all of those thousands of Rebel soldiers he would just happen to know this particular one? I saw the officer's sword was already gone and figured if nothing was said about me taking a shot at him the better off it would be for me. I up real quick like and volunteered to bury that Rebel, Lieutenant Blue, so Jakey said his name was. We buried him near a farm house and the Colonel went and told those folks there his name and where he was located. Jakey was sure his mother would want to come and carry her boy back home to bury with his family.

This is the "old" model, similar to the one awarded Pvt. Thomas Evans for capture of an enemy flag.

After awhile I went down back to where the cavalry had charge of the prisoners and looked up that Rebel who had been carrying the flag. I had to search all around and finally found him, and do you know he had shrunk down to where he was the same size as me! We talked awhile and he was a Virginia boy and very polite talking. I gave him some hardtack since he hadn't eaten much that day, and some greenbacks because he was out of money and would probably be going to a prison camp where food would be hard to get. We shook hands and he said he would look me up after the war, but he never did, and I guess he didn't make it through. He was a brave man to stand there holding only that flag in the thick of all of that fighting. I wish he was here today so I could shake his hand. I don't have any hard feelings against those boys, they only fought for what they believed was right, the same as we did.

Old Colonel Jakey Campbell was right pleased I had captured that flag, and said so in his report about the battle. He ended up with it coming from a Tennessee regiment instead of a Virginia regiment, but the Lieutenant got it straightened out. My Christmas present that year was the Medal of Honor for capturing that flag. I may be the only man to win such an honor that had to sign for it with an 'X.' I was right pleased with the whole affair and that is how a poor boy from Wales who can't read or write captured a Rebel flag down in Virginia during the big war and won a medal."

On February 9, 1864, Thomas Evans enlisted in Company D, 54th Regiment Pennsylvania Volunteer Infantry, at Huntington, Pennsylvania. His enlistment was for three years and he received a $300 bounty payable by installments. Mr. Evans was aged 40 years, had gray eyes, dark hair, a dark complexion, was 5 feet 7 inches high, and indicated his birthplace was Wales. His enlistment form was signed with an 'X.'[1]

The 54th Regiment was organized at Harrisburg, Pennsylvania, on February 27, 1862, to serve a period of three years.[2] Its structure consisted of 13 companies, Companies A through M, and the Colonel of the Regiment was Jacob M. Campbell.[3]

On June 5, 1864, a little remembered battle was fought at Piedmont, Virginia. Two armies totaling 16,000 men fought desperately with a decisive victory for the Union Army and near annihilation of the Confederate forces. The Confederate defeat resulted in the loss of 1,000 prisoners alone, and caused the Confederate Government at Richmond to initiate frantic efforts to regain control of the Shenandoah Valley. During the battle, the Confederate leader, Brigadier William E. 'Grumble' Jones, was killed by a minie ball through the head as he waved his hat in an attempt to cheer his men on to victory.[4] Another Confederate casualty was Lieutenant Monroe Blue, commanding Company K of Harper's Reserves, who was then on a recuperative leave of absence. Blue had been seriously wounded during April, 1864, when, with 30 men, he had attacked a camp of 300 Union cavalry in hope of stampeding their horses.[5] Lieutenant Blue was found dead on the battlefield by Colonel Jacob M. Campbell of the 54th Pennsylvania, who had known him prior to the war. Colonel Campbell wrote the mother of the dead officer the following letter:

Staunton, Va., June 7th 1864

Mrs. Blue:—

Respected Madam:—

I feel it to be my unpleasant duty to communicate to you, the sad intelligence of the death of your son, Lt. Monroe Blue. He fell upon the battle field at Piedmont, Augusta Co. Va. on Sunday the 5th inst. In passing over the battle field after the battle, I recognized him among the dead. On his person I found the enclosed paper confirming me in my belief as to his identity. It was all that was in his pockets. If he had any thing else, it had been taken from him.

I took the body & buried it by itself, placing a board at the head with his name upon it. The grave is about two hundred yards from the house of Martha Rankin, — in the woods, & about fifty feet from the road — on the right hand side, as you come from Port Republic, to Piedmont & about fifty feet from a fence running at right angles to the road. He is wrapped in a blanket, with a brown blanket over him — a (illegible) cloth being first placed over the face — a board on top of the blanket, with earth on the board, — his wound was from a bullet, through the neck, from right to left, & must have been instantly fatal. I am thus minute, so there may be no mistaking the grave, presuming you would desire to remove the remains. I gave Mrs. Rankin his name on a slip of paper, & told her where the grave was.

I regret exceedingly that I could not give him a more proper burial, but lack of time, as well as material rendered it impossible, as there were, alas, too many to bury on that sanguinary field. Still more do I regret to announce the death of your gallant son, who, however we may differ in sentiment, I am pleased to accord all those noble traits of character, which must endear him to a mothers heart, — a sister's love, — and the high esteem of all his friends and acquaintences.

He fell near the same spot, and the same moment, with his brave commander, Maj. General Jones, gallantly contending for the palm of victory.

Permit me Madam to condole with you in this your affliction, and to unite with you in prayer, that we may all meet in the world above, where war and strife will never enter.

Very Respectfully
your obt. servant
J.M. Campbell
Col. 54th Reg. Penna. Vols.

It would be a satisfaction to know that you received this. A line to my address, '2nd Brigade Sullivan's Division via Martinsburg' would reach me.[6]

Lieutenant Blue's body was never removed and he lies buried on the battlefield. A courageous officer, he is remembered as one of the South's fallen heroes.

The 54th Pennsylvania suffered numerous casualties in the battle. Two men were killed, twenty nine severely wounded, and forty men suffered slight wounds that did not require removing them from duty.[7] The number of wounds resulting from the unusual bayonet charge is not recorded. Contrary to popular belief very little bayonet fighting occurred in Virginia, the battle at Piedmont being a rare exception. For comparison purposes the records of the Medical Director of the Army of the Potomac show that from May 1, 1864 to July 31, 1864 a period that included the Wilderness, Spotsylvania Court House, Cold Harbor, the initial assaults and the Crater at Petersburg, and other miscellaneous engagements only thirty seven bayonet wounds were treated by army surgeons.[8] During the entire Civil War period of 1861-1865, the Surgeon General reported treating only 400 cases of wounds from bayonets of which 30 proved fatal.[9] Rifle fire was so powerful and accurate troops were simply unable to approach close enough to the enemy to use the bayonet.

In his official report of the engagement Colonel Campbell wrote: 'Private Thomas Evans, of Company D, wrested the colors from the color bearer of a Tennessee regiment, sending the color bearer to the rear.'[10] This mistake was corrected by the commander of Company D when he directed the following letter to Colonel Campbell for indorsement through official channels to the War Department at Washington, D.C.:

'Camp of 54th Regt, Penna, Vols,
infty Harrisonburg Va,
Oct 4th 1864

Col

I have been informed that Enlisted men Who have captured battle flags etc, from the Enemy during this campaign have been rewarded by the Government, Therefore I would respectfully call your attention to the fact that Thomas Evans, a Private of my Company (D) of the 54th Regt, Penna, Vols, Infty, did at the battle of Piedmont Va, June 5th 1864, during the charge made by the 2d Brigade, Capture the Color Bearer & the Colors of the 45th Va Infty, a Rebel Regiment.

Private Thomas Evans is a good & brave Soldier and for his Meritorious Conduct, I would respectfully recommend him to your favorable consideration.

I am Very Respectfully
Your Obdt. Servt.
George W. Gageby, 2d Lieut.
Comp D, 54th Regt, pa, Vols, Infty
Comdg Comp'[11]

To J. Thoburn
Col Comdg
1st Div, Dept W. Va.

This letter was endorsed through the chain of command and on November 26, 1864 Private Thomas Evans was awarded the Medal of Honor.[12] The medal did not make an appearance until December when a package arrived at the 54th Pennsylvania and the receipt following was executed:

'Recd one pkge. from Adjt Genls Office December 3, 1864.

his
Thomas X Evans
mark

Private Co D Penn Vols.
Witness J. B. Dunlop
1 Lieut Comdg
Co H 54 Vol.'[13]

The honor of winning a medal was not lost on Private Evans and securing the assistance of a more literate comrade he dispatched the following letter:

'Army of the Shenandoah
Dec 4th 1864

E.D. Townsend
Sir

Your form of Nov 26th accompanied by the medal of honor awarded me has been received, and I hereby tender my heartfelt thanks to the Donor for testifying in such a manner the appreciation of what was merely a Soldier's duty.

Very respectfully
your Obt Servt
Thomas Evans
Co D 54th Regt Pa Vols Inft.'

Private Evans served actively through the remainder of the war and was the only member of the 54th Regiment to be awarded the Medal of Honor. The entire regiment was captured on April 9, 1865 during the final days of the war at Rice's Station, Virginia and marched as a unit to Camp Parole, Annapolis, Maryland.[15] Private Evans was officially paroled on April 17, 1865 and discharged at Camp Parole by War Department Special Order no. 77 dated May 31, 1865. The 54th Pennsylvania was deactivated July 15, 1865 at Harrisburg, Pennsylvania.[16] Private Evans' meticiously kept service records show one interesting error; at discharge he was still 40 years of age. Officially, he had not aged during his 15 month enlistment.

FOOTNOTES

[1] Evans, Thomas, military service records, National Archives, Washington, D.C.

[2] *Official Army Register of the Volunteer Force of the United States Army,* Adjutant General's Office (Washington, 1865), Part III, p. 863.

[3] *Annual Report of the Adjutant General of Pennsylvania for the Year 1866,* Adjutant General's Department (Harrisburg, 1867), Vol I, p. 261.

[4] Edward A. Pollard, *The Lost Cause* (E.B. Treat & Co., New York, 1867), p. 531.

[5] *The War of the Rebellion, A Compilation of the Official Records of the Union and Confederate Armies* (Washington, 1880-1901). Hereafter referred to as OR, Series I, Vol XXXIII, p. 309.

[6] Marshall M. Brice, *Conquest of a Valley* (University Press of Virginia, 1965), pp. 145, 146.

[7] *Ibid.*, p. 79.

[8] OR, op. cit., Vol XXXVI, Part I, p. 265.

[9] *The Medical and Surgical History of the War of the Rebellion,* Surgeon General United States Army (Washington, 1883), Part III, Vol II, p. 686.

[10] OR, op. cit., Vol XXXVII, Part I, p. 119.

[11] Evans, op. cit., records group 500276.

[12] *Medal of Honor Recipients,* 1863-1963, Subcommittee on Veterans Affairs, Washington, 1964, p. 432.

[13] Evans, op. cit., records group 501170.

[14] *Ibid.*, records group 506308.

[15] Fredrick H. Dyer, *A Compendium of the War of the Rebellion,* (Reissue), Morningside (Dayton, Ohio, 1978), p. 1592.

[16] Evans, op. cit., military service records.

SIGNALS

Dear Editors:

Enclosed is eighteen dollars for a year subscription for THE KEPI.

I was introduced to your magazine by my daughter. My interest has been rekindled in the Civil War period after reading your three issues from cover to cover. I was so enthused, I wrote to your unclassifieds for more information. Their response was prompt and several had personal notes enclosed.

Keep up the good work.

Sincerely yours,
Mary Jackson
Latham, New York

Dear Editors:

I wish to thank you for your first rate article in the August/September issue by Thomas J. Evans — "Carrying Isaiah Home." Also, I would like to know where to write for more information about the John S. Mosby Society's periodical "The Partisan Review." Once again Thank You.

Darrell W. Stridh
Whittier, California

[*Letters of inquiry may be sent to THE KEPI, and will be forwarded to Mr. Evans.*]

Dear Editors:

Excellent magazine! Especially the story on my alma mater — V.M.I.

You caught my attention. A check for a subscription is enclosed.

Is it possible to purchase the prior back issues? If so, please advise the price.

Much luck and much success.

Don DeLuca
Pittsburgh, Pennsylvania

[*Back issues are available for $3.50 each. Please allow six weeks for delivery, unless first class delivery is desired. If so, please include 54¢ extra per issue. April, June/July, August/September and October/November 1983 issues are available.*]

Dear Editors:

Enclosed you will find U.S. postal money order in the amount of $18.00 for a years subscription of THE KEPI magazine.

I enjoyed the sample copies and look forward to getting the magazines during the coming year.

Looking forward to the August-September issue and thanking you for your kind attention, I remain

Sincerely,
Robert T. Millikin, Sr.
Richmond, Virginia

Dear Editors:

Really enjoyed your first two issues, looking forward to more. Please accept this money order for a one year subscription.

Thanks
Michael W. O'Brien
SPECIALTY JEEPARTS
San Francisco, California

Dear Editors:

I was very pleased with my first copy of THE KEPI. I can only hope that others are equally pleased and that your venture will be a success. Also, I wish to thank you for the personal note in this Computer Age. Personal contacts seem to be fewer and fewer.

Enclosed you will find a Bank Money Order for twenty one dollars in payment for the next issues.

May you and THE KEPI thrive and grow.

Sincerely,
William R. Ross, M.D.
Winnipeg, Manitoba
Canada

"PARDON FROM THE ADVERSARY THAT YOU DESPISE"

by

Everett K. Cooper

After four years of bitter, fratricidal war, the transition of Southern survivors to reconstructed civilians was to be an arduous campaign. The three major surrenders of Confederate Army commands — Lee at Appomattox, Johnston at Durham Station and Kirby Smith at Galveston — and the numerous other surrenders did not of themselves automatically reinstate those veterans to the benefits of United States citizenship. It was not to be as simple a task as clipping CSA buttons (for those fortunate enough to have such) from what passed as a uniform to convert soldiers into civilians. The terms of surrender dealt only with the military aspects, placing the soldiers on a "parole" status, in which they promised not to resume military duties until exchanged for a Yankee prisoner of comparable rank. The restoration of citizenship was to be a civilian process, conducted once the war was officially over. The vastness of this restoration was indicated by the official numbers of surrendered Confederate forces, which, of course, did not include those who refused to surrender.[1]

Surrendered Confederate Forces	
Army of Northern Virginia	27,805
Army of Tennessee	31,243
Gen. Richard Taylor's command	42,293
Army of Trans-Mississippi	17,686
Others	55,196
Total	174,223

Those men already languishing in confinement as prisoners of war would face delay in their release, as piecemeal surrenders of field armies did not apply to them. For some, the horrors of prison life would continue for several months after Appomattox. Thus, the end of armed conflict meant the beginning of the humiliation of defeat. The victors would endeavor to make their subjugation of the South complete by requiring leaders and influential citizens to humble themselves by requesting a pardon. For some, this was too bitter a price to pay, and they fled the country.

Beginning in July of 1862, the U.S. Congress passed the Confiscation Act. Its prime purpose was to legalize the seizure of property belonging to supporters of the rebellion. The thirteenth section of this legislation stated: "The President is hereby authorized at any time hereafter, by proclamation, to extend to any person who may have participated in the existing rebellion in any state or part thereof, pardon and amnesty, with such exceptions and at such time and on such conditions as he may deem expedient for the public welfare." Along with the Constitutional authority for presidential reprieves and pardons, this law would become the legal basis for President Lincoln, and later President Andrew Johnson, to grant amnesty* and pardon to ex-Confederates.[2]

The first utilization of this authority was on December 8, 1863, when Abraham Lincoln issued his "Proclamation of Amnesty and Reconstruction." This offer, made after the successful military campaigns of 1863, was a plan to restore Unionist government in the Southern states and weaken the Confederates' will to fight. The only requisite for pardon was to take and subscribe to an oath to support and defend the Constitution of the United States, and to abide by that oath. However, certain groups of individuals were excluded from this pardon offer, as would be the pattern in subsequent amnesty offers. This proclamation was more a weapon of psychological warfare than were later proposals, released after the Southern armies had laid down their arms. While Lincoln excluded only six groups of Southerners from pardon, he made no provision for overcoming the exclusions.[3]

The next general offer of amnesty was made by President Andrew Johnson on May 29, 1865, at the time of the last surrender of a major Confederate army. This proclamation was issued at a time when bitter animosities ran high. But despite the strong personal disinclination of the ex-Confederates towards seeking amnesty, it was made a matter of everyday economics. Without this official forgiveness, it would be impossible to have confiscated holdings restored, buy or sell property, enter into contracts, or perform many other ordinary legal transactions. The nonrepatriated Confederate would also be excluded from the political rebuilding of the South. While the spirit of the "Unreconstructed Rebel" was good for the soul, it was impractical. Nevertheless, there were many who would continue their resistence, as exemplified

*Amnesty — a general pardon for offenders by a government, especially for political offenses.

in the popular verse, "O I'm a Good Old Rebel," written by Major Innes Randolph of Virginia:

"....

I only wish we'd won.
And I don't want no pardon
For anything I done."

The tone of President Johnson's first proclamation, dated May 29, 1865, and the text of the required oath (See Appendix I), were similar to Lincoln's wartime proclamation. The exclusions were increased to fourteen, but with the provision "that special application may be made to the President for pardon by any person belonging to the excepted classes; and such clemency will be liberally extended as may be consistent with the facts of the case and the peace."[4]

During the 1865-1866 period, when the mass of pardon requests were presented, over 80% were approved. The President was deluged with petitions and petitioners, including many who made pilgrimages to his office to personally plead their cases. Andrew Johnson preferred to be involved in each case, but he did not anticipate the continuing daily burden on his time this would require. Nor did he anticipate the surrogate involvement with the appearance of "pardon brokers" to intercede for the humbled petitioners. The pardon brokers, whose fee was $150, though some cases ranged as high as $500, became active in the Southern state capitals, as well as in Washington. The activities of the brokers in obtaining pardons served a useful purpose, but nearly developed into a national scandal of corruption.[5]

A third amnesty proclamation on September 7, 1867, reduced the number of classes excluded from pardon. This was followed by a fourth amnesty proclamation issued on July 4, 1868, reflecting the progress of reconstruction by further reducing the number of exclusions. (See Appendix II) At this time, those unpardoned consisted of only a handful of Southerners "who may be under presentment or indictment in any court of the United States having competent jurisdiction upon a charge of treason or other felony." This reduced the unforgiven to an elite group upon whom a special revenge was desired, such as Jefferson Davis, John C. Breckinridge and Robert E. Lee.[6]

Andrew Johnson's final "General Amnesty Proclamation," issued on December 25, 1868, gave absolution and forgiveness to all by the lame-duck President. The proclamation of a pardon "unconditionally and without reservation" at this time was little more than politic. A few months earlier, the Fourteenth Amendment to the U.S. Constitution had been passed. Section 3 of this amendment disqualified from holding office any person who, as a member of Federal or state government, had taken an oath to support the Constitution, and afterward violated that oath by engaging in the rebellion. This exclusion could be removed only by a two-thirds vote of Congress. With the passage of the Fourteenth Amendment, Congress assumed an active role in the process of forgiveness. (See Appendix III) It was estimated that approximately 18,000 were subject to exclusion by the provisions of Section 3, but Congress made wholesale approvals, and reduced the number to about 750.[7]

In 1876, Congress considered another amnesty bill which would have allowed the remaining 750 excluded by Section 3 to merely appear before a local U.S. judge, swear to a simple oath and be restored to complete citizenship. Representative James G. Blaine, an active presidential aspirant, "waved the red shirt" when he introduced a modification to the bill, reviving the sectional animosities. His request was to add only a few words to the bill: "With the exception of Jefferson Davis." This, he argued, was based on the fact that President Davis bore the sole responsibility for the horrors of Andersonville. To overcome resistance and aid in passage of the bill, Jefferson Davis wrote to the Chairman of the Amnesty Committee. "...I have no claim to pardon, not having in any way repented, or changed the convictions on which my political course was founded..." The legislation was passed with the change requested by Congressman Blaine. Finally, on June 6, 1898, Congress repealed the disability imposed by Section 3 of the Fourteenth Amendment, thirty-three years after termination of the armed conflict. Jefferson Davis was dead, and, since the end of the war, had not sought political office.[8]

What administrative processing and "paper work" were required for the Southerner to be restored to U.S. citizenship? An example of the simplest process would be the case of Sergeant John Kipps, 4th Virginia Infantry, who was surrendered at Appomattox. The surrender terms between Grant and Lee required that:

"Rolls of all the officers and men be made in duplicate, one copy to be given to an officer to be designated by me (Grant), the other to be retained by such officer or officers as you (Lee) may designate. The officers to give their individual paroles not to take up arms against the Government of the United States until properly exchanged, and each company or regimental commander to sign a like parole for the men of their commands."

In two days, Sergeant Kipps was given a slip of paper which allowed him to return to his Virginia home. The printed pass stated:

Appomattox Court House
April 10, 1865

THE BEARER, Sgt. John Kipps of 4th Virginia Infantry, a Paroled Prisoner of the Army of Northern Virginia, has permission to go to his home, and there remain undisturbed.[9]

Despite the agony at Appomattox, all this simple piece of paper procured for Sergeant Kipps and other veterans was permission to go home — they were still Confederate soldiers. A month and a half after Appomattox, President Johnson issued his first amnesty and pardon proclamation. If Sergeant Kipps or any other ordinary Confederate veteran chose to start the reconstruction process, he had only to go to the closest Provost Marshal or county court house and swear his oath. He would then be given a signed copy of the oath, which would serve as his "new citizenship" paper.

A second example would be the case of a civilian who did not serve in the Confederate military or government, such as Mr. Allen Coward, a 57-year-old farmer from Galveston County, Texas. The final surrender of a major Confederate military command occurred on June 2, 1865, at Galveston. The occupation troops came ashore a few days later, and, on July 21st, the Federally sponsored governor, A.J. Hamilton, arrived at Galveston. Citizens began to line up at Provost Marshal offices and the county offices of appointed officials. There, the process of swearing and recording the amnesty oath took place. As a farmer of modest resources and not a holder of political office, Mr. Coward would be quickly restored to U.S. citizenship.

What of an individual on the exception list, as described in the amnesty and pardon proclamations? The case of Confederate General John B. Gordon is typical of an excluded army officer. "On September 15 (1865) Gordon went to Fulton County (Georgia) Court House, in Atlanta, and signed an oath of amnesty before Ordinary Daniel Pittman. With it he forwarded a letter to President Johnson begging for pardon for his part in the war. This must have been a bitter dose for Gordon, partly because Andrew Johnson was one of the few men whom he thoroughly despised; but more especially because he felt that he had done nothing to ask pardon for. But he swallowed his pride, as Lee and others did, to do what he believed was best for the country."[10]

Perhaps the classic example of a simple, straight-forward request for a parole was that of General Robert E. Lee's. His letter, forwarded to President Johnson through General Grant on June 13, 1865, shortly after Johnson's proclamation, read:

"Sir: Being excluded from the provisions of the amnesty and pardon contained in the proclamation of the 29th ult., I hereby apply for the benefits and full restoration of all rights and privileges extended to those included in its terms. I graduated at the Military Academy at West Point in June, 1829; resigned from the United States Army, April, 1861; was a general in the Confederate Army, and included in the surrender of the Army of Northern Virginia, April 9, 1865. I have the honor to be very respectfully, Your obedient servant, R.E. Lee."

The full processing of General Lee's request was never made. He would, however, be covered by the General Amnesty of December 25, 1868.[11]

The reason for Lee's prompt request for pardon was his awareness that he would be an example for others. General P.G.T. Beauregard sought counsel from Generals Lee and Johnston. In a letter to Lee, dated September 1, 1865, Beauregard expressed the thought of many: "It is hard to ask pardon of an adversary that you despise." Both Lee and Johnston recommended that patriotism and duty required he apply for pardon. On September 16th, Beauregard swore his oath of fidelity before the Mayor of New Orleans. This oath was forwarded with a letter, which stated:

"In taking up arms during the late struggle (after my native State, Louisiana, had seceded) I believed, in good faith, that I was defending the constitutional rights of the South against the encroachment of the North. Having appealed to the arbitration of the Sword, which has gone against us, I accept the decision as settling finally the questions of secession & slavery — & I offer now my allegiance to the Govt. of the United States, which I promise, truly and faithfully, to serve & uphold hereafter, against all external or internal foes." To his signature, he added "late Genl. C.S.A." Perhaps Beauregard's letter, like that of Lee's, did not admit of any wrongdoing. No action was taken on the Creole's request, even after personal follow-ups by himself and others.[12]

The usual method for excepted individuals to apply for a pardon was to prepare a letter of request with a signed amnesty oath, and submit it to the governor of the state. The letter could be a direct statement, as that made by Lee, or something elaborate, such as the letter submitted by Joseph Davis, elder brother of Jefferson Davis. He had printed endorsements from more than twenty-five leading political figures, accompanied by six pages of additional signatures on a petition. Like many others, Mr. Davis needed to make a strong plea for pardon so he could recover his valuable plantation, which the U.S. government had seized and occupied. The governor had the power to reject any of the submitted pardon requests. When the papers were forwarded to Washington, they were reviewed and processed by the Attorney General's office. President Johnson had appointed as Pardon Clerk in the Attorney General's office a former Confederate Colonel, Matthew F. Pleasants. The application was then forwarded to the President, who, if he approved the request, would issue a large, impressive document, a "warrant" for a pardon, which was sent to the requestor. (See Appendix IV) The warrant, signed by the President, granted "full pardon and amnesty for all offences by him committed" with certain stipulations for its acceptance. If the requestor was in agreement, he was to "notify the Secretary of State, in writing, that he has received and accepted the foregoing pardon." A printed form was provided for this acceptance. Then the Secretary of State would issue to the requestor another impressive document, with seal, to which was attached the signed acceptance form from the requestor.

For some, even this organized program for repentance and forgiveness was further complicated. Border state Maryland clandestinely contributed many volunteers to support the Southern cause in military, naval and government functions, the greatest number being those who served in the Army of Northern Virginia. The Appomattox surrender agreement allowed each Confederate to be released on parole and return to "his home and remain there undisturbed." The returning Marylanders found they were required to register immediately with the Provost Marshal to receive another piece of paper which would allow them to be "undisturbed." They quickly learned that General Grant was contravened by an opinion of the U.S. Attorney General, who ruled that Maryland veterans "had no right to come to places which were their homes prior to their going into the rebellion." These ex-Confederates were arrested, and released only if they were willing to take the oath of allegiance. But, once released, they were required to go "north where they were tolerated." The President's May 29, 1865 proclamation eased that harsh action, though the exception groups were still required to register with the Provost Marshal.

In the numerous surrenders of field and garrison commands, many were given a parole, en masse, but chose to depart the country for foreign shores. The pain and hostility of four years of war, a surrender they did not support, and the degradation of requesting a pardon were more than they cared to endure. At the end of the war, there were 299 active generals on the Confederate Army rolls. Approximately 32, or more than 10%, fled the country. Most would return within a 3 year period. Six governors or former governors of Southern states also left the country. (See Appendix V) The healing effect of time, continued liberalization of the amnesty regulations, love of country and the restoration of their citizenship lured them back. The temporary exiles satisfied their consciences that they were unreconstructed rebels. A few had the satisfaction of never asking for pardon. This spirit was epitomized by Jefferson Davis in an address to the Mississippi Legislature in 1884. "It has been said that I should have applied to the United States for a pardon, but repentance must precede the right of pardon and I have not repented. Remembering as I must all which has been lost, disappointed hopes and crushed aspirations, yet I deliberately say, if it were to do all over again, I would again do just as I did in 1861."[13]

In recent years, under the continued pressure of Southern Congressmen, the President and the U.S. Congress completed the task of restoring the rights of full citizenship to Jefferson Davis and Robert E. Lee. The application made by Lee for pardon under the May 29, 1865 amnesty and pardon proclamation was never acted upon by the government. Lee failed to send a signed oath of allegiance with his letter requesting pardon. The oath was sent a few months later, but the two documents were never joined and processed. Secretary of State W.H. Seward, under the impression that the oath had already been recorded, gave the letter to a friend as a souvenir.[14] The general amnesty of December 25, 1868, restored citizenship to Lee, excepting the right to hold office as detailed in Section 3 of the Fourteenth Amendment. This limitation was voided by Congressional legislation in 1898, twenty-eight years after his death. In August, 1975, Congress enacted, and the President approved, the elimination of any restrictions on his citizenship, retroactive to June 13, 1865, the date of Lee's letter requesting pardon. This left Jefferson Davis as the last unpardoned Confederate.[15]

Jefferson Davis never asked for amnesty or pardon, though the last amnesty proclamation of December 25, 1868, restored his basic rights, except those prohibited by the Fourteenth Amendment. To ask forgiveness would have been an admission of wrong, not only for Davis, but for the entire Confederate South. Davis refused to do this. To "wipe the slate clean" of the last wound of the war, Congress passed, and President Jimmy Carter signed, on October 17, 1978, the law by which the "legal disabilities placed upon Mr. Jefferson F. Davis are hereby removed and that Mr. Jefferson F. Davis is posthumously restored to full rights of citizenship, effective December 25, 1868." President Carter commented at the signing that it "officially completes the long process of reconciliation that has reunited our people following the tragic conflict between the states."[16]

FOOTNOTES

[1] Clement A. Evans, **The South After the War**, p. 501.

[2] John H. Franklin, **Reconstruction After the Civil War** (Chicago, 1961), p. 16.

[3] John H. Franklin, **Reconstruction After the Civil War;** See also: Michael Perman, **Reunion Without Compromise** (Cambridge University, 1973), pp. 122-125.

[4] Edward McPherson, **The Political History of the United States of America During the Period of Reconstruction April 15, 1865 — July 15, 1870** (New York, 1972).

[5] Hodding Carter, **The Angry Scar** (Garden City, 1959), p. 73.

[6] **Ibid.**

[7] Edward McPherson, **The Political History of the United States of America During the Period of Reconstruction;** See also: James M. McPherson, **Ordeal by Fire,** p. 516.

[8] **Calendar of the Jefferson Davis Postwar Manuscripts in the Louisiana Historical Association Collection** (New Orleans, 1943), Letters 77 — 80, 82, 83, 228.

[9] Philip Van Doren Stern, **Soldier Life in the Union and Confederate Armies** (New York, 1961), p. 369.

[10] Allen P. Tankersley, **John B. Gordon — A Study in Gallantry** (Atlanta, 1955), p. 229.

[11] Hodding Carter, **The Angry Scar;** See also: **Citizenship For General Lee,** Lincoln Herald (Summer 1979), pp. 94-101.

[12] T. Harry Williams, **P.G.T. Beauregard — Napolean in Gray** (Baton Rouge, 1955), pp. 313-318.

[13] Hudson Strode, **Jefferson Davis — Tragic Hero** (New York, 1964), p. 469.

[14] Charles Bracelen Flood, **Lee — The Last Years** (Houghton Mifflin Co., Boston, 1981), p. 100.

[15] **Citizenship for General Lee.**

[16] U.S. Public Law 95-466, October 17, 1978, 95th Congress, "To restore posthumously full rights of citizenship to Jefferson F. Davis." Approved October 17, 1978.

APPENDIX I

Amnesty Oath

I, ——, do solemny swear, in the presence of Almighty God, that I will henceforth faithfully support, protect, and defend the Constitution of the United States and the Union of the States thereunder; and that I will in like manner abide by and faithfully support all laws and proclamations which have been made during the existing rebellion, with reference to the emancipation of slaves: So help me God.

APPENDIX II

Amnesty Exclusion Classes

	Exempted From Amnesty	
	From	To
Military Causes:		
All who are or shall have been military or naval officers of Confederate government above rank of Colonel (Army) or Lieutenant (Navy)	Dec. 8, 1863	Sept. 7, 1867
As above but minumum rank increased to Brigadier General (Army) or Captain (Navy)	Sept. 7, 1867	July 4, 1868
All who resigned commissions in the Army or Navy of the U.S. and afterwards aided the rebellion (wording modified 5/29/1865 to say resignation was "to evade duty in resisting the rebellion")	Dec. 8, 1863	Sept. 7, 1867
All Confederate military and naval officers who were educated at the Military Academy or Naval Academy	May 29, 1865	Sept. 7, 1867
All who engaged in destruction of U.S. commerce on high seas; all persons who made raids into the U.S. from Canada or engaged in destroying commerce on lakes and rivers that separate Canada from U.S.	May 29, 1865	Sept. 7, 1867
All who engaged in any way in treating colored persons, or white persons in charge of such, otherwise than lawfully as prisoners of war, and which persons may have been found in the U.S. service as soldiers, seamen or other capacity	Dec. 8, 1863	July 4, 1868
Civil Causes:		
All who are or shall have been pretended civil or diplomatic officers or otherwise domestic or foreign agents of the C.S.A.	Dec. 8, 1863	July 4, 1868
All who left seats in the U.S. Congress to aid the rebellion	Dec. 8, 1863	Sept. 7, 1867
All who served as governors of states while members of the Confederate States	May 29, 1865	July 4, 1868
All who left judicial positions under the U.S. to aid the rebellion	Dec. 8, 1863	Sept. 7, 1867
General Causes:		
All who voluntarily participated in the rebellion and has taxable property over $20,000	May 29, 1865	Sept. 7, 1867
All who have been or are absentees from the U.S. for the purpose of aiding the rebellion	May 29, 1865	Sept. 7, 1867
All persons who currently are in military, naval or civil confinement or bond as prisoners of war or offences of any kind	May 29, 1865	Dec. 25, 1868
All who took the oath for amnesty under Dec. 8, 1863 proclamation or thereafter and violated the oath	May 29, 1865	Sept. 7, 1867
All persons who were engaged directly or indirectly in the assassination of Pres. Lincoln or in any plot or conspiracy connected therewith	Sept. 7, 1867	July 4, 1868

APPENDIX III

Fourteenth Ammendment, Section 3, U.S. Constitution

That no person shall be a Senator or Representative in Congress, or elector of President and Vice-President, or hold any office, civil or military, under the United States, or under any State, who having previously taken an oath as a member of Congress, or as an officer of the United States, or as a member of any State Legislature, or as an executive or judicial officer of any State, to support the Constitution of the United States, shall have engaged in insurrection or rebellion against the same, or given aid or comfort to the enemies thereof. But Congress may, by a vote of two-thirds of each House, remove such disability.

General Amnesty Law of May 22, 1872:

That all political disabilities imposed by the third section of the fourteenth article of ammendments of the Constitution of the United States are hereby removed from all persons whomsoever, except Senators and Representatives of the Thirty-six and Thirty-seventh Congresses, officers in the judicial, military, and naval service of the United States, heads of departments, and foreign ministers of the United States.

APPENDIX IV

Presidential Pardon Documents

(Document size $11^3/_8$"x$17^7/_8$")

ANDREW JOHNSON
PRESIDENT OF THE UNITED STATES OF AMERICA

To all whom these presents shall come, greeting:

Whereas________________________

of________________________by taking part in the late rebellion against the Government of the United States, has made himself liable to heavy pains and penalties; AND WHEREAS, the circumstances of his case render him a proper object of Executive clemeny; NOW, THEREFORE, BE IT KNOWN, that I, ANDREW JOHNSON President of the United States of America, in consideration of the premises, divers other good and sufficient reasons me thereunto moving, do hereby grant to the said________________________a full pardon and amnesty for all offences by him committed, arising from participation, direct or implied, in the said rebellion, conditioned as follows:

1st. This pardon to be of no effect until the said________________________shall take the oath prescribed in the Proclamation of the President, dated May 29th, 1865.

2d. To be void and of no effect if the said________________________shall hereafter, at any time, acquire any property whatever in slaves, or make use of slave labor.

3d. That the said________________________first pay all costs which may have accrued in any proceedings instituted or pending against his person or property, before the date of the acceptance of this warrant.

4th. That the said________________________shall not, by virtue of this warrant, claim any property or the proceeds of any property that has been sold by the order, judgment, or decree of a court under the confiscation laws of the United States.

5th. That the said________________________shall notify the Secretary of State, in writing, that he has received and accepted the foregoing pardon.

IN TESTIMONY WHEREOF, I have hereunto signed my name and caused the Seal of the United States to be affixed.

Done at the City of Washington, this____day of______A.D. 186—, and of the Independence of the United States the________________________

(S E A L)

Andrew Johnson

By the President:

/s/ William H. Seward Secretary of State

(Document size 7⁵/₈"x9¾")
Honorable William H. Seward
Secretary of State

Sir;

I have the honor to acknowledge the receipt of the President's Warrant of Pardon bearing date____________________186—, and hereby signify my acceptance of the same, with all the conditions therein specified.
I am, Sir,
Your obedient servant,

____

(Document size 7¾"x10")

United States of America
Department of State

To all to whom these presents shall come, Greeting:

I CERTIFY, That the document hereunto annexed is a true copy of the original on file in this Department.

IN TESTIMONY WHEREOF, I, WILLIAM H. SEWARD,

Secretary of State of the United States, have hereunto subscribed my name and caused the seal of the Department of State to be affixed.

Done at the City of Washington, this________________day of________________A.D. 186—, and of the Independence of the United States of America the Ninetieth.
(S E A L)

William H. Seward

APPENDIX V

Confederates in Exile*

Generals:

Daniel W. Adams	Fled to England and Canada; died in Canada
Henry W. Allen	Also served as Governor of Louisiana at end of war; fled to Mexico and died in Mexico
Hamilton P. Bee	Fled to Mexico; returned in 1876
John C. Breckinridge	Fled to Cuba, Europe and Canada; returned 1869
William H. Carroll	Fled to Canada in 1863; died in Canada
John B. Clark	Fled to Mexico; returned
Jubal A. Early	Fled to Mexico and Canada; returned
Daniel M. Frost	Fled to Canada in 1863; returned
Birkett D. Fry	Fled to Cuba; returned in 1868
William P. Hardeman	Fled to Mexico; returned
James M. Hawes	Fled to Mexico; returned
Alexander T. Hawthorn	Fled to Brazil; returned in 1874
Thomas C. Hindman	Fled to Mexico; returned in 1868
William L. Jackson	Fled to Mexico; returned
Edwin G. Lee	On duty in Canada; returned 1866
Danville Leadbetter	Fled to Mexico and Canada; died in Canada 1868
Hylan B. Lyon	Fled to Mexico; returned in 1866
John B. Magruder	Fled to Mexico; returned before 1871
James P. Major	Fled to France; returned
Mosby M. Parsons	Fled to Mexico; died in Mexico in 1865
William Preston	Fled to Mexico, England and Canada; returned in 1866
Sterling Price	Fled to Mexico; returned in 1866
Robert V. Richardson	Fled to Europe; returned
Roswell S. Ripley	Fled to England; returned
Joseph O. Shelby	Fled to Mexico; returned
Henry H. Sibley	Fled to Europe and Egypt; returned after 1873
James E. Slaughter	Fled to Mexico: returned in 1868

Edmund K. Smith	Fled to Mexico; returned
Walter H. Stevens	Fled to Mexico; died in Mexico
Robert A. Toombs	Fled to Europe; returned in 1867
John G. Walker	Fled to Mexico; returned
Cadmus M. Wilcox	Fled to Mexico; returned
Governors:	
Edward Clark (Texas)	Fled to Mexico; returned
Isham G. Harris (Tenn.)	Fled to Mexico and England; returned
Thomas O. Moore (La.)	Fled to Mexico; returned
Charles S. Morehead (Ky.)	Fled to Mexico
Pendleton Murrah (Texas)	Fled to Mexico; returned
Thomas C. Reynolds (Mo.)	Fled to Mexico; returned in 1868

*Based on biographical information listed in Warner **Generals in Gray**, Boatner **The Civil War Dictionary**, and Davis **Confederate Exiles** American History Illustrated June 1970.

CONCURRENT EVENTS

FRANCE

On the 14th inst. the Emperor reviewed the troops and the National Guard at Paris.

It was the first time the guard had been reviewed by the Emperor since it was reconstituted after the **coup d'etat.** The spectacle was a magnificent one, and the crowd of spectators was immense.

Several nominations to the Legion of Honor took place.

(New-York Daily Tribune, Thursday, August 28, 1862.)

Another Attempt on the Life of the Marquis Wielopolski.

Warsaw, Aug. 16, 1862.

Yesterday evening, as the Marquis Wielopolski was driving out in an open carriage, a lithographic printer, named Jean Rjoutsa, made a desperate attempt to stab him with a poignard. The weapon, however, only struck the lining of the carriage. It is said the blade was poisoned.

Zarozinski, the man who attempted to assassinate the Grand Duke Constantine, has been condemned to be shot. The sentence has been submitted to the Emperor for approval.

(New-York Daily Tribune, Thursday, August 28, 1862.)

SYRIA

It was reported that a conference would be held at Paris in relation to Syria.

The Porte had notified the Western Powers of the conclusion of peace between the Druses and Maronites. He promised, nevertheless, to act with the utmost rigor against the authors of the massacres.

It was reported that the Porte had protested against French intervention, and that the expedition was stopped.

(The New York Herald, Monday, August 6, 1860.)

ENGLAND

THE QUEEN ON NEUTRALITY.

The Queen, in a speech proroguing Parliament, says: "Civil war, which for some time has been raging in America, has unfortunately continued in unabated intensity, and the evils with which it has been attended have not been confined to the American continent: but her Majesty, having from the outset determined to take no part in the contest, has seen no reason to depart from the neutrality which she has steadily adhered to."

(Harper's Weekly., August 30, 1862.)

"A few hours before his death, in a long interview with his trusted and honored friend, Schuyler Colfax, he [Abraham Lincoln] stated that he wished to give the Rebel leaders an opportunity to leave the country and escape the vengeance which seemed to await them here."

(The Field, the Dungeon, and the Escape, 1865, p. 326.)

MEDICAL MANUAL

"Status of the American Medical Profession in 1860"

by
Robert Kamer, M.D. and
Samuel Jackson, M.D.

The art of medicine in America was just coming into its own as the country readied itself for Civil War. The established Medical colleges and universities, along with the new ones, created by the demand of an expanding population, were educating young physicians in the theory of disease and surgical intervention, pharmacology and anatomical structure. Microscopes revealed to the doctor the mystical world hidden to his naked eye, case histories and statistics were studied and "distinctive features," such as *chemical analyses, operative surgery and Practical Anatomy,*[1]—were demonstrated. Teaching hospitals offered "clinical lectures daily, except Sunday, on Medicine, Surgery and Obstetrics, for which ample material is furnished in the lying-in wards and general hospital under the same roof."[2] Other schools promised that "Especial Attention will be given to preparing students for the Medical Staff of the Army and Navy,"[3] and private practioners offered instruction in such varied specialties as: "Medicine and Surgery of the Eye and Ear,"[4] "Urinary Pathology" (where arrangements were made by which specimens of the most important urinary deposits occuring in the several Hospitals of this city can be obtained),[5] "Medical Jurisprudence"[6] and "Physiology and Microscopic Anatomy."[7] Fees for these medical courses ranged from $5.00 for the "introduction in urinary pathology"[8] to $10.00 for a course of lectures on Surgical Anatomy,[9] to $100 for a full session of medical instruction with matriculation, demonstrators' and graduation fees extra.[10]

Despite these advances in the fields of medical education in the eastern part of the country, the westward migration caused some schools and colleges to send forth "a great deal of crude material."[11] These doctors, responding to the urgent demands of a widely extended population, could not wait to become thoroughly educated, but had to rely on "an active mind and a strong constitution"[12] to accumulate the rest of their knowledge in the field. In the profession, however, as doctors discussed the future, optimism was high. They felt that, despite inefficiency in the present system, instruction had advanced and become better adapted "to the necessity of the times and the spirit of the age,"[13] and the large body of medical men in the United States would be a credit to any nation of the old world.

Medical practioners were becoming more aware of their responsibilities, and began speaking out against their peers who were less than scrupulous with the patients they treated. Of particular concern in 1860 was the practice of doctors, who were "in the habit of exclusively patronizing some particular apothecary."[14] The collusion arose when a physician sent a prescription to a druggist, then later received a percentage of the amount paid for it. "This percentage seems to vary from fifteen to fifty per centum of the receipts."[15] If the inducement, therefore, was to get as much money as possible into the hands of the apothecary, the doctor prescribed large doses and changed the remedy frequently. "Esprit De Corps," who wrote the charge against his contemporaries, pleaded with them to give up this vice, which he acknowledged was brought about by the fact that many patients did not pay for prescribing and that only by extortion could the doctor be compensated. His advice (for those patients "not proper objects for professional charity"),[16] was to refuse them attention. He concluded "To the Censors of Medical and pharmaceutical societies the subject commends itself, as one deserving their scrutiny,"[17] aware that the "trucking for lucre" was greater than the relief of human suffering."[18]

Pursuant to the advance of American medical education and ethics was the demand for professional journals. Prior to the Civil War, much of the literary contributions to this country were imported from Europe: "The Lancet," "The British and Foreign Medico-Chirurgical Review," and "Braithwaite and Ranking" were all extensively republished here. American periodicals were attempted, but "languished for want of pecuniary and literary support."[19] Looking for a uniquely American perspective, the practitioners of the medical sciences called for editors to forego the re-

Medical College of Georgia.

publishing of prominent foreign magazines (which cost them little or nothing to reprint), and encourage the build-up of native material. They also wished to see the physicians, surgeons and professors record their observations and write them into articles of "home literature." By patronizing them in well-conducted journals, this would do much toward developing the medical mind of the country. This, in turn, would aid the original thinkers and prove America the new leader in the medical field.

These physicians of 1860 little realized that in less than a year they and their collegues would be drawn into bloody civil war, some of them joining the ranks as common soldiers, but most as Army and Navy surgeons. The education they received in the classrooms of New York, Pennsylvania, Boston, Virginia, and Georgia would be quickly put to the test, and, perhaps more importantly, their moral courage and originality would be stressed to the utmost. Before sterilization, cross-contamination, viruses, blood transfusions or pharmacology were well understood, these men, and not infrequently women, cared for the wounded soldier, and relying on "an active mind and a strong constitution," saved thousands of lives that otherwise would have perished. In 1860, the feeling was optimistic for American medicine: five years later, these dedicated people had proven that American ingenuity and courage were indeed equal to any.

FOOTNOTES

[1] New York Medical College.

[2] Long Island College Hospital, Brooklyn, New York.

[3] Jefferson Medical College, Philadelphia, Pennsylvania.

[4] Given by Dr. Turnbull, Howard Hospital.

[5] Given by Dr. John W. Lodge, Philadelphia, Pennsylvania.

[6] New York Medical College.

[7] College of Physicians and Surgeons, New York.

[8] Given by Dr. Lodge, Philadelphia, Pennsylvania.

[9] Philadelphia School of Anatomy.

[10] Long Island College Hospital, Brooklyn, New York.

[11] "The Status of American Medical Literature" from *The Medical and Surgical Reporter*, (Philadelphia: 1860), p. 16.

[12] *Ibid*, p. 17.

[13] *Ibid*, p. 17.

[14] "Dishonorable Mercenary Arrangements Between Physicians and Apothecaries," letter written by "Esprit de Corps," from *The Medical and Surgical Reporter*, (Philadelphia: 1860), p. 19.

[15] *Ibid*, p. 19.

[16] *Ibid*, p. 20.

[17] *Ibid*, p. 20.

[18] *Ibid*, p. 20.

[19] "Status," p. 17.

WHIMSEY

SHOWER OF BRIMSTONE.— At and about New Brighton, Beaver County, on Saturday morning last, between seven and eight o'clock, they had a shower of rain, which brought with it considerable quantities of flour of sulphur of brimstone. We learn it extended beyond Beaver; how much further we do not know. That it was genuine sulphur was indicated by its odor, and the readiness with which it ignited. The source of this phenomenon is a queston for the enquiring. Did it come from the Copper Works of Pittsburgh, the volcanoes of the Sandwich Islands, or from a place that is sometimes delicately hinted at by Theologians?

(*The Saturday Visiter*, Pittsburgh, April 2, 1853.)

Charles Incledon was proverbial for the coolness with which he regarded the turbulence of an audience. He always listened to the "storm" with the utmost nonchalance and occasionally addressed the noisy tenants of Olympus. One evening he was prevented from singing by a dire conflict in the most classical part of the house; and after pacing the stage for some time, "nursing his wrath to keep it warm," he pulled his watch deliberately from his fob, and thus addressed them: 'Ladies and gentlemen, if you would contrive to finish this row in a quarter of an hour, I would esteem it as a particular favor. I'm engaged to sup with a friend at half past eleven, and I have very little time to spare." This good-natured rebuke had the desired effect, and the belligerent parties "grounded arms" immediately.

(*Harper's Weekly*, August 30, 1862.)

The law is a pretty bird, and has charming wings, 'Twould be quite a bird of Paradise if it didn't carry such a terrible bill.

(*Harper's Weekly*, August 30, 1862.)

MILITARY REVIEW

Denmark Vesey's Revolt by John Lofton. 294 pp., index. The Kent State University Press, Kent, Ohio, 44242. 1983.

When did the Civil War begin? With the election of Abraham Lincoln? The secession of South Carolina? The firing on Fort Sumter? In *Denmark Vesey's Revolt*, author John Lofton argues that it was the abortive slave uprising of 1822, in Charleston, South Carolina, that set the stage for the eventual disruption of the Union.

Denmark Vesey was a black youth of African or West Indian ancestry, who was living on the island of St. Thomas when he was sold to Captain Joseph Vesey in 1781. Vesey, in turn, sold him to the plantation owners of St. Domingue, who returned him three months later as being "unsound and subject to epileptic fits." He became the seaman's personal slave (thus the name), and spent two years on his chattel ship, before settling in Charleston, South Carolina, with his master.

By a stroke of providence, Denmark won $1,500 in a lottery and purchased his freedom, in the year 1799, when he was about 33 years old. He learned the carpentry trade, and supported himself so successfully that at one point his value was estimated at over $8,000.

Influenced by the American Revolution, which had promised freedom for *all* men, the spirit of the French Revolution and the remarkable achievements of the blacks on Haiti, who had won their freedom from slavery by the sword, Denmark Vesey (who, as a free Negro, enjoyed few of the rights of the white population) plotted emancipation for South Carolina's slaves. By the turn of the century, he had learned that abolitionists in the Federal government meant well, but seemed powerless to aid the suffering blacks; armed resistance, in the manner of a patriot fighting for liberty, seemed the only answer. Gathering around him trusted aides from the abundant slave population, the plan for July 14, 1822, was to take Charleston by surprise, arm the loyal blacks from the gunshops and arsenal, and "leave not one white skin alive," sparing neither women, children nor ministers. It was a desperate, daring, thoroughly well-prepared idea. Slaves would be brought from the plantations, and together with those in the city, would fight for freedom. Timetables and assignments were prepared; but it was not to be. Word of the intended "servile insurrection" was passed on to the city authorities by slave owners who had been forewarned by their own slaves. Amidst near panic, the governor and the local officials alerted the militia, called for help from Washington and began arresting any Negros whose names came to be associated with the intended violence. Slowly, the network of Denmark's soldiers collapsed, and he was captured and identified as the leader of the plot. Not a single white man's blood had been shed, but the state was thrown into a panic from which it would nor recover until the end of the Civil War. Denmark Vesey and his subordinates were hung for their crimes, and South Carolina initiated a series of laws against the Negro that would span the next forty years.

John Lofton argues convincingly that the fear instilled in the white population after Vesey's revolt separated them from the rest of the country. Paranoid that another black man would raise the slaves, the influential plantation owners saw that strict anti-Negro legislation was passed. The Negro Seaman Act of 1822, calling for free black sailors of any nation to be jailed while their ships were in the harbor, set South Carolina directly against the seafaring nations of the world, and the laws of the Federal government. Its tariff nullification efforts were directly related to the slavery issue, and as early as 1826, a Senator from South Carolina declared: "The very day the unhallowed attempt is made [to interfere with our domestic concerns] by the authorities of the federal government, we will consider ourselves as driven from the Union."

In 1964, when this book was first published, the author wrote in his preface that the technique of his work was to examine the life of one slave and prove that he and the black community he lived in helped shape the destiny of their race. By 1983, in the preface to the paperback edition, John Lofton's thoughts were more well-defined. He explained that his objectives were to interweave Captain Vesey's life and experiences into that of Denmark's to examine the effects of culture on his development, to describe the insurrection and its place in history and to follow-up on its aftermath, which led ultimately to the Civil War. Mr. Lofton has done all of this, and more. Skillfully piecing together Denmark Vesey's life from meager evidence, he presents the story of insurrection *in context with its times*. He has made Charleston at the turn of the 18th Century come alive, much in the fashion of a good storyteller. Throughout the text, he has supplemented the passage of years with letters and newspapers of the day, while including statistics on population trends, politics and economy. To tell the tale of Denmark Vesey to a modern audience, this background is essential. It raises a man and his motives from preconceived ideas, and allows him, as well as the individuals who comprised that society, to be judged impartially by the reader.

Denmark Vesey's Revolt is a psychological study of the 18th and 19th centuries; it is unfortunate that in the 1983 preface we must face that same critical look at our own times. In 1976, when a portrait of Denmark Vesey was displayed in Charleston, there were many who were outraged. To them, Vesey was a murderer and a racist, akin to Hitler, Attila the Hun and Herod. He was a man to be deprived of his rightful place in history. For blacks honoring him in Charleston, he was a hero, a liberator sent by God. Mayor Riley referred to Vesey as a "man who was ahead of his time." After reading *Denmark Vesey's Revolt*, one's impression is more that he was a man *of his times*. If history had judged him harshly, and consigned him to an insignificant place, perhaps the reason is, as the character of Captain Henry Wirz said in the 1960 play, "The Andersonville Trial:" "You are all the victors here, and you make up the morality for the losers."

S.L. Kotar

Concurrent Events

NO OYSTERS.—For some reason, accounted for either by the presence of the Yankees in oyster waters, or the presence of the holidays in Richmond, oysters have entirely disappeared from the market and the saloons, and none have been obtained since Saturday last.

Some dealers account for the scarcity from the fact that the oyster dredgers have dropped the "tongs," and gone off to enjoy the holidays. As a consequence, no adequate supply need by looked for in the Richmond market until after the setting in of the new year.
(Daily Richmond Examiner, Wednesday Morning, December 31, 1862.)

Articles of War

An Act For Establishing Rules and Articles For The Government Of The Armies Of The United States.

ART. 58. All public stores taken in the enemy's camp, towns, forts, or magazines, whether of artillery, ammunition, clothing, forage or provisions, shall be secured for the service of the United States; for the neglect of which the commanding officer is to be answerable.

ART. 59. If any commander of any garrison, fortress, or post shall be compelled, by the officers and soldiers under his command, to give up to the enemy, or to abandon it, the commissioned officers, non-commissioned officers, or soldiers who shall be convicted of having so offended, shall suffer death, or such other punishment as shall be inflicted upon them by the sentence of a court-martial.

ART. 60. All sutlers and retainers to the camp, and all persons whatsoever, serving with the armies of the United States in the field, though not enlisted soldiers, are to be subject to orders, according to the rules and discipline of war.

ART. 61. Officers having brevets or commissions of a prior date to those of the regiment in which they serve, may take place in courts-martial and on detachments, when composed of different corps, according to the ranks given them in their brevets or dates of their former commissions; but in the regiment, troop, or company to which such officers belong, they shall do duty and take rank both in courts-martial and on detachments which shall be composed of their own corps, according to the commissions by which they are mustered in the said corps.

ART. 62. If, upon marches, guards, or in quarters, different corps of the army shall happen to join, or do duty together, the officer highest in rank of the line of the army, marine corps, or militia, by commission, there on duty or in quarters, shall command the whole, and give orders for what is needful to the service, unless otherwise specially directed by the President of the United States, according to the nature of the case.

ART. 63. The functions of the engineers being generally confined in the most elevated branch of military science, they are not to assume, nor are they subject to be ordered on any duty beyond the line of their immediate profession, except by the special order of the President of the United States; but they are to receive every mark of respect in which their rank in the army may entitle them respectively, and are liable to be transferred, at the discretion of the President, from one corps to another, regard being paid to rank.

GENERAL STORE

BROWN'S METAL TOP LAMP CHIMNEY.

AN INVENTION OF RARE MERIT!

Long Wanted, and Found at Last.

A KEROSENE LAMP CHIMNEY,

EASILY CLEANED AND WHICH WILL NOT CRACK OR BREAK FROM SUDDEN CHANGES OF TEMPERATURE.

The CHIMNEY in common use breaks from unequal expansion and contraction, caused by the upper portion becoming much more highly heated than the lower portion. This constant liability of the Chimney to be broken by being suddenly cooled or "over-heated," is entirely removed by a unique and beautiful invention—BROWN'S METAL TOP LAMP CHIMNEY, which, being made in accordance with the simple law of expansion and contraction, is found to be almost "fire-proof," as far as heating by the flame of the lamp is concerned.

The clear, positive superiorities of the NEW LAMP CHIMNEY over the old style are

1. It will not crack or break from sudden changes of temperature, or "over-heating."

2. The top can be instantly removed and the glass wiped perfectly clean in a moment *without wetting*.

3. It is shorter than the common chimney, therefore less liable to accidents and more convenient for carrying about the house; it is of elegant proportions and a decided ornament to the lamp, which will be found to burn with a clear, steady and beautiful light.

It is really the ideal of perfection in a lamp chimney, and after using one for an evening you will wonder that you could so long tolerate the old expensive style.

A "BROWN'S METAL TOP LAMP CHIMNEY" IS WARRANTED, *with ordinary care, to last just as long as the lamp on which it is used.*

Price (with a small neat shade for common size), Zinc top, 25 cents; Brass, 31 cents; Silver-Plated, 50 cents.

Dealers will be furnished, per Express, with a *sample box* containing one half doz. chimneys, assorted styles and sizes, with circulars, colored show cards, price list, etc., on receipt of $1 25.

BROWN'S METAL TOP LAMP CHIMNEYS are manufactured only by the

NEW LAMP CHIMNEY MANUFACTURING CO.,
MILTON BRADLEY, Treasurer.

HOLIDAY

AND BIRTHDAY GIFTS FOR CHILDREN.

The most useful and beneficial are the Patent Spring Rocking Horses, invented by Jesse A. Crandall. They are not dangerous, like the common rocker, and can not be easily upset. They expand the chest, and learn the rider to sit correctly.

Send stamp inclosed for illustrated circular and price list. Also, a large assortment of Sleighs, Propellers, Drums, etc., etc.

000 L. TIBBALS, No. 480 Broadway, New York.

DR. MUNROE,

SURGEON DENTIST,

OF RICHMOND, VA.

IS NOW AT THE EAGLE HOTEL.

DR. M. will insert *Mineral, Porcelain* and *Animal Teeth*, upon the most modern improved scientific principles. ALSO, Clean, File, Separate and Extract, as well as regulate Children's, and remedy defects in Adults Teeth, or every variety of operation, to make healthy, preserve and beautify the Teeth.

July 8—3t

MOWER,
AND
COMBINED REAPER AND MOWER.

Manny's Patent,

WITH WOOD'S IMPROVEMENT.

THESE Machines can be had, for the Western part of New-York State, (except the counties of Alleghany, Cattaraugus, Chautanque and Erie,)—the States of Virginia, Maryland—for District of Columbia and Western Vermont—of the proprietor at Hoosick Falls, N. Y.,

WALTER A. WOOD,
Manufacturer and Proprietor.

The above Machines are warranted capable of cutting from 10 to 15 acres of Grass or Grain per day, as well as is done with scythe or cradle. Price of Combined Machine, with 2 Scythes and extras, delivered on cars at Hoosick Falls...$125

Mower, with 2 Scythes and extras,........... 110

April 19—wtf

JAMES R. JOHNSON

OFFERS FOR SALE,

At his Jewelry Store on Main Street, 3 doors south of the corner leading to Mayo's Bridge,

AN EXTENSIVE ASSORTMENT OF

Rich Jewelry and Fancy Goods,

New Book by Henry Ward Beecher.

EYES AND EARS.

ONE HANDSOME VOLUME. 12MO. PRICE $1 25.

(Copies in Fine Cloth Extra Gilt, with Gilt edges, $1 75.)

A new volume by Mr. Beecher "*to inspire a love of Nature, an enjoyment of rural occupations, to form a kindly habit of judging men and events, to enliven the tedium of sickness, to while away an hour with innocent amusement,*" will be sure to attract the attention of every household.

Sent *post paid* to any address on receipt of price by the publishers, TICKNOR & FIELDS,
135 Washington St., Boston.

PENSIONS, BOUNTIES, BACK PAY, PRIZE MONEY, &c, procured by SOMES, BROWN, & THRUSH, 2 Park Place, N. Y.

N.B. Enclose stamp for our book.

G. WESTINGHOUSE & CO.

CONTINUE the manufacture of Threshing Machines Clover Cleaners, Wood Saws, &c., at Central Bridge Schoharie Co., N. Y.

We have improved our Thresher and Cleaner, (and for which we have obtained a Patent last year,) which works superior to anything of the kind in use, and has given entire satisfaction where used.

Our Horse-Power, Thresher and Separator, has the name of being the best machine in use, where known. Those wanting machines will be more likely to get them when wanted by ordering them early, as we shall not be able to make more than 100 of them this season. Last year we did not supply the demand by a large number, being unable to get them out in time.

Further information given on application by mail otherwise.

G. WESTINGHOUSE & CO.
Central Bridge, N. Y.

May 3—w22,24,26,28,30 32—m3t

Grover & Baker's

CELEBRATED NOISELESS

SEWING MACHINES

For Family and Manufacturing Use,

495 BROADWAY, NEW YORK.

The GROVER & BAKER S. M. CO. now offer, in addition to their splendid stock of Machines making the celebrated Grover & Baker stitch, new and superior Shuttle or "Lock-Stitch" Machines, adapted to all varieties of sewing. Much more Simple, Durable, Noiseless and Perfect than any "lock-stitch" machines heretofore in use. 3380

Wesson's Breech-Loading Rifle.

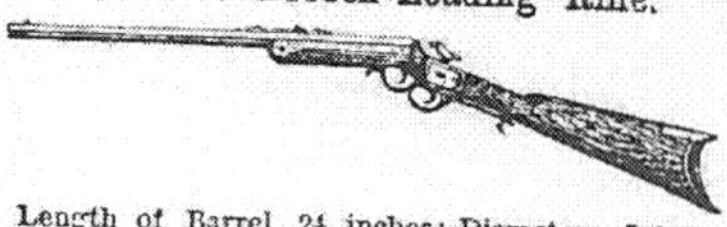

Length of Barrel, 24 inches; Diameter of Bore, 32-100 of an inch; weight only 6 pounds.

This is the best Rifle yet invented, its great superiority consisting of rapidity and facility of Loading, and being used with a metallic Cartridge, there is no escape at the breech, and cannot possibly be loaded improperly; shoots with perfect accuracy; can be used all day without cleaning, and will not heat with the most rapid firing.

J. W. STORRS,
Sole Agent, 256 Broadway.

Also Agent for Smith & Wesson's Revolvers and Cartridges.

313-38o

MACKENZIE'S

PATENT CANTERING HORSES—The most delightful out-door exercise for Boys' and Girls that could be desired. Run rapidly over the ground by the weight of the rider. Send for an illustrated circular.

377-9 S. WM. SMITH, No. 498 Broadway.

Capturing the flag.

THE KEPI
P. O. Box 4542
North Hollywood, California 91607

BULK MAIL

THE KEPI

FEBRUARY-MARCH 1984 $3.50

"A New Concept in Civil War Reporting"

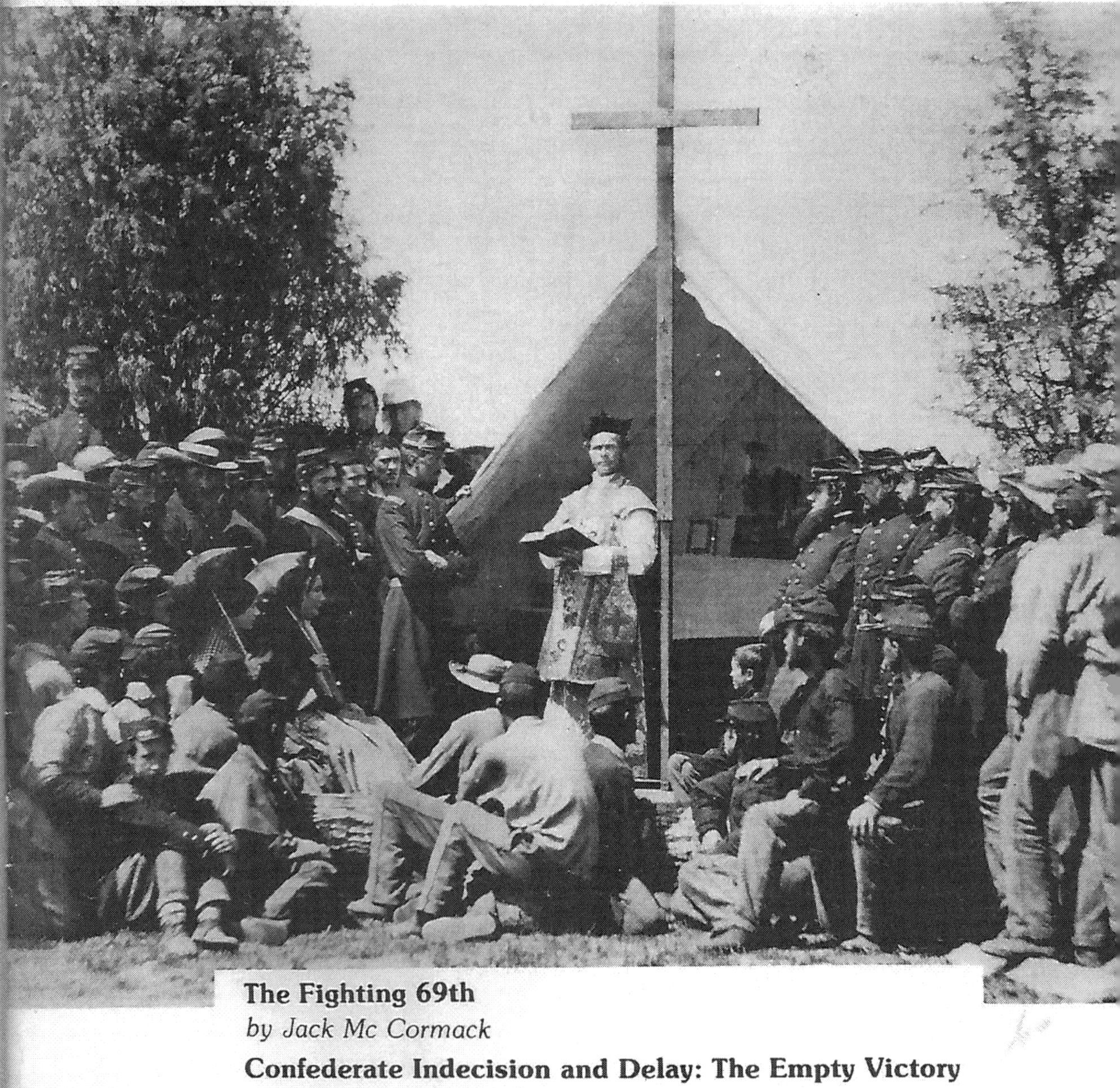

THE KEPI

"A New Concept in Civil War Reporting"

FEBRUARY-MARCH 1984 **VOLUME II, NUMBER 1**

THE KEPI is published bi-monthly for $3.50 per issue, or $18.00 per year. Canadian single copy, $4.00; subscription, $21.00 per year. United Kingdom single copy, $4.50; subscription, $28.00 (Air Mail) per year. Correspondence to the editors may be directed to: **THE KEPI**, P.O. Box 4542, North Hollywood, CA. 91607. Unless expressly stated, **THE KEPI** neither endorses nor is responsible for the content of advertisements appearing in its pages. **THE KEPI** requests authors to query with specific topics before submitting articles, and assumes no responsibility for unsolicited manuscripts not accompanied by return postage. Work not fully documented and footnoted where applicable will not be considered.

All inquiries should be directed to the above address. **THE KEPI** reserves the right to edit and copyright all letters. Please allow 6 weeks for delivery of first issue.

ILLUSTRATIONS

Cover shows Fr. Thomas Mooney offering Sunday Mass to the 69th NYSM; Col. Corcoran stands with arms folded. Courtesy Library of Congress.

Inside front cover is an engraving of Brig.-Gen'l. Michael Corcoran from *Harper's Weekly*, August 30, 1862.

Photographs from "Indecision and Delay" courtesy Library of Congress.

Photographs from "Black Day at Buena Vista" courtesy of the author.

Cartoons on pages 27, 33 and back cover from "Yankee Notions," February, 1859.

THE REGULARS

THE IRREGULARS

GENERALLY SPEAKING

The first issue of our second year features an article by Jack Mc Cormack, entitled "The Fighting 69th.' This New York infantry regiment, comprised primarily of Irish immigrants, responded to President Lincoln's call for volunteers in April, 1861, and fought from Bull Run to the final days of 1865. Nearly decimated after four years of war, this brave group of men earned their motto, "Never Retreat From The Clash Of Spears." Mr. Mc Cormack is Professor of History at Delaware County Community College in Medina, Pennsylvania, and President of the Pennsylvania State Education Association. He has written articles for Civil War Times Illustrated, British History Illustrated *(now* British Heritage) *and* Mankind. *He is a regular contributor to* The Irish Edition *newspaper, given tours and conferences at the Gettysburg Battlefield, and has served as editor of the Little Big Horn Associates'* Research Review.

Vernon H. Crow has written our second feature article, "Confederate Indecision and Delay: The Empty Victory at Chickamauga." After the loss of Vicksburg and the defeat at Gettysburg in early July, 1863, the Confederate government faced the dire problem of being divided into thirds with the impending loss of Chattanooga. A two week conference was held in Richmond in late August and early September by the Confederate high command to determine what course of action was to be taken. This fourteen day period, characterized by misinterpretation of events, lack of communication and an inability to act promptly, led to the fruitless victory at Chickamauga. While historians have blamed General Bragg for the situation in Tennessee, Mr. Crow presents a look at the failures in Richmond that directly led to the Confederate losses in the center.

Mr. Crow has written several articles for Civil War Times Illustrated, *and is the author of "Storm in the Mountains," a thoroughly documented account from MCI Press, describing William H. Thomas' Legion of Cherokee Indians and white mountaineers from Tennessee and North Carolina.*

Accompanying this article is TIMELINE, a new feature designed to give KEPI readers a chronological background and guideline for selected articles, and REPURCUSSIONS, which adds a critical look at Longstreet's attempts to transfer west, and the delays encountered at the August 24—September 7 meetings in Richmond.

Thomas J. Evans' narrative, "Black Day at Buena Vista" recounts the adventures of several Indiana men who volunteered for service in the Mexican War, and later joined the ranks of the Union and the Confederate armies. These citizen soldiers, who had fought side by side in the late 1840's were separated by political differences in the early 1860's but their friendships were to transcend these differences as their paths crossed during the Civil War. The story of these messmates is a valuable look at the importance of integrating the times prior to and during the Civil War, for a complete understanding of the times.

Another new feature in this issue is Paul Kallina's FAMILY BIBLE. This serial will provide a month-by-month list of the birthdays of the notable men and women of the Civil War. It is fascinating to note how young the great leaders were, or whose birthdays fell close together, and is an example of the human ties that bind us all. Also by Mr. Kallina is a short work, entitled, "Memorandum of Books Sent to Winder's Building," listing engineering texts in 1862. Paul Kallina is a free lance researcher, specializing in Civil War records in the National Archives. He discovered the memorandum while researching for the museum at Fort Ward, a Civil War site in Alexandria, Virginia.

Rounding out the issue is LEXICON, the Articles of War and some Valentine's Day humor from 1859. We wish you all a happy St. Patrick's Day, as well, and add a birthday remembrance for Abraham Lincoln, a man whose faith and courage will always be cherished.

S.L. Kotar *J.E. Gessler*
Editors-in-General

Head-Quarters of THE KEPI
1st February, 1984

GENERAL ORDERS
No. 5
by Command of the Editors-in-General

Coming in the April-May issue of THE KEPI is an article on one of the tragedies of the war. In the southern Appalachian Mountains of North Carolina, sectionalism ran high. While some of the men went to war, many stayed behind or deserted the Army to become guerilla partisans. Hatreds based on divisions of wealth, kinship and politics often caused these ruthless men to kill and rob innocent civilians, and in January, 1863 when Confederate troops were sent to punish these men, thirteen were illegally executed. "Massacre in Shelton Laurel" explores the incident and the men involved in it.

Also featured is "The Affair at Fairfax Court-House," the exciting story of Ranger John S. Mosby's capture of Union general Ed Stoughton, containing information never before published.

19th Century life is explored in TRACTS with an article on Sabbath school teachers, THE GENTLEMAN AGRONOMIST is presented, and the Articles of War continue.

"The Fighting 69th"

by
Jack Mc Cormack

Riamh Nar dhruid O sbairn lann proclaimed the new green flag presented to the "First Regiment, Irish Brigade," better known as the 69th Regiment, New York Volunteer Infantry. The ancient Gaelic words echoed a hope for the nation on that November day of 1861: *Never Retreat from the Clash of Spears.* – Only once in its history was the 69th forced to deny its motto. In Burnside's mad frontal assault upon the impregnable Confederate position on Marye's Heights, Fredericksburg, Virginia, December, 1862, the 69th fell back, having come closer to the formidable stone wall than any other Union unit.[2] The regiment had been staggered by 53.7% casualties.[3]

The actual beginnings of the regiment go back to 1851, when New York State reorganized its militia. Strangely enough, the 69th began its career as the 9th New York State Militia in what is perhaps best described as the infiltration of that regiment by Irish rebels, who hoped to use it as a training mechanism for the eventual war to drive the English from their homeland. Although the 69th was formed in 1851, its character was more solidified in 1858, when most of the remaining Irish elements of the 9th were transferred to it.[4] In that same year, the Fenian Brotherhood was formed in the United States for the purpose of gaining Irish independence. It numbered members of the 69th in its ranks, including Colonel Michael Corcoran.[5] In the so-called Quarantine War of 1858, the 69th did duty at the quarantine station on Staten Island when the good burghers of that island decided that they didn't want the hospital to grace their pleasant shores. In 1859, following militia tradition, it elected as its new colonel, Michael Corcoran, proprietor of Hibernian Hall, a local New York City pub. Lest his current occupation be misunderstood, he was the son of a retired career British Army officer, and had served in the Irish police before emigrating to America. He worked his way through the ranks of the regiment and was considered an excellent officer.[6] That is, he was until the affair concerning "Baron Renfrew."

It seems that the gentleman in question was Queen Victoria's son and heir, Edward, Prince of Wales. New York society was intent upon honoring him and the New York City militia regiments were ordered to parade before him. Corcoran and his Irish republicans refused to obey. The colonel was relieved of his command, and measures were taken to court-martial him and disband his regiment for "insulting" the prince. Fate intervened, and South Carolina seceded from the Union in December, 1860. She was followed by others. Suddenly, the country needed Corcoran and his Irishmen. The charges were dropped, and the regiment girded for war.[7]

The 69th responded to President Lincoln's call for three months volunteers and left New York amidst great ceremony on 23 April. Mrs. Maria Lydig Daly, wife of prominent Irish-American jurist, Judge Charles P. Daly, presented the regiment with its national color. At first, the regiment guarded a railroad line in the vicinity of Annapolis, Maryland, for about a week, then left to help defend the nation's capital, which was virtually surrounded by Confederate sympathizers. While in the District of Columbia, it occupied the buildings of Georgetown College. It was here that it received reinforcements led by a man whose name would be permanently interwoven with the destiny of the regiment—Captain Thomas Francis Meagher. Shortly after Meagher's zouave company joined it, Corcoran's regiment occupied Arlington Heights across the Aqueduct Bridge over the Potomac River.[9]

At this spot took place the first official flag-raising over Union earthworks in the "Old Dominion" at Fort Corcoran on 30 May. The honor went to Captain James Cavanaugh of C Company of the 69th. For the next six weeks, the usual boring routine of the soldier prevailed as the regiment drilled itself into shape. In June, the 69th was attached to the brigade commanded by a relatively unknown officer by the name of William Tecumseh Sherman. In due time, it was marching toward its destiny along the banks of a small Virginia creek called Bull Run.

At about noon, 21 July 1861, Sherman's Brigade moved across Bull Run to the attack with the 69th NYSM in the lead. After a severe struggle, outnumbered by their foes, Sherman ordered a withdrawal. Colonel Corcoran formed an irregular hollow square to guard against increasing Rebel cavalry activity.[10] In the disorderly retreat of the Union Army, Corcoran was captured when his horse was killed under him.[11] His regiment marched back to Fort Corcoran in better order than other Union units after the rout at Bull Run. It had lost more men killed and wounded than any other element of Sherman's Brigade.[12] Subsequently, Corcoran was held as a hostage for William Smith, a Confederate privateersman from the brig, *Jefferson Davis*. The United States authorities were trying Smith on charges of piracy, a hanging offense. Fortunately for all concerned, the Southern sailor was judged to be a prisoner of war.[13] In due time, Corcoran was exchanged and formed the Corcoran Legion, a brigade of New York Irishmen.[14]

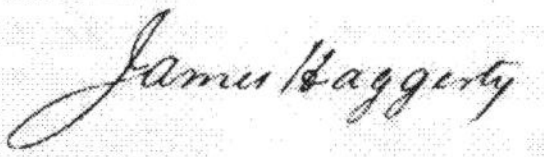

Lt. -Col. James Haggarty, 69th NYSM, who was killed at First Bull Run. Courtesy, USAMHI.

The 69th NYSM was mustered out of service shortly after Bull Run, its time having expired. Captain Meagher promptly received permission from the Secretary of War to raise a brigade of Irishmen for the Union cause. Meagher ended up as the commander of what became one of the Union Army's most famous units, the Irish Brigade. The newly formed 69th New York Volunteer Infantry was the first regiment of the brigade, and it came from a nucleus of the old regiment which had fought at Bull Run. Its colonel was Robert Nugent. On 18 November, the patriotic ladies of New York presented the regiment with its colors: the Stars and Stripes of America and the previously mentioned green standard of Erin. On the same day, the 69th was blessed by Archbishop John Hughes of New York, and left for Virginia.[15]

Life was pleasant at Camp California, but Christmas away from home for the first time was lonely. Fiddle and bagpipe music helped to restore flagging spirits. Homesick soldiers sang "The Girl I Left Behind Me" and "Home, Sweet Home," as well as patriotic Irish airs. Soon, the bell of the chapel tent called the men to Midnight Mass.

One of the celebrants of this divine service was the colorful chaplain of the 69th, Father Thomas Ouellet, S. J. of St. John's College (now Fordham University). He was a small French-Canadian with a volatile Gallic temper. Woe to the soldier who ignored his religious duties when the swarthy Jesuit was around. He once upbraided those who didn't attend Mass after the wearying retreat from the "Seven Days" fighting by characterizing them as "coffee coolers" and "skeddadlers." The men of the 69th always claimed that their chaplain had originated these army terms. Yet, none complained when Ouellet would be discovered on many a battlefield, prowling around at night with a lantern seeking to console the wounded and dying of both armies.[16]

On 13 March 1862, Meagher's Brigade was assigned to the First (Richardson's) Division of the Second (Sumner's) Corps of the Army of the Potomac.[17] The Peninsula Campaign opened shortly after this. The 69th was shipped to the James River Peninsula aboard the *Ocean Queen*, which, ironically, may have been one of the ships which brought survivors of the horrendous Irish famine of the mid-1840s to the States.[18] Certainly, the description of the ship given by 88th New York Infantry chaplain, Fr. William Corby,[19] would fit one of those "coffin ships." He reported the troop transport as having "plenty of ocean but not much queen!"[20] It was during this campaign that the Irish Brigade decided to enliven the usual army *ennui* by sponsoring a "Day at the Races." There were two horse races for the officers, followed by what must surely have been the *piece de resistance*, a mule race ridden by the drummer boys. The enlisted men were scheduled to have a Gaelic football (similar to soccer) game. Following the day's activities, an evening of theatricals was planned. Alas, before all could be accomplished, the soldiers were to be actors in a more deadly drama: the Battle of Fair Oaks.[21]

The regiment also took an active part in the fighting at Gaines' Mill, Savage Station, Glendale and Malvern Hill. At Gaines' Mill, the Irish soldiers came to the aid of the hard-pressed Fitz John Porter. Captain Felix Duffy's company did yeoman work in halting, at the points of their bayonets, a panic among Porter's teamsters and stragglers. The rest of the regiment helped to stiffen the resistance of Porter's troops which stopped the pursuing Confederates. At Savage Station, the combative command of Nugent participated in the storming, capture and spiking of two artillery pieces of a Virginia battery. The next day, at Glendale, the regiment had the very trying experience of supporting Hazard's Union Battery. This unpleasant duty consisted of lying near the "redlegs" as they worked their guns, fervently praying that enemy counter-battery fire would not hit themselves.[22]

The next day, 1 July 1862, the 69th participated in the slaughter at Malvern Hill, where the Union Army turned and viciously snarled back at their Confederate attackers. Nugent led the 69th in the van of the Irish Brigade's fierce counterattack upon the advancing Confederates. As the brigade came up, General Daniel Butterfield seized the beloved green flag and led the regiment toward the rapidly advancing Rebel lines. The charging Irish poured a galling oblique fire into the attacking gray legions, stemming their advance.[23] The price paid was a heavy one, 155 casualties.[24] Among the dead was Sergeant Haggarty, whose brother, Lt. Col. Haggarty was killed at First Bull Run with Corcoran's 69th NYSM.[25] Private Peter Rafferty, B Co., won the Congressional Medal of Honor at Malvern Hill. He had been wounded, refused an order to go to the rear and was wounded again. He was finally captured and disabled for life.[26]

Another sterling example of the spirit of the regiment was provided by Lieutenant John H. Donovan of D Co. Whilst charging the enemy, he was shot through the right eye, and left for dead. The next morning, Confederate Generals Hill and Magruder visited the captured Union officers, asking for their sidearms. Lt. Donovan replied that he had given them to his servant to return to his regiment for him. Said Hill: "I think from the apparent nature of your wound, you won't have much need of them in the future." Donovan replied, "I think differently...I have one good eye left yet and I will risk that in the cause of the Union. Should I ever lose that, I'll go it blind!" The Confederate officers were duly impressed with the *esprit* of the "Fighting Irish."[27]

While brigade commander Meagher headed north to recruit, the 69th spent several weeks recuperating at Harrison's Landing. It later moved to cover the retreat of Pope's defeated Federal Army from Second Bull Run. The respite proved to be all too brief. The contending armies were being inexorably drawn toward the most sanguine single day's battle of the war, soon to take place along the hitherto peaceful banks of the meandering Maryland creek called the Antietam.

Captain Jack Gosson of the 69th began the campaign in rather startling manner. On a fine September eve, he decided to sleep at the top of a haystack. Come morning, satisfied with his comfortable perch, he slid down the stack onto the chest of a fellow officer who exclaimed, "...you scoundrel; who the devil are you?" "And who the hell are you?" snapped Gosson, and he emphasized his point with a good swift kick to the other. The latter replied with a string of oaths. The Irish officer suddenly paled when the other man got up and turned to face him. It was Israel B. Richardson, his division commander! Fortunately, after profuse apologies, and probably some thick blarney, the captain was able to take the chill off the morning for both with some good whiskey from his canteen.[28]

At 9:30 A.M., 17 September 1862, the 69th forded the Antietam Creek under the command of Lt. Col. James Kelly. It acted as the right flank of Richardson's Division. As such, it charged over the crest of the hill near the Piper House. For 100 yards, it was exposed to a galling fire from the 2nd and 4th North Carolina Regiments posted in a sunken road, forever afterwards called "The Bloody Lane." General Meagher went down, knocked unconscious when his horse was killed under him. General Richardson was mortally wounded. The green flag of the regiment went down in the swirling *melee*. In fact, this time it was the eighth regimental color bearer to fall. Captain James Mc Gee raised the flag again. Within seconds the staff was shattered by still another Minie ball. As Mc Gee reached down to snatch the fallen banner once more, his kepi was shot off. Col. Kelly went down. The firing was so furious, men's muskets seared their hands. Finally, having exhausted their ammunition, Meagher's Brigade was relieved in excellent battlefield order by Caldwell's.[29]

The cost had been high. The 69th lost ten officers and 186 enlisted men killed and wounded.[30] Bloody Lane was filled with ninety-six casualties from the two gallant North Carolina regiments. Colonel Tew of the 2nd had been killed and every officer of the 4th was killed or wounded.[31] The 69th New York had earned the title that would forevermore follow it: "The Fighting Sixty-Ninth."

After Antietam, the 69th moved with the army to Bolivar Heights, above Harper's Ferry. It was fortunate that the men of the regiment could not foresee the future. Its greatest trial was only months away. One thing that they did know for certain was that they were not very happy with the decision to replace army commander, George B. Mc Clellan. In fact, several 69th officers handed their resignations to Gen. Meagher. He refused to accept them. It was not only politics; Mc Clellan was a Democrat, and so were the Irishmen. However, far more important than that — Mc Clellan never wasted the lives of his men. In fact, at their beloved general's leave-taking, the entire Irish Brigade broke ranks to swarm around their leader. The men of the 69th and the whole Second Corps of the Army of the Potomac would have cause to remember their new commander. He would be the one to send them against the Stone Wall at Fredericksburg.[32]

The 69th marched toward their destiny along the banks of the Rappahannock River in November. On their way, they were accosted by a haggard woman, screeching at them. It reminded the men of the wailing of the banshee, and many wondered for whom the evil spirit wailed. The Irish Brigade camped at Falmouth just across from Fredericksburg. While these events were transpiring, the battle-shredded flag of the regiment was sent north with Captain Mc Gee to be replaced with a new manner. The latter never arrived in time for the coming battle.[33]

On the morning of 12 December, the 69th crossed the Virginia river on a pontoon bridge. They were cheered on by the 14th Brooklyn, while the band of Hawkin's Zouaves (9th New York) struck up the 69th's anthem, "Garryowen." Not quite so festive was the presence of professional embalmers, who passed out their calling cards among the advancing soldiers. The men settled down in Fredericksburg to await further developments. The night was one of bitter and penetrating cold.[34] While the Irish Brigade waited to go into action on Saturday the 13th, Gen. Meagher had the men of his command place sprigs of green boxwood on their kepis. Out Hanover Street marched the proud brigade in column of fours, as Confederate shells from the Washington Artillery of

New Orleans rained about them. The Irishmen followed in support of Zook's Brigade. Next came the words that presaged the awful slaughter: "Irish Brigade, Advance. Forward, Double Quick. Guide Center."[35]

Obstructing fences were thrown down with Col. Nugent, himself, pulling off rails. Above the troops, half-hidden by the smoke of battle, staunchly stood the fearsome stone wall. Still further up Marye's Heights were the well-placed pieces of the Washington Artillery.[36] Crammed into the position behind the stone wall were the two Rebel infantry brigades of Cobb (Georgia) which contained a number of Irishmen and Kershaw (South Carolina). Meanwhile, the 69th passed its way through the melancholy remnants of the earlier attackers. One can only imagine the thoughts which must have gone through their minds as they passed the dead, dying and wounded of the first waves. Several times the Irishmen hit the impregnable gray lines, with fearful losses. Oh, that McClellan would have led the army that day! The "Fighting 69th" added to its reputation on that sad December day. Captain Saunders, its fourth commander of the day, finally brought it from the blood-stained field. The regiment had entered the fray with nineteen officers and 219 enlisted men. Sixteen of its officers and 112 of its men were casualties.[37] At least one soldier of the 69th retained his sense of humor. Confederate General James B. Longstreet found the Irishman lying about twenty feet from a Confederate cannon. "Are you hurt my lad," asked the solicitous enemy officer. His answer was unexpected: "It's the truth I am telling you, General, I have seen better days." Later that same evening the wounded Union soldier died.[38] The second regimental Medal of Honor winner emerged from the debacle at Fredericksburg. Private Timothy Donoghue of B Co. received it for rescuing a wounded officer despite his own wound.[39] Ironically, the cornfield over which the 69th had advanced belonged to Colonel Marye of the Confederate Army. It seems that during the dreadful famine years of Ireland, the entire corn crop of that same field had been sent to the Emerald Isle to relieve the distress there.[40] Now Irish blood nourished the field and repaid the debt.

Perhaps the highest tribute came from a man who would lead a similar forlorn charge within eight months time. In a letter to his wife. Confederate General George Pickett wrote, "The brilliant assault on Marye's Heights of their Irish Brigade was beyond description. Why, my darling, we forgot they were fighting us, and cheer after cheer at their fearlessness went up all along our lines."[41]

Fittingly, what really amounted to a gigantic Irish wake followed the Battle of Fredericksburg. It seems that a celebration had been planned earlier in honor of the arrival of the new green flags for the New York regiments (63rd, 69th and 88th) of the brigade. Meagher decided to hold the affair anyway, despite the brigade's heavy losses. The "Death Feast" was held on 15 December in a Fredericksburg theater. Twenty-three generals were present. The grief-stricken Meagher made an unfortunate (though understandable) speech in which he bitterly assailed political ger erals. His words were to doom his later efforts to get permission to recruit for his sadly depleted regiments. Indeed, the losses were so great that the officers of the three New York regiments refused to receive their new colors until they were strong enough to defend them.[42]

At last, however, the survivors could relax in winter quarters at Falmouth. In fact, they enjoyed a signal stroke of good fortune. The Second Corps did not take part in Burnside's other piece of stupidity, the "Mud March" of mid-January, 1863.[43] Moreover, the Irishmen were even able to get in their beloved St. Patrick's Day festivities.

As usual, the good saint's day began with a splendid Military Mass. The feature of the day was the "Grand Irish Brigade Steeple-chase." Over ten thousand soldiers arrived to watch the great horse race plus the other events of the day which included a weight-throwing competition, a half-mile run, a half-mile hurdles race, a greased pig contest, a 500-yard sack race and a wheel-barrow race. There were competitions for the best Irish step-dancers in jigs, reels and hornpipes. The evening saw theatricals and recitations. The day was a huge success with new army commander, "Fighting Joe Hooker," toasting his Celtic hosts.[44]

On 3 May, the 69th took a relatively minor part in the Chancellorsville Battle. After spending some time guarding United States Ford, the regiment moved to the field of action in support of the 5th Maine Battery, which was being cut to pieces by enemy shelling, and was now threatened by advancing Rebel infantry. Its commander, Captain Leppine, went down, as did Lieutenant Kirby. Soon, only one gun was working, served by a corporal and a private. Several men of the 69th aided the 116th Pennsylvania to pull off the battery's five guns. They then helped to cover the army's retreat.[45]

Just subsequent to the Chancellorsville campaign, the 69th was saddened by the resignation of brigade commander Meagher. The fiery officer had left in protest against the shabby treatment accorded his request (never acknowleged) to recruit the "poor vestige and relic" of the Irish Brigade to full strength.[46] There wasn't much time to fret, for Gettysburg was just around the corner.

The greatly reduced regiment was committed to battle on 2 July near the Wheatfield. It had been consolidated into a two-company battalion under Captain Richard Moroney. Before charging, all of the men of the Irish Brigade witnessed a most extraordinary scene: a benediction and general absolution. Before going into the fight, the men were briefly halted while Fr. Corby addressed the soldiers as they stood at "Order Arms." After exhorting all to do their duty, the chaplain gave a general absolution to those intently gathered before him as Corps Commander Winfield S. Hancock and his staff looked on. The 69th then plunged into the swirling fray. After capturing several Confederates, the regiment was forced to quickly withdraw. However, on the next day, they helped to repel the all-out Rebel attempt to break through the Union

Statue of "Fair Catch" Corby at Gettysburg. After the Civil War chaplain Corby served as President of Notre Dame University. Courtesy of the author.

had begun by the attackers as well as the defenders.[50]

Spottsylvania sapped the strength of the already seriously weakened regiment. The 69th attacked across 200 yards of clearing before they broke into the first line of Confederate troops. A second line was captured in desperate hand-to-hand combat. As Lt. King described it to his brother: (The 69th) "had to rush at them through a storm of grape and rifle-balls. Men dropped in scores... (Capt.) Blake led us through a shower of balls... He then took the regimental flag (the green one) which was the only color on that part of the line...and stood there waving it and calling us, 'Come on, boys, and I will show you how to fight.' ...till he was struck in the knee." Blake actually was mortally wounded. Shortly after King sent this letter home, he, too, received his death wound.[51] Besides the two officers, the regiment lost 122 others as Spottsylvania.[52]

The 69th lost another forty-one men in the struggle at Cold Harbor (well-named!) It then pitched headlong into the Petersburg campaign. Fighting was so severe that Capt. Moroney was transferred from the 69th to command the decimated 63rd New York. He eventually became the officer commanding the Irish Brigade replacing vice Colonel Kelly, who was killed in action. Forty-three more men were lost by the 69th in the siege of Petersburg. Then came the shock: the depleted Irish Brigade was merged with Carroll's, to form the Consolidated Brigade.[53]

The Irish regiment took part in an exhausting night march to Deep Bottom, where it helped assault the Confederate lines. It then returned to the Petersburg lines. In August, the 69th again successfully charged, at Deep Bottom. Here, the enmity between division commander Francis C. Barlow and the Irishmen worsened when he accused them of "breaking" in the engagement. Just as his staff had absorbed this criticism, up rode Adjutant Smith of the 69th, covered with sweat and battle grime, to report that the enemy's works had been taken by the brigade![54] The regiment was not so successful in the desperate combat along the Weldon

lines.[47]

Shortly after Gettysburg, the new, manifestly unfair, draft act was implemented in New York City. Since it permitted the purchase of a draft exemption upon payment of the then prohibitive sum of $300, or the buying of a substitute to take one's place, the urban poor resented the law. This discontent boiled over into severe rioting in the city. The duty of enforcing the unpopular legislation devolved upon the provost marshal of the city, Col. Robert Nugent of the 69th, still recuperating from a wound received at Fredericksburg. Col. Nugent's home was one of those burned in the riots. It took a week to quell the disorders. The soldiers of the 69th were at once angered, sympathetic and embarrassed by the fact that most of the rioters were Irish.

During the remainder of 1863, the regiment took part in the pursuit of Lee's Army of Northern Virginia. It was in on the Mine Run and Bristoe Station campaigns. Toward the close of the year, the regiment reenlisted practically to the man! It was among the first units of the Army of the Potomac to do so. Those who did so were granted veterans' bonuses and thirty days' leaves. The weary soldiers, headed by Capt. Moroney, arrived in New York on 2 January 1864, and were welcomed by a small detail of the 69th New York National Guard. They were saddened by the fact that their ill and very pro-Union archbishop, John Hughes, died during the first week of the month.[48] The general indifference of the war-weary citizens was more than made up for by the enthusiastic reception tended by their comrades in arms. Gen. Meagher welcomed the men in a great reception at Irving Hall on 16 January. Sergeant-Major O'Driscoll presided over the festivities. A fitting toast was proposed by Col. Brewster of the Excelsior (New York) Brigade: "The Irish Brigade; what there is left of it." Gen. Corcoran was also toasted; the old commander of the original 69th had died as the result of an unfortunate accident.[49]

During January and February, the regiment was finally recruited up to minimum strength. Col. Nugent resumed command. The new brigade leader was Brigadier General Thomas Smyth, a native of County Cork. On the evening of 3 May, the 406 officers and men of the 69th broke camp for the trying campaigns to come. At the Wilderness, 5-7 May, it lost 13% of its numbers. That the war had taken a new turn is evident from such official reports as "...threw up a line of breastworks..." and "...threw up entrenchments..." Incessant trench warfare

Colonel Robert Nugent, commander of the 69th New York Volunteer Infantry and later leader of the Irish Brigade. He was wounded at Fredericksburg and was Provost Marshal of New York during the draft riots. Courtesy, National Archives.

Railroad, near Reams Station, Va. After thrice repelling Rebel attacks on their lines, the 69th got caught in a nasty struggle when a fourth Southern effort dislodged a neighboring Union brigade, allowing the 69th to be outflanked with a loss of fifty-two men, most of these captured.[55] Despite this loss, the regiment helped celebrate the third anniversary of the formation of the Irish Brigade in early September. Gen. Meagher was on hand for the occasion.[56]

On 30 October, one of those baffling mysteries of war happened. Most of the picket line of the 69th disappeared along with part of that of the 111th New York. This took place near Fort A. Hays, opposite Confederate Fort Davis. Apparently, an enterprising force of about 200 Confederates, dressed partly in blue uniforms, infiltrated the Union lines between the 69th and the 111th. The Rebels "gobbled-up" the Yank pickets until an alert 69th sergeant of G Company informed Lt. Murtha Murphy. The latter ordered his pickets to open fire on the intruders. In the meantime, the enemy had made off with 168 (including 141 new recruits-which may partly explain the great success of the raid) members of the 69th, and eighty-two from the 111th-a nice bag! Some of those captured did manage to escape back to the Union lines.[57] It was not one of the 69th's more auspicious moments!

Not long afterwards, the Irish Brigade was reconstituted, and Robert Nugent once more commanded it.[58] The winter was spent in the entrenchments opposite the Confederate lines. The last particularly severe fighting in which the 69th took part occurred at Hatcher's Run, 25 March 1865. The Confederates of General William Mahone vigorously struck the Union works. Interestingly enough, Mahone's Division contained a large number of Irishmen. In this sharp fighting, the regiment suffered nine dead and eighty-five wounded. For a while, it was touch and go before the audacious Rebs were repulsed. The 69th had used up sixty rounds of ammunition per man before the attack ended. Captain Mulhall, leading the

Irish Brigade Monument at Gettysburg at "The Loop." At the base lies the Irish Wolfhound, symbol of devotion, and mascot to the 69th Regiment. Courtesy of the author.

regiment's skirmishers, fell severely wounded at the beginning of the engagement. He then lay for over an hour between the contending forces before he was rescued by his comrades.[59]

Fortunately for this decimated Union regiment, the end was almost near. The unit broke camp on 29 March to pursue the harassed Army of Northern Virginia as it pulled back in a frantic effort to escape Union encirclement. The Confederates were overtaken at Sutherland Station of the Southside Railroad, where two pieces of artillery and some 600 prisoners were taken by the brigade. One of the last wartime duties of the 69th was to destroy a captured Confederate supply train of 284 wagons.[60] Two days later, Gen. Robert E. Lee surrendered the gallant remnants of his command.

The 69th New York Volunteer Infantry proudly marched in the Grand Review of the Union Armies held in Washington on 23 May 1865. It had good cause to be proud. It had lost more men in action than any other New York regiment, and it ranked sixth in battle casualties of all Union Army regiments. Its total losses of 1084 exceeded its top strength at any single time.[61] Indeed, it had justly earned the sobriquet, THE FIGHTING SIXTY-NINTH.

FOOTNOTES

[1] Augustine I. O'Flynn, "The Fighting 69th, Irish Soldiers of New York, in *The Irish Voice*, Summer, 1954, p. 34.

[2] William Miller Owen, "A Hot Day on Marye's Heights" in *Battles and Leaders of the Civil War* (New York: Thomas Yoseloff, Inc., 1956), III, p. 98. See also by the same author, *In Camp and Battle With the Washington Artillery of New Orleans* (Boston: Ticknor & Co., 1885), p. 187.

[3] William F. Fox, *Regimental Losses in the American Civil War* (Albany, NY: Albany Pub. Co., 1898), p. 35.

[4] George A. Hussey, *History of the Ninth Regiment* (New York: Veterans of the Regiment, 1889), pp. 3-7.

[5] Patrick D. O'Flaherty, *History of the Sixty-ninth Regiment of the New York State Militia, 1852-1861* (Ann Arbor, MI: University Microfilms, 1981), p. 95, p. 164. Doctoral dissertation.

[6] *Ibid.*, pp. 200-201.

[7] W. F. Lyons, *Brigadier-General Thomas Francis Meagher* (New York: D. J. Sadlier & Co., 1870), p. 75.

[8] Maria Lydig Daly (Harold E. Hammond, ed.), *Diary of a Union Lady, 1861-1865* (New York: Funk & Wagnalls Co., 1962), p. 15.

[9] Michael Cavanaugh, *Memoirs of Gen. Thomas Francis Meagher* (Worcester, MA: Messenger Press, 1892), pp. 376-377, pp. 379-380.

[10] *War of the Rebellion, Official Records of the Union and Confederate Armies* (Gettysburg, PA: National Historical Soc., 1972; reprint). Series I, Vol. II, Reports of Col. Sherman and Capt. Kelly, pp. 368-372. Hereafter cited as *OR*.

[11] Michael Corcoran, *Gen. Corcoran's Captivity* (Philadelphia: Barclay & Co., 1862), p. 24.

[12] *OR*, Ser. I, Vol, II, Report of Gen. Tyler, p. 351.

[13] Corcoran. See also: Charles P. Daly, *Are the Southern Privateersmen Pirates?* (New York: James B. Kirker, 1862) and William D. Kelley, *United States vs. William Smith* (Philadelphia: King & Baird, n.d.).

[14] One of those captured with Corcoran was Pvt. James Mc Kay Rorty who escaped two months later. O'Flaherty, 296. Rorty was killed at Gettysburg commanding B Battery, 1st New York Light Artillery, as it helped to repel Pickett's attack.

[15] Cavanaugh, pp. 425-428.

[16] William Corby, *Memoirs of Chaplain Life* (Chicago: La Monte & O'Donnell & Co., 1893) pp. 299-303.

[17] Francis A. Walker, *History of the Second Army Corps* (New York: Charles Scribner's Sons, 1886), pp. 3-5.

[18] Ira. A. Glazier, ed., *The Famine Immigrants* (Baltimore: Genealogical Pub. Co., 1983), I, xxvii.

[19] Fr. Corby served as President of Notre Dame University after the war. Due to a campus statue (similar to the one at Gettysburg) showing him raising his hand in benediction, he is known irreverently among football conscious ND students as "Fair Catch" Corby!

[20] Corby, p. 39.

[21] David Powers Conyngham, *The Irish Brigade and Its Campaigns* (New York: William Mc Sorley & Co., 1867), pp. 144-157.

[22] *Ibid.*, pp. 182-209.

[23] *OR*, Ser. I, Vol. II, Part 2. Report of Gen. Meagher, 72-75.

[24] Fox, p. 204.

[25] *OR*, *op. cit.*

[26] *Medal of Honor, 1863-1968* (Washington: Government Printing Office, 1968), p. 208. Hereafter cited as *MH*.

[27] Conyngham, pp. 219-220.

[28] *Ibid.*, pp. 293-294.

[29] *Ibid.*, pp. 303-308. *OR*, Ser. I, Vol. 19, Part I. Reports of Gen. Hancock, pp. 275-283; Gen. Meagher, pp. 293-295; Maj. Cavanaugh, pp. 296-298.

[30] *OR*, *op. cit.*, Return of Casualties in the Union Forces at the Battle of Antietam, Md., p. 192.

[31] *Ibid.*, Reports of Col. Bennet, pp. 1047-1048; Maj. Gen. D. H. Hill, p. 1029.

[32] Conyngham, p. 324.

[33] *Ibid.*, pp. 330-337.

[34] St. Clair Mulholland, *The Story of the 116th Regiment, Pennsylvania Volunteers* (Philadelphia: F. Mc Manus, Jr. & Co., 1903), pp. 35-37.

[35] *Ibid.*, pp. 44-45. Conyngham, pp. 341-343.

[36] Owen, *In Camp and Battle...*, pp. 178-194.

[37] *OR*, Ser. I, Vol. XXI, Report of Gen. Hancock, pp. 226-232.

[38] G. Wilfred Pearce, "Stories of the Irish Brigade" in *Journal of the American Irish Historical Society* (New York: AIHS, 1911), Vol. X, p. 360.
[39] *MH*, p. 77.
[40] John Dooley, *John Dooley, Confederate Soldier: His War Journal* (Georgetown, D.C.: Georgetown University Press, 1945). pp. 116-117.
[41] George E. Pickett, *The Heart of a Soldier* (New York: Seth Moyle, Inc., 1913), p. 66.
[42] Cavanaugh, pp. 471-473, pp. 33-34 (Appendix).
[43] Walker, pp. 199-200.
[44] Conyngham, pp. 372-382. Corby, pp. 138-145.
[45] *OR*, Ser. I, Vol. XXV, Part 1, Reports of Gen. Hancock, pp. 311-316; Maj. Mulholland, pp. 124-130. Mulholland, pp. 93-101. Walker, pp. 245.
[46] Lyons, pp. 180-183.
[47] *OR*, Ser. I, Vol. XXVII, Part 1, Report of Lt. Smith, 388-390. Mulholland, pp. 124-130, pp. 407-408.
[48] Daly, p. 274.
[49] *OR*, Ser. I, Vol. XXIX, Part 1, Report of Capt. Richard Moroney, pp. 263-264. See also Conyngham, pp. 424-437.
[50] *OR*, Ser. I, Vol. XXXVI, Part 1, Reports of Maj. Richard Moroney and Capt. Robert H. Milliken, pp. 393-396.
[51] *Ibid.*, Conyngham, pp. 451-453.
[52] *Ibid.*, p. 137. See also Fox, p. 204.
[53] *OR*, Ser. I, Vol. XL, Part 1, pp. 349-350. See also Frederick H. Dyer, *A Compendium of the War of the Rebellion* (New York: Thomas Yoseloff, 1959), I, p. 288.
[54] Conyngham, p. 466-473.
[55] *Ibid.*, pp. 476-477. See also *OR*, Ser. I, Vol. XLII, Part 1, p. 129.
[56] Corby, pp. 264-268.
[57] *OR*, Ser. I, Vol. XLII, Part 1, Reports of Nelson A. Miles, C. D. Mac Dougall, Murtha Murphy, Richard Moroney, Robert Milliken, pp. 255-260, p. 290 and Part 3, Correspondence of G. K. Warren, pp. 439-440.
[58] *Ibid.*, Part 3, General Orders No. 1, Robert Nugent, pp. 476-477 and Special Orders No. 275, Sept Carncross.
[59] *OR*, Ser. I, Vol. XLVI, Part 1, Report of James J. Smith, pp. 200-201. Report of Charles Smart, p. 696.
[60] *Ibid.*, Report of James J. Smith, 728-731. See also Conyngham, pp. 517-526.
[61] Fox, p. 204. See also Conyngham, pp. 597-599, and Dyer, III, pp. 1430-1431.

GENERAL CORCORAN

On page 545 (of *Harper's Weekly*) we present a portrait of BRIGADIER-GENERAL MICHAEL CORCORAN, whose release from captivity has been one of the most interesting events of the past week.

General Michael Corcoran was born in Ireland, but came to this country at a very early age. He engaged in various employments here, and at the time the war broke out was employed as a clerk in the New York Post-office. He had always evinced a fondness for soldiering, and had risen, by his attention to drill and his devotion to the interests of the regiment, from the rank of private to that of Colonel of the Sixty-ninth regiment New York State Militia. He first rose into notice when the Prince of Wales was here, by refusing to turn out his regiment in honor of the nation's guest. His conduct at the time was the subject of severe animadversion, and he was even court-martialed for it. It will suffice to say that for that offense no one will be disposed to censure him now. If Prince Albert Edward were to come here to-morrow, there is not a Colonel in the service who would willingly pay honor to the heir to the throne of a country which has treated us as England has done during the past year. At the outbreak of the war Colonel Corcoran was one of the first Colonels who r ported a regiment ready; and early in April, 1861, the gallant Sixty-ninth marched down Broadway 1300 strong. They were stationed for some time on Arlington heights, where they left a substantial token of their visit in the shape of Fort Corcoran. Their term of service expired on the 20th July, and they were then entitled to return to their homes. But a battle was imminent, and after a heart-stirring appeal from Colonel Corcoran, the entire regiment decided to see it out. At Bull Run they behaved with gallantry, and won the praise of their General. Unfortunately, in the retreat the Colonel got separated from his men, and was taken prisoner. This was thirteen months ago. During that time General Corcoran has endured privations of all kinds, in prison at Richmond, Charleston, Columbia, and Salisbury—suffering slights and insults beyond all imagination; badly fed, badly lodged, badly treated by brutal jailors; constantly deluded by false promises of liberty which were always broken. And now the gallant fellow is back with us again, and ready and eager to fight once more under the glorious old Flag. At his reception at Washington he said:

SOLDIERS AND CITIZENS.— For this kindly greeting on my arrival I return you my most fervent thanks. Such cheers I have not heard for thirteen months, nor is it possible for such a cheer to come from any other people than those actuated by the principles which move the men I see before me, marching, as they are, to the maintenance and support of the outraged laws of this glorious Union. There is nothing of the yell of the tiger or hyena in those cheers. They remind me of the cheers I heard on the Fourth of July at Salisbury, North Carolina, for the maintenance of the glorious institutions under which we have lived. [Immense applause.] Massachusetts has always done well, and taken a great part in all the struggles of the past. It does my heart good to see that she has come forth with redoubled vigor with the best of her children, who will, I hope, strike such a blow as will crush the miscreants that have dared to raise the flag of this infamous rebellion. [Cheers, and cries "We will, we will!"] I hope to hear a good account of the men who are now before me. I know I shall. I hope that New York will send forth her thousands to your Massachusetts' thousand, and I am satisfied that, with the hearty co-operation of the other States, we shall soon put an end to this rebellion. Gentlemen, I had hoped the whole matter would have been settled ere now; but I do not regret that I am now here and shall be permitted to take my musket in my hand, if in no other position, and strike one more blow against our enemies. I hope the number that has been called for by the President of the United States will come forward voluntarily and enroll themselves in their country's cause; but I shall not regret if some are obliged to be drafted, for I want a few of the fault-finders—[Cheers and cries of "Good, good!"]— some of the stay-at-home military critics to have an opportunity of displaying their imaginary genius. I shall not be sorry either, for I want to see the lukewarm men, the men who remain at home and mind their business, and after peace is restored by the labor and the blood of their more patriotic neighbors, will be the first to take advantage of every thing good in this country they are unworthy of and unwilling to sustain the hour of her need. I want those men made to fill up the ranks, and shall not regret if there is no other way of getting them to draft them. [Cries of "Good, good!"] I am glad to find on my return here that the Government has adopted vigorous measures for the prosecution of the war, and that Congress has clothed the Executive with ample power to do so. I trust the President implicitly. I would see him invested with unlimited power to crush the rebellion. The rebels have given to their chief leader dictatorial powers, and we must meet force with force. If the power of the loyal people is put forth now the rebellion will be broken and the Union sustained and restored. Gentlemen, I did not intend to make a speech—[Cries of "Go on, go on!" and cheers]— and will reserve any further remarks I may have to make to a future occasion.

The public learned with satisfaction that the President, in consideration of his captivity and his heroism, has appointed him a Brigadier-General, his commission to date from the Battle of Bull Run.

[From *Harper's Weekly*, New York, Saturday, August 30, 1862.]

CONCURRENT EVENTS

A meeting for the relief of Ireland.—We learn from the Post that on Monday evening last, an immense meeting was held at the Odeon for the above named object.

Addresses were delivered by the Rev. Messers. O'Flaherty and O'Reily and Isaac H. Wright, Esq. Mr. O'Flaherty laid on the table $79 collected in his congregation at Salem. Mr. Wright proposed a series of resolutions explanatory of the causes of the present distressing condition of the Irish peasantry, and they were adopted. The collections were then commenced, and six hundred and thirty dollars were paid down before the audience. A few minutes after the wardens had spread throughout the house, a gentleman of this city requested that it might be announced that if $500 should be raised with in a fortnight, he would contribute $25 to the fund. In less than half an hour after this announcement was made more than $500 had been piled upon the table.

[*Boston Cultivator,* Boston, Massachusetts, December 13, 1845.]

UNION PRISONERS IN CLOSE CONFINEMENT

A number of Union prisoners from General Pope's army had arrived in Richmond. They were put "in solitary confinement, in conformity with instructions, and the rule, as enforced, will not be departed from under any consideration whatever." Among the prisoners was Brigadier-General Prince and thirty-four officers.

(**Harper's Weekly.,** August 30, 1862.)

FAMILY BIBLE

(N = North, S = South.)

FEBRUARY

Feb. 2, 1803 — Gen. Albert Sidney Johnston, Washington, Kentucky. (S)
Feb. 3, 1793 — Capt. Charles S. McCauley, commandant of Norfolk Navy Yard (1861), Philadelphia, Pennsylvania. (N)
Feb. 3, 1805 — Gen. Samuel R. Curtis, near Champlain, New York. (N)
Feb. 3, 1807 — Gen. Joseph E. Johnston, Cherry Grove, Prince Edward County, Virginia. (S)
Feb. 3, 1811 — Horace Greeley, *New York Tribune* editor, Amherst, New Hampshire. (N)
Feb. 6, 1813 — Gen. Joseph R. Anderson, Botetourt County, Virginia. (S)
Feb. 6, 1832 — Gen. John B. Gordon, Upson County, Georgia. (S)
Feb. 6, 1833 — Gen. James Ewell Brown Stuart, Patrick County, Virginia. (S)
Feb. 6, 1834 — Gen. William Dorsey Pender, Edgecombe County, North Carolina. (S)
Feb. 6, 1836 — Lt. Com. Thomas O. Selfridge, Charlestown, Massachusetts. (N)
Feb. 7, 1817 — Gen. Leroy P. Walker, also first Secretary of War (1861), Huntsville, Alabama. (S)
Feb. 7, 1826 — Lt. Com. James E. Jouett, near Lexington, Kentucky. (N)
Feb. 8, 1817 — Gen. Richard S. Ewell, Georgetown, District of Columbia. (S)
Feb. 8, 1820 — Gen. William T. Sherman, Lancaster, Ohio. (N)
Feb. 9, 1826 — Gen. John A. Logan, Jackson County, Illinois. (N)
Feb. 11, 1811 — Capt. Benjamin F. Sands, U.S.N., Baltimore, Maryland. (N)
Feb. 11, 1812 — Vice President Alexander H. Stephens, Crawfordsville, Georgia. (S)
Feb. 12, 1809 — President Abraham Lincoln, Hardin County, near Hodgenville, Kentucky. (N)
Feb. 13, 1812 — Adm. Samuel P. Lee, Fairfax County, Virginia. (N)
Feb. 14, 1817? — Frederick Douglass, abolitionist, Tuckahoe, Maryland. (N)
Feb. 14, 1824 — Gen. Winfield S. Hancock, Montgomery Square, Pennsylvania. (N)
Feb. 15, 1797 — John Bell, ante-bellum politician, near Nashville, Tennessee. (S)
Feb. 15, 1835 — Gen. Alexander S. Webb, New York City. (N)
Feb. 16, 1823 — Gen. John D. Imboden, near Staunton, Virginia. (S)
Feb. 17, 1824 — Gen. William F. "Baldy" Smith, St. Albans, Vermont. (N)
Feb. 17, 1837 — Gen. Francis J. Herron, Pittsburgh, Pennsylvania. (N)
Feb. 18, 1805 — R.A. Louis M. Goldsborough, Washington, D.C. (N)
Feb. 18, 1817 — Gen. Lewis A. Armistead, New Bern, North Carolina. (S)
Feb. 21, 1800 — Gen. John H. Winder, Commissary General of Prisons, Somerset County, Maryland. (S)
Feb. 22, 1819 — Com. John N. Maffitt, aboard an immigrant ship en route from Ireland to America. (S)
Feb. 24, 1811 — Sen. Edward D. Baker, colonel in command at Ball's Bluff, London, England. (N)
Feb. 24, 1836 — Winslow Homer, artist, Boston, Massachusetts. (N)

Feb. 25, 1807 — George A. Trenholm, financier and Secretary of the Treasury (1864-65), Charleston, South Carolina. (S)
Feb. 26, 1846 — William F. Cody, guerrilla and scout, Scott County, Iowa. (N)
Feb. 27, 1823 — Gen. William B. Franklin, York, Pennsylvania. (N)
Feb. 28, 1825 — Gen. Quincy A. Gillmore, Black River, Lorain County, Ohio. (N)
Feb. 29, 1784 — Gen. John E. Wool, Newburgh, New York. (N)
Feb. 29, 1844 — Pvt. Albert Sieber, Grand Duchy of Baden, Germany. (N)

MARCH

Mar. 1, 1820 — George Davis, Attorney General (1864-65), New Hanover County, North Carolina. (S)
Mar. 2, 1793 — Samuel Houston, Governor of Texas (1859-61), near Lexington, Virginia. (N)
Mar. 2, 1824 — Gen. Henry B. Carrington, Wallingford, Connecticut. (N)
Mar. 2, 1828 — Gen. Jefferson C. Davis, Clark County, Indiana. (N)
Mar. 2, 1829 — Gen. Carl Schurz, Liblar-am-Rhein, Germany. (N)
Mar. 4, 1826 — Gen. John Buford, Woodford County, Kentucky. (N)
Mar. 6, 1820 — Gen. Horatio G. Wright, Clinton, Connecticut. (N)
Mar. 6, 1831 — Gen. Philip H. Sheridan, Albany, New York. (N)
Mar. 7, 1832 — Gen. Orlando M. Poe, Navarre, Ohio. (N)
Mar. 8, 1799 — Simon Cameron, Secretary of War (1861-62), Maytown, Pennsylvania. (N)
Mar. 8, 1821 — Gen. Morgan L. Smith, Mexico, New York. (N)
Mar. 8, 1841 — Lt. Col. Oliver Wendell Holmes, Jr., Boston, Massachusetts. (N)
Mar. 10, 1818 — George W. Randolph, Secretary of War (1862), Monticello, near Charlottesville, Virginia. (S)
Mar. 11, 1812 — James Speed, Attorney General (1864-65), Farmington, Kentucky. (N)
Mar. 12, 1817 — Capt. John L. Worden, U.S.N., Mt. Pleasant, Westchester County, New York. (N)
Mar. 14, 1815 — Gen. Jerome B. Robertson, Woodford County, Kentucky. (S)
Mar. 14, 1816 — Gen. Montgomery D. Corse, Alexandria, Virginia. (S)
Mar. 14, 1823 — Gen. Roswell S. Ripley, Worthington, Ohio. (S)
Mar. 14, 1833 — Gen. John S. Marmaduke, Arrow Rock, Missouri. (S)
Mar. 16, 1822 — Gen. John Pope, Louisville, Kentucky. (N)
Mar. 16, 1828 — Gen. Patrick R. Cleburne, County Cork, Ireland. (S)
Mar. 18, 1782 — John C. Calhoun, Long Canes Settlement, near Abbeville, South Carolina. (S)
Mar. 20, 1812 — Gen. George B. Crittenden, Russellville, Kentucky. (S)
Mar. 20, 1830 — Gen. Eugene A. Carr, Concord, Erie County, New York. (N)
Mar. 20, 1835 — Lt. Com. John G. Grimes, Hillsboro, New Hampshire. (N)
Mar. 22, 1817 — Gen. Braxton Bragg, Warrenton, North Carolina. (S)
Mar. 22, 1824 — Gen. William H.C. Whiting, Biloxi, Mississippi. (S)
Mar. 22, 1834 — Gen. Francis A. Shoup, Laurel, Indiana. (S)
Mar. 23, 1818 — Gen. Don Carlos Buell, near Marietta, Ohio. (N)
Mar. 25, 1818 — Gen. Isaac I. Stevens, Andover, Massachusetts. (N)
Mar. 25, 1827 — Com. Stephen B. Luce, Albany, New York. (N)
Mar. 26, 1813 — Gen. Thomas W. Sherman, Newport, Rhode Island. (N)
Mar. 28, 1818 — Gen. Wade Hampton, Charleston, South Carolina. (S)
Mar. 29, 1829 — Gen. Robert E. Rodes, Liberty, Virginia. (S)
Mar. 30, 1825 — Gen. Samuel B. Maxey, Tompkinsville, Kentucky. (S)

Mar. 31, 1810 — Com. James Alden, Portland, Maine. (N)
Mar. 31, 1823 — Mary Boykin Chesnut, *A Diary from Dixie*, Pleasant Hill, near Camden, South Carolina. (S)

Addendum to FAMILY BIBLE

FEBRUARY

Feb. 5, 1838 — Fr. Abram Ryan, poet, "The Conquered Banner," Hagerstown, Maryland. (S)
Feb. 6, 1824 — Gen. Nathan D. "Shanks" Evans, Marion, South Carolina. (S)
Feb. 10, 1818 — Isham G. Harris, governor of Tennessee (1857-62), Tullahoma, Tennessee. (S)
Feb. 15, 1797 — John Bell, near Nashville, Tennessee. (S)
Feb. 16, 1832 — Gen. Camille de Polignac, Millemont, France. (S)
Feb. 18, 1829 — Gen. Alfred Mouton, Opelouses, Louisiana. (S)
Feb. 19, 1821 — Gen. Francis P. Blair, Jr., Lexington, Kentucly. (N)
Feb. 20, 1805 — Angelina Grimke, abolitionist, Charleston, South Carolina. (N)
Feb. 27, 1831 — Edward A. Pollard, editor, *Daily Richmond Examiner*, Alta Vista, Nelson County, Virginia. (S)

MARCH

Mar. 6, 1810 — John B. Jones, *A Rebel War Clerk's Diary*, Baltimore, Maryland. (S)
Mar. 17, 1777 — Chief Justice Roger B. Taney, Calvert County, Maryland. (S)
Mar. 21, 1824 — Peter W. Alexander, newspaper reporter, Elberton, Georgia. (S)
Mar. 24, 1834 — Maj. John Wesley Powell, Mount Morris, New York. (N)
Mar. 25, 1830 — Col. Alfred L. Rives, Chief of Engineering Bureau, Paris, France. (S)
Mar. 26, 1817 — Gen. Herman Haupt, chief of military railroads, Philadelphia, Pennsylvania. (N)
Mar. 28, 1820 — William H. "Bull Run" Russell, reporter, London *Times*, Lilyvale, County Dublin, Ireland.
Mar. 29, 1812 — John Letcher, governor of Virginia (1860-64), Lexington, Virginia. (S)

The McGraw-Hill Encyclopedia of World Biography. 12 vols. New York: McGraw-Hill Book Company, 1973.

BIBLIOGRAPHY

Biographical Directory of the United States Executive Branch, 1774-1971. Ed. Robert Sobel. Westport, Connecticut: Greenwood Publishing Company, 1971.
Dictionary of American Biography. 20 vols. New York: Charles Scribner's Sons, 1928-1936.
The National Cyclopaedia of American Biography. 57 vols. New York and Clifton, New Jersey: James T. White & Company, 1892-1977.
Wakelyn, John L. *Biographical Dictionary of the Confederacy*. Westport, Connecticut: Greenwood Press, 1977.
Webster's American Military Biographies. Ed. Robert McHenry. Springfield, Massachusetts: G. & C. Merriam Company, 1978.

"Confederate Indecision and Delay: The Empty Victory at Chickamauga"

By
Vernon H. Crow

About August 23, 1863, General Robert E. Lee, commanding the powerful Army of Northern Virginia and military adviser to President Jefferson Davis, was summoned to Richmond for a lengthy period of strategic conferences with the Confederate high command. The situation that faced the Richmond government in the South's struggle for independence was dire: the Confederacy had been split into two by the loss of the Mississippi River at Vicksburg; Tennessee had almost been given up; and in Virginia Lee's army was still too weak from the battle of Gettysburg to resume the offensive. This partitioning of the South at Vicksburg created a new Confederate left in Mississippi. To oppose General Ulysses S. Grant and the Federals there was a small army under General Joseph E. Johnston. In the Confederate center lay General Braxton Bragg's Army of Tennessee at Chattanooga threatened by General William S. Rosecrans and the Union Army of the Cumberland. Lee's army on the Rapidan held the Southern right opposite General George G. Meade and the Army of the Potomac in Virginia.

It was obvious to Lee, Davis, Adjutant General Samuel Cooper and Secretary of War James A. Seddon that the Confederacy had to act. To remain on the defensive spelled certain defeat for the young nation. Fortunately for the Confederates on the left, Grant had dispersed his troops after his victory at Vicksburg. On the right, Meade did not seem overly anxious to start another "on to Richmond" campaign. Only Rosecrans in the center seemed ready, at the time, to make a move.

Lee favored reinforcing his army and striking a blow at Meade, while Cooper thought it best to strengthen Bragg from Johnston's army and attack Rosecrans. Bragg had recommended sending some of his troops to Johnston, while the latter seemed to have no plan at all.[1] The debate lasted two weeks until finally, on August 31, 1863, Lee wrote to his senior lieutenant, General James Longstreet, commander of the famous First Corps, to prepare the Army of Northern Virginia for a general offensive.[2] This order must have displeased Lee's "Old War Horse" who had advocated sending reinforcements from Lee's army to Tennessee as early as January 1863, and again in April. On September 2nd, Longstreet wrote to his chief: "I do not know that we can reasonably hope to accomplish much here by offensive operations, unless you are strong enough to cross the Potomac. If we advance to meet the enemy on this side, he will, in all probability, go into one of his many fortified positions; these we cannot afford to attack." Longstreet must have known the effect this statement would have on Lee. Without Longstreet's enthusiastic support, Lee would be reluctant to go north again. Besides, Lee knew that his lieutenant was right. Meade was a good defensive soldier and would probably retire at Lee's advance into strong fortifications.

Longstreet in his reply to Lee, however, saw his chance to urge a transfer west. Said Longstreet: "I know but little of the condition of our affairs in the west, but am inclined to the opinion that our best opportunity for great results is in Tennessee."[4] Longstreet then proposed to put Lee on the defensive in Virginia with two corps, and send the third to Bragg to "destroy Rosecrans' army."[5]

On the fourth of September, Lee wrote Longstreet and apparently asked how long it would take to move a corps to Tennessee. The next day, Longstreet replied that he didn't know, but again took the opportunity to renew his support for a westward movement. Then "Old Pete" added an idea that must have been wandering his head for months: "I will say more; I think that is time that we had begun to do something in the west, and I fear if it is put off any longer we shall be too late. If my corps cannot go west, I think that we might accomplish something by giving me [three] brigades, and putting me in General Bragg's place, and giving him my corps."[6]

One wonders how the resulting battle of Chickamauga might have turned out if Davis had chosen Longstreet to command the Confederate center. Actually, Davis had wanted Lee to go west and assume command, a position Lee politely refused. Said Lee to the president: "...the duty could be better performed by the officers already in that department."[7] Davis eventually came to agree with Lee: Longstreet and his First Corps would be transferred to the Army of Tennessee, but Bragg would keep his command. The final decision to send Longstreet to Tennessee was probably made by Davis on September 6th, because that evening Lee wrote to the president that he had given the order to the quartermaster-general to transport Longstreet's troops west.[8] The plan which had taken fourteen days to formulate called for two of Longstreet's experienced divisions, commanded by Major Generals Lafayette McLaw's and John Hood—nine veteran brigades of about 12,000 arms—along with Porter Alexander's battalion of artillery of twenty-six guns, to be carried entirely by train to Chattanooga via the Bristol-Knoxville line, a distance of about 540 miles.[9] The first trains began arriving at Orange Court House on September 9th. That night, or the next, Longstreet visited the tent of his chief. As he bid Lee farewell and made his way to his horse, the commanding general followed his most trusted lieutenant. As Longstreet prepared to mount, Lee said: "Now, General, you must beat those people out in the West."[10]

Longstreet withdrew his foot from the stirrup and came to what he called a "respectful position" and promised: "If I live; but I would not give a single man of my command for a fruitless victory."[11]

Lee's "Old War Horse" did not know it at the time, but Union Major General Ambrose Burnside, the goat of Fredricksburg, already had begun another campaign that would lead to Longstreet's "fruitless victory."

For more than two years Abraham Lincoln, the Federal president, had urged general after general to invade and occupy East Tennessee. There were political as well as military reasons why Lincoln wanted the region held. Most important to the Northern president was that the vast majority of the people there were loyal to the Union. East Tennessee, bounded by the Smoky Mountains on the east and the Cumberland Range on the west, was also the gateway to Georgia and Alabama as well as the back door to Virginia. It contained the East Tennessee and Virginia Railroad which stretched from Knoxville to Lynchburg, linking the Confederate right with the center and the left. At the point where upper East Tennessee touched Virginia and Kentucky, was Cumberland Gap, considered by both sides to be a major invasion route for armies marching north or south, as well as a virtual bastion when fortified. Lincoln gave the job of capturing East Tennessee to General Ambrose Burnside following his removal from command of the Army of Potomac. Burnside had been assigned the Department of the Ohio along with his famous Ninth Corps. The campaign against Vicksburg, however, required the use of his veteran troops and as a result Burnside resisted his superiors' proddings to move on East Tennessee.[12] After Vicksburg was captured by Grant, and the Ninth Corps sent back to Kentucky, Burnside decided it was safe to move on Knoxville. He organized the Twenty-third Corps from troops in the department and placed Major General George L. Hartsuff in command.[13] Burnside's movement into East Tennessee was to be made in concert with that of Rosecrans, who was to march on Bragg, both to begin their campaigns on August 16th.[14]

Burnside's troops were garrisoned south of Lexington, Kentucky, and were ordered to march over the Cumberland Range, west of Cumberland Gap. To oppose Burnside, the Confederates had about eight thousand troops, under General Simon Buckner, scattered from below Knoxville to Cumberland Gap.[15]

From the inception of Burnside's invasion, Bragg and Buckner thought his objective was Rosecrans' army. Though they knew that Burnside's force was moving on East Tennessee they believed he would join the Army of the Cumberland near Chattanooga. Even Lee thought so.[16] The movement was discovered on August 19th by Confederate Brigadier General John Frazer, who commanded 2,000 men at Cumberland Gap. On August 23rd, Buckner wrote Bragg: "Burnside's main column will move by Jamestown. It is expected to cooperate with Rosecrans' left. Alone I can do little against him."[17] The next day he telegraphed General Cooper in Richmond: "Is it not possible to re-enforce me?"[18] Buckner pleaded for help from General Sam Jones, commanding the Confederates in southwest Virginia and whose main responsibility was to protect the important salt works there.

Buckner's situation was difficult. He had a 200-mile front to defend and his troops were scattered guarding railroad bridges as well as garrisoning Cumberland Gap. In addition, as Buckner pointed out to Cooper, he had forty-three gaps in his front to patrol, "through nearly all of which cavalry and infantry can pass."[19]

As Burnside's movement developed, Bragg ordered Buckner to concentrate his forces south of the Tennessee River at Loudon, thirty

Gen. A. E. Burnside.

Gen. Rosecrans.

Gen Longstreet.

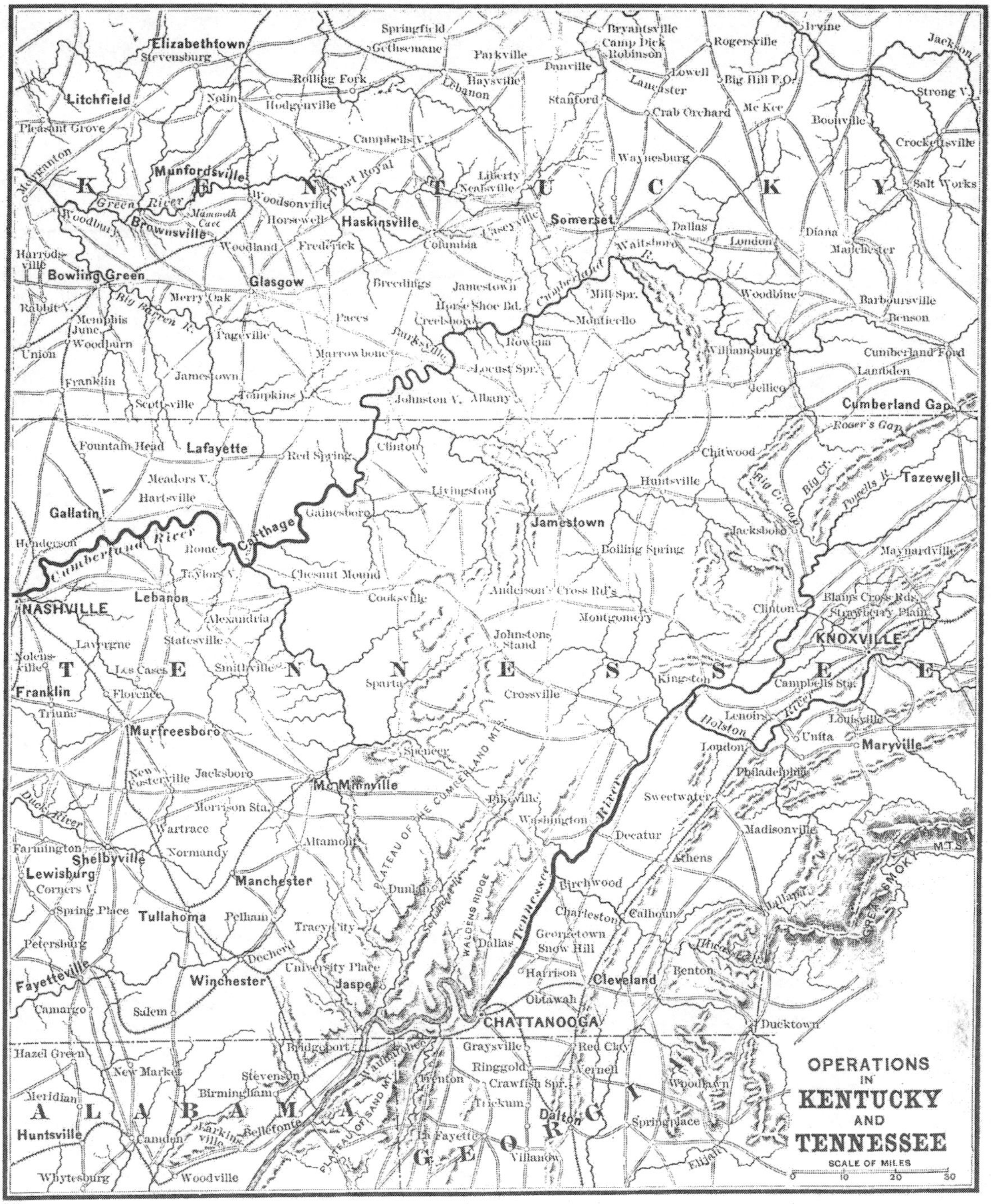

miles from Knoxville. Government stores and the railroad stock were removed from Knoxville creating what General John C. Vaughn called "a perfect panic." As the Confederates evacuated Knoxville, the loyal Southerners became alarmed. One Confederate wrote to President Davis: "Our people are desponding very much. East Tennessee should be held at all hazards, as we cannot afford to lose it and its advantages...if we cannot hold East Tennessee we cannot hold any State."[20]

Burnside descended on East Tennessee with 15,000 men in five columns all directed to converge on Montgomery, Tennessee, about forty-five miles west of Knoxville, but it was no small feat.[21] Using pack mules and horses to carry most of the burden, the command climbed the mountains, but the going became even too rough for the beasts. "In several instances, the animals utterly failed to drag the artillery up the acclivities, and their places were filled by men, who, with hands upon the drag ropes and shoulders to the wheels, dragged or lifted the guns, caissons and wagons from height to height."[22] The march was remembered by Burnside with pride considering "the great difficulties in the way of steep rugged mountains, bad roads, and short forage."[23] Captain Orlando M. Poe, chief engineer at the time for the Twenty-third Corps, found his maps to be "perfectly worthless, nay worse, as they only serve to mislead."

Poe's defective maps, however, did not prevent the heads of the two infantry columns from reaching Montgomery at the same time on August 30th, after a march of 112 miles. Eight miles outside of Montgomery, on the road to Kingston, Burnside sent Brigadier General John W. Foster forward with his cavalry to capture Knoxville.[24] Foster's troopers entered the town to the cheers and wild enthusiasm of its citizens. Not all Knoxville, however, was happy to see the Union cavalrymen. One Confederate sympathizer wrote in her journal: "Yesterday was the first day of September 1863. Long shall it be remembered among us. It was the day when we beheld for the first time our most dreaded foe, the Yankees, so long expected throughout East Tennessee."[25]

By September 3rd, Burnside occupied Knoxville, but found that Cumberland Gap was still in Confederate hands. Immediately he ordered Brigadier General James M. Shackelford—the man who had captured the dreaded Confederate raider John Morgan—to take his cavalry to the south side of the gap. Before leaving Kentucky, Burnside had sent Colonel John De Courcy with a brigade of cavalry to invest the gap from the north (Kentucky) side. De Courcy found the gap too heavily fortified by Frazer, who had been ordered by Buckner to defend it "to the last," an order Frazer was ready to obey. Still, the Union cavalrymen found the position too strong to be carried, and so Shackelford sent word back to Burnside that it could not be reduced by the force with him. Burnside got Shackelford's message on the seventh and set out immediately for Cumberland Gap with Colonel Samuel A. Gilbert's infantry brigade, arriving there after an extraordinary march of sixty miles in fifty-two hours.[26]

With De Courcy blocking his front, and Burnside now in command of the force in his rear, Frazer agreed to surrender the Confederate garrison on September 9th.

Ironically, Burnside's capture of Cumberland Gap came on the same say day that Longstreet troops were boarding trains at Orange Court House for their transfer to Tennessee. Burnside's capture of East Tennessee forced Longstreet's corps to change its route from an estimated four-day trek of 540 miles through Knoxville to an 843-mile, exhausting seven-and-a-half-day trip via Wilmington and Charlotte and Augusta, and then by a single track of road to Atlanta. It was an uncertain railroad with different gauges and unreliable engines.[27]

Later, Longstreet would write that the success of the plan was "thought from the first to depend upon its prompt and vigorous execution."[28] Not only was speed a major consideration, but so was secrecy. It is no wonder that when the Northern newspapers published Longstreet's detachment on September 9th, Lee was furious. "I fear," said Lee, "that there has been great imprudence in talking on the part of our people."[29]

On the same day that Longstreet's troops left their camps on the Rapidan, Rosecrans' forces occupied Chattanooga, which Bragg evacuated on the eighth. Buckner, ordered down from East Tennessee, joined the Army of Tennessee, and the three corps retired on La Fayette, Georgia. But on the seventeenth, Bragg crossed the Chickamauga River with an eye to bringing Rosecrans to battle, which he succeeded in doing on the nineteenth and twentieth. With the help of Longstreet's troops, which had just arrived, Bragg won a decisive victory. But his victorious troops did not pursue the retreating enemy, thus allowing them time to reach Chattanooga. Once safely behind the extensive fortifications, Rosecrans was virtually untouchable. Only by protracted seige operations could Bragg now hope to reap the fruits of Chickamauga. The Confederate line was brought up to the heights around the city, but before a single Union soldier had lost his life from starvation, the famous "Cracker line" was established, and supplies were smuggled into the city. The Confederates lost both their chance to crush Rosecrans, and regain the strategically important city; the months of indecision and delay had finally culminated in the empty victory that Longstreet feared.

Bragg has taken the major share of the blame from historians because of his lack of leadership in following up the victory. At least one historian, however, has put part of the blame on Longstreet for failing to order Rosecrans' retreat cut off until it was too late.[30] Others claim the troops were too tired and there were no reserves, while some point to officers who failed to carry out orders from the commanding general.

Perhaps the blame for a lost opportunity must also be shared by the Confederate high command in Richmond. It took two weeks of conferences and meetings for Davis, Cooper and Seddon to agree to detach Longstreet's corps and reinforce the Confederate center. To add to this costly delay, apparently Bragg was not aware that he was to be reinforced by troops from Lee's army until September 13th.[31] Even then, he did not know which troops were coming, nor how many. It wasn't until the fifteenth that Bragg found out three of Longstreet brigades were at Atlanta on their way to him.[32] One has to wonder why no official word was sent by the authorities in Richmond. If Bragg had known the extent of Longstreet's reinforcements, he might have delayed his crossing of the Chickamauga until Longstreet was up.

Another contributing factor to the emptiness of Chickamauga was the miscalculation of Burnside's movement. As late as September 13th or 14th, Lee still thought that his objective was Chattanooga. Said Lee: "Had I been aware that Knoxville was the destination of General Burnside, I should have recommended that General Longstreet be sent to oppose him..."[33] Lee had been lead to believe that the Union force in East Tennessee was on its way to join the Army of the Cumberland by dispatches he was getting through the War Department and President Davis, although in defense of the War Department, it was getting most of its information from Bragg, Buckner and Jones. Not only was the Confederate high command misinformed and lacking in a good intelligence gathering system, but it was slow to act on what information it had. When Buckner pleaded for help from Jones in southwestern Virginia, it preferred to defend the salt works. When Lee finally intervened, Richmond then ordered Jones to defend East Tennessee.[34] Perhaps the whole Chickamauga campaign might have had a different outcome if Seddon and Cooper had sent Jones with his 3,000 troops immediately to Buckner's assistance and ordered Bragg to let Buckner's Third Corps retire up the East Tennessee and Virginia Railroad behind the Holston River, perhaps to Strawberry Plains. Frazer's force at the gap was ample to keep De Courcy at bay. The railroad would have remained open, and Longstreet's troops could then have been sent through Bristol and combined with Buckner. That general would have been able to fight a delaying action until Longstreet arrived. "Old Pete," ready to fight in September, might have driven Burnside from Knoxville then and there. After driving off Burnside, Longstreet could have moved to Bragg's right.

Longstreet was critical of Cooper and Seddon in his memoirs when commenting on the loss of Cumberland Gap and Knoxville. "These facts," he wrote, "were known to the Richmond authorities at the time of our movements, but not to General Lee or myself until the move was so far advanced as to prevent recall."[35] His point was that the high command withheld information that might have altered both his and Lee's thinking. But it is doubtful if Seddon or Cooper intentionally withheld information from Longstreet. Their inability to interpret facts, their lack of communication to Bragg about Longstreet's transfer and their lack of prompt support for Buckner evidences a confused command structure. For example, in July, the War Department abolished Buckner's Department of East Tennessee, and placed his troops under Bragg as Buckner's Third Corps; but Buckner was to maintain administrative control of East Tennessee, reporting directly to Richmond. Buckner was naturally confused. How could he serve in Bragg's army and yet report to Richmond?[36]

Moreover, when it became ap-

Major General Simon Buckner

parent to Bragg that Frazer was about to be bottled up in Cumberland Gap, he ordered the Gap evacuated, but again the high command interfered, allowing Jones to reverse the order, resulting in the loss of 2,000 men.[37]

After the war, Porter Alexander wrote that it was actually the delay in accepting Longstreet's strategy that cost the South a greater victory at Chickamauga. Speaking of the battle, the famous artillery officer remarked: "This was consequently but another bloody and fruitless victory to be followed by a terrible defeat in a few weeks when the enemy's reinforcements had joined."[38] Alexander put the blame squarely on Richmond: "It is first to tell of the dilatory consideration and slow acceptance of the proposed strategy which should have been decided upon even before Lee's army was again south of the Potomac, and every subsequent movement planned to facilitate it."[39]

Perhaps most prophetic of all was Lee's letter to Davis on September 14th: "Should General Longstreet reach General Bragg in time to aid him in winning a victory, and return to this army, it will be well but should he be detained there without being able to do any good, it will result in evil."[40] Evil did result, but as much as historians like to blame Bragg for the empty victory at Chickamauga—and he certainly deserves his share—the high command in Richmond must share it, too.

FOOTNOTES

[1] Thomas L. Connelly, *Autumn of Glory* (Baton Rouge: Louisiana State University Press, 1971), p. 102, hereinafter cited as *Connelly;* Stanley F. Horn, *The Army of Tennessee* (Norman: University of Oklahoma Press, 1953), p. 239.

[2] *War of the Rebellion, Official Records of the Union and Confederate Armies* (Washington: Government Printing, 1881), in 70 volumes in 128 parts, Series I, 29, Part 2, p. 693, hereinafter cited as *O.R.* All references are to Series I.

[3] *Ibid.*

[4] *Ibid.*, p. 694.

[5] *Ibid.*

[6] *Ibid.*, p. 699. See also James Longstreet, *From Manassas to Appomattox* (ed. by James I. Robertson, Jr., Bloomington: Indiana University Press, 1960) p. 435, hereinafter cited as *Longstreet.*

[7] *O R* , 29, Part 2, p. 701.

[8] *Ibid.*, pp. 700-01.

[9] Edward Porter Alexander, *Military Memoirs of a Confederate* (New York: Scribner's, 1907) p. 448, hereinafter cited as *Alexander.*

[10] *Longstreet*, p. 437.

[11] *Ibid.*

[12] *O R* , 30, Part 2, p. 545.

[13] *Ibid.*, p. 548.

[14] *Ibid.*, pp. 545, 548.

[15] *Ibid.*, 30, Part 4, p. 526.

[16] *Ibid.*, 29, Part 2, p. 706.

[17] *Ibid.*, 30, Part 4, p. 540.

[18] *Ibid.*, p. 546.

[19] *Ibid.*, p. 526.

[20] *Ibid.*, p. 560.

[21] *Ibid.*, 30,Part 2, p. 548; Augustus Woodbury, *Major General Ambrose E. Burnside and the Ninth Army Corps* (Providence; Sidney S. Rider & Brother, 1867), pp. 306-07, hereinafter cited as *Woodbury.*

[22] *Woodbury*, pp. 305-06.

[23] *O R* , 30, Part 2, p. 548.

[24] *Ibid.*

[25] Mary U. Rothrock (ed.), *The French Broad-Holston Country* (Knoxville: East Tennessee Historical Society, 1946, p. 138.

[26] *O R* , 30, Part 2, pp. 548-49.

[27] *Alexander*, p. 449; *Connelly*, p. 191.

[28] *Longstreet*, p. 436.

[29] *O R* , 29, Part 2, pp. 719-20.

[30] *Connelly*, p. 227-28.

[31] *Ibid.*, p. 191.

[32] *Ibid.*

[33] *O R* , 29, Part 2, p. 720.

[34] *Ibid.*, 30, Part 4, pp. 565-66.

[35] *Longstreet*, p. 436.

[36] *O R* , 30, Part 4, pp. 512, 561-62; *Connelly*, p. 158.

[37] *Ibid.*, 30, Part 2, p. 604.

[38] *Alexander*, p. 448.

[39] *Ibid.*

[40] *O R* , 29, Part 2, p. 721.

TIMELINE

After Shiloh in April, 1862, the Confederate Army retreated to Tupelo, Mississippi; not until August did newly appointed Braxton Bragg begin a campaign to regain lost land and initiative. Opposed by Don Carlos Buell at Nashville, the two armies marched through Tennessee to Kentucky and back again, culminating in October with the Battle of Perryville. Immediately afterwards, the Confederates retreated, via Cumberland Gap, to Knoxville. The recovery of the Gap, which, until their advance, had been held by the Federal commander G.W. Morgan, was an important acquisition for Bragg. Cumberland Gap controlled access from Tennessee to Virginia, and was vital for Confederate communications and the transportation of supplies from the rich Tennessee Valley to Virginia. It also controlled the vital railway link, without control of which Rebel forces would be obliged to make long detours when passing from east to west.

The Confederate newspapers considered Bragg's invasion of Kentucky a failure and critized him for it, but President Davis took no action when he visited Tennessee in December. Later in the month, General Rosecrans, who then commanded the Union army at Nashville, advanced his troops toward Murfreesboro, and Bragg moved up to meet him. On the last day of 1862, after the Battle of Murfreesboro, or Stones' River, Bragg telegraphed Richmond, "God has granted us a happy new year." On January 2nd, his attempt to dislodge Rosecrans was defeated, causing him to retreat toward Tullahoma. By this action, Bragg lost the confidence of his Corps commanders; the situation became so acute, Davis ordered General Joe Johnston, the overall commander of the department, to go from his headquarters in Mississippi to Tennessee to determine whether Bragg should be removed. Johnston's decision to leave Bragg in command did little to improve the morale of the army.

Nothing major occured in Tennessee until June, 1863, when Rosecrans began moving his army from Murfreesboro toward Bragg, in order to push him back beyond the Tennessee River. This he succeeded in doing, and by July, Bragg had retreated across the river, toward Chattanooga, the vital hub of the railroad system in the middle of the Confederacy.

On the far left of army operations stood Vicksburg, Mississipi. The Confederates controlled the great river south of the city, and Union forces held the north. Throughout 1862, and into 1863, numerous attempts were made by Federal forces under Grant and Sherman to reduce Vicksburg by water and land. All met with little success until the spring of 1863, when the Federals passed the shore batteries and gained a foothold south of Vicksburg. Crossing to the east shore, Grant's men wasted little time marching north to Jackson, which they took easily, as Johnston had retreated northward. Turning to Vicksburg, the victorious troops defeated General Pemberton's men in the Battle of Champion Hill, or Baker's Creek, and the Battle of Big Black River. Pemberton then retreated to Vicksburg, and was subsequently seiged by Grant. At the same time, Union troops under Nathanial Banks seiged Port Hudson, the second most critical Confederate position on the Mississippi.

In Virginia, after the Battle of Chancellorsville in May, 1863, the Federal army ceased its aggressive advancements, and General R.E. Lee, feeling a Confederate offensive in the north would relieve Vicksburg more successfully than by his sending troops west, plotted strategy. The Army of Northern Virginia moved into Federal territory, and by July was gathering around Gettysburg, Pennsylvania. The Union troops, under General Meade, also moved into position near Gettysburg.

The following chronological list of events details particular incidents that led to "the empty victory at Chickamauga."

July 1, 1863

Pennsylvania

Ewell's forces encounter the enemy and push them back in the first days fighting at Gettysburg.

Mississippi

Union forces explode a mine under the Confederate works, leaving a huge crater.

Tennessee

Rosecrans positions himself at Tullahoma as Bragg retreats to Chattanooga.

July 2nd

Pennsylvania

The second day of fighting at Gettysburg.

July 3rd

Pennsylvania

The third day of fighting at Gettysburg sees the defeat of the Confederate Army.

Mississippi

Pemberton meets Grant to discuss the surrender of Vicksburg.

July 4th

Pennsylvania

The ANV retreats from Gettysburg.

Mississippi

Vicksburg is formally surrendered.

July 5th

Mississippi

The Confederates at Vicksburg are paroled.

July 8th

Louisiana

Port Hudson surrenders, leaving the entire Mississipi under Federal control.

July 9th

Louisiana

General Gardner formally surrenders Port Hudson.

July 12th

Virginia

The ANV prepares to cross the Potomac into Virginia. Lee writes Sam Jones to take his force to Winchester.

July 13th

Virginia

The ANV crosses into Virginia. In Richmond, Davis assigns D.H. Hill the command of General Hardee, and orders him to report to Bragg.

July 16th

Mississippi

Johnston pulls out of Jackson because of fear that Sherman's superior numbers will defeat him.

Virginia

Lee writes to Davis that after the fall of Vicksburg, some site must be selected on the MIssissippi, fortified and the rest of the troops in Mississippi left free to campaign.

July 17th

Mississippi

Bragg proposes to reinforce Johnston for an attack on Grant. (Johnston will reply: "it is too late.")

July 19th

Tennessee

D.H. Hill reports to Bragg; he will later write that Bragg was gloomy, despondent, prematurely gray and nervous.

August 8th

Virginia

Lee offers to resign command of the Army of Northern Virginia.

August 11th

Virginia

Davis writes Lee he will not accept Lee's resignation.

Mid-August 1863

Virginia

Longstreet writes directly to the Secretary of War (without notifying Lee), outlining a general transfer of troops west, from the ANV to Bragg.

He suggests defensive tactics in Virginia, and a concentration of troops in Tennessee, so Rosecrans could be crushed. He soon after informs Lee of his communication to Seddon, and Lee remains opposed to his plan.

August 16th

Tennessee

Rosecrans moves toward the Tennessee River, his goal being the occupation of Chattanooga. Burnside also begins a march from Lexington, Kentucky, toward eastern Tennessee.

August 19th

Tennessee

Burnside's movement on East Tennessee is discovered by General John Frazer, at Cumberland Gap.

August 21st

Tennessee

Rosecrans reaches the Tennessee River.

August 22nd

Virginia

Lee writes Davis that if Meade doesn't move, he is planning to attack him.

August 23rd

Virginia

Lee is summoned to Richmond for a military strategy conference.

Mississippi

Buckner writes Bragg that Burnside will move by Jamestown, and is expected to cooperate with Rosecrans' left. Alone, he warns, he can do little against him.

August 24th

Virginia

Lee arrives in Richmond for Day 1 of the conferences, after first writing Davis: "Nothing prevents my advancing now but the fear of killing our artillery horses." The troops borrowed from Sam Jones must also be returned to him today. He will have many long, private sessions with Davis.

Mississippi

Buckner telegraphs Cooper, asking for reinforcements to fight Burnside. He also pleads for help from Sam Jones.

August 25th

Virginia

Day 2 of Confederate High Command strategy session.

August 26th

Virginia

Day 3 of Confederate High Command strategy session.

August 27th

Virginia

Day 4 of Confederate High Command strategy session.

August 28th

Virginia

Day 5 of Confederate High Command strategy session.

August 29th

Tennessee/Georgia

McCook's Corps, from Rosecrans' army, crosses the river at Caperton's Ferry, approximately 35 miles below Chattanooga. Thomas' Corps also crosses without opposition.

Virginia

Day 6 of Confederate High Command strategy session.

August 30th

Virginia

Day 7 of Confederate High Command strategy session.

August 31st

Virginia

Day 8 of Confederate High Command strategy session. Lee writes to Longstreet: "I have wished for several days past to return to the army, but have been detained by the President. He will not listen to my proposition to leave tomorrow...I can see nothing better to be done than to endeavor to crush his Army while in the present condition." He orders the ANV to prepare for offensive operations.

September 1, 1863

Virginia

Day 9 of Confederate High Command strategy session.

Tennessee

Rosecrans' army continues to cross the Cumberland, largely unopposed. A vanguard of Federal cavalry enters Knoxville.

September 2nd

Virginia

Day 10 of Confederate High Command strategy session. Longstreet answers Lee's letter of August 31st by re-stating his belief that "I am inclined to the opinion that our best opportunity for great results is in Tennessee."

Tennessee

Buckner evacuates Knoxville and Burnside enters without opposition. This places Bragg in a serious position, and the authorities in Richmond realize he must be reinforced.

September 3rd

Virginia

Day 11 of Confederate High Command strategy session. Lee leaves Richmond to inspect the enemy's position from Clark's Mountain.

Tennessee

Burnside sends the cavalry to check Cumberland Gap, which is still in Confederate hands.

September 4th

Virginia

Day 12 of Confederate High Command strategy session. Lee is still at Orange Court House. He writes his wife he believes the enemy in his front has been sending off troops to reinforce Rosecrans, but has been getting up others. Lee also writes Longstreet, asking him how long it would take to move a corps to Tennessee.

Tennessee/Georgia

Rosecrans finishes crossing the river.

September 5th

Virginia

Day 13 of Confederate High Command strategy session. Longstreet writes to Lee that he doesn't know how long it would take to move a corps to Tennessee, but warns "if we put it off any longer we shall be too late." President Davis offers command of the Army of Tennessee to Lee, but the matter is left open.

Tennessee/Georgia

Rosecrans, believing Bragg is retreating south, divides his army into three groups, in order to move as quickly as possible through widely spread mountain gaps.

September 6th

Virginia

Day 14 of the Confederate High Command strategy session. Lee, in turn, offers Longstreet the opportunity to go west. Longstreet stipulates he must be given time to gain the confidence of the men, and if a victory is achieved, it must be followed up. Lee also decides it would be better to leave command of the troops in Tennessee to those officers familiar with the troops. In Richmond, the final decision is made to send troops west, and orders are given to the Quartermaster General to prepare transportation. Troops are to be carried entirely by train to Chattanooga, via the Bristol-Knoxville line, approximately 540 miles. Lee determines to leave tomorrow for the army.

Tennessee/Georgia

Bragg decides to evacuate Chattanooga.

September 7th

Virginia

Lee leaves Richmond to return to the army.

Tennessee/Georgia

After receiving a message that Cumberland Gap could not be taken by cavalry, Burnside sets out immediately with Gilbert's infantry brigade. Rosecrans' army closes in on Chattanooga: McCook crosses Lookout Mountain to Alpine; Thomas crosses McLemore's Cove, and Crittenden is in Wills Valley.

September 8th

Tennessee/Georgia

Bragg evacuates Chattanooga and withdraws toward Lafayette, Georgia.

September 9th

Tennessee

The Confederates surrender Cumberland Gap. Federal forces occupy Chattanooga.

Virginia

The first cars begin arriving at Orange Court House to take Longstreet's troops west. Before the First Corps reaches Richmond, news reaches them that the Gap has been lost. This necessitates a re-routing of Longstreet, via Wilmington, Charlotte and Augusta, to Atlanta, a trek of 843 miles and 7½ days. In Richmond, amid broad hints from the War Department that Longstreet would be named to succeed Bragg, there are a myriad of problems as to the specific troops he is to take with him. Lee writes Davis that from information received, it appears Burnside plans to form a junction with Rosecrans. This being the case, he ought to be attacked as soon as possible. Lee than stages a spectacular review of the 2nd Corps, ANV.

September 10th

Georgia

Bragg orders an attack on Rosecrans' troops that fails to materialize. The Federals pull back to regroup.

Arkansas

Little Rock falls, posing a serious threat to the Trans-Mississippi area.

Virginia

Longstreet's troops reach Richmond. Lee writes that if he were stronger, he could drive Meade to Washington, but with the absence of the First Corps, he may have to retreat to Richmond.

September 11th

Georgia

Bragg orders a second attack that does not come off.

Virginia

Lee stages a grand review of the 3rd Corps, ANV.

September 12th

Georgia

Rosecrans works at reuniting his forces. At dusk, the first of Longstreet's men arrive in Atlanta.

September 13th

Georgia

Bragg plans another assault that is not carried out.

Virginia

Lee withdraws across the Rapidan and Meade occupies Culpeper Court House.

September 14th

Virginia

Lee writes Davis that he learned by papers today that Rosecrans occupied Chattanooga on the 9th and that Burnside had marched upon Knoxville. He says Bragg was either misinformed, or Burnside changed his plans. Had he known, he would have sent Longstreet to oppose him. He fears Burnside will force Sam Jones back, and thus aid the advance of Meade.

Georgia

Longstreet's men leave in the evening on the Western & Atlantic Railroad to Bragg.

September 15th

Georgia

Longstreet's men continue their journey from Atlanta to where Bragg's forces are waiting for them.

September 17th

Georgia

Union troops gather around Lee and Gordon's Mill, and Bragg determines to turn their left flank, get behind them and cut off their retreat to Chattanooga.

September 18th

Georgia

The first of Longstreet's forces arrive. Bragg's attack does not come off. Dense woods hinder each army from sizing up the other.

September 19th

Georgia

Intermittent heavy fighting as Bragg concentrates on the Union left. In the afternoon, Longstreet arrives with the bulk of his troops, and doesn't find Bragg until 11 P.M.

September 20th

Georgia

After a fierce contest, Rosecrans fears his army has been defeated, and races back to Chattanooga. Thomas holds the Federal left and turns back attack after attack, until night, when he retires in good order. (The Battle of Chickamauga.)

September 21st

Georgia

The Confederates do not follow up their victory, allowing Rosecrans to organize defenses in Chattanooga.

September 22nd

Tennessee/Georgia

Bragg cancels an attack on the Federals and occupies the commanding heights. A seige is begun that the Union forces eventually break. They maintain control of their military goal, Chattanooga, and the Confederates are left with an "empty victory," and a concentrated army with nowhere to go.

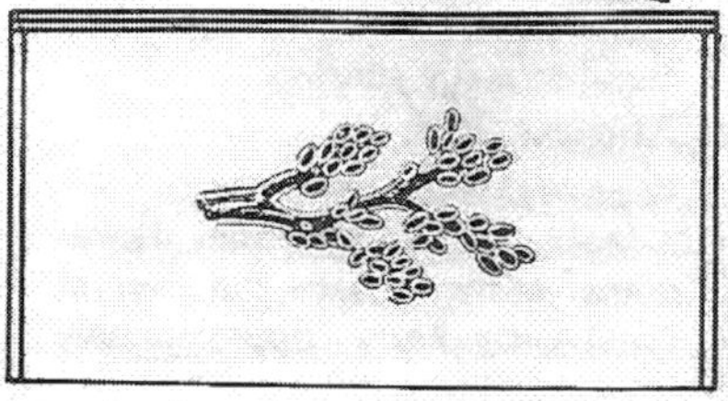

REPURCUSSIONS

The critical days of August 24 through September 7, 1863, marked a turning point in the war. The Confederate high command, composed of President Jefferson Davis, Secretary of War James A. Seddon, Adjutant & Inspector General Samuel Cooper and General Robert E. Lee delayed two weeks before acting to help Bragg's army. Their indecision proved fatal, because in those precious fourteen days Knoxville fell, and Cumberland Gap was lost: troops sent from Virginia to Tennessee would have to travel hundreds of miles out of their way to reach their destination. When they finally arrived, the command structure, notoriously weak and inefficient, collapsed. The entire attempt to salvage the middle of the Confederacy was inept, and its failures foreshadowed the death of the new nation.

The problems facing Davis' government were not new. As soon as the lines were drawn, strategic points became apparent. Vicksburg, on the left of the Confederacy, was vital because of its controlling position on the Mississippi. Chattanooga, in the center, was the hub of the railroad network, and Richmond, on the right, was the seat of government.

The paranoia over protecting the national capitol applied equally to North and South and the symbolic achievement of capturing Washington and Richmond became obsessive. This virtual monomania dictated the course of events, and the war in the East assumed gigantic proportions. While Lincoln went through many commanders of the Army of the Potomac, Davis had, after Johnston's wounds disabled him early in 1862, only one: R.E. Lee. Lee was determined to drive the Federals out of Virginia and pursue a campaign in enemy territory. His reluctance to transfer any portion of the Army of Northern Virginia stalled attempts to aid the western troops.

In the spring of 1863, General James Longstreet was invited to stop in Richmond and confer with the Secretary of War. His opportunity came in early May, when the 1st Corps was hurrying to Fredericksburg in answer to Lee's urgent summons. After paying his respects, Longstreet noted that Seddon was "deeply considering the critical condition of Pemberton's army at Vicksburg,"[1] and had a plan for sending Longstreet's Corps there. The general replied that the concentration of Confederate troops should be at Tullahoma, where they could confront Rosecrans. Neither course of action was adopted because, Longstreet felt, of "the difficulty of withdrawing the force suggested from Gen. Lee's army."[2] On May 7th, President Davis wrote Pemberton, stressing the necessity of holding both Vicksburg and Port Hudson, and promised, "You may expect whatever it is in my power to do for your aid."[3] Shortly thereafter, Longstreet presented his idea for a transfer west to Lee. "The idea seemed to be a new one to him, but he was evidently seriously impressed with it."[4] They discussed this plan several times over the coming month, with Lee's chief objection to it "the fact that if adopted, it would force him to divide his army."[5] Lee preferred a campaign into Maryland and Pennsylvania, and convinced the authorities in Richmond that this would accomplish the same result as sending his troops west, and would require less time and preparation. Lee's opinion carried a great deal of weight, and as military advisor, his arguments swayed the decision. The ANV reorganized, and after a long month of inaction, moved north.

The beginning of July saw the defeat of the Confederate Army at Gettysburg and its retreat south. On July 4th, the second disaster of the month occurred when Vicksburg surrendered, and within a week Port Hudson fell. These three events should have necessitated an immediate conference in Richmond to determine how the beleaguered and scattered armies might be put to best use. Without the Mississippi River, Tennessee and its pathway to Virginia, Cumberland Gap, took on new significance. Defenses there should have demanded the priority of the Confederate government, but precious time was allowed to slip away.

In mid-August, Longstreet wrote to the Secretary of War (without notifying Lee), outlining a general transfer of troops from east to west. The ambitious general didn't mention whether or not he received a reply, but shortly thereafter, on August 24th, Lee was invited to Richmond. Curiously, Longstreet was not. The same day, General Buckner telegraphed Cooper, asking for reinforcements to help fight Burnside. He also pleaded for help from General Sam Jones (who was receiving his troops back from Lee's army, where they had been on loan since July 12th), but did not get any. Jones' troops in Western Virginia could have been sent to aid either Buckner or Bragg (as Lee suggested on June 25th[6] and later, on September 23rd,[7] although at neither time did Lee have authority over those men), but they were not. Instead, the high command let this important little force remain where it was, while they debated the possibilities for action.

During the two weeks Lee was in Richmond, he had "long private conferences"[8] with Davis. It is apparent from his correspondence during this period that his mind was still set on an offensive in the East. On August 24, 1863, while at Orange Court House, he had written to Davis: "Nothing prevents my advancing now but the fear of killing our artillery horses."[9] His opinion did not change after eight days of meetings, as he wrote Longstreet on August 31st to prepare the ANV for an offensive. By admitting "I have wished for several days past to return to the Army, but have been detained by the President. He will not listen to my proposition to leave tomorrow..."[10] it is evident that for some time he had considered his input to the talks in Richmond at an end: he could "see nothing better to be done than to endeavor to bring Gen. Meade out and use our efforts to crush his Army while in the present condition."[11] D.S. Freeman's comment, in his Pulitzer prize-winning biography, *R.E. Lee,* that Lee's "strict conception of the subordination of the military arm of the civil government kept him from offering suggestions regarding other armies than his own, except when his views were sought by the President or by the Secretary of War,"[12] seems strangely out of keeping with the situation. Lee was called to Richmond for the very purpose of discussing strategy. (Davis' letter to Lee, dated August 24, 1863, states: "For some days I have hoped to be able to visit you, wishing to consult with you on military questions of a general character. Events in the south and west continue to detain me here."[13]) It appears from Lee's letters that his mind was made up before he ever went to the capitol and, like the critical decision-making period after Chancellorsville, could see nothing better than to attack the enemy in his front. Mr. Freeman also notes that the conferencees were well aware that Burnside would probably compel Buckner to evacuate Knoxville.[14] This he did, the vanguard of Federal soldiers entering the city on September 1st, seven days after the conference had begun.

General Samuel Cooper believed the best opportunity for success lay in reinforcing Bragg from Johnston's army. Leonidas Polk believed a concentration against Rosecrans was the soundest policy,[15] and Bragg himself had suggested, on July 17th, that he reinforce Johnston for an attack on Grant. Johnston's succinct reply "it is too late" remained his sole contribution to the pressing Confederate problems.

Decision-making did not come quickly in Richmond, either. Facing the very real possibility of having the Confederacy divided into thirds with the fall of Chattanooga and a march by Rosecrans through Georgia, the decision was again made to move against the Union Army in Virginia. On August 31st (two days after Rosecrans crossed the river below Chattanooga), Lee notified Longstreet to prepare the ANV for an offensive. On September 2nd, Longstreet bluntly replied that unless Lee felt strong enough to cross the Potomac, the best opportunity for great results was in Tennessee.[16] That same day, Burnside entered Knoxville without opposition, which necessitated yet another look at the conditions in the West. Davis himself concluded that now Bragg must be reinforced from Virginia, and Lee agreed, with misgivings.[17]

On September 4th, Lee wrote to Longstreet again, strangely asking how long it would take to move a corps to Tennessee. At this point, it might have been wiser to summon his lieutenant to him, or to have met him in Richmond, where the transfer of troops, as well as the specific units he would take, could be discussed at length, without losing time. This wasn't done, however (perhaps because Davis had intimated to Lee that he wished *him* to command the men sent west), and on the following day. Longstreet replied that he did not know how long it would take to move a corps to Tennessee.

On September 6th, the same day Bragg decided to evacuate the hub of Confederate railroads, orders were given to the Quartermaster-General to transport Longstreet's men to Chattanooga. The plan was to carry them by train on the Bristol-Knoxville line, a distance of about 540 miles. Back in Richmond, Lee wrote Davis a letter that sounded a bit like an apology: "In my conversation with you on this subject when the question was proposed, I did not intend to decline the service if desired that I should undertake it..."[18]

Lee's obvious reluctance to go made Longstreet the compromise choice. The two stipulations he claimed to have presented to Lee when the opportunity was finally offered clearly indicated "Old Pete" had more than a brief trip west on his mind. The first condition he claimed to have made was that he must be given time to gain the confidence of his own men. Longstreet seemed to have believed the rumors that were circulating in the capitol regarding his assuming command of the entire department of Tennessee. His second stipulation, that any victory be exploited, goes hand-in-hand with his desire to superceed Bragg. Unfortunately for Longstreet's ambition, Lee had already expressed his opinion when he turned down the command for himself: it had better be left to those officers already in that department.[19] Even if Lee had agreed to both conditions, neither were his to grant. No one, from Davis on down, could promise the general enough time to become familiar with any troops, and by leaving Bragg in command, no number of orders could induce an officer in charge of a battle to follow up a victory. That could only be done as circumstances warrented. As it turned out, Longstreet could not even be promised a direct route to Chattanooga. On September 7th, as the first cars arrived at Orange Court House, Cumberland Gap was surrendered. This development was the obvious result of losing Knoxville. The fact that Burnside's movements were consistantly misinterpreted as thinking he was going to Rosecrans evinced a fatal flaw in the information gathering network of the Confederacy. No spies, no cavalry and no general officer in the vicinity was able to advise Richmond in which direction the Federal Army was moving. (Reports were received from outside sources, however. On August 29th, G.A. Henry wrote to Seddon from Lexington, Virginia, that passengers from East Tennessee said Burnside was approaching East Tennessee from Kentucky. This report was forwarded to Davis.[20]) As late as September 7th, Lee thought Burnside was planning to join Rosecrans, and only learned by newspapers on the fourteenth that he had marched on Knoxville. Perhaps an even greater error in communication was the fact that it wasn't until mid-September that Bragg found out Longstreet was on his way to him.

When the troops from Virginia finally reached their destination, the Confederate concentration paid off. Despite the fact that confusion reigned before and during the Battle of Chickamauga, a major victory was achieved when Rosecrans was defeated. Unfortunately for the gray, he fell back on Chattanooga unmolested, and gained the extensive fortifications there. Bragg was unable to starve him out, and succeeded only in antagonizing Longstreet and D.H. Hill. Bragg had been an unpopular leader as early as December 1862, when Confederate news-

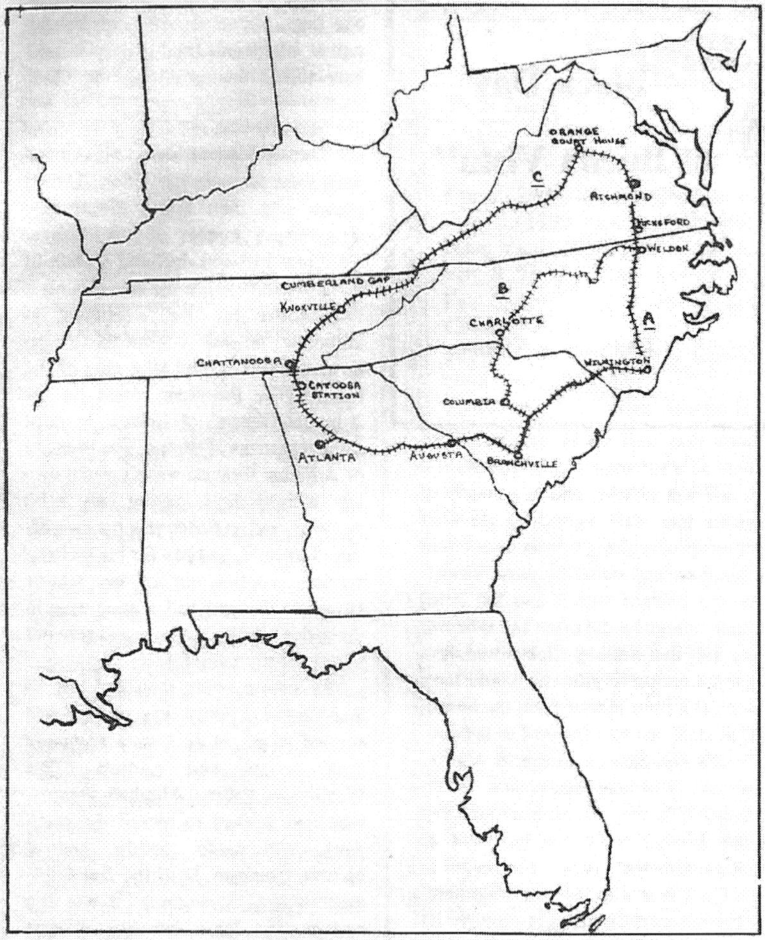

Routes marked A and B show the two routes taken by Longstreet's troops. Route marked C shows the original plan of travel through Cumberland Gap to Knoxville. (A=705 miles; B=775 miles; C=540 miles)

papers complained bitterly about his handling of the expedition into Kentucky. After his retreat from Murfreesboro in January, his own officers expressed a lack of confidence in him, and a month later, General Polk wrote directly to Richmond, requesting an immediate substitution for Bragg. Davis dispatched Johnston to review the situation, and relieve him, if necessary, but after briefly visiting Tennessee, the department commander returned to Mississippi without taking action. Davis himself had visited Tennessee during December of 1862, and gained Lee's compliment of having inspired the troops to victory at Murfreesboro. He did not, however, alter Bragg's authority, and thus seven months later, Braxton Bragg became the scapegoat for the loss of Chattanooga, and for Longstreet's "empty victory."

FOOTNOTES

[1] James Longstreet, "Lee in Pennsylvania" from *Annals of the War* (Philadelphia: The Times Publishing Co., 1879), pp. 415-16.
[2] *Ibid*, p. 416.
[3] John Bowman, *The Civil War Almanac* (New York: Bison Books Corp., 1982), p. 146.
[4] Longstreet, "Lee," p. 416.
[5] *Ibid*, p. 416.
[6] Clifford Dowdey, *The Wartime Papers of R.E. Lee* (Boston: Little, Brown and Co., 1961), p. 531.
[7] *Ibid*, pp. 602-3.
[8] Douglas Southall Freeman, *R.E. Lee* (New York: Charles Scribner's Sons, 1943), Vol. III, p. 165.
[9] Dowdey, *Wartime Papers*, p. 594.
[10] *Ibid*, p. 594.
[11] *Ibid*, p. 594.
[12] Freeman, *Lee*, p. 165.
[13] *War of the Rebellion, Official Records of the Union and Confederate Armies* (Washington: Government Printing Office, 1897), Vol. LI, Part 2, p. 759.
[14] Freeman, *Lee*, p. 164.
[15] *OR*, XXIII, Part 2, pp. 932-33.
[16] *OR*, XXIX, Part 2, pp. 693-94.
[17] Jefferson Davis, *The Rise and Fall of the Confederate Government*, p. 428.
[18] Dowdey, *Wartime Papers*, p. 596.
[19] *Ibid*, p. 596.
[20] *OR*, LI, Part 2, pp. 760-61.

Decidedly Not.

Valentine Dealer.—Well, I suppose you want something in the comic line?

Young Gentleman (who is deeply in love).—Do I look like a person that wanted anything comic?

Echoes of War

"Black Day at Buena Vista"

by

Thomas J. Evans

[This interview was given to Tom Evans by James William Lusk at his Walnut Ridge farm, Washington County, Indiana, in July, 1952.]

"That big cypress tree growing here by the porch...is the only cypress tree in Washington County, and may be the only one this far north in the United States, for all I know. Now, that tree has an unusual history. The early settlers in this area used to have to use the rivers for travel because of the poor condition of the roads in those days. They would build them a big raft out of logs and load it up with bacon, dried fruit, honey and live hogs and chickens, and float it down the river to New Orleans, Lousiana, to market. They could even sell the logs in the raft for lumber after they arrived there. When they had sold everything out, a group of them would get together and walk back home. They went in groups because of the river robbers, who would let them float down the river without bothering them, but on the way back, those robbers would ambush anyone they could for their money. That is why the men would get together and walk back for mutual protection. They went well-armed, and there was many a would-be robber who ended up shot and thrown in the river.

"Well, my Grandfather, Robert Lusk, and some local men had floated a raft down to New Orleans, and had to walk back to Indiana. When they left the city, he cut a good walking staff from a cypress tree to help him along and to kill any water moccasin snakes that happened to get in his way. When he finally arrived home here, he stuck that staff up in the lawn right there as a souvenir of his trip. Now it turned wet weather and in a couple of weeks that staff started to get little green sprouts along it, and it took root right there and started to grow. Grandpa was pleased with it and left it for shade though he did often say if he had any idea that walking stick would have taken a notion to grow he would have stuck it a little farther from the house. That staff just had the will to survive.

"It's also human nature to want to survive. Now take Joe Jackson, for instance. You won't remember him, but your brother will, for he went to school with Joe's twins. Joe fought in the Civil War with the 53rd Regiment, if I recall correctly. He got captured by the Confederates and put in the Andersonville, Georgia, prison camp. That was a real hell hole, and many a Union Army boy starved to death in there. Joe told me once that he was starving, when he had the good luck to make a meal from a half cup of peas that gave him strength and saved his life. Now those peas had already passed undigested through one man before Joe had got hold of them. But Joe had the will to survive, and he lived to come back here and become a successful farmer. He did say, though, that he never had cared much for peas since Andersonville.

"Well, today I want to tell you a little about the Mexican War, and about a few of the boys from around here who fought in it. It's tradition that the State of Indiana troops did not fight well in Mexico, and most people say the less said about the matter the better. Now that is simply not true because the fault of our boys retreating at Buena Vista is the responsibility of one man, a colonel, who ordered the retreat, afterwards tried to deny it, and later turned traitor during the Civil War.

"During May of 1846, Mexico invaded the southern part of the United States with their army, killing and wounding a number of our citizens. Congress met and declared a state of war existed between the two republics and voted for the President to authorize 50,000 volunteers to go down to Mexico and take care of the matter. The President called on the Adjutant General of Indiana to raise three regiments of troops. The position of Adjutant General wasn't much of a job in those days, paying only $100 per year, and he had to rent his own office, furnish it, and pay for his fuel and writing supplies out of his salary. However, Indiana had a good man in the office, and he got busy and sent out a notice for volunteers.

"Of course every man and boy in Washington County was excited and wanted to go. In no time a company was raised and named 'The Washington Rifles.' Abraham Dennis, who had worked to recruit the company, was made captain, and it became Company D of the 2nd Regiment Indiana Volunteers. It was the custom then for a few boys to get together and form a mess, it saved each one from having to carry cooking equipment, a good idea when there was a lot of marching to be done.

"The mess I want to tell you about was made up of 2nd Lieutenant Thomas C. Parr, Sergeant Andrew H. Ratts, and Private Horace N. Attkisson and David Colglazier. Things were more democratic in the army in those times, and when Tom Parr started putting on airs about being an officer, the boys voted him the honor of carrying their skillet and cooking pot for a month. In June the company was drawn up in ranks in town, and a pretty girl presented them with a flag, and each soldier with a pocket testament. Then after they were fired up with a lot of patriotic speeches, mostly from men who didn't plan to go, they marched off to whip the Mexicans.

"The company marched down to New Albany, where the men were issued a flintlock musket, equipment, and 40 cartridges. Then they got aboard a river boat and rode down to New Orleans, rested up a bit, then went on transport ships down to the mouth of the Rio Grande River in Mexico. They got there in August and set up camp and started training. The boys in the Regiment did not eat well on those transports, but Tom Parr slipped his friends plenty of fruit from the officers' mess where the eating had been a lot better. Now, Joseph Lane had been elected colonel of the 2nd Regiment; he was later governor of Oregon, senator from that state, and ran for vice president, but was defeated. The President felt he was a capable officer and promoted him to brigadier general. They had to have an election to replace him, and William A. Bowles, who had raised a company and was made captain, campaigned, and barely won. Billy Bowles was pretty much of a politician with no military background, and there was a lot of dissatisfaction with his winning.

"Come December, the 2nd Regiment, which had been training and doing guard duty broke camp and marched into the interior of Mexico. By February of 1847, they had located Mexican General Santa Anna, with 20,000 troops, awaiting the little U.S. Army of 7,000 men near a ranch named 'Buena Vista.' General Zachary Taylor, commanding the U.S. forces, lined his men up in a good defensive position with the 2nd Regiment holding the left flank. On the 23rd of February, the Mexican Army attacked their position. The boys in Company D were in line out in the open, and the Mexican artillery was shooting cannon balls into the Regiment and giving them a deadly fire of grape shot and cannister balls in the flanks. The Mexican Infantry massed in their front, and they exchanged musket fire until the Mexicans made a charge. Lt. Tom Parr was killed instantly, and Sergt. Andy Ratts was seriously wounded. Suddenly Col. Bowles shouted, 'Cease firing and retreat!' Now, the boys didn't want to do it, and he had to shout it several more times before they would pay attention and start backing up. Andy said he wasn't able to go with them and for them to leave him there. Well, his mess mates would have none of that, for they had already lost Tom, and they hadn't a mind to lose Andy, too. Dave threw Andy over his shoulder like a sack of feed, and carried him to the rear while Horace, who was a crack shot, covered them with his gun. Dave said later he kind of got the idea Horace was using him and Andy for bait, because every time one of those Mexican cavalry lancers would ride over to spear them, why Horace would blow him right out of the saddle. Those Mexican lancers killed a lot of the wounded boys; one fellow, Chapman by name, was speared a dozen times by those lancers while he was trying to help a wounded friend off the battlefield. He must not have had anybody like Horace looking after his back. And those lancers caught a wounded captain being hauled off the field in a wagon and murdered him, robbed him, and stripped off his uniform for souvenirs. Andy was lucky the boys were able to carry him back to a field hospital where he was safe.

"After they had left Andy at the hospital, Horace and Dave discovered the Mexican lancers had broken through the lines and were trying to capture the Army's supply train there in the rear, so they joined up with those members of the 2nd Regiment that had rallied and fought them off. Then they reformed in companies and marched back to the front line and fought until the Mexicans fell back. General Taylor, 'Old Rough and Ready' the men called him, was quick to use his own artillery when he saw the Mexican soldiers were bunching up.

'Double-shot them guns and give'em hell, Bragg!' he shouted to Captain Braxton Bragg, who afterwards became a Confederate general during the Civil War. Now newspapers can't print that kind of language, and they reported him as shouting, 'A little more grape, please, Captain Bragg.' Anyway, that statement won him the presidency later.

"The boys never saw any more of Col. Bowles that day, as he and a few men had fallen in with Jefferson Davis' Mississippians and fought along with them. The Mexicans finally fell back after near whipping the U.S. Army, and next day it was learned the Mexicans had retreated during the night. It had been a bloody battle, the Mexicans had 2,000 men killed, and the U.S. Army lost about 750 boys. Company D had Tom Parr and three other boys killed, and Andy Ratts and six others wounded so bad they had to be taken to the hospital. But what really got everyone upset was that Col. Bowles, in his official report of the battle, wrote that the Regiment had broke and retreated without orders, and he could rally only a few men to fight with the Mississippi troops.

"Well, the damage was done, and everybody thought the 2nd Regiment had turned tail and run out on their comrades. Then Lt. Spicely, who later became a Union Army general during the Civil War, spoke up that he had heard Col. Bowles order the retreat, and he wanted an investigation of the matter. There was a Court of Inquiry held, and a dozen or so officers and men all testified they had heard Col. Bowles give the order to retreat, not just once, but several times. General John Wool who was head of the court came right out and told Col. Bowles he had disgraced the 2nd Regiment and Indiana, and that he was unfit to command. Captain Abe Dennis of Company D was right put out about the matter and challenged Col. Bowles to a duel. Cooler heads prevented it from taking place, which was probably a good thing for Billy Bowles, because Abe was not a man to be trifled with. The newspapers here in Indiana made it hot for Billy, but nothing was ever done to him officially. He was left in command until the 2nd Regiment was returned to the States, and discharged at New Orleans in the spring.

"Well, the United States won the war at the cost of 12,000 lives lost, and for every boy that was killed by a bullet, ten had died by disease. Dave had a bad spell with his bowels, and it didn't look like he would make it for a time, but he was strong and recovered in time to come home with Company D. They carried Tom Parr's body back with them, had a grand funeral, fired a volley over his grave, and marked it with a tall flat tombstone inscribed that he had been killed at Buena Vista. The rest of the killed were buried in Mexico, and Tom probably would have been, too, if it hadn't been for his mess mates.

DAVID COLGLAZIER, CO. D., 2 IND. INF., MEX. WAR

"The folks were glad to see the boys back home again from the war, and they had all kinds of welcoming back festivities, with long speeches, and a huge barbecue out at the edge of town. Then it was over, and time for the boys to start being civilians again. The old time, until Andy Ratts decided he liked what he had seen down in Mexico ed what he had seen down in Mexico and went down to Texas to try his hand at ranching. Dave Colglazier started farming and bought three hundred acres in the county. He liked to raise stock and was one of the best judges of cattle in all of Indiana. He never did get completely over the illness from the war, and some folks think that is what eventually caused his death. He is buried over in the town cemetery not very far from Tom Parr and has 'MEX.WAR' carved on his white Government grave stone.

"It wasn't long before people began to suspect there was going to be another war over State's rights, slavery, and other issues. Zachary Taylor, who had been elected president after the war for what he had accomplished down in Mexico, had completed his term of office, and was a sick old man when the talk of rebellion started. He raised right up off his sick bed and wanted to take an army down South and settle the matter right away. Now his daughter had married Jefferson Davis, who became President of the Confederacy, and his son, Richard Taylor, served as a general in the Confederate Army. I guess you could say his family did not entirely agree with his views.

"The Civil War started with the Confederates firing on Fort Sumter, and soon the States were caught up in the conflict. Andy Ratts was in Texas, and he joined the Texas Militia and supported the Confederacy. Horace Attkisson was as ready to fight for the Union as before, and when they raised the first company of volunteers in the county, he was elected first lieutenant. That company eventually became Company G, 13th Regiment Indiana Volunteer Infantry. Horace proved to be an outstanding officer because of his previous military experience, and he soon came back to the county to recruit a company of his own. This was Company C, 50th Regiment Indiana Volunteer Infantry. The Regiment was sent down in Tennessee to guard railroad bridges and supplies from the local guerillas and the Confederate Army.

"Now, Confederate General John Hunt Morgan decided he would conduct one of those big raids for which he became famous. He and his men rode through Tennessee, burning houses and confiscating horses, and they trapped Horace and twenty of his men in a log blockhouse guarding the railroad at Edgepoint Junction. There were a thousand Confederate cavalrymen in Morgan's command, and they attacked the blockhouse figuring Horace would surrender right away. Well, that wasn't the way of it at all. Horace said right off that the Confederates could never run a bunch of crack shots like them out of that blockhouse. Horace had several cases of ammunition, and plenty of food and water, and he had no intention of surrendering.

"One of the boys up and says, 'Captain, there are fifty of them for every one of us!'

Horace replied, 'That is their hard luck. That means each of you has fifty targets for their one!'

"So the boys all had a good laugh about that and felt better, particularly because Horace didn't seem upset about the unfavorable odds and wasn't the least afraid. During the first attack, he was all over the blockhouse firing his pistol through the slits in the walls and encouraging everybody. When the Confederates ended the attack the boys got to looking around and saw none of them had been shot while there were quite a few Rebels laying around outside on the ground. That got them to feeling pretty good and raised their confidence in Horace's leadership.

"As they were getting ready for the next attack they figured would be coming, a Confederate officer mounted on a horse casually rode out on a knoll about five hundred yards away and stopped. It appeared as if he was writing in a notebook and sketching the blockhouse. He was a former newspaper reporter who had come down from the North to join up with General Morgan, and the boys later thought he may have been making notes for a little newspaper he was trying to publish for the Confederate troops.

'One of you squirrel hunters drop that officer,' Horace says.

"One of the boys drew a bead on him and squeezed off a shot, and that officer threw up his notebook and fell dead from his horse. The next issue of that newspaper carried his own obituary. Now the boys generally weren't much on shooting officers except during breaks in the serious fighting. The unwritten rule was it was best to shoot the privates since they had rifles, and were trying to kill you. The more privates you shot, the better your chances of surviving the battle. Officers were usually only armed with a pistol and a sword, and couldn't do a man much damage. From what I have heard the Civil War veterans say, I am convinced that most of the officers that got killed in the fighting during the battles accidently stopped a bullet that was meant for some poor private.

"Now, the shooting of that officer really stirred the Confederates up, just like throwing a rock at a hornet's nest, and with a yell they started their second attack. The shooting was something fierce, but the boys felt secure inside the blockhouse, and Horace was jumping all over the place shouting that everybody was doing a good job and to make every shot count. After about half an hour, the Confederates drew off and quieted down again. The boys all checked themselves, and outside of some minor splinter damage and some bruises from ricocheting bullets, all twenty of them were in good shape and still able to shoot.

"Then here rides out a Confederate holding a white flag, and shouts, 'Flag of truce! Hold your fire!'

"It was clear the Confederates wanted to talk, so Horace spruced up a bit, took some white gloves out of his saddlebags, and put on the sword he hardly ever carried if he could get out of it.

'I am going to confuse them a little bit,' says Horace.

"Then, with his sergeant behind him, he strutted right out the door just like he was on dress parade. He marched up to where the fellow on horseback had been joined by a Confederate captain.

'Captain,' Horace says, 'I accept your offer of surrender. Have your men stack their arms in the field on your left and move on over to the area on the right. I will issue paroles, and there will be liberal terms for all.'

"The Captain was taken back and stammered, 'The General's compliments, Sir. He is requesting you to surrender because your situation is hopeless. Our next attack will surely wipe you out.'

'Well,' says Horace, trying to look surprised, 'my compliments to your General. However his attacking is going to end, for the way my boys are shooting, he will soon have only as many men left as I have. Then it will be our turn to attack if the relief column from our regiment doesn't get here by then and spoil things. Give your General my regards and tell him we will be waiting for your attack.'

"With those words Horace saluted smartly, then he and the sergeant marched back into the blockhouse. Horace told the boys what he had done, and they thought it was a pretty good joke. However, they guessed the Confederates probably hadn't appreciated the humor in Horace's little act and would be getting kind of mad when they got to thinking on it. So the third attack started with a cavalry charge, and the boys emptied a few saddles before the Confederates decided cavalry shouldn't be attacking infantry who were so well protected and full of fight. Then they threw in dismounted troops and went at it hot and heavy for about an hour before they finally realized there was no way they were going to get Horace and his boys out of that blockhouse. They called a halt to the fighting and continued on their raid leaving behind eight dead and about twenty men too seriously wounded to move. There was no telling how many more wounded were able to ride away, because the boys saw several who had to be held up on their saddles. The boys came out and gave a cheer for Horace, and they later said those logs in the blockhouse had so many bullets in them a man could have burned them and collected a fortune from selling the lead.

"The Union Army was right pleased with Horace, and said so, and promoted him to be a major. The 50th Regiment marched on down into Arkansas, and fought Confederate General Nathan Bedford Forrest's army, and some of Forrest's cavalry captured their wagon train during the fighting. As soon as Horace heard about it he shouted out that cavalry couldn't hold that train, and for some boys to follow him. I guess he remembered that from down in Mexico. Anyway, Horace led an infantry charge and captured all of those wagons right back before the Confederates knew what was going on. Then Horace had the bad luck to be captured.

"Horace was taken down to Texas and put in a prison camp. There were officers in there that knew him, and they couldn't understand the way he suddenly changed. Oh, he was polite as ever to all, but he didn't mingle with the other officers, and he didn't join any of the groups that spent their time planning an escape. The folks in there were beginning to think maybe he was sick or that being captured had broken his old fighting spirit. Then one morning when they awoke, Horace was gone! Completely disappeared! Nobody could figure out how he had escaped. They tried to cover up for him, but at roll call the Confederates found out he wasn't there. The Confederates announced they would have him back in no time, for it was impossible to escape from that prison camp. They set a pack of dogs kept for that purpose after him. Well, those dogs lit right in and picked up his scent. They went barking along for a couple of miles to a clump of trees and then were quiet. When the Confederates arrived, they discovered a horse had been tied there waiting for Horace. They couldn't figure out how any Yankee cavalry could be that far South, and there weren't any Union sympathizers in the area that they knew about.

"Now, the Confederates had another trick up their sleeves. They had a couple of Indians they used for scouts, something the U.S. Army never got around to doing until the Indian Wars. So these Indian scouts rode all around the prison telling how they intended to catch Horace and scalp him and bring his scalp back to show the other prisoners. They said they would catch up with Horace in no time, and they did, too, for about a week later, a Confederate cavalry patrol discovered what the buzzards had left of them. They had gotten so interested in following Horace's trail up a dry rocky river bed, they forgot to be cautious and probably never did see him sitting in the shade of some big rocks on the bank waiting patiently for them with a carbine and pistol. Each of those Indians had been shot neatly between the eyes, they reported. Horace rode on leisurely back to the Union Army lines, and made a report on all the troop movements he had seen while he was escaping.

"After the war was over, Horace told how he had managed to get away. One of the main guards there at the prison was his old mess mate from the Mexican War, Andy Ratts! Even though Andy was a Confederate, he had to help his old comrade who wouldn't leave him on the battlefield at Buena Vista when he was wounded. They had never let on they knew each other, and were careful not to be seen talking. Andy fixed Horace up with a good fast horse, guns, money, maps, food and a canteen. One dark night he slipped Horace out and sent him on his way to the Union Army lines. Horace just seemed to disappear! No one ever suspected just how Horace had managed to pull it off until he told. Andy was a guest at Horace's home many times after the war. Andy eventually died down South, I heard, and was buried beneath a Confederate tombstone, although if he had wanted he could have had a Government stone due to his Mexican War service.

"Horace was promoted to lieutenant colonel in the army by the time the war was over. He came home to stay, and was well thought of by all, and respected throughout the county. The local veterans gave him a nice funeral when he died, and fired a volley over his grave. He had ordered his own monument, not a government tombstone, and it has a sword decorating it to show he was in the military service. His family had a nice verse carved on it, and I suppose he is in a fitting resting place for a hero.

"Well, Colonel Bowles had returned to his home over at French Lick when the Mexican War had ended. His first wife died, and he remarried a widow woman, from Mississippi, if I recall correctly, who had inherited a huge plantation and a large number of slaves. They decided to continue to live at French Lick, and spent a lot of time on river boats going back and forth to look after the operation of the plantation. Pretty soon Colonel Bowles began to consider himself a planter, and to talk about the need for the South to secede from the Union.

"When the Civil War started, Colonel Bowles decided he didn't want to serve in the Confederate Army, and that he would continue to live in Indiana. He didn't have to worry about serving in the Union Army because of his past war record and the enemies he had made then who were now high-ranking officers. What he did do was become a major general in the Knights of the Golden Circle, an organization of traitors, that helped recruit soldiers for the South, forwarded money and dispatches to Richmond, Virginia, and generally acted as spies. The Governor of Indiana got wind of what was going on and infiltrated the organization with his own detectives. Soon he knew as much as the leaders about what was going on inside that secret society.

"When the detectives had collected enough evidence for conviction, he had Colonel Bowles and the other leaders of the Knights arrested, and it was decided they would be court-martialed by a military court, since there was a war going on. The court found Colonel Bowles guilty and sentenced him to be hanged at a military camp outside of Indianapolis. Several local folks thought it was a fair verdict and long overdue, and several of the former soldiers who had served under him were planning to attend the ceremony, when the President decided to commute his sentence to life imprisonment at hard labor. They sent him to the federal penitentiary in Ohio to serve his sentence. When the Civil War ended, it was decided a lot of the court-martial trials of civilians should have been conducted in the civil courts, and a lot of the sentences that had been handed out were suspended. Colonel Bowles' case was one of these, and he was released from prison about the middle of 1866.

"His wife's plantation was ruined during the war by some of his enemies in the Union Army, and the slaves were all set free. Colonel Bowles decided to return to French Lick, and lived out the rest of his days. He died peacefully in bed, and how that old rascal escaped being shot over the years is beyond me. And so that is the story of a black day at Buena Vista, the man who caused it, and four county boys who fought there."

Colonel William A. Bowles

OBITUARY

Indianapolis (Indiana) Journal, April 2, 1873.

Colonel Wm. A. Bowles died at his residence in French Lick, Orange County, on Friday last, in the seventy-eighth year of his age. Colonel Bowles, though not a brilliant man, possessed some remarkable traits of character, and at one time exercised a powerful influence in southern Indiana, where he has resided for twenty-five or thirty years past.

Prior to the Mexican War he was somewhat of a politician and had been elected to fill various positions in public life. Upon the opening of that war he raised a company for one year's service and it became a part of the Second Indiana Regiment. On the appointment of Colonel Joe Lane, who was elected first colonel of the Second Regiment, to a brigadiership by President Polk, Captain Bowles became a prominent candidate for the vacant colonelcy. The vacancy was not filled, however, until the arrival of the regiment at Brazos Island, where the election took place, and Captain Bowles suceeded in defeating Captain W.L. Sanderson, who was the opposing candidate. As commandant of the regiment Colonel Bowles proved a failure, as he was wanting in those pecular traits of character that are required for success. He was in command of the regiment at the battle of Buena Vista and by his miserable bungling, if not cowardice, the regiment and State were disgraced. A well-known gentleman now residing in this city, who was a captain in Colonel Bowles's regiment, severely denounced his conduct in that engagement to his superiors, and branded Bowles as a coward. His friends charitably attributed his course to military ignorance or error, but those who had the best means of knowing maintain to this day that he showed the white feather most unmistakably. At any rate, from that time to the end of the service Colonel Bowles was without influence, and the regiment which he commanded saw no further service, by which it could wipe out the stigma attached to it.

On his return from Mexico Colonel Bowles retired to his home and until the beginning of the late War of the Rebellion busied himself with his private affairs. Having a large interest in property in the South, which had been secured by his last marriage, he took sides with the Southern people in the war and became a prominent member of the Knights of the Golden Circle. His connection with that remarkable passage in the State's history will be well remembered. There was abundant evidence to prove that he was associated with other members of the Democratic party to liberate the rebel prisoners at this city, sieze the arsenal and military stores here and revolutionize the State.

In September, 1864, Bowles was arrested, along with Harrison S. Dodd, L.P. Milligan, Andrew Humphrey, John C. Walker and others, tried by a military commission in this city on the charge of treason and sentenced to death. His sentence was commuted by President Lincoln to imprisonment for life, and Bowles was placed in a military prison in Ohio. Afterward the case was carried to the Supreme Court and, after a protracted argument, the finding of the military commission was set aside and Colonel Bowles discharged from custody. He returned to his home in Orange County, where he has since lived a quiet and obscure life. The deceased was a man of amiable private character and was highly esteemed by those of whom he chose to make friends.

ACKNOWLEDGEMENTS

Mrs. Edith Dalton, Historical Researcher, Salem, Indiana. Mr. Herman Jackson, Superintendent, Crown Hill Cemetery, Salem, Indiana.

VERIFICATION SOURCES:

Bowles, William A.

"Captain Bowles Elected Colonel Of Second," *Indiana Democrat*, August 28, 1856. "We learn from the Army that Captain W. A. Bowles has been elected colonel of the Second Regiment to fill the vacancy occasioned by the promotion of Colonel Joseph Lane."

"Letter From A Soldier Of The Second Regiment," *Indiana Sentinel*, May 9, 1847. Letter indicates Col. Bowles gave the order to "Cease firing and retreat" during the battle of Buena Vista.

Lavender, David, *Climax at Buena Vista*, (New York: Lippincott & Co., 1966), p. 191. Describes Col. Bowles standing in the rear as shouting "Cease fire and retreat" three times before his men would obey.

Ibid., p. 168. Indicates at the time of his commission Bowles had known nothing whatsoever of military science, and during more than half a year in Mexico made not the least effort to learn.

Smith, Justin H., *The War with Mexico*, (New York: Macmillan, 1919), p. 557. Describes Bowles as personally brave, but he had been away much of the time and did not understand his orders.

Reeder, Russel Potter, *The Story of the Mexican War*, (New York: Merideth Press, 1967), p. 105. States Bowles' order to retreat was not the result of cowardice but that he was green and confused.

War of the Rebellion, Official Records of Union and Confederate Armies (Washington: Government Printing, 1881), Series I, Vol XXXIX, Part 2, p. 237. Bowles identified as major general in the Knights of the Golden Circle.

OR, Series II, Vol VII, p. 340. Bowles accused of enlisting men in the Rebel Army and secreting Rebel Officers.

OR, Series II, Vol VIII, p. 548. Bowles sentenced to be hanged for treason.

Ibid., p. 583. Sentence commuted to life prison at hard labor.

Ibid., p. 896. Bowles released from prison on the President's orders on April 10, 1866.

Attkisson (AKA: Atkisson, Atkinson), Horace N.

Perry, Oran, Adjutant-General, *Indiana in the Mexican War*, Indianapolis, Buford Press, 1908, p. 376. Identified as Private, Co. D. 2nd Ind. Inf.; Mustered—June 20, 1846; Discharged—June 23, 1847.

Terrell, W. H. H., *Adjutant General's Report— Indiana*, Indianapolis, 1865, Vol II, p. 103. First Lieutenant Horace N. Attkisson, Co. G, 13thReg. Ind. Vols. promoted to Captain in 50th Reg. Ind. Vols., Dec. 6, 1861.

Ibid., p. 491. Horace N. Attkisson commissioned Major on Sept. 8, 1862, and Lieutenant Colonel on Nov. 19, 1863.

Ibid., p. 491. Horace N. Attkisson commissioned Major on Sept 8, 1862, and Lieutenant Colonel on Nov. 19, 1863.

Ibid., p. 500. Mentions Capt. Attkisson's battle at Edgefield Junction, Tenn.

OR, Series I, Vol XVI, Part 2, p. 454. Capt. Attkisson inspires Union soldiers at Edgefield Junction, Tennessee.

OR, Series I, Vol. XVII, Part 1, pp. 581-583. Maj. Attkisson recaptures 50th Ind. wagon train following loss to Forrest's cavalry.

OR, Series I, Vol XLI, Part 3, p. 318. Major Attkisson identified as a prisoner just escaped from Tyler, Texas reports on Confederate troop disposition near Ft. Smith.

Parr, Thomas C.

Perry, *op. cit.*, p. 378. Identified as 2nd Lt., Co. D, 2nd Reg. Ind. Inf.; Mustered—June 20, 1846; Died—Feb. 23, 1847.

"Funeral Of Lieutenant Parr," *Salem Republican*, July 16, 1847. 2000 citizens attended funeral of Lt. Parr, killed at Buena Vista.

Stevens, Warder W., *Centennial History of Washington County Indiana*, Indianapolis, Bowen & Co., 1916, p. 295. Describes funeral ceremony performed by returning veterans.

Colglazier, David.

Perry, *op. cit.*, Identified as Private, Co. D, 2nd Reg. Ind. Inf.; Mustered—June 20, 1846; Discharged—June 23, 1847.

Stevens, *op. cit.*, p. 1015. Provides family history.

Ratts, Andrew H.

Perry, *op. cit.*, Identified as Sergt., Co. D. 2nd Reg. Ind. Inf.; Mustered—June 20, 1846; Discharged—June 23, 1847.

Stevens, *op. cit.*, p. 286. Relates Ratts' aid to Maj. H. N. Attkisson in escape from prison camp where Ratts was a Confederate guard.

Miscellaneous

"Captain Kinder, Second Regiment," *State Journal*, April 20, 1847. Identifies Captain T. B. Kinder, Co. B, 2nd Regiment, wounded at Buena Vista, as being murdered, robbed, and stripped by Mexican lancers while being removed from the battlefield in a wagon.

Reeder, *op. cit.*, p. 228. Describes Private Giles Chapman, Co. K, 2nd Reg., as sacrificing his life while trying to carry off a wounded comrade when he received 11 lance thrusts through his body.

Thomas, Edison H., *John Hunt Morgan and his Raiders*, (Lexington: University Press, 1975), p. 51. Describes death of Confederate Lt. Gordon E. Miles of Morgan's command, former newspaper reporter from New York, who was killed at Edgefield Junction, Tenn.

Wilcox, Cadmus M., General., *History of the Mexican War*, (Washington, D.C.: Church News Pub. Co., 1892), p. 561. Provides Mexican War statistics—100,000 men served. Of this number, 120 officers and 1,400 men fell in battle or died of wounds; 100 officers and 10,800 men perished by disease.

The retreat of the 2nd Regiment may be found in *Senate Executive Document*, 30th US Congress, 1st Session, Official reports and subreports of the Battle of Buena Vista and its aftermath. For frank and interesting opinions of Col. Bowles by participants, see Scribner, B. F., *A Campaign in Mexico*, Philadelphia, 1850, pp. 60-62, and Carelton, James H., *The Battle of Buena Vista*, New York, 1848, pp. 59-62.

Argumentum ad Hominem.

Strict old gent.—What's the matter, boy?

Boy.—Boo, hoo, hoo! I've lost all my marbles.

Old gent.—Now you see the consequences of playing marbles on Sunday.

SIGNALS

Dear Editors:

I would like to subscribe to your magazine THE KEPI for the year. I've enjoyed the sample issues you've sent me.

I have enclosed a check of $18.00 for the subscription.

Thank you,
Mark H. Daniels
Pawling, New York

Dear Editors:

Just received my first issue of your magazine this month, and I would like to congratulate you and your staff on its scholarly content, no-frills format, and unbiased choice and presentation of material.

Please send guidelines regarding the submission of articles.

Sincerely,
Thomas M. Martin
Norfolk, Virginia

Dear Editors:

As you see, I am working on my first 1984 issue. Thought you might want to use some of my "SHADES OF SHERMAN"!

Your KEPI is a fine magazine!

Sincerely,
Stan Schirmacher
Editor, Sons of Sherman's
March to the Sea
Tempe, Arizona

Dear Editors:

Enclosed is a check for $7.00 as payment for two back issues of THE KEPI. Please send me the first two issues published, Vol. I, Numbers 1 & 2, the April and June/July issues.

I have read the next two issues cover to cover and enjoy everything about your magazine. I anxiously await receiving the earlier issues. Thank you.

Sincerely,
Glen A. Noto
Bergenfield, New Jersey

[First class postage for Vol. I, Numbers 1—4, is 54¢. First class postage for issue Vol. I, Number 5, and all subsequent issues is 71¢.]

LEXICON

AGUE-TREE, á'gue-trēē. s. Sassafras.

AHIGHT, à-hīte'. ad. Aloft, on high.

AIRING, āre'ing. s. 410. A short jaunt.

BUGBEAR, bug'bāre. s. A frightful object; a false terrour.

BUGLE, bū'gl. A shining bead of black glass.

BURR, bŭr. s. The lobe or lap of the ear.

CHIROMANCY, kīr'rò-măn-sē. s. 353, 519. The art of foretelling the events of life, by inspecting the hand.

CIRURGEON, kī-rŭr'jē-ŭn. s. 353. One that cures ailments not by internal medicines, but outward applications, now written *Surgeon;* a surgeon.

EGOTISM, ē'gō-tĭzm. Too frequent mention of a man's self.

Contrary to my own judgement I have made the *e* in the first syllable of this word long, because I see it is uniformly so marked by all the Dictionaries I have seen: but I am much mistaken if analogy does not in time recover her rights, and shorten this vowel by joining it to the *g*, as if written *eg-o-tism;* not because this vowel is short in the Latin *ego,* (for the English quantity has very little to do with the Latin,) but because the word may be looked upon as a simple in our language, and the account is on the antepenultimate syllable. Mr. Elphinston, whose opinion on this point is of greatest weight, makes the

A TABLE OF SIMPLE AND DIPHTONGAL VOWELS REFERRED TO BY THE FIGURES OVER THE LETTERS IN THIS DICTIONARY

ENGLISH SOUNDS	FRENCH SOUNDS
1. à The long slender English *a*, as in fàte, pàper, etc.	é in *fée, épée.*
2. á the long Italian *a*, as in fá, fá-ther, pa-pá, mam-má.	*a* in *fable, rable.*
3. â. The broad German *a*, as in fâll, wâll, wâ-ter.	*â* in *âge, Châlcus.*
4. ă. The short sound of the Italian *a*, as in făt, măt, măr-ry.	*a* in *fat, matin.*
1. è. The long *e*, as in mè, hère, mè-tre, mè-dium.	*i* in *mitre, epitre.*
2. é. The short *e*, as in mét, lét, gét.	*e* in *mette, nette.*
1. ì. The long diphtongal *i*, as in lìne, tì-tle.	*ai* in *laique, naif.*
2. í. The short simple *i*, as in pín, tít-tle.	*i* in *innée, titrée.*
1. ò. The long open *o*, as in nò, nòte, nò-tice.	*o* in *globe, lobe.*
2. ó. The long close *o*, as in móve, próve.	*ou* in *mouvoir, pouvoir.*
3. ô. The long broad *o*, as in nôr, fôr, ôr; like the broad â.	*o* in *or, for, encor.*
4. ŏ. The short broad *o*, as in nŏt, hŏt, gŏt.	*o* in *hotte, cotte.*
1. ù. The long diphthongal *u*, as in tùbe, cùbe, cù-pid.	*iou* in *Cioutat, chiourme.*
2. ú. The short simple *u*, as in túb, cúp, súp.	*eu* in *neuf, veuf.*
3. û. The middle or obtuse *u*, as in bûll, fûll, pûll.	*ou* in *boule, foute, poule.*
ôi. The long broad ô, and the short i, as in ôil.	*oi* in *cycloide, heroique.*
ôû. The long broad ô, and the middle obtuse û, as in thôû, pôûnd.	*aoû* in *Aoûte.*
Th. The acute or sharp *th*, as in *th*ink, *th*in.	
TH. The grave or flat TH, as in THIS, THAT.	

When *G* is printed in the Roman character, it has its hard sound in *get, gone, &c. as go, give, geese, &c.; when it has its soft sound it is spelled in the notation by the consonant J,* as giant, ginger, *ji-ant, jin-ger.* The same may be observed of *S:* the Roman character denotes its hard sound in *sin, sun, &c.; its soft sound is spelled by z* as *rose, raise, &c.; roze, raze,* &c.

first vowel short.

EXARTICULATION, ĕks-ăr-tĭk-ū-lā'shŭn. The dislocation of a joint.

To FRIBBLE, frĭb'ble. v. n. 405. To trifle.

To FROUNCE, fròùnse. v. a. To frizzle or curl the hair.

To IRK, ĕrk. v. a. 108.

This word is very expressive; it comes from the Islandick *yrk*, work. It is only used impersonally, and signifies to disgust, as, It irks me, I am weary of it.

JIGOT, jĭg'ŭt. s. 166. A leg; as, a Jigot of mutton.

MAIDMARIAN, māde-māre'yăn. s. A kind of dance.

MASTLIN, mĕs'lĭn. s. Mixed corn, as wheat and rye.

OYES, ō-yĭs'. s. Is the introduction to any proclamation or advertisement given by the publick criers. It is thrice repeated.

This word, like several others, has been changed by the vulgar into something which they think they understand. It is derived from the old French imperative *Oyez*, Hear ye! but is now universally heard in courts of justice like the affirmative adverb *yes*, preceded by the long open *o*.

PSYCHOLOGY, sī-kŏl'ō-jē. s. 513. The doctrine of the soul or mind.

To NARRATE, năr'rāte. v. a. 91. To relate, to tell.

Dr. Johnson says this word is only used in Scotland; but as it is regularly derived from the Latin *narro*, and has a specifick meaning to distinguish it from every other word, it ought to be considered as a necessary part of the language. To *tell* seems to imply communication in the most general sense; as, to *tell* a *story*, to *tell* a *secret*, *&c. To relate*, is to tell at some length, and in some order, as, to *relate the particulars of a transaction;* but to *narrate* seems to relate a transaction in order from beginning to end; which often becomes insipid and tiresome. Hence the beauty of Pope's —— *narrative* old age:

"The poor, the rich, the valiant, and the sage,

"And boasting youth, and *narrative* old age."

ROUSE, ròùze. v. n. A dose of liquor rather too large.

RUDESBY, rŏŏdz'bē. s. An uncivil turbulent fellow. *Obsolete.*

SEASURGEON, sē'sŭr-jŭn. s. A chirurgeon employed on shipboard.

SENNIGHT, sĕn'nĭt. s. 144. The space of seven nights and days, a week.

To SPITCHCOCK, spĭtsh'kŏk. v. a. To cut an eel in pieces and broil it.

STARSHOOT, stăr'shŏŏt. s. An emission from a star.

To WHITTLE, hwĭt'tl. v. a. 405. To make white by cutting; as boys are said to Whittle a stick when they cut off the bark and make it white.

From "A Critical Pronouncing Dictionary and Expositor of the English Language" by John Walker, published by Collins and Hannay, New York, 1825.

MEMORANDUM OF BOOKS SENT TO WINDER'S BUILDING

By
Paul Kallina

American military men of the Civil War era admired the French army. They studied—and often copied—its ways and traditions. Their dependence on French military science is exemplified in the memorandum reproduced below. The books listed there were sent from the United States Military Academy to army headquarters at Winder's Building, Washington, D.C.

The English titles in the text of the memorandum are as they appear in the original. Translations of the French will be found in the footnotes.

The memorandum was found at the National Archives in the Records of Detached Engineer Officers (Defenses of Washington, 1861-1866), subject folder of correspondence regarding forts, Records of the Office of the Chief of Engineers, Record Group 77, Entry 557.

Received Washington D.C. July 3d 1862 of W.C. Gunnell Engineer &c—the following books belonging to Engr Department—Chief Engineer—U.S.M.A.—&c

Dictionaire de l'armée de terre par Bardin[1] — 7 Volumes
Aide mémoire to the military Sciences Vol I — A to F
J. Claudel: Formules et tables[2]
Cours sur le tracé et la construction des batteries de toute espèce[3] with Plates
Haillot: Ponts militaires[4]
Ecole de Sapes[5]
Naval & military dictionary of the French language—Burn
Monge: Application de l'analyse à la géomêtrie[6]
Fleming & Tibbins Dictionary — 2 Volumes
Tables & Formulae by Capt. T.J. Lee[7]
Palsey: Elementary fortifications — 2 Volumes
Birago's Recherches sur les équipages des ponts militaires.[8]

J.D. Kurtz
Capt & Asst

FOOTNOTES

[1] Bardin's Army Dictionary.
[2] J. Claudel: Formulas and Tables.
[3] Course in the planning and construction of batteries of every type.
[4] Haillot: Military Bridges.
[5] Sapper Training.
[6] Monge: Applying analysis to geometry.
[7] The full title of this work is *A collection of Tables & Formulae, useful in Geodesy and Practical Astronomy.* The author, Captain Thomas J. Lee of the Topographical Engineers, was born in France. See George W. Cullum, *Biographical Register of the Officers and Graduates of the United States Military Academy,* (New York: D. Van Nostrand, 1868), I, (1802-1840), pp. 362-363.
[8] Birago's Research on military bridging equipment.

MILITARY REVIEW

"*A German in the Yankee Fatherland, The Civil War Letters of Henry Kircher,*" edited by Earl J. Hess, The Kent State University Press, Kent, Ohio, 44242; 1983 list price $19.50. 178 pages, Bibliography, Index, Illustrations.

This book details the letters written home by a young German-American. Nineteen-year-old Henry A. Kircher, born in Illinois to German natives who immigrated to the United States in 1833, joined the 9th Illinois Infantry shortly after President Lincoln's call for troops on April 15, 1861. After the expiration of his three month term of service, he left the unit, apparently because of ethnic conflicts. By August, he joined Company A, 12th Missouri Infantry, that had, by the end of September, a ratio of more than nine immigrants to every one native-born soldier, and over 84% of the foreigners were German.

The 12th Missouri saw service throughout the Western campaigns, participating at and around Vicksburg, and later with Sherman in Tennessee and Georgia. Kircher worked his way up to captain, and was wounded three times on November 27, 1864, while acting as Hooker's vanguard, which was chasing Bragg's army south after Missionary Ridge. His right arm and left leg were amputated, and Kircher was sent home. Recovering from his wounds, he was elected Circuit Court Clerk, and later took over his father's hardware firm.

Kircher's Civil War letters, written in German and primarily to his parents, offer a firsthand perspective of the war. His comments on officers, and other regiments are an interesting commentary on life in the Western Army, and add to the knowledge of how life was perceived by a soldier. Especially revealing was his accurate knowledge of Grant's campaign strategy prior to Vicksburg, which he described in a letter of April 25, 1863.

The letters divulge a dislike of the

Negro and the idea of black soldiers: "...I, like most, don't like the idea of fighting side-by-side with a Negro. That's asking a little too much...to see him fighting with weapons in his hand at my better side, hey now, no siree bob." In 1863, when the government decided that Negroes would be armed as soldiers, Kircher wrote home that the policy was presented carefully to the whites, and added, "It seemed just as if he or the government didn't have the courage simply to command and to require obedience. By God, the soldiers have sworn to be loyal to the government, and will be loyal, with few exceptions."

Kircher's opinion as to why the German regiments had better health was that they had better stomachs, had no sweet tooth, cooked their food thoroughly and were cleaner. Throughout his correspondence, he praised German troops and German officers, but his ethnic heritage and upbringing, contrary to the author's preface, didn't seem more acute than that of a Virginian praising a Virginian unit, or a New York regiment preferring a native son officer. The author additionally writes in the preface that Kircher's struggle against combat-related disabilities is inspirational. While Mr. Hess does devote some of the last chapter to the latter part of the Illinoisian's life, the picture seems sketchy and incomplete. More detail on his post-war readjustments, whether or not he used an artificial leg, what other problems he overcame would help to keep him in perspective.

The most severe criticism about "A German in the Yankee Fatherland" is the exclusion of Kircher's diary. This he kept throughout his term of service, and wrote in English, the language he would have to use outside his group of German-speaking comrades. While it is interesting to note his fluency in his parents' tongue, it is much more relevant to read his thoughts as he would have expressed them to an English-speaking comrade. Very often, acceptance is based on the ability to communicate, which in this instance meant accurate verbalization in English. If, as Mr. Hess states in the preface, Kircher's English improved between 1861 and late 1863 (a fact that is not immediately evident, due to limited diary inclusions), it is extremely important the reader be able to gauge this. And, while the author skillfully ties together the sequence of events between letters home by using very brief excerpts from the diary, there is too much valuable information left untouched. In the bibliography, it is mentioned that a revealing diary quote on regimental discipline exists, but the author found no place for it in the book. It would have been preferable to include extensive quotes from the four volumes of diary (or perhaps do the book on the diary, and annotate the letters), not only to give the researcher more information on the war, but also to aid in understanding Kircher himself.

As it stands, "A German in the Yankee Fatherland" provides a good firsthand account of the Civil War as seen by a young soldier. It is well-edited, very readable and a useful addition to those studying Vicksburg and the Western theatre; unfortunately, it is just too short.

S.L. Kotar

"Field Artillery Weapons of the Civil War," by James C. Hazlett, Edwin Olmstead and M. Hume Parks, with a Forward by Harold L. Peterson; Newark: University of Delaware Press, Associated University Presses, Inc., 440 Forsgate Drive, Cranbury, New Jersey, 08512; 1983 list price $45.00 332 pages, Tables, Appendixes, Glossary, Index, Bibliography, Notes, Illustrations.

The Company of Military Historians sponsored this volume "as a useful reference in American military history." The word "useful" seems to be an understatement after reading this authoritative, well-researched work. The late Harold L. Peterson not only wrote in his Forward that "...the authors have done a tremendous job in sorting out and locating the types, sizes, and variations of these historic cannon," but he made free to them the resources of the National Park Service to aid them in their labors. Countless others provided assistance to make this the most complete study of classification of Civil War fieldpieces, both Federal and Confederate. The highly detailed tables, specific diagrams, numerous photographs, lengthy appendixes and glossary are minute, yet extensive. Attention to nomenclature where confusion often reigned is most helpful, and careful tracing of the history of pieces is quite enlightening. I do not doubt that the enormous amount of work put into this book will render it the bible for all students of artillery, and that it will remain so for some time to come.

J.E. Gessler

Articles of War

An Act For Establishing Rules and Articles For The Government Of The Armies Of The United States.

ART. 64. General courts-martial may consist of any number of commissioned officers, from five to thirteen, inclusively; but they shall not consist of less than thirteen where that number can be convened without manifest injury to the service.

ART. 65.* Any general officer commanding an army, or Colonel commanding a separate department, may appoint general courts-martial whenever necessary. But no sentence of a court-martial shall be carried into execution until after the whole proceedings shall have been laid before the officer ordering the same, or the officer commanding the troops for the time being; neither shall any sentence of a general court-martial, in the time of peace, extending to the loss of life, or the dismission of a commissioned officer, or which shall, either in time of peace or war, respect a general officer, be carried into execution, until after the whole proceedings shall have been transmitted to the Secretary of War, to be laid before the President of the United States for his confirmation or disapproval, and orders in the case. All other sentences may be confirmed and executed by the officer ordering the court to assemble, or the commanding officer for the time being, as the case may be.

ART. 66. Every officer commanding a regiment or corps may appoint, for his own regiment or corps, courts-martial, to consist of three commissioned officers, for the trial and punishment of offenses not capital, and decide upon their sentence. For the same purpose, all officers commanding any of the garrisons, forts, barracks, or other places where the troops consist of different corps, may assemble courts-martial, to consist of three commissioned officers, and decide upon their sentences.

ART. 67. No garrison or regimental court-martial shall have the power to try capital cases or commissioned officers; neither shall they inflict a fine exceeding one month's pay, nor imprison, nor put to hard labor, any non-commissioned officer or soldier for a longer time than one month.

ART. 68. Whenever it may be found convenient and necessary to the public service, the officers of the marines shall be associated with the officers of the land forces, for the purpose of holding courts-martial, and trying offenders belonging to either; and, in such cases, the orders of the senior officer of either corps who may be present and duly authorized, shall be received and obeyed.

ART. 69. The judge advocate, or some person deputed by him, or by the general, or officer commanding the army, detachment, or garrison, shall prosecute in the name of the United States, but shall so far consider himself as counsel for the prisoner, after the said prisoner shall have made his plea, as to object to any leading question to any of the witnesses, or any question to the prisoner, the answer to which might tend to criminate himself; and administer to each member of the court, before they proceed upon any trial, the following oath, which shall also be taken by all members of the regimental and garrison courts-martial:

"You, A.B., do swear that you will well and truly try and determine, according to evidence, the matter now before you, between the United States of America and the prisoner to be tried, and that you will duly administer justice, according to the provisions of 'An act establishing Rules and Articles for the government of the armies of the United States,' without partiality, favor, or affection; and if any doubt should arise, not explained by said Articles, according to your conscience, the best of your understanding, and the custom of war in like cases; and you do further swear that you will not divulge the sentence of the court until it shall be published by the proper authority; neither will you disclose or discover the vote or opinion of any particular member of the court-martial, unless required to give evidence thereof, as a witness, by a court of justice, in a due course of law. So help you God."

And as soon as this said oath shall have been administered to the respective members, the president of the court shall administer to the judge advocate, or person officiating as such, an oath in the following words:

"You, A.B., do swear, that you will not disclose or discover the vote or opinion of any particular member of the court-martial, unless required to give evidence thereof, as a witness,

*Whenever a general officer commanding an army, or a colonel commanding a separate department, shall be the accuser or prosecutor of any officer in the army of the United States, under his command, the general court-martial for the trial of such officer shall be appointed by the President of the United States.

The proceedings and sentence of the said court shall be sent directly to the Secretary of War, to be by him laid before the President, for his confirmation or approval, or orders in the case.

So much of the sixty-fifth article of the first section of "An act for establishing rules and articles for the government of the armies of the United States," passed on the tenth of April, eighteen hundred and six, as is repugnant hereto, shall be, and the same is hereby, repealed. — **Act 29th May, 1830, Sects.** 1, 2, 3.

by a court of justice, in due course of law; nor divulge the sentence of the court to any but the proper authority, until it shall be duly disclosed by the same. So help you God."

ART. 70. When a prisoner, arraigned before a general court-martial, shall, from obstinacy and deliberate design, stand mute, or answer foreign to the purpose, the court may proceed to trial and judgement as if the prisoner had regularly pleaded not guilty.

ART. 71. When a member shall be challenged by a prisoner, he must state his cause of challenge, of which the court shall, after due deliberation, determine the relevancy or validity, and decide accordingly; and no challenge to more than one member at a time shall be received by the court.

ART. 72. All the members of a court-martial are to behave with decency and calmness; and in giving their votes are to begin with the youngest in commission.

ART. 73. All persons who give evidence before a court-martial are to be examined an oath or affirmation, in the following form:

"You swear, or affirm (as the case may be), the evidence you shall give in the cause now in hearing shall be the truth, the whole truth, and nothing but the truth. So help you God."

ART. 74. On the trials of cases not capital, before courts-martial, the deposition of witnesses, not in the line or staff of the army, may be taken before some justice of the peace, and read in evidence; provided the prosecutor and person accused are present at the taking of the same, or are duly notified thereof.

ART. 75. No officer shall be tried but by a general court-martial, nor by officers of an inferior rank, if it can be avoided. Nor shall any proceedings of trials be carried on, excepting between the hours of eight in the morning and three in the afternoon, excepting in cases which, in the opinion of the officer appointing the court-martial, require immediate example.

ART. 76. No person whatsoever shall use any menacing words, signs, or gestures, in presence of a court-martial, or shall cause any disorder or riot, or disturb their proceedings, on the penalty of being punished at the discretion of the said court-martial.

ART. 77. Whenever any officer shall be charged with a crime, he shall be arrested and confined in his barracks, quarters, or tent, and deprived of his sword by the commanding officer. And any officer who shall leave his confinement before he shall be set at liberty by his commanding officer, or by a superior officer, shall be cashiered.

ART. 78. Non-commissioned officers and soldiers, charged with crimes, shall be confined until tried by a court-martial, or released by proper authority.

ART. 79. No officer or soldier who shall be put in arrest shall continue in confinement more than eight days, or until such time as a court-martial can be assembled.

ART. 80. No officer commanding a guard, or provost marshal, shall refuse to receive or keep any prisoner committed to his charge by an officer belonging to the forces of the United States; provided the officer committing shall, at the same time, deliver an account in writing, signed by himself, of the crime with which the said prisoner is charged.

ART. 81. No officer commanding a guard, or provost marshal, shall presume to release any person committed to his charge without proper authority for so doing, nor shall he suffer any person to escape, on the penalty of being punished for it by the sentence of a court-martial.

ART. 82. Every officer or provost marshal, to whose charge prisoners shall be committed, shall, within twenty-four hours after such committment, or as soon as he shall be relieved from his guard, make report in writing to the commanding officer, of their names, their crimes, and the names of the officers who committed them, on the penalty of being punished for disobedience or neglect, at the discretion of a court-martial.

ART. 83. Any commissioned officer convicted before a general court-martial of conduct unbecoming an officer and a gentleman, shall be dismissed the service.

ART. 84. In cases where a court-martial may think it proper to sentence a commissioned officer to be suspended from command, they shall have power also to suspend his pay and emoluments for the same time, according to the nature and heinousness of the offense.

ART. 85. In all cases where a commissioned officer is cashiered for cowardice or fraud, it shall be added in the sentence, that the crime, name, and place of abode, and punishment of the delinquent, be published in the newspapers in and about the camp, and of the particular State from which the offender came, or where he usually resides; after which it shall be deemed scandalous for an officer to associate with him.

ART. 86. The commanding officer of any post or detachment, in which there shall not be a number of officers adequate to form a general court-martial, shall, in cases which require the cognizance of such a court, report to the commanding officer of the department, who shall order a court to be assembled at the nearest post or department, and the party accused, with necessary witnesses, to be transported to the place where the said court shall be assembled.

From "Revised Regulations for the Army of the United States, 1861," Philadelphia, J.P. Lippincott & Co., pp. 509-512.

SENSIBLE CONCLUSION.

MR. BLAWKINS (*who is tired of a life of single blessedness*).—That's the thing! That will settle the business at once! I never could screw my courage up to the sticking point to propose; but that would melt the heart of any female in the world. Here goes; Hurrah for ST. VALENTINE! Darn the expense! What's a man without a wife, and how is a man to get a wife without a Valentine?

THE KEPI
P. O. Box 4542
North Hollywood, California 91607

BULK MAIL

THE KEPI

APRIL-MAY 1984 $3.50

"A New Concept in Civil War Reporting"

The Affair at Fairfax Court-House

by Thomas J. Evans

Massacre in Shelton Laurel

by Vernon H. Crow

Hitherto Taboo

by Robert Kamer, M.D. and Samuel Jackson, M.D.

THE KEPI

"A New Concept in Civil War Reporting"

APRIL-MAY 1984 — VOLUME II, NUMBER 2

THE KEPI is published bi-monthly for $3.50 per issue, or $18.00 per year. Canadian single copy, $4.50; subscription $23.00 per year. United Kingdom single copy, $4.50; subscription $28.00 (Air Mail) per year. Correspondence to the editors may be directed to: **THE KEPI**, P.O. Box 4542, North Hollywood, CA, 91607. Unless expressly stated, **THE KEPI** neither endorses nor is responsible for the content of advertisements appearing in its pages. **THE KEPI** requests authors to query with specific topics before submitting articles, and assumes no responsibility for unsolicited manuscripts not accompanied by return postage. Work not fully documented and footnoted where applicable will not be considered. All inquiries should be directed to the above address. **THE KEPI** reserves the right to edit and copyright all letters. Please allow 6 weeks for delivery of first issue.

ILLUSTRATIONS

Cover: Mosby in uniform; Brig.-Gen. Edwin H. Stoughton, Commander of 2nd Vermont Brigade; plaque outside Gunnel house where Stoughton was captured.

Illustrations from *"Massacre"* courtesy of the author.

Picture of J.S. Mosby courtesy of Terri Singleton.

Photographs for Tracts from *"The Child's Illustrated Scripture Question Book."*

Back cover: courtesy of Charles Lowe.

THE REGULARS

THE IRREGULARS

GENERALLY SPEAKING

The cover story in this issue, written by Thomas J. Evans, concerns John S. Mosby's famous capture of General Ed Stoughton. Considered by many to be the partisan's crowning achievement, "The Affair at Fairfax Court-House" presents information never before published, introduces Sarah Lake, the brave woman who helped save the wounded Mosby's life through great personal sacrifice; the story also puts an end to the myth that Antonia Ford was a hardcore spy for the Confederacy.

Our second feature, "Massacre at Shelton Laurel" by Vernon H. Crow, covers the sensitive topic of how a regular army unit dealt with guerilla opposition in the mountains of North Carolina. After constant feuding between irregular Union sympathizers and local Confederates, the latter wrote to Governor Zeb Vance for relief. Vance, in turn, applied to the army, and finally General Henry Heth ordered indigenous troops into the area. Their pre-existing animosities resulted in the deaths of thirteen civilians, as the soldiers ruthlessly captured any male Federalists they could find and executed them without trial or appeal. Mr. Crow supplies the background leading to this incident, traces the Confederate units in the area, and follows the killings and their tragic consequences.

"Hithertoo Taboo," by our medical advisors, Robert Kamer and Samuel Jackson, covers the (a) fascinating, (b) embarassing, (c) humorous ideas and "misconceptions" of the sexual ideas held by our 19th Century ancestors.

Continuing is Family Bible, *with the birthdates and birthplaces of Civil War notables who were born in the months of April and May. Charles A. Lowe, author of "John Jeffers' Battle for the Union and Against the Pension Bureau," describes how his great-grandfather fought to receive what he considered a fair compensation for his disabling wound received at Gettysburg. Mr. Lowe, a retired Weymouth police officer and stringer correspondent for many Greater Boston newspapers, started researching his family background and discovered the interesting and very human story of his Civil War ancestors, proving to the reader of the 20th Century that, indeed, "governments never change!"*

Completing the issue are our regular features: Concurrent Events, Whimsey and the Articles of War.

S.L. Kotar J.E. Gessler

Editors-in-General

Head-Quarters of THE KEPI
1st April 1984

GENERAL ORDERS
No. 6
by Command of the Editors-in-General

The June/July issue of THE KEPI features an article on the "Jackson flag" and its introduction to the Southern soldiers who would bravely fight under it. Employing unpublished source material, "To Raise A Flag" sheds a humorous new light on why the Second Corps, A.N.A. enjoyed the unveiling of the Confederacy's third flag.

Also featured is an article on the deadly consequences of smuggling along the Virginia/Maryland border. Following the career of a young Southerner, "A Smuggler and a Guerilla" vividly recounts his experiences bringing goods in and out of the Confederacy and his associations with Mosby's men, John Mobberly and "Sink" Sinclair.

"Shoes" covers the invention of the McKay Stitcher and the shoe production in Abington, Massachusetts, that greatly aided the Union war effort.

The regular/irregulars appear, as well as an article on breaking the Federal code at Vicksburg, and Family Bible continues.

Massacre In Shelton Laurel

by
Vernon H. Crow

Clash of irregulars and soldiers.

Two of Mary Allen's children were ill with the fever, dreaded scarlet fever. Mrs. Allen was caring for her son, Romulus, age six, and her four-year-old daughter, Margaret, at her home in Marshall, the seat of Madison County in western North Carolina, while her husband, Colonel Lawrence M. Allen, was off with his regiment, the Sixty-fourth North Carolina, in Tennessee. But, by January 1863, the women of Laurel Valley, a poor mountainous section of the Tarheel state—the backwoods of the Confederacy, as some called this part of southern Appalachia—were used to such hardships. Mary Allen was cheered somewhat by the appearance of Captain John Peek, her kinsman, who commanded Company C of her husband's regiment, home on leave.[1] With times so difficult, it was good to have Peek near and to hear the news of her husband and other relatives who were in the Southern army.

During the night Captain Peek was awakened by a disturbance outside the Allen house. As he ventured out to investigate he was suddenly shot in the right arm. Although the wound was serious—shattering his limb—he would live to report that he had been shot by deserters from his own regiment.

While Peek lay bleeding, the attackers entered the Allen house and began to ransack the place. Mrs. Allen pleaded with the "Marshall raiders," as they were later styled, that her children were gravely ill, but they paid no attention to her. They broke open trunks and bureaus, taking clothes, shoes and money. One account claims that they entered the dying children's rooms and stole their clothing.[2]

The raiders, who numbered about fifty, struck other homes, too, and apparently burned several dwellings and stole salt and other provisions. The little town of Marshall, which had less than twenty buildings, including the old county courthouse and an out-of-date jail, was throughly sacked.

The Marshall raid had far-reaching effects, eventually leading to the massacre of thirteen people by soldiers who claimed to be acting on orders. In a war supposedly filled with chivalry and honor, how could such a thing have happened?

Shortly after the raid, rumors began to spread in western North Carolina and East Tennessee that it signaled the beginning of a general Unionist rebellion. There was one report that organized troops would soon be arming and drilling in Madison County. The governor of North Carolina, Zeb Vance, called out the militia and appealed for help from General Henry Heth, under whose command the western part of North Carolina fell. Heth, headquartered at Knoxville on the other side of the mountains that separated Tennessee and North Carolina, immediately ordered Brigadier General William G. M. Davis to put down the uprising.

Davis sent Captain William B. Nelson with a company of cavalry into Shelton Laurel, the local name for Laurel Valley, and ordered Colonel William H. Thomas to "furnish 200 men" and join him at Greeneville, Tennessee. Thomas, whose North Carolina legion included several hundred Cherokee Indians, assembled his force of whites and Cherokees and met Davis at the railroad depot. Then the command marched to Warm Springs, north of Marshall, in the Carolina mountains. From there Thomas put out reconnaissance units on both sides of the French Broad River. About this time Captain Nelson returned and reported that he had a brush with the tories, killing twelve and capturing twenty. Davis passed along the news of Nelson's skirmish with the insurgents to Heth on January 20, but told his chief that he thought the rumors of a Union uprising exaggerated. But said Davis: "I think the attack on Marshall was gotton up to obtain salt, for want of which there is great suffering in the mountains. Col. [L.M.] Allen's Sixty-fourth North Carolina Regiment and the men of his command are said to have been hostile to the Laurel men and they to the former for a long time—a kind of feud existing between them." Heth passed Davis' letter on to Governor Vance and dismissed the Laurel troubles.[3]

While Davis was at Warm Springs he received word that Heth had ordered the Sixty-fourth to join up with his force. Believing the country quiet, Davis wrote Governor Vance and recommended that the militia be sent home. He then returned to Greeneville leaving Thomas and Allen's regiment to operate in western North Carolina. Thomas's orders were to scour the counties south of Madison and "arrest all deserters and recusant conscripts." Davis also sent three companies of the Sixty-fourth under Major W. N. Garrett into the counties adjacent to the mountains to "pursue and arrest every man...of known bad character..." By February 20, Thomas had completed his mission and returned to East Tennessee, but the Sixty-fourth remained in the mountains.[4]

Lawrence Allen was twenty-eight when the Civil War began. The son of a wealthy farmer, Allen married Mary Peek in 1853, and their union produced three children. He was elected clerk of the superior court of Madison County, took a leading role in politics and was a frequent delegate to state Democratic party conventions. In July 1861, he organized a company of Madison volunteers for the Second Battalion North Carolina Infantry, and served for nearly one year in that command. However, Allen wanted more glory and fame. In May of 1862, he received permission to raise the Eleventh Battalion Infantry which was recruited to a full regiment—the Sixty-fourth—by July, when Allen was elected colonel.[5] Captain B.T. Morris, who headed Company E in Allen's regiment, described his chief as "not an attractive man—rather otherwise—but was chosen leader because he was known to be brave and fearless. Fighting was expected and his men had the utmost confidence in him."[6] Morris spoke for most of the officers, many of whom were related to Allen, and perhaps for most of the enlisted men. But a good number of privates had gone into the army because of the draft. As soon as the command was ordered to East Tennessee, many mountaineers deserted at the first opportunity, returning to hide in the hundreds of secret places known only to their families.

It was not long before western North Carolina became a haven for deserters from both armies, North and South. In Madison County, perhaps the most pro-Union County of the state, a boy in blue could find shelter. Local Unionists, or "tories," combined with the deserters in creating bloody guerrilla warfare, which resulted in raids like the one in Marshall. The families of Confederate soldiers were forced to submit to numerous outrages. Livestock was killed for no other reason than it was owned by a Confederate sympathizer. Houses were burned, crops ruined, clothes and money stolen, and the soldiers themselves were hunted, especially if they traveled in small units. From behind rocks and laurel thickets the "bushwhackers" would wait to ambush the enemy. The loyal Southern populace responded with their own guerrilla tactics, and committed similar acts of violence against the tories and bushwhackers when caught, and against pro-Union families when not. Beatings and torture were often used to make those still at home reveal the hiding place of a Union man. Murder also became a common occurrence. The condition of western North Carolina became so desperate that Governor Vance wrote General Heth on January 21, 1863, and pleaded: "I hope you will not relax until the tories are crushed."[7]

Lawrence Allen, more than anyone else, wanted to crush the tories after he heard of the Marshall raid. But Allen had been placed on suspension for "crime and drunkenness" as one newspaper claimed, although he may have been under suspension for reporting a soldier present who was actually absent from the Sixty-fourth. Allen requested, and got, an interview with General Heth, taking along Lieutenant-Colonel James A. Keith, his second-in-command.[8]

At thirty-five, Keith was tall and lean, with black hair and beard, and blue eyes. One of the men who served under him remembered Keith as being "intrepid and fearless."[9] The son of a Baptist minister, Keith was an ambitious man who had become one of the wealthiest men in Madison by the time of the Civil War. He, like Allen, was also a resident of the Laurel Valley, and Keith shared his chief's intense hatred for the Unionists of his region.

During the interview, the two colonels pleaded with Heth to let them punish the Marshall raiders. Phillip Shaw Paludan, in his excellent study of the Shelton Laurel affair, takes Heth to task for not realizing that Keith was "a dangerous man, a leader of what the army called partisan rangers, of men who made their own rules of warfare..."[10] Paludan also points out that Keith "...was a man with first-hand experience in guerrilla fighting, a man with scores to settle. Not only was it likely that Keith did not know the rules [of warfare], but he had powerful reasons for not wanting to know. As a result, Heth's orders took on major importance."[11] The meeting with General Heth pleased both North Carolinians. The general gave his permission to send the Sixty-fourth into the mountains to search and destroy the tories, and did not object to letting Allen accompany the command. Keith later claimed that Heth gave him verbal orders to take no prisoners. "I want no reports from you about your course at Laurel. I do not want to be troubled with any prisoners and the last one of them should be killed..."[12] Satisfied that their general understood the mission, Allen and Keith left Heth and prepared the regiment for the journey to Laurel Valley. The regiment was split into two commands: Allen

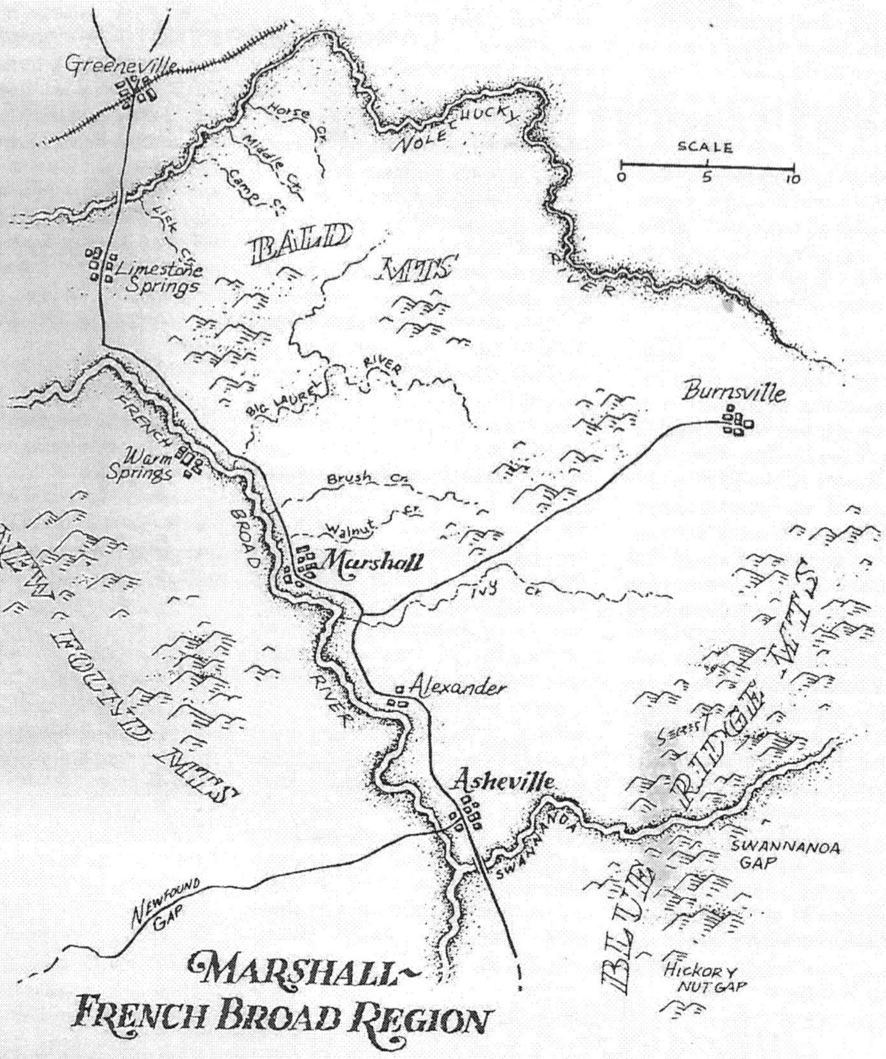

Map of the Shelton Laurel Valley, North Carolina.

took charge of one, Keith of the other.

The two columns trudged their way through the snow and up the Tennessee side of the mountains, but as they wound their way down into North Carolina they encountered scattered gunfire as snipers attempted to pick off the graycoated soldiers. In running gun battles, Allen's troops were able to kill eight of the bushwhackers. At Bill Shelton's, Allen bumped into a force of about fifty tories, who offered stiff resistance, but were driven off leaving six Unionists dead. Allen decided to camp at Shelton's and wait for Keith to come up. Sometime during the night, Allen received word that his six-year-old son had died and his daughter was near death. Accompanied by several soldiers, Allen set out for his home in Marshall, ducking sniper fire as he and his men rode through Shelton Laurel. Allen reached home before noon and found his grieving wife and dead boy's body. During the day, his little daughter, Margaret, died. The next day Colonel Allen, embittered more than ever, buried his children and returned to his command.[13]

While Allen tended to his family, James Keith began to sweep the valley, hunting for anyone who either had taken part in the Marshall raid, or might know a raider's name. At least three tories came in and surrendered voluntarily, and were hustled off to Asheville by Keith's acting adjutant-general, W.H. Bailey.

Keith turned ruthless, interrogated everyone, and committed numerous savage and outrageous acts. Captain Daniel Ellis, the famous Federal guide who led many Unionists out of the mountains to Northern lines and who interviewed an eye-witness, wrote later: "Some of these poor defenseless women were then stripped of their clothing, they were tied up, and their naked backs were lashed until the blood trickled down to their feet, while the balance of them were hung up by the neck until they were nearly dead. They were then taken down, and their tormentors then left them tied until their feet and hands were frozen to such a degree that their nails all came off."[14] Ellis's account may have been exaggerated, but it is certain that several women were whipped by Keith's men.[15]

By Friday, January 16, Keith rounded up at least fifteen prisoners, most taken quietly at their homes in Shelton Laurel, and who ranged in age from Davey Shelton, thirteen, to fifty-six-year-old James Shelton. At least seven of the prisoners were related and bore the name Shelton, one of the earliest families in the Laurel Valley. When the war came, most of the Sheltons remained with the old flag, as did many poorer western North Carolinians. The less affluent mountaineers did not favor abolition so much as they resented the more wealthy landowners. The mountainous nature of the far western counties created an isolationist society which led to suspicion of one's neighbors and resulted in family antagonisms. Sometimes the war was little more than a good excuse to rekindle old feuds and quarrels—to punish those who had insulted family or friends. The desire for revenge was certainly on Keith's mind, as well as Allen's as the latter returned to his command.

For reasons which are not clear, Keith and Allen kept their prisoners closely guarded for three days, perhaps discussing their fate. Sometime during Sunday night, two prisoners escaped into the blackness, probably sealing the fate of the others. When their escape was discovered in the morning, Keith told the men to get ready to march, they were going to be taken to Knoxville. The prisoners were marched along a road for a distance of three miles, then turned into a hollow and to "a sink-hole about three hundred yards from the main road, where they stopped."[16] There are several versions of what happened next. One account claims Keith gave the men and boys ten minutes to pray. They were then put into a line and tied together. Another version reported that the prisoners were taken five at a time and lined up in front of a squad of soldiers. At least one of the men begged for mercy, but Keith ignored his pleas and, instead, ordered his men to fire. However, despite the animosity that obviously existed between the Sixty-fourth and the Laurel men, the soldiers refused to obey. Keith shouted at them that if they did not fire they would take the prisoners' place. Again Keith ordered the squad to fire. This time the muskets reported, and four of the prisoners died instantly. Another lay wounded "mortally shot in the bowels and while he was writhing in agony and praying to God for mercy a soldier mercilessly and brutally shot him in the head with his pistol."[17] Daniel Ellis claimed the pistol was fired by Keith.

This wretched scene was repeated twice more as the other eight prisoners were similarly shot by the firing squad. Young Davey Shelton was wounded in the second round and begged to be spared, but he was dragged back into place and shot to death. The thirteen bodies were then buried together in a shallow grave and allowed to remain there for three weeks. Later the families were given permission to remove them for proper burial.

Following the killings, the Sixty-fourth remained in Madison County for several weeks hunting bushwhackers and tories until it was ordered back to Tennessee. Meanwhile, word spread quickly in the mountains of the massacre; Governor Vance heard rumors of it in Raleigh on February 2. On the ninth of the month, he wrote Augustus S. Merrimom, the state solicitor for the western district, demanding an investigation. Merrimom, later a

Artist's depiction of atrocity.

United States senator from North Carolina, searched for witnesses and apparently found people willing to talk. By February 16, he reported to the governor that he had interviewed someone "who got his information from some of the guilty parties."[18] Merrimom added: "I know not what you intend doing with...[them]..., but I suggest they are all guilty of murder."[19] Two weeks later, the state's attorney sent more details to the governor, revealing that the killings were ordered by Keith and "that probably eight of the thirteen killed were not in the company that robbed Marshall and other places. I suppose they were shot on suspicion."[20] Merrimom also learned the names of the victims and some of their ages, but not the names of the soldiers who fired the shots. "One thing is certain," he concluded, "thirteen prisoners were killed—shot without trial or any hearing whatever and in the most cruel manner."[21]

Merrimom's reports did not land on deaf ears. Governor Vance, himself a western North Carolinian, was upset at the news. Said Vance: "I was fearful in the great excitement prevailing among our people that the misguided people of Laurel might be dealt too harshly with, and warned the officers to be civil and just."[22] When newspaper editor William W. Holden, North Carolina's postwar governor, heard of the Laurel massacre he visited Vance. "I called at his office to get the facts," said Holden. "Governor Vance was very indignant. His form dilated as he said, 'This was done by Colonel Keith of the 64th N.C. and I will write to Mr. Seddon, Secretary of War at Richmond, to have him courtmartialed...' " Vance promised Holden, "I will follow him (Keith) to the gates of hell or hang him."[23]

Vance attempted to keep his promise. On February 28 he sent Seddon the facts and wrote: "I desire you to have proceedings instituted at once against [Keith], who if the half be true is a disgrace to the service and to North Carolina."[24] Vance also wrote to General Davis at Knoxville for an investigation of the murders.

If Vance, Merrimom and others were outraged at the killing of prisoners of war, there were those who thought that, while Keith may have acted rashly, he acted properly. Captain B.T. Morris of the Sixty-fourth defended the action. Said Morris: "When an officer finds himself and men bushwhacked from behind every shrub, tree or projection on all sides of the road only severe measures will stop it. Keith and Allen were fighters—soldiers. Their first duty was self protection, protection of their people from midnight marauders."[25]

Bushwhacking in the mountains of North Carolina.

SIXTY-FOURTH REGIMENT.
B. T. Morris, Captain, Co. E.

Many of the loyal Confederate citizens in the mountains thought the Shelton Laurel men got exactly what they deserved. One who called himself "Elbert" wrote Holden's Raleigh *North Carolina Standard*: "I am sure [the slain men and boys] have received nothing more than justice according to the enormity of their crimes."[26] Unionist Daniel Ellis found even the women of the area unsympathic to the murder of pro-Union men. Once Ellis stumbled across the skeletons of three men who had been left hanging from a tree. As he was looking at the remains, an old woman appeared in the road. Their conversation quickly revealed that she was a "rebel." The Union guide, dressed in civilian clothes, asked if the woman felt any pity for the hanged men. "No, none whatever," was her response. "That is the very way that men ought to be treated whenever they are caught running away to Old Abe. All of them ought to be killed...that is not a strange sight to me...I have seen...them kneel to pray before being shot."[27]

Despite the local justifications for the killings, Governor Vance continued to press for an investigation. Confederate War Secretary Seddon received Vance's letter on March 5, but turned the matter over to General Daniel S. Donelson, new commander of the Department of East Tennessee, as Heth had been transferred to Lee's army in Virginia. General Davis also forwarded Vance's request to Donelson. Meanwhile, Allen and Keith and the Sixty-fourth left Shelton Laurel and returned to the other side of the mountains.

Zebulon Baird Vance, wartime governor of North Carolina.

Vance heard nothing more of the Shelton Laurel affair from the army until May, and then what he heard smacked of a cover-up. He wrote Seddon on May 18 and demanded to know if the rumors were true that Keith had been brought "before a court-martial and honorably acquitted by producing an order for his conduct...I have also been officially notified of [Keith's] resignation."[28] Vance was desirous of a copy of the court-martial so he could prosecute Keith in the civil courts. "Murder," Vance concluded, "is a crime against the common law in this State and he is now subject to that law."[29]

Seddon replied a few days later that Keith's resignation was accepted on May 15, and that Keith claimed he could prove his orders came directly from Heth. The War Department had submitted Keith's defense to the general who admitted "...he told Keith that those found in arms ought not be treated as enemies, and in the event of an engagement with them to take no prisoners as he considered that they had forfeited all such claims, but [General Heth] denies in strong terms the making use of any remarks which would authorize maltreatment of prisoners who had been accepted as such..."[30] In other words, the guerrillas were not to be taken prisoner: if engaged, they were to be killed. However, once a guerrilla was accepted as a prisoner of war he was to be treated under the rules of war. True, a soldier of Heth's background and education understood this, but it was asking more than could be expected of civilian-soldiers like Keith, considering the circumstances of the Laurel expedition, to fully understand the military code. Heth's failure to clarify this distinction made the general partially to blame for the Shelton Laurel massacre.

The story did not end with Keith's resignation. Vance attempted to have him arrested by Confederate troops, but the colonel's friends in Madison hid him for two years. After the war he was finally captured by Union troops and spent twenty-six months in jail. He was finally brought to trial in December 1868 on one count of murder and found not guilty. However, he was quickly rearrested and indicted on two more charges of murder. While awaiting trial, Keith decided that he would never be safe again in North Carolina. On February 21, 1869, he escaped from his jail in Asheville and was never heard from again. Rumors placed him in Arkansas, probably given credence because Lawrence Allen and his family soon relocated there.[31]

The Shelton Laurel massacre never became an issue in the Civil War, at least not like the Fort Pillow massacre. Maybe it was because only thirteen people were killed, or perhaps it happened in too remote an area. It may not have become

(continued on page 22)

The Affair at Fairfax Court-House

by
Thomas J. Evans

J.S. Mosby after the war.

The fact that Richmond was the capital of the Confederacy has long given the impression that all Virginians supported secession: nothing could be further from the truth. As soon as the question of leaving the Union arose, representatives from the Western counties of Virginia expressed their desire to remain loyal to the United States. When the government and later the citizens voted to join the Confederate States, the nonslave-owning mountaineers elected a new governor, and in 1863 became the new state of West Virginia. This loyalty to the Union also applied to the people in the northern counties along the Virginia-Maryland border. When hostilities commenced between the two governments, the result was a continuous and bloody war throughout northern Virginia. The "Border War," as Union General Henry Halleck labeled it, was conducted continuously by partisans who destroyed roads, burned bridges and captured wagon trains.[1] The region abounded in partisans, spies, secret agents and sympathizers of both sides.

The Union presence was felt strongly enough in northern Virginia that Captain Samuel C. "Quaker Sam" Means recruited the Independent Loudoun (Va.) Rangers at Waterford, Virginia. This became the only organized military unit from the state that fought in the Union Army during the Civil War. Across the Potomac River, in Maryland, Captain Henry A. Cole, organized a company of hard-riding cavalry that crossed the border at will, often joining with Captain Means' Rangers to harass the Confederate forces.[2]

The Confederacy had its ardent defenders, also. Captain Elijah "Lige" A. White's "Comanches" were soon battling with Captain Means' Rangers and the Union cavalry, as were the forces of Colonel John Imboden and Captain Harry Gilmore.

Serving as a partisan was a dangerous business. Partisan leader John "Sweet Johnny" Mobberly harassed the Union Army until his death in an ambush.[3] William "French Bill" Loge was captured, court-martialed, and hanged at Harper's Ferry.[4] "Pony" Ormsby was executed by a firing squad at Vienna, and John "Hanse" McNeill was shot in the back by one of his own men, and died

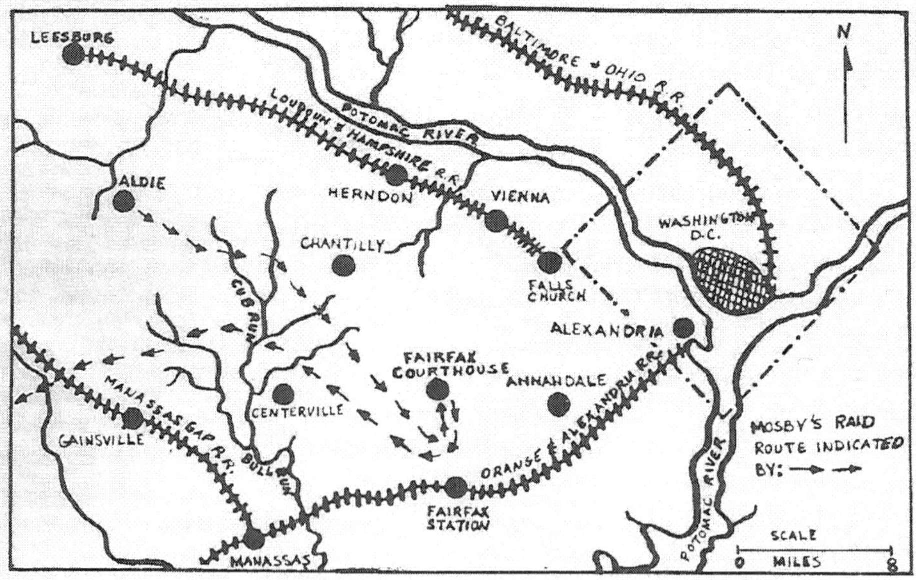

Map of Northern Virginia, showing the route of Mosby's raid.

at Harrisonburg, Virginia.[5] The odds were slim for a partisan to reach old age in Virginia.

When the Second New York Cavalry unexpectedly swooped down upon Beaver Dam Railroad Station during the summer of 1862, they captured a lone Confederate courier asleep on the platform. An enemy of his during the war, who would later became President of the United States, described him as: "Slender, not tall, wiry, and looks as if he could endure any amount of physical exercise."[6] A comrade would write of him: "He was rather a slouchy rider, also did not seem to take any interest in military duties. He had been but seldom at our drills before starting, and we all thought he was rather an indifferent soldier [however] Mosby distinguished himself in every affair we had."[7] And so the name of John Singleton Mosby, the captured Confederate courier, became known to the Federal authorities for the first time when he was taken to the Old Capitol Prison in Washington, D.C., as a prisoner of war.

Prior to his capture, Mosby had served in Captain William E. "Grumble" Jones' cavalry company, part of the famous 1st Virginia Cavalry Regiment. Mosby had seen little of the action at First Manassas, where he was engaged in scouting, but was later made Adjutant and First Lieutenant of Jones' regiment when Jones received a larger command. When elections replaced Jones with Fitzhugh Lee, Mosby lost his appointment and became a scout for General James E.B. "Jeb" Stuart. Mosby greatly admired Stuart, and wrote of him: "In his works on the outposts Stuart soon showed that he possessed the qualities of a great leader of cavalry. He never had an equal in such service."[8]

Following his exchange during August, 1862, Mosby served with Stuart at Second Manassas, Sharpsburg, Fredricksburg, and on the Dumfries Christmas Raid. Then, with Stuart's assistance, he became a partisan leader on January 24, 1863, a date that would become a black day for the Union Army. Mosby and a small contingent of men crossed the Rappahannock River into northern Virginia to begin his partisan campaign. He wrote: "I think that Stuart was the only man in the Army of Northern Virginia, except two or three who accompanied me and knew me well, who expected that I would accomplish anything."[9] Mosby immediately began to scout the area closely, enlist a few dedicated supporters, and seek an opportunity to carry the war to the Union forces protecting Washington, D.C. In early March, 1863, Mosby conducted a raid that many historians believe to be his greatest military achievement.[10] It was certainly his most visible military action.

Fairfax Courthouse was a two-story brick building that contained the county government and records of Fairfax County, Virginia, prior to the war. It was located approximately fifteen miles from the nation's capitol, and over the years was surrounded by a village containing a number of substantial dwellings and business establishments. On May 31, 1861, it was the scene of an opening conflict of the Civil War when a company of Union cavalry made a reconnaissance through the area. Here Confederate Captain John Quincy Marr became the "First to fall for the rights of the South" as his gravestone would proudly announce, and Lieutenant-Colonel Richard S. Ewell, destined to become a Confederate Lieutenant-General, was wounded in the shoulder during the skirmish. The leader of the Union cavalry company, Lieutenant Charles H. Tompkins, received the Medal of Honor for his personal valor, and would also become a general officer during the war.[11]

By March, 1863, the Union Army had been in possession of Fairfax Courthouse for approximately two years; a sutler had established his store directly in front of it to serve the soldiers stationed in the town and in the surrounding area. A number of Union Army officers were billeted in the nearby residences. The Edward R. Ford residence, owned by a local merchant of the town, had received officer "guests" since early in the war. Their comforts were seen to by Ford's attractive daughter, Antonia, who liked to play at being a spy.

Antonia Ford was a woman of twenty-five, with brown eyes and hair, possessed a pleasing personality, and was always ready to engage any willing Union officer in conversation. During the Battle of First Manassas, she had passed some minor information to the Confederate Army, and the fun-loving Jeb Stuart had commissioned her his honorary aide. He had, in fact, drawn an elaborate document which read:

"TO ALL WHOM IT MAY CONCERN, KNOW YE:

That reposing special confidence in the patriotism, fidelity and ability of Antonia J. Ford, I, James E.B. Stuart, by virtue of the power vested in me as brigadier general in the Provisional Army of the Confederate States of America, do hereby appoint and commission her my honorary aide-de-camp, to rank as such from this date.

She will be obeyed, respected and admired by all lovers of a noble nature. Given under my hand and seal at the headquarters of the Cavalry Brigade at Camp Beverly the 7th October, A.D. 1861, and the first year of our independance.

(Signet ring seal) (Signed) J.E.B. Stuart[12]

(X true copy) (Signed) L.L. Lomax."

Antonia was quite proud of her appointment, kept it hidden beneath the mattress of her bed, and was quick to show it to all loyal to the Southern cause.

Stopping at the courthouse occasionally was a tall man dressed in black, the Reverend John D. Read, pastor of the First Baptist Church in nearby Falls Church. He ministered to the needs of all sick and wounded soldiers, Union and Confederate alike, and tried to help the poor blacks in the county to learn to read and write. No notice was taken of the comings and goings of this good man, but it was noticed by some that he exhibited a particular interest in any wounded Confederate, and would often ask leading questions about military matters. A few people good-naturedly commented if the Reverend Read wasn't such a good preacher, he just might have been mistaken for a spy.[13]

The Dr. William Presley Gunnel residence, a two-story brick house located near the courthouse, was the headquarters of Brigadier-General Edwin H. Stoughton, who commanded a section of the Washington Defenses protecting the nation's capital. Official orders directed: "Brigadier Generals are authorized to establish their headquarters at the points most convenient to their commands."[14] Ed Stoughton had determined the settlement around Fairfax Courthouse was convenient to his Second Vermont Brigade, although it was camped four miles distant at Fairfax Station. In the opposite direction on the road to Vienna was located the camp of his cavalry. Ed Stoughton fought hard and played hard, and because there was little fighting during the cold, muddy winter, he intended to relax where there was considerably more comfort than that afforded by a cold tent.

Edwin H. Stoughton had been appointed to the U.S. Military Academy from the state of Vermont, and was commissioned a second-lieutenant upon his graduation on July 1, 1859. Recognizing a war was forthcoming and that he could receive more rapid promotion in the volunteer services, he resigned from the regular army and was appointed colonel of the Fourth Vermont Volunteer Regiment on September 21, 1861. The regiment was part of the Army of the Potomac, and Ed Stoughton led it during the Virginia Peninsular Campaign. The regiment fought valiantly at the Siege of Yorktown, Lee's Mills, the battles of Williamsburg and Savage Station, and the Operations before Richmond.[15] His soldiers liked to say there was enough fight in Ed Stoughton to go around twice with some left over. His record being excellent, he was appointed Brigadier-General, U.S. Volunteers on November 5, 1862. He assumed command of the Second Vermont Brigade, then engaged in the defense of Washington. At twenty five years of age, Ed Stoughton was moving rapidly upwards through the ranks of the Union Army, and it was predicted he had a bright future in the military service.

Stoughton was not without his problems. His commission expired by Constitutional Limitation on March 4, 1863, and after that date he was technically not in command of the brigade. In his opinion, the bureaucrats in the War Department in Washington were not moving very fast in his behalf. Colonel Sir Percy Wyndham, commanding the First New Jersey Cavalry, was not effective in controlling the local Southern partisans. Inflexible in his tactics, this English soldier of fortune continued to use antiquated methods, involving large bodies of cavalry trying to catch small numbers of partisans. Wyndham's commanding officer of the post and headquarters detachment, Lieutenant-Colonel Johnstone, had moved his wife into a nearby house and spent more time with her than on his duties. His provost marshall, the hard-drinking but conscientious Lieutenant L.L. O'Connor, was not doing all that Stoughton thought should be done concerning security measures in town. On March 1, 1863, at the bottom of a message to the headquarters of his commanding general, Stoughton had added: "Last night, about 9 o'clock, while I was at headquarters, at the station (Fairfax Station), a man, undoubtly a spy was at the courthouse, dressed as a captain. He interrogated all my servants minutely respecting the troops in the vicinity, asking if I kept my horse saddled in the night, and other suspicious questions."[16]

The Gunnel house today. Stoughton's bedroom was on the second floor left.

The arrival of Ed Stoughton's mother and sister for a visit was a cheering influence. He decided to have a champagne party in their honor on the evening of March 8, 1863, and set about making arrangements with his customary zeal. He made provision for his family to stay at the Ford home, believing the lovely Antonia to be a spy, but satisfied she would gain little information of military value from the women. He planned to forget his problems with the commission, his officers, the Confederate Army, and the local partisans, and to enjoy himself for the evening. He sent party invitations to all of his officers (for he was not one to hold a grudge over a little thing, such as not performing one's duty), and to all the local young females.

The weather on Sunday, March 8, was disagreeable. Snow was on the ground, and a cold rain started falling early that afternoon. But lights blazed from the Gunnel house, lighting the way for guests, and the party was soon underway. It was past midnight before the guests began to make polite exits. Colonel Wyndham had been called to Washington in the afternoon and could not attend. His adjutant felt he should at least take advantage of the absence to use Wyndham's warm bed, and proceded to do so. Lieutenant O'Connor received word of problems with the pickets on the Vienna road, and with a number of oaths, collected his reserves and rode off through the rain to investigate the matter. Seeing his mother and sister to their lodgings, Stoughton retired to his upstairs bedroom. He was soon sound asleep, satisfied his party had been a success. His adjutant briefly cleaned up from the party and retired. Only the marching of the guards disturbed the black night surrounding the courthouse.

On the afternoon of March 8, John S. Mosby and a party of twenty-nine men started from Aldie to Fairfax Courthouse. Aldie was a favorite rendezvous for partisans because numerous roads entering the town and nearby mountains offered ready op-

Marker in front of the Gunnel house.

portunity for escape should the need arise. Mosby had been interested in Fairfax Courthouse for some time because a Britisher, Sir Percy Wyndham, Commanding Officer of the First New Jersey Cavalry, was stationed there. Colonel Wyndham, exasperated by Mosby's recent partisan tactics, had denounced him as a horse-thief, a serious accusation in those times, even during war. Jeb Stuart, aware of Mosby's outrage, had urged the capture of Wyndham in his own camp as an embarrassment to both Wyndham and the Union Army. Mosby knew his plan was dangerous, but as he later explained: "The safety of the enterprise lay in the novelty; nothing of the kind had ever been done before."[17] Mosby and his men made little noise as they rode through the night, armed only with revolvers. Mosby believed the saber to be of limited value against gunpowder, and the rattling of the carbine and saber would have made it impossible to surprise an enemy.[18]

Mosby had received advance information from local farmers of an opening in the Union defense lines where his party could enter undetected. He spent a good deal of personal time in scouting the enemy and visiting farmers friendly to his cause. He would appear suddenly in a farmyard, scoop up any small child present, place it in front of him and ride it around on his horse. He allowed the older children to lead the horse as the little one sat alone in the saddle.[19] The farmers appreciated Mosby's interest in their children, and were quick to respond to requests for information. They showed him the little known cow paths and back roads that would prove useful for his purposes. The farmers' wives were also pleased to see Mosby, because he was quick to pay for any meals eaten by himself, his men or his prisoners. He always paid twenty-five cents for a meal, using Union or Confederate money, depending upon who controlled the territory at the time.[20]

Mosby's small party penetrated the Federal lines between Chantilly and Centerville, both heavily garrisoned with Union soldiers. He knew that once inside, his group would be mistaken for Federal cavalry. Army pickets permitted them to pass within sight of their outposts because they had no reason to believe enemy forces were in their rear and moving inside their own lines.[21]

About 2 A.M., pickets guarding the area near Fairfax Courthouse heard cavalry riding in on the road from Fairfax Station where the Vermont Infantry camp was located. They felt no cause for alarm however, and continued their slow marching in the rain and darkness. The horsemen divided into small groups and went about their business. Pickets soon looked down the barrels of big Colt revolvers and heard the command: "Surrender, you Damn Yankee!" Mosby quietly took possession of the headquarters area without a single shot being fired. After quickly grilling the telegraph operator on duty, he had a man cut the lines, breaking all contact with the outside. He discovered General Stoughton was at the Gunnel house, and was determined to capture him. Silently, his men performed their assigned tasks. One group went to the stables and selected the best horses there. Another group found the residence occupied by Wyndham, and finding him gone, pulled his adjutant from the colonel's bed and made him a prisoner.

Mosby and several men proceeded to the Gunnel house, banged on the door, and shouted that they carried important dispatches for the personal attention of General Stoughton. When the door was opened, Mosby rushed in and silenced the staff officer on duty by thrusting his pistol into his side. The best description of what happened next are Mosby's own words, written to a friend in Richmond shortly after the incident. "I walked into his room with two of my men and shaking him in bed said, 'General, get up!' He rose and rubbing his eyes, asked what was the meaning of all this. I replied, 'It means, sir, that Stuart's cavalry are in possession of this place, and you are my prisoner!' "[22]

Mosby knew he could not linger, and departed an hour after his arrival. Leading his men and prisoners slowly back down the road toward the camp of the Vermont Infantry, Mosby believed no one was aware of what was happening. At the edge of the village the window of a house was suddenly lifted, and an authoratative voice demanded to know what the cavalry was doing at that late hour.

After the extreme tension of the past hour, Mosby's men could not help laughing in relief as the owner of the voice suddenly realized he was not addressing friendly troops. Lieutenant-Colonel Johnston, the post commander, was spending the night with his wife when the horsemen's noise disturbed them. Johnstone raced for the back door, ran through the yard, and dived under the outhouse. Additionally, he had the misfortune to lose his nightshirt in the process. Two of Mosby's men had broken open the door of the house, but were attacked by Johnstone's wife, thus delaying them until the colonel had made good his escape. With no time to search, Johnstone was left in his hiding place.

Guide to Mosby's midnight raid for the present student.

Upon leaving the village, Mosby ordered a sharp turn and headed directly towards Centerville. He knew there were several thousand Union soldiers stationed there — no one would expect him to ride in that direction.[23] The group rode rapidly through the darkness, and again passed undetected through Union lines.

A short time later, a wet Provost Marshall O'Connor returned from his disagreeable duty to find that during his absence Mosby had captured most of the headquarters he was responsible for. He immediately dispatched the following message to higher head quarters:

"Provost Marshall's Office,
Fairfax Court-House, Va.
March 9, 1863 — 3.30 a.m.

Captain Mosby, with his command, entered this town this morning at 2 a.m. They captured my patrols, horses, etc. They took Brigadier-General Stoughton and horses, and all his men detached from his brigade. They took every horse that could be found, public and private, and the commanding officer of the post, Colonel Johnstone, of the Fifth New York Cavalry, made his escape from them in a nude state by accident. They searched for me in every direction, but being on the Vienna road, visiting outposts, I made my escape.

L.L. O'Connor
Lieutenant, Provost Marshall

P.S. All our available cavalry forces are in pursuit of them."[24]

After Mosby's party had departed, Lieutenant-Colonel Johnstone came out of his hiding place, scrubbed down with a horse brush, collected all immediately available cavalry, and led the chase for the partisans. "Outhouse" Johnstone (as he was afterwards nicknamed by his troops) galloped off in the direction opposite Mosby's route. All pursuit by the Union Army proved to be unsuccesful.

By the time Mosby entered the safety of the Confederate lines, he had captured Stoughton, two captains, thirty enlisted men and fifty eight horses.[25] It was a remarkable accomplishment! Mosby personally took Stoughton to Culpeper and turned him over to one of the general's former West Point classmates, Confederate General Fitz Lee. Wrote Mosby: "Lee came out of his tent and welcomed General Stoughton...as a long lost brother. He took him into the tent to give him a drink and left me out in the rain."[26]

The Affair at Fairfax Court-House, as Union Army records named the incident, was officially over: Mosby's great raid would be long remembered and affect the lives of many people.

Northern newspapers were outraged over Mosby's raid. Commented the *New York Herald:* "The capture of Brigadier-General Stoughton, with his headquarters guard and fifty-five horses, by a marauding party of rebels, within our own lines, is regarded by military authorities here as deeply mortifying and disgraceful."[27] Equally angry, *Frank Leslies Illustrated Newspaper* added: "It is said Mr. Lincoln was more annoyed at the loss of the horses than of the Brigadier-General, of whom he said he could make a dozen any day, but a horse was beyond his power of creation."[28] These articles reflected the general feelings of the North on learning of the raid, and it was believed the military men at Fairfax Courthouse were not doing the job expected of them.

Brigadier-General Stoughton was placed on a prisoner of war status, and transported to the Confederate States Military Prison at Richmond, Virginia. Arriving there on March 12, his escort thoughtfully checked him into a suitable room at the plush Ballard Hotel for the night. The prison warden, on learning of this unusual action, descended on the hotel at 1.00 A.M. and transported the general to his own establishment, generally regarded as neither comfortable nor prestigious.[29] Poor Ed Stoughton was still having a problem getting an undisturbed night's sleep. Fortunately, he did not have long to languish in Richmond. He was exchanged on May 7, 1863, arriving at Annapolis, Maryland, with another captured general officer and some two hundred officers of lesser rank.[30] As could be expected, someone was needed to blame for the fiasco at Fairfax Courthouse: Ed Stoughton's commission was not renewed. He entered the legal profession, practicing in New York City until 1868, died on Christmas Day at the age of thirty one years. Of his Civil War service, it can be said he fought hard and he played hard, and that his misfortune was to be captured while playing.

Secretary of War Edwin M. Stanton was not satisfied with the furor arising from Mosby's raid and asked General Lafayette C. Baker's Secret Service to investigate the incident. Suspicion soon came to rest upon Antonia Ford. A female agent posing as a Secessionist traveling south stopped at the Ford home. She soon won over the gullible Antonia, who proudly showed her commission from Jeb Stuart. Antonia was taken into custody shortly thereafter, and confined in the Old Capitol Prison at Washington, D.C., along with her father and several other residents of Fairfax County.

Jeb Stuart, realizing he had contributed to Antonia's imprisonment, wrote Mosby: "I wish you would send me whatever evidence you may be able to furnish me of Miss Ford's innocence of the charges of having guided you in your exploits at Fairfax so that I can insist upon her unconditional release."[31] Unfortunately, Stuart did not possess the influence to negotiate for her release, and he was not in the position to "insist" on it. Antonia was ultimately released through the efforts of Union Major Joseph C. Willard, a former Provost Marshall of Fairfax, and an admirer of Antonia from the time he was billeted in her home during the early days of the war. After a whirlwind courtship, the dedicated rebel married her Union Army officer suitor in Washington on March 10, 1864, a year and a day after Mosby's raid. Seven years later she died, leaving one child, a son, Joseph, who would become one of Virginia's most effective lieutenant governors.

In the year 1900, long after the war had ended, Mosby indicated in a letter to a friend that Antonia was innocent of assisting him in any manner, and referred to her son: "There is not a particle of truth in the statement she ever acted as a spy — or that I ever did. I refer to Antonia Ford's being arrested on the charge of aiding me — of her being put in prison — & getting her revenge by marrying the provost marshall — Result — Joe Willard — She was as innocent as Abraham Lincoln — Now don't you see the obligation that Joe Willard is under to me? If it hadn't been for me there never would have been a Joe Willard. I made him. Don't you think he ought to pay me an annuity?"[32]

There was one man who believed Mosby could be captured and he concentrated his efforts to accomplish this. Reverend John D. Read had numerous confidential sources of information, and he soon provided Mosby's location to Brigadier-General G.A. De Russy then commanding the Headquarters, Defense of Washington, South of the Potomac. In an act of sheer stupidity, General De Russy dispatched a message to Washington, naming his source: "A Mr. Reed, resident of that place (Falls Church), says he knows that Mosby's headquarters are only about 5 miles from Falls Church, where he had about 40 men."[33] Soldiers like to be mentioned in dispatches and reports because it is often an aid to promotion: spies want no mention at all, for to be anonymous is to survive.

Shortly thereafter, a company of Mosby's men raided Falls Church capturing five Union Army pickets and the Reverend Read. He was taken into a dark pine thicket, made to kneel, and a pistol placed to the back of his head. The official report of Major John Birdsall, Thirteenth New York Volunteer Cavalry, who pursued the partisans, describes what they found. He wrote: "There is no doubt concerning the murder of Mr. Reed, as the surgeon, who has made an examination of the body, states that the skull at the base of the brain is blown to atoms, and the flesh about the wound is filled with powder, as if the pistol had been placed close to the head."[34] Spying was a dangerous business.

For years following the tragic event, children in Northern Virginia chanted the following verse as they jumped rope:

"Isn't any school,
Isn't any teacher;
Isn't any church,
Mosby shot the preacher."[35]

Few of them realized the historical significance of their song.

For John S. Mosby, the Fairfax Courthouse raid provided the recognition he needed to launch his war record as a partisan leader. On March 25, 1863, he received his commission as Captain of Partisan Rangers in a letter from Jeb Stuart, who also thanked him for the present of General Stoughton's saddle and inquired if he could catch that "unprincipled scoundrel Wyndham" without being caught himself.[36] Mosby immediately expanded his small command and initiated a campaign of partisan military operations believed by many to be unequaled to this day. The Union Army made special efforts to capture him, and ordered Colonel Charles R. Lowell of the Second Massachusetts Cavalry to "Go where you please in the pursuit of Mosby..."[37] The effort came to no avail, and following the great cavalry battle at Brandy Station, Union Brigadier General Alfred Pleaston decided a bribe was the only method of stopping Mosby. Pleaston wrote to headquarters: "Ask the general how much of a bribe he can stand to get Mosby's services."[38] He received in reply: "If you think your scheme can succeed in regard to Mosby, do not hesitate as to the matter of money."[39] Again, the effort came to nothing. Mosby could not be caught or bought!

Mosby was a hero to the hard-pressed people of the South, who supported him during the hardships inflicted on them throughout the war. An example of this support was the critical wounding of Mosby at Lakeland, the residence of Ludwell Lake, a prominent Virginia farmer. Mosby and Ranger Tom Love were dining on spare ribs and biscuits late one evening with Lake and his two daughters, Sarah and Landonia. Landonia's husband had ridden with Mosby until his capture. She had just finished telling of her visit to him in the Union prison camp at Point Lookout, Maryland, and how she had smuggled him a twenty dollar gold piece hidden inside an apple, when the sound of Union cavalry was heard.

Spotting horses tied in front of the Lake home, a passing detachment of drunken New York Cavalry surrounded the house. Looking into a window, Corporal Kane saw a Confederate officer, and fired one shot at him through the window glass. The bullet clipped a button from the vest of Ludwell Lake before striking Mosby in the lower abdomen.

"Oh, My God, I'm shot! Give me something fast!" Mosby exclaimed.

Sarah Lake took the bonnet from her head and handed it to Mosby to staunch his bleeding. The lights had gone out in the confusion, and when the troopers rushed in to establish order, they found a Confederate cavalryman ready to surrender. Nearby, a Confederate officer, who identified himself as Lieutenant Johnson, 6th Virginia Cavalry, lay on the floor, bleeding from the stomach. Mosby had daubed blood from the wound on his lips to make it appear he was bleeding internally, usually a sign of impending death. He had been able to shed and hide his uniform coat and high riding boots in the short time after he had been shot, therefore was not recognized. Believing "Lieutenant Johnson" to be mortally wounded, the Union cavalrymen left him on the floor to die, and departed with their captured prisoner, Ranger Love. As they left, a cavalryman noticed Mosby's plumed slouch hat and took it for a souvenir.

It was not safe for Mosby to remain at Lakeland, so he was moved in an oxcart driven by a slave, "Uncle" Daniel Strother, to the nearby residence of Aquilla "Quilly" Glasscock for medical treatment. His uniform and high boots had been sent ahead to lessen any chance of his being identified. As Mosby was being loaded in the oxcart, his bare feet protruded from under the quilts he was wrapped in, and Sarah Lake removed the only pair of shoes she owned and slipped them on his feet. The following day, when the Federals discovered the hat taken on the previous night may have belonged to Mosby, they rushed back to the Lake home. During four hours of intense interrogation, Sarah Lake stood barefoot outside in the yard in falling sleet. She never wavered in her statement that the wounded officer was not Mosby, that she had never seen him before, and that she had no idea where he was now. It was a dedicated Southern woman's way of supporting the cause she believed in.[40]

According to an interview with Virgil Carrington Jones, Civil War author and Mosby authority, if Mosby had died gloriously for the Southern cause, he would have been recognized today as a distinguished Southern hero.[41] However, he survived his wound to fight again, to disband his Rangers at the finish of the war (he never formally surrendered), and to call for a time of healing and cooperation with the Union he had fought so hard to destroy. He became a "Black Republican," and backed reconciliation. Asked once by a minister if he believed in hell, he replied anyone who voted Republican and lived in Virginia had to believe in it. John Singleton Mosby served his state and country as a soldier, lawyer and statesman. At the time of his death, his military prowess was largely unrecognized, and many of his former Rangers refused to speak of him. For the "Grand Old Rascal," the raid on Fairfax Courthouse was his finest hour.

ACKNOWLEDGEMENTS

Kim B. Holien, U.S. Army historian, Washington, D.C.

Horace H. Mewborn, Jr., historian and Mosby collector, Fairfax County, Virginia.

FOOTNOTES

[1] *War of the Rebellion: A Compilation of the Official Records of the Union and Confederate Armies*, Series I, Vol. XXIX, Part 2, p. 397.

[2] Thomas J. Evans, "'Quaker Sam,' Captain Samuel C. Means, Co. A, Independent Loudoun Rangers," *North-South Trader*, IX-5:20-27, July-August, 1982.

[3] *OR*, Series I, Vol. XLVI, Part 3, p. 590.

[4] *OR*, Series I, Vol. XLIII, Part 2, pp. 721, 717.

[5] Virgil Carrington Jones, *Gray Ghosts and Rebel Raiders* (New York: Holt, Rinehart and Winston, 1956), pp. 305, 307.

[6] Ulysses S. Grant, *Personal Memoirs* (New York: Charles L. Webster & Co., 1886) Vol. II, p. 142.

[7] William Willis Blackford, *War Years with JEB Stuart* (New York: Charles Scribner & Sons, 1945), p. 14.

[8] John Singleton Mosby, *The Memoirs of Colonel John S. Mosby* (Bloomington, Indiana: University Press, 1959), p. 31.

[9] John Singleton Mosby, *Mosby's War Reminiscences* (New York: Dodd, Meade and Co., 1887), pp. 39-40.

[10] Albert E. Castel, "The Guerilla War, 1861-1865," *Civil War Times*, Special Issue: 17, October, 1974.

[11] Thomas J. Evans, "The Skirmish at Fairfax Courthouse, Virginia," *North-South Trader*, X-4; 8-12, May-June, 1983.

[12] Jeanne Johnson Rust, *A History of the Town of Fairfax* (Fairfax, Virginia, 1960), p. 54.

[13] Thomas J. Evans, "For God and Country — The Reverends Read," *North-South Trader*, VIII-3:13-17, March-April, 1981.

[14] *OR*, Series I, Vol. LI, p. 975.

[15] Maj. Gen. George W. Cullem, *Biographical Register of the Officers and Graduates of the US Military Academy* (Cambridge: The Riverside Press, 1891), p. 729.

[16] *OR*, Series I, Vol. XXV, Part 2, p. 114.

[17] Mosby, *Memoirs*, p. 72.

[18] Adolphus "Dolly" E. Richards, "Mosby's 'Partisan Rangers,'" *Famous Adventures and Prison Escapes of the Civil War* (New York: The Century Co., 1898), p. 114.

[19] From Numerous interviews by the author.

[20] *Fairfax County, Virginia — A History*, Fairfax County Board of Supervisors, Fairfax, Virginia, 1978, p. 357.

[21] John Singleton Mosby, "A Bit of Partisan Service," *Battles and Leaders of the Civil War* (New York: Thomas Yoseloff, Inc., 1956), Vol. III, p. 150.

[22] Rust, *History*, p. 49.

[23] Mosby, *Memoirs*, pp. 176, 177.

[24] *OR*, Series I, Vol. XXV, Part 1, p. 43.

[25] *Ibid*, Part 2, p. 856.

[26] Rust, *History*, p. 49.

[27] *New York Herald*, March 10, 1863, p. 1.

[28] *Frank Leslie's Illustrated Newspaper*, March 28, 1863, p. 2.

[29] *OR*, Series II, Vol. V, pp. 847, 848.

[30] *Ibid*, p. 567.

[31] *OR*, Series I, Vol. XXV, Part 2, p. 858.

[32] *The Partisan Review*, The John S. Mosby Society, Fairfax, Virginia, Winter-Spring, 1983, p. 1.

[33] *OR*, Series I, Vol. XXIX, Part 2, p. 26.

[34] *OR*, Series I, Vol. XLIII, Part 2, pp. 414, 415.

[35] Thomas J. Evans, "In Search of the Railroad Camp," *North-South Trader*, X-1:20-25, 32, November-December, 1982.

[36] *OR*, Series I, Vol. XXV, Part 2, p. 858.

[37] *OR*, Series I, Vol. XXVII, Part 3, p. 74.

[38] *Ibid*, p. 72.

[39] *Ibid*.

[40] Interview with John Gott, great-grandson of Sarah Lake. (Mr. Gott has in his possession Sarah's blood-stained bonnet and the bullet-shattered glass window pane from Lakeland.)

FAMILY BIBLE

Birthdays of Civil War Notables

by
Paul Kallina

(N = North. S = South.)

APRIL

Apr. 1, 1801—Capt. William F. Lynch, C.S.N., Norfolk, Virginia. (S)
Apr. 1, 1823—Gen. Simon B. Buckner, Hart County, near Munfordville, Kentucky. (S)
Apr. 2, 1802—Prof. Dennis Hart Mahan, dean of the faculty, U.S.M.A. (1838-1871), New York City. (N)
Apr. 2, 1812—Rev. John N. Waddell, army chaplain, Willington, South Carolina. (S)
Apr. 2, 1814—Gen. Henry L. Benning, Columbia County, Georgia. (S)
Apr. 3, 1798—Adm. Charles Wilkes, New York City. (N)
Apr. 4, 1792—Rep. Thaddeus Stevens, Danville, Vermont. (N)
Apr. 4, 1802—Dorothea Dix, Superintendent of Army Nurses, Hampden, Maine. (N)
Apr. 4, 1806—Claiborne F. Jackson, governor of Missouri (1860-1861), Fleming County, Kentucky. (S)
Apr. 4, 1831—Gen. Edward C. Walthall, Richmond, Virginia. (S)
Apr. 4, 1844—Cpl. James Tanner, Richmondville, New York. (N)
Apr. 4, 1853—Thomas "Tad" Lincoln, Springfield, Illinois. (N)
Apr. 5, 1825—Gen. David R. Jones, Orangeburg District, South Carolina. (S)
Apr. 5, 1839—Capt. Robert Smalls, U.S.N., Beaufort, South Carolina. (N)
Apr. 5, 1856—Booker T. Washington, *Up From Slavery*, Hale's Ford, Franklin County, Virginia. (N)
Apr. 6, 1828—Gen. Charles W. Field, Airy Mount, Woodford County, Kentucky. (S)
Apr. 7, 1822—Gen. Gershom Mott, Lamberton, New Jersey. (N)
Apr. 9, 1812—Gen. Randolph B. Marcy, Greenwich, Massachusetts. (N)
Apr. 10, 1806—Gen. Leonidas Polk, Raleigh, North Carolina. (S)
Apr. 10, 1807—Gen. Benjamin F. Kelley, New Hampton, New Hampshire. (N)
Apr. 10, 1810—Rep. Willis B. Machen, Caldwell County, Kentucky. (S)
Apr. 10, 1817—Gen. John C. Robinson, Binghamton, New York. (N)
Apr. 10, 1827—Gen. Lewis Wallace, Brookville, Indiana. (N)
Apr. 10, 1847—Pvt. Joseph Pouletzes (later Pulitzer), Mako, Hungary. (N)
Apr. 11, 1837—Col. Elmer E. Ellsworth, Malta, Saratoga County, New York. (N)
Apr. 12, 1777—Henry Clay, The Slashes, Hanover County, Virginia.
Apr. 12, 1805—Capt. Theodorus Bailey, U.S.N., Chateaugay, New York. (N)
Apr. 12, 1831—Gen. Grenville M. Dodge, Danvers, Massachusetts. (N)
Apr. 14, 1820—Gen. Harry T. Hays, Wilson County, Tennessee. (S)
Apr. 14, 1828—Adalbert Volck, artist, Augsburg, Germany. (S)
Apr. 14, 1842—Lt. Adna R. Chaffee, U.S. Army, Orwell, Ohio. (N)
Apr. 15, 1821—Joseph E. Brown, governor of Georgia (1857-1865), Pickens District, South Carolina. (S)
Apr. 15, 1821—Commander Catesby ap Roger Jones, Fairfield, Virginia. (S)

Apr. 16, 1808—Caleb B. Smith, Secretary of the Interior (1861-1863), Boston, Massachusetts. (N)
Apr. 16, 1816—Gen. Edward Johnson, Salisbury, Virginia. (S)
Apr. 17, 1806—William Gilmore Simms, ante-bellum writer, Charleston, South Carolina. (S)
Apr. 17, 1809—Gen. Philip St. George Cocke, Fluvanna County, Virginia (S)
Apr. 17, 1837—John P. Morgan, war profiteer, Hartford, Connecticut. (N)
Apr. 18, 1815—Beriah Magoffin, neutralist governor of Kentucky (1859-1862), Harrodsburg, Kentucky. (S)
Apr. 20, 1807—John Milton, governor of Florida (1861-1865), Jefferson County, Georgia. (S)
Apr. 20, 1808—Napoleon III, Emperor of France (1852-1870), Paris, France.
Apr. 20, 1824—Gen. Alfred H. Colquitt, Walton County, Georgia. (S)
Apr. 20, 1827—Gen. John Gibbon, near Philadelphia, Pennsylvania. (N)
Apr. 21, 1809—Sen. Robert M. T. Hunter, Secretary of State (1861-1862), Essex County, Virginia. (S)
Apr. 21, 1816—Gen. and Sen. Louis T. Wigfall, Edgefield, South Carolina. (S)
Apr. 22, 1792—Capt. Uriah P. Levy, U.S.N., Philadelphia, Pennsylvania. (N)
Apr. 22, 1815—Rev. Thomas W. Humes, East Tennessee Relief Association, Knoxville, Tennessee. (N)
Apr. 22, 1831—Gen. Alexander M. McCook, Columbiana County, Ohio. (N)
Apr. 22, 1832—Col. Edward E. Cross, Lancaster, New Hampshire. (N)
Apr. 23, 1791—Pres. James Buchanan, Mercersburg, Pennsylvania.
Apr. 23, 1813—Sen. Stephen A. Douglas, Brandon, Vermont. (N)
Apr. 23, 1815—Andrew G. Curtin, governor of Pennsylvania (1860-1866), Bellefonte, Center County, Pennsylvania. (N)
Apr. 24, 1807—Gen. Charles F. Smith, Philadelphia, Pennsylvania. (N)
Apr. 26, 1814—Commander James H. Strong, Canandaigua, New York. (N)
Apr. 26, 1817—Lord Lyons, British minister to the United States (1858-1865), Lymington, England.
Apr. 27, 1791—Samuel F. B. Morse, telegraph inventor, Charlestown, Massachusetts.
Apr. 27, 1822—Gen. Ulysses S. Grant, Point Pleasant, Ohio. (N)
Apr. 27, 1835—Gen. John M. Corse, Pittsburgh, Pennsylvania. (N)
Apr. 28, 1815—Gen. Andrew J. Smith, Bucks County, Pennsylvania. (N)
Apr. 29, 1815—Gen. Abram Duryee, New York City. (N)
Apr. 29, 1820—Gen. Henry W. Allen, governor of Louisiana (1864-1865), Prince Edward County, Virginia. (S)

MAY

May 2, 1837—Capt. Henry M. Robert, U.S. Army, Robertville, Jasper County, South Carolina.
May 3, 1816—Gen. Montgomery C. Meigs, Augusta, Georgia. (N)
May 4, 1818—Edwin De Leon, diplomat, Columbia, South Carolina. (S)
May 6, 1801—Gen. George S. Greene, Apponaug, Rhode Island. (N)
May 6, 1812—Maj. Martin R. Delany, Charles Town, Virginia. (N)
May 6, 1825—Gen. Joseph Bailey, Pennsville, Ohio. (N)
May 7, 1818—Juliet Ann Opie Hopkins, hospital administrator, Woodburn, Jefferson County, Virginia. (S)
May 7, 1826—Varina Howell Davis, The Briers, near Natchez, Mississippi. (S)
May 8, 1824—William Walker, filibuster adventurer, Nashville, Tennessee. (S)

May 8, 1826—Pilot Horace E. Bixby, Chief of Union River Service, Geneseo, New York. (N)
May 8, 1829—Moses Embree Milner, "California Joe" of Berdan's Sharpshooters, Standford, Kentucky. (N)
May 8, 1833—Gen. Frank Wheaton, Providence, Rhode Island. (N)
May 8, 1835—Augusta Jane Evans, novelist and nurse, Columbus, Georgia. (S)
May 9, 1800—John Brown, abolitionist, Torrington, Connecticut. (N)
May 9, 1844—Belle Boyd, spy, Martinsburg, Virginia. (S)
May 10, 1813—Montgomery Blair, postmaster general (1861-1864), Franklin County, Kentucky. (N)
May 10, 1838—John Wilkes Booth, near Belair, Maryland. (S)
May 12, 1806—Gen. James Shields, Altmore, County Tyrone, Ireland. (N)
May 12, 1840—Lt. Charles W. Read, C.S.N., Yazoo County, Mississippi. (S)
May 13, 1830—Zebulon B. Vance, governor of North Carolina (1862-1866), Buncombe County, North Carolina. (S)
May 15, 1802— Gen. Isaac R. Trimble, Culpeper County, Virginia. (S)
May 15, 1819—Gen. Thomas L. Crittenden, Russellville, Kentucky. (N)
May 15, 1837—Col. Garnett Andrews, Washington, Georgia. (S)
May 16, 1801—William H. Seward, Secretary of State (1861-1869), Florida, Orange County, New York. (N)
May 16, 1824—Gen. Edmund Kirby-Smith, St. Augustine, Florida. (S)
May 17, 1812—Gen. Joseph W. Revere, Boston, Massachusetts. (N)
May 18, 1798—Gen. Ethan A. Hitchcock, Vergennes, Vermont. (N)
May 18, 1815—Rep. Thomas S. Bocock, Speaker of the House (1862-1865), Buckingham County, Virginia. (S)
May 19, 1820—Margaret Junkin Preston, poet, Milton, Pennsylvania. (S)
May 19, 1812—Gen. Felix K. Zollicoffer, Maury County, Tennessee. (S)
May 21, 1822—Gen. Dabney H. Maury, Fredricksburg, Virginia. (S)
May 22, 1804—Commodore John W. Livingston, New York City. (N)
May 22, 1819—Gen. Richard L. T. Beale, Hickory Hill, Virginia. (S)
May 23, 1824—Gen. Ambrose E. Burnside, Liberty, Indiana. (N)
May 24, 1810—Capt. Melancton Smith, U.S.N., New York City. (N)
May 24, 1819—Victoria, Queen of England (1837-1901), Kensington Palace, London.
May 25, 1816—Gen. Henry Hopkins Sibley, Natchitoches, Louisiana. (S)
May 26, 1806—Adm. Henry K. Thatcher, Thomaston, Maine. (N)
May 26, 1835—Gen. Edward Porter Alexander, Washington, Georgia. (S)
May 27, 1819—Julia Ward Howe, author "Battle Hymn of the Republic," New York City. (N)
May 27, 1819—Commander Thomas H. Stevens, Middletown, Connecticut. (N)
May 27, 1823—Gen. John G. Foster, Whitefield, New Hampshire. (N)
May 27, 1837—James B. "Wild Bill" Hickok, scout and spy, Troy Grove, La Salle County, Illinois. (N)
May 28, 1818—Gen. Pierre G. T. Beauregard, St. Bernard Parish, Louisiana. (S)
May 28, 1830—Gen. George L. Hartsuff, Tyre, Seneca County, New York. (N)
May 28, 1838—Gen. Basil W. Duke, Scott County, near Georgetown, Kentucky. (S)
May 29, 1810—Gen. Erasmus D. Keyes, Brimfield, Massachusetts. (N)
May 29, 1824—Gen. Cadmus M. Wilcox, Wayne County, North Carolina. (S)
May 30, 1812—Gen. John A. McClernand, Breckinridge County, near Hardinsburg, Kentucky. (N)

(Continued on page 24)

CONCURRENT EVENTS

In 1643, at Ipswich, beans were ordered to be used in voting — the white denoting yea, the black nay. From this originated the old phrase, "You don't know beans;" as no one was allowed to vote who did not know a white bean from a black one.
[*Boston Cultivator*, Boston Massachusetts, December 13, 1845.]

ROBBERIES BY CHLOROFORM —
A Dunkirk barber has been sentenced to fifteen years' imprisonment for robbing a man of fourteen hundred dollars while under the influence of chloroform, which the accused had administered. The penalty, says the New-York Times, is a severe one, but not more so than the nature of the offence demanus. Robberies by the aid of chloroform are becoming of almost daily occurrence, and unless the courts and juries take care to visit with severest punishment offencers who resort to the use of this subtle agency, special legislation will soon be required to meet such cases. There is a progress in crime as well as in everything else, and the employment of chloroform as an accomplice in robbery is one of the latest refinements. We question whether it ought not to be met with penalities quite as terrible as is the crime of arson in the first degree, for the circumstances under which these crimes are perpetrated are similar in the most important respects. In each case the crime is committed when its victim is unconscious of what is going on about him, and has not the slightest opportunity to defend his person and property; and in each case his life is seriously periled. If this new species of crime, already so alarmingly prevalent, be not met by stern and fearful retribution, robberies will be multiplied tenfold, for the use of chloroform involves little risk to the rogue, unless it be the risk of the halter.
[From *The Country Gentleman*, Albany, New York, May 24, 1855.]

John Jeffer's Fight For the Union And Against the Pension Bureau

by
Charles A. Lowe

John Jeffers was typical of the immigrant who came to the United States in the mid-19th Century. Leaving Ireland in 1850 because he was hungry, he settled in Milford, Massachusetts, went to school and learned the "cordwainer's trade:" bootmaking. Like most newcomers who stayed within their ethnic group, Jeffers married an Irish girl from County Mayo, the place he originated from. At twenty-two years of age, he and Bridget Hardman were married at St. Mary's of the Assumption Catholic Church by Rev. Fr. Michael O'Brien. They eventually had four children, a girl and three boys.

In June of 1861, responding to the call of President Abraham Lincoln for volunteers to suppress the rebellion of the Confederacy, Jeffers put aside his tools and bid farewell to Bridget and the children. He journeyed to Yonkers, New York, and enlisted in the Union Army as an infantryman. Captain Coggswell mustered him into Co. G., 40th New York Volunteers, the Mozart Regiment. His company commander was Lt. Augustus Keene. The unit became part of the First Division (Major-General David Birney), Third Corps (Major-General Daniel Sickles).

Beginning with the siege of Yorktown on April 5, 1862, until Gettysburg, July 3, 1863, they were involved in nineteen major engagements, including Fair Oaks (twice), Williamsburg, Malvern Hill, Second Bull Run, Fredericksburg and Chancellorsville. Six hundred and thirty were killed, wounded, missing or taken prisoner in that fourteen-month span.

At the Peach Orchard, July 2, 1863, on the second day of the three-day holocaust at Gettysburg, Jeffers took a minie ball through the left side of his chest that shattered the breast bone, punctured a lung and exited the right side. After emergency battlefield medical treatment, he was sent by ambulance to the General Hospital at Alexandria, Virginia. From there, he was transferred to McDougall General Hospital, Ft. Schuyler, New York Harbor, and in June of 1864 to DeCamp General Hospital, David's Island, New York Harbor.

The casualty sheet, signed by F.M. Montague, noted that "J. Jeffers, Private, Co. G, 40th Regiment of Infantry of the New York Volunteers, wounded, lungs, severe," dated July 2, 1863. Dr. W. Webster, the Assistant Surgeon of the Division, who apparently operated on Jeffers, stated, "gunshot wound of the lower end of the sternum (breast bone) received in action at Gettysburg, July 2, 1863. He is two-thirds (2/3) disabled. Physically unable to do duty with Third Corps."

Jeffers was finally discharged from DeCamp in 1864, and awarded a pension of $8.00 per month. He went home to Milford, and in November of that year, he and Bridget traveled to Boston and became United States citizens, being sworn by N.J. Clifford, Clerk of the United States District Court for Massachusetts.

Unable to work, he found the meager pension too small to support his family; he reenlisted in the Veteran's (Invalid) Reserve Corps, December 1864 for three years for garrison duty. Again, he was in and out of hospitals for his wound, malaria (contracted along the Potomac River) and bronchitis. He was again discharged in November 1865.

Seeking to increase his pension, he engaged Attorney George G. Parker of Milford to represent him. Many petitions were submitted to the Pension Bureau, accompanied by affidavits from neighbors, attesting to his infirmity as well as physical exams taken in Worcester, Massachusetts, but nothing worked.

Examining Physician John G. Metcalf of the Pension Bureau remarked at Worcester, "he was wounded as above, by a bullet entering the left chest and coming out the right side, fracturing the sternum. Wounds closed. Is emaciated, has cough, percussion dull over the left chest, front and back. Present disease of the lungs resulted from gunshot wound, disability is permanent and equal to the loss of hand or foot." Again, this made no favorable impression on the Bureau.

On March 17, 1873 (St. Patrick's Day), after Jeffer's neighbors David Burke and Patrick Fitzsimmons went to Worcester District Court to present affidavits that "Jeffers certainly was an invalid," Justice Charles A. Dewey awarded a pension increase to $15.00 per month.

Ten years later, on September 12, 1883, Pension Case #32055, the Acting Secretary of the Interior approved an increase to $24.00 per month. The approval was by W.L. Josslyn, and countersigned by W.W. Dudley, Commissioner of Pensions. By that time Jeffers was living with his daughter, Sarah, in Hopedale. She was a cook and domestic for the Draper family in Boston and Hopedale, barons in the dry-goods manufacturing business. He died in Robbury, Massachusetts, at the age of sixty-nine.

Jeffers spent twenty years struggling to obtain the pension he felt he deserved. He fought for his adopted country, even though he was not a citizen. He knew hard times in Ireland, in Milford, in the service and after the war. An ungrateful nation reluctantly acceded to his needs. Governments never change!

Bibliography

Family records
New York State Archives
National Archives

SIGNALS

Dear Editors:

I ran across a cartoon today that reinforced my belief that America is losing its history. The cartoon shows a young schoolboy talking to a teacher. He says, "The Civil War...was that a mini-series?"

I feel the cartoon is representative of the ignorance pervading this country of our own past. We must become more aware of the forces that have shaped our United States. Secession, civil rights, and states rights are still as current issues today as they were 120 years ago.

Your magazine I feel is an excellent tool showing both the military and civilian life of that time. I hope your magazine is a success, even in this television age. It shows that period as more than an ugly war, but of a survival of a young country, that, to me at least, a mini-series couldn't capture.

Best luck with your magazine. Keep up the good work.

Betsy J. Bennett
West Jordan, Utah

Dear Editors:

Your magazine is truly a joy to read. I have received the last two issues but not the first three. I would like to order these back issues: April, June/July and August/September. Enclosed is a check for $10.50 and I will be looking forward to receiving these. Keep up the good work.

Paul Kearley
Ceres, CA.

Dear Editors:

I have found your articles on medicine very enlightening. They have not only amazed and amused me, they have helped me appreciate the whys and wherefores of the 19th Century citizen.

Phillip York
NYC, New York

Dear Editors:

I have just finished reading the Feb/March issue and felt I must comment on it. I found "Timeline" to be an excellent supplement to Mr. Crow's article on the events leading to Chickamauga. His presentation of Longstreet's empty victory was authoritative and well-written, while "Repurcussions" was thought-provoking and, I must admit, a bit disturbing. It presented those critical days in a new perspective, and proves to me the absolute need for a magazine such as THE KEPI. I subscribe to other CW publications, but never in my long association with them, can I remember being as excited with a single issue as I am with this.

John Toliver
Malibu, CA.

(Continued from page 10)

an issue with the Northern press because of the repulsion widely felt in North Carolina. As for Keith's defense that he was acting under orders, Augustus Merrimom, the future North Carolina senator, spoke for most loyal Confederates when he aptly rejected it in his February 16 letter to Governor Vance: "I do not suppose [Keith and Allen] had any order to do so barbarous a deed, but if they had the order [it] was void absolutely, no matter by whom issued."[32]

FOOTNOTES

[1] Phillip S. Paludan, *Victims: A True Story of the Civil War* (Knoxville: University of Tennessee, 1981), p. 85.

[2] *War of the Rebellion, Official Records of the Union and Confederate Armies* (Washington: Government Printing, 1881) in 70 volumes in 128 parts, Series I, Vol. XVIII, p. 853, hereinafter cited as *O.R.* All references, unless stated, are to Series I; Paludan, *Victims*, op. cit.

[3] *O.R.*, op. cit.; William H. Thomas to Sarah J. Thomas, January 28, 1863, William H. Thomas Papers, Western Carolina University, Cullowhee, North Carolina.

[4] *O.R.*, Vol. XVIII, pp. 810-811; Vol. XXIII, Part 2, p. 644.

[5] Abstracts of Lawrence M. Allen's service file, Record Group 109, Confederate War Records, National Archives, Washington D.C.

[6] Walter Clark, (ed.) *Histories of the Several Regiments and Battalions from North Carolina in the Great War 1861-65* (Raleigh: E.M. Uzzell and Co., 1901) in 5 volumes, III, p. 661, hereinafter cited as *NC Regts.*

[7] *O.R.*, op. cit., 854; John G. Barrett, *The Civil War in North Carolina* (Chapel Hill: University of North Carolina, 1963), pp. 185-192, 197-199; Georgia L. Tatum, *Disloyalty in the Confederacy* (Chapel Hill: University of North Carolina, 1934, pp. 107-135.

[8] Paludan, *Victims*, p. 85; Allen's service file, op. cit.

[9] *NC Regts.*, op. cit.

[10] Paludan, *Victims*, p. 86.

[11] *Ibid.*

[12] *O.R.*, Ser. II, Vol. V, p. 956.

[13] Paludan, *Victims*, p. 93.

[14] Daniel Ellis, *Thrilling Adventures of Daniel Ellis* (New York: Harper & Brothers, 1867), pp. 113-115.

[15] *O.R.*, Vol. XVIII, p. 893.

[16] Ellis, *Thrilling Adventures*, p. 410.

[17] *O.R.*, Vol. XVIII, p. 881; Paludan, *Victims*, p. 97; Ellis, *Thrilling Adventures*, p. 441.

[18] *NC Regts.*, III, pp. 660-61; *O.R.*, Vol. XVIII, pp. 867, 881.

[19] *O.R.*, Vol. XVIII, p. 881.

[20] *Ibid.*, p. 893.

[21] *Ibid.*

[22] *Ibid.*, p. 867.

[23] William W. Holden, *Memoirs of W.W. Holden* (Durham: Seeman Printing, 1911), pp. 27-28.

[24] *O.R.*, Ser. II, Vol. V, p. 841.

[25] *NC Regts.*, III, p. 661.

[26] Raleigh. *North Carolina Standard*, April 23, 1863.

[27] Ellis, *Thrilling Adventures*, pp. 187-188.

[28] *O.R.*, Ser. II, Vol. V, pp. 843, 858, 952.

[29] *Ibid.*

[30] *Ibid.*, p. 956.

[31] Paludan, *Victims*, pp. 111-115.

[32] *O.R.*, op. cit., p. 839.

[redacted]

Tracts

The Sabbath-School Teacher

by
Rev. Allen Kelley

Religion in the 19th Century dominated every aspect of life. It was a strict moral code for behavior, with punishments and rewards clearly laid out. Affiliation with a religious denomination and a specific church was necessary for economic and social reasons, as well as spiritual ones. A man was judged by the company he kept, and his advancement often depended upon his beliefs. A stranger coming to work in a new town would literally be forced to join a parish or remain an outcast. His intentions were suspect until he worked his way into the mainstream of religious society. Frequently entire towns were inclined toward one particular faith, and the stranger who joined an "opposing" religion usually had to keep within that circle for his social contacts.

Attending church on Sunday very often meant committing the whole day to worship. Services in the morning and evening, prayer groups and the religious instruction of the young occupied the hours of the first (or seventh) day. The benevolent employment as a Sabbath School teacher was often layman's work, done by volunteers. This "twice blessed" occupation (blessing the teacher as well as the taught),[1] was considered as the one next to the ministry in importance, and one which demanded the warmest affections of the heart and the most vigorous exercise of the mind.[2] The Sabbath School was where children were taught divine truth, it being felt their youthful minds were more susceptible to these teachings than at other times.[3]

Scholars were encouraged to commit to memory passages from the Scriptures, and answer a set of questions afterward. A typical study for a class at Eastertime in 1860 might have been Lesson 44, "The Resurrection of Christ," Matthew 28:1-10.

1. In the end of the sabbath as it began to dawn toward the first *day* of the week, came Mary Magdalene and the other Mary to see the sepulchre.

2. And, behold, there was a great earthquake: for the angel of the Lord descended from heaven, and came and rolled back the stone from the door, and sat upon it.

3. His countenance was like lightning, and his raiment white as snow:

4. And for fear of him the keepers did shake, and became as dead *men*.

5. And the angel answered and said unto the women, Fear not ye: for I know that ye seek Jesus, which was crucified.

6. He is not here: for he is risen, as he said. Come, see the place where the Lord lay.

7. And go quickly, and tell his disciples that he is risen from the dead; and, behold, he goeth before you into Galilee; there shall ye see him: lo, I have told you.

8. And they departed quickly from the sepulchre with fear and great joy; and he did run to bring his disciples word.

9. And as they went to tell his disciples, behold, Jesus met them, saying, All hail. And they came and held him by the feet, and worshipped him.

10. Then said Jesus unto them, Be not afraid: go tell my brethren that they go into Galilee, and there shall they see me.

1. After the crucifixion, what was done with the body of Jesus?

Ans. It was taken down and buried in a tomb.

2. Who performed this service?

Ans. Joseph and Nicodemus.

3. Where did they lay the body?

Ans. In Joseph's own new tomb which was hewn out in the rock.

4. How was the sepulchre secured?

Ans. A great stone was rolled against the door.

5. How was the stone secured?

Ans. it was sealed.

6. How was the tomb guarded?

Ans. By a watch, probably, of sixty Roman soldiers.

7. When did Mary Magdalene and the other Mary go to the tomb?

Ans. In the end of the Sabbath, as it began to dawn toward the first day of the week. 1.

8. Is not the sabbath on the first day of the week?

Ans. Yes, since the resurrection of Christ, but before that, it was on the seventh day.

9. What had taken place at the sepulchre?

Ans. There had been an earthquake, and the angel of the Lord had descended from heaven, rolled back the stone from the door and sat upon it. 2.

10. What was his appearance?

Ans. His countenance was like lightning, and his raiment white as snow. 3.

11. How were the keepers, or soldiers, affected on seeing the angel?

Ans. For fear of him they did shake and became as dead men. 4.

12. What did the angel say to the women?

Ans. Fear not ye, for I know that ye seek Jesus which was crucified. 5.

13. What else did he say?

Ans. He is not here; for he is risen from the dead as he said. Come, see the place where the Lord lay. 6.

14. What did he tell them to do?

Ans. To go quickly and tell his disciples that he is risen from the dead. 7.

15. Where did he tell them to go to meet Jesus?

Ans. Into Galilee; there shall ye see him; lo, I have told you.

16. What did the Marys do?

Ans. They departed quickly, with fear and great joy, and ran to tell his disciples. 8.

17. As they went, who met them?

Ans. Jesus met them, saying, All hail. 9.

18. What did the women do?

Ans. They held him by the feet, and worshipped him.

19. What did Jesus say to them?

Ans. Be not afraid; go tell my brethern that they go into Galilee, there shall they see me. 10.

20. Did they meet him there?
Ans. Yes; he was seen by the eleven disciples, and above five hundred brethren at once. Matt. 28:16; 1 Cor. 15:6.

Lo! Jesus dies, and God in wrath
Enwraps the earth in gloom!
He lives! Victorious over death,
And rises from the tomb.

As Jesus rose, so will he raise
My scattered, sleeping dust
Clothed with eternal beauty, for
The dwellings of the just.

His children, all who love the Lord
And in his word delight,
Though they were dead, shall live again
Immortal in his sight.

This active frame of mine snall die
And mingle with the earth;
But Christ shall keep and form it, for
The resurrection birth.[4]

Then (as now), ministers of the gospel had to encourage Christians to serve the Lord by teaching Sabbath school. Numerous excuses were given as to why a person could not teach. Then (as now), answers were always available. If, for example, a parishioner turned down the offer to teach because "I am not a professor of religion," the 1860's reverend might answer, "You ought to be," then add if you profess religion, it is your duty to teach, even if it be to instruct in how to read the Word of God. "Not good enough?" Teaching will make you better. "Too old?" You have more experience to assist you. "Married?" Do you cease to be the Lord's by becoming a husband or wife? "You wish the Sabbath for yourself, to rest and meditate?" The soul finds some of it's sweetest rest in works of mercy.[5]

THE RESURRECTION OF CHRIST.

Sabbath School was not only for the well-to-do, or the white children. Although in many states it was unlawful to teach Negros to read, that was one of the things black students were instructed in at Sabbath School. Enlightened teachers wished to help their students learn so that they might read the Bible. Classes for slaves and freemen were always well-attended, but suffered from a lack of instructors. In some instances, students had to be turned away, or gave up attending for lack of teachers. Those who defied the law or their "peers" to spread the Word of God benefited not only their pupils, but mankind, as well.

The Sabbath School teacher taught the Scriptures, sang hymns, taught reading, answered questions, and "baptized with the holy influences of the word and spirit of God...lead many souls to him who is waiting to receive them."[6]

FOOTNOTES

[1] Rev. William Nevins, *"Practical Thoughts,"* American Tract Society (New York: 1836), p. 55.
[2] Henry Holt, *"The Child's Illustrated Scripture Question Book,"* Henry Holt (Boston: 1860), p. 3.
[3] *Ibid*, p. 4.
[4] *Ibid*, pp. 158-61.
[5] Nevins, pp. 54-57.
[6] Holt, p. 3.

(Continued from page 20)

May 31, 1810—Horatio Seymour, governor of New York (1862-1864), Pompey Hill, Onondaga County, New York. (N)
May 31, 1813—Sen. Albert Gallatin Brown, Chester District, South Carolina. (S)
May 31, 1818—John A. Andrew, governor of Massachusetts (1860-1866), Windham, Maine. (N)
May 31, 1819—Walt Whitman, West Hills, Long Island, New York. (N)
May 31, 1834—Anthony Burns, fugitive slave, Stafford County, Virginia. (N)
May 31, 1837—Gen. William H. Fitzhugh "Rooney" Lee, Arlington, Virginia. (S)
May 31, 1837—Gen. Stephen D. Ramseur, Lincolnton, North Carolina. (S)

BIBLIOGRAPHY

Biographical Directory of the United States Executive Branch, 1774-1971. Ed. Robert Sobel. Westport, Connecticut: Greenwood Publishing Company, 1971.
Dictionary of American Biography. 20 vols. New York: Charles Scribner's Sons, 1928-1936.
The National Cyclopaedia of American Biography. 57 vols. New York and Clifton, New Jersey: James T. White & Company, 1892-1977.
Wakelyn, John L. *Biographical Dictionary of the Confederacy.* Westport, Connecticut: Greenwood Press, 1977.
Webster's American Military Biographies. Ed. Robert McHenry. Springfield, Massachusetts: G. & C. Merriam Company, 1978.

ONE MORE BIRTHDAY

[In adding Margaret Junkin Preston to "Family Bible," I noticed that we had forgotten to add her sister, Elinor Junkin Jackson, to our last installment. Ellie was born March 6, 1825, in Milton, Pa., and was the fourth child (second daughter) of George Junkin and Julia Rush Miller. In the fall of 1848, the Junkin family moved to Lexington, Virginia, where Rev. Junkin became president of Washington College. In 1851, a young army major became professor of Natural and Experimental Philosophy at the Virginia Military Institute, and on August 4, 1853, Thomas Jonathan Jackson married Elinor Junkin.— The Editors.

NOSTRUMS AND NOTIONS

Hitherto Taboo

by
Robert Kamer, M.D., and
Samuel Jackson, M.D.

Humans have always been fascinated by sexual prowess, whether it be their own, or that of others. While it was often laughed about or whispered over in men's clubs or women's get-togethers, very little was known about sexual *dysfunction* in the mid-19th Century. While the use of microscopes and a more thorough knowledge of anatomy and physiology were available to the Medical man of the time, little emphasis was put on educating or helping the public at large with their "unspeakable problems." Cures for impotence, sterility, barrenness, loss of sexual power, drive or enjoyment were sought in tonics, powders and "regimens."

Few physicians felt comfortable explaining or treating diseases of the genitals. Parents often had difficulty discussing sex with their children and often the subject was postponed until a marriage was arranged. Then it fell into the realm of ministers, who gave counciling to the young couple. This moralistic, usually stern lecture did little to educate either the man or the woman in what was going to happen after the wedding festivities were over and they were left on their own. Often, the only contact these newly marrieds would have had with procreation or the sexual function would be from their experiences with playmates.

In the 1850's, Dr. Frederick Hollick began giving a series of Popular Lectures on Sexual physiology (to Ladies and Gentleman separately) that he hoped would familiarize the public mind with such matters, and "point our the real source of numerous physical and moral evils under which society labors."[1] He used Anatomical Models, Diagrams and the new creation, a "Manikin," to demonstrate to his audiences how the human body worked (and for those who didn't know, what it looked like!). His discussions on the *female system*, the *male system* (with all it's derangements), midwifery, venereal diseases and generative organs soon brought the masses to his door. The "Medical *Old Fogies*,"[2] however, were quite shocked at what he had to say, and tried to stop him. They only served to bring Dr. Hollick more publicity, and "the would-be monopolists of knowledge became truly pitiable objects of public scorn and contempt."[3]

One of the topics young ladies were introduced to in books and lectures appearing in the 1850's was the fascinating discovery that the outline of the future being always existed in the egg, even before it is impregnated. The male principle was necessary, however, to give it life.

Both sexes were warned against *Continence*, or the voluntary abandonment of sexual indulgences by those who were capable of and who had a desire for them. Continence, it was taught, drove persons to the most singular Mental eccentricities and sometimes even to moral perversity. Physical derangements of the Genital and Urinary organs, inflammation of the brain or spinal marrow, a wasting of the flesh and fever occured, especially when the person was abstaining because of "mistaken religious views."[4] Sex, young audiences were reminded, was the foundation of marriage, and statistics proved that married persons lived longer than single ones: of single men between 25-45, twenty-eight out of one hundred would die before age 60, while only eighteen out of one hundred married ones would. At 80 years of age, only three single men survive, compared to nine married men.

A typical "case from the notebook" used to illustrate the evils of continence concerned a Lawyer, naturally very amative, who had made up his mind not to marry until he had amassed a fortune. His troubles began with nightly emissions, which sometimes affected his mind. His memory eventually began to fail, and also his powers of application. He suffered fits of languor and depression, and when advised he should marry before his sexual powers became completely impared, he determined instead to try a *Nervous stimulant*. This didn't help, and he was on the verge of suicide, when he came under the care of a physician who understood about *male generative organs.* He immediately diagnosed the source of the entire problem as *seminal losses* (passing of semen in the urine), and *caurterized* the Ducts. This was said to have removed the irritability, and within six months the man was able to marry with physiological propriety. His virility was restored, although inferior to what it would have been had he married earlier. His mind recovered, also, but never so that he could concentrate as before.

19th Century lecture-goers were taught that the connection between the Sexual Organs and the *Brain* was of overwhelming importance, not only for their physical well-being, but as influencing their moral tendencies as well. Citing the "obvious" principle of Animal Physiology that no action could occur without *Nervous power*, and that every idea was stimulated by *Nervous fluid*, and organ deficient in this mysterious power would decay. If one organ be overworked, it exhausts the nervous system, and the entire body becomes imperfect. If, for example, a man indulged in *sexual excesses*, he would soon find all his organic functions failing, because he was drawing off too much energy from one area to another. *Sexual abuses* in young persons also resulted in stunted growth, weakly developed organs, liability to Disease and premature old age.

Masturbation was another topic for the genital lecturer. Stating that a very large portion of the human race was guilty of this excess, it was predicted "the great majority and great portion of the evils and suffering of society are produced by it."[5] The object of this vice was to "excite by artificial means, the sexual feeling. This is accomplished by friction, of the most sensitive parts of the genitals, usually with the hand, but often by other means, assisted by a highly excited state of the imagination."[6] The immediate effects of this "self abuse" were lassitude and depression of spirits. This was followed a short time later by other symptoms, many of which were never suspected to have arisen from this cause. As with Continence, the memory would begin to fail and the mind could not be directed to anything for any length of time, leading to insanity and idiocy. This drain on the nervous energy (caused by an intensely excited imagination without any natural or appropriate associations) lead to constitutional degeneracy and affected both the individual *and his offspring*, which was formally looked upon "as a mysterious dispensation of Providence."[7]

From The Casebook, Dr. Hollick told of a patient who suffered from *Retarded development* due to Masturbation. This young man, full of animal spirits, was introduced to the habits of self abuse by a female servant and her playmates. Owing to his temperament, he soon gave himself up to the habit, but kept it secret from his parents. His nurse did not inform on him, although knowing of his occupation, because it often relieved her of the trouble of watching him. At six years of age, his habit became a perfect furor, and he found himself growing weak and irritable. At fourteen, when he first developed semen, he indulged more than ever until he became so childish of mind he was thought to be Consumptive. He also discovered, to his horror, that at twenty-one, his Sexual Organs had not grown since age fourteen, nor had his voice deepened. He pleaded *to be made a man.* A regular course of diet, daily exercise, bathing and friction of the skin were prescribed, while his Genital Organs were treated with the *Cougester* (used daily with shampooing) and the *Aphrodisiac Remedy.* After two years of treatment, the individual was fully capable of the duties of Married Life, but not so frequently as persons of his age. Unfortunately, he was killed in a steamboat accident out West, or he would have been a remarkable instance of what proper Medical and Hygienic means could accomplish.[8]

Delving into the matter of masturbation further, parents were warned that no idea was more false than the doctrine of delicacy and reserve in the treatment of this habit. The 1848 report on Idiocy presented to the Massachusetts Senate stated: "There need be no fear of weakening virtue by letting it look upon such hideous deformity as this vice presents. Virtue is not salt or sugar to be softened by such exposure; but the crystal or diamond that repells all foulness from its surface."[9] It was suggested that children should be raised with habits of modesty and reserve; after the infant leaves its mother's arms, it should sleep in a bed by itself; after adolescence, children should never be permitted to sleep together. Boys should be taught to shrink sensibly from any unnecessary exposure of person before each other, and be trained in the habits of delicacy and self-respect. The philosophy rampant amongst boys away at school, that masturbation was a meritorious thing to which they should all aim at was to be rooted out, and, if necessary, all academies closed.

Second only to the fountain of youth is Man's quest for the perfect aphrodisiac. This Medical Remedy, which acts upon the sexual organs so as to preserve or restore their powers[10] has been sought all over the world. Some act as powerful *stimulants*, giving temporary power, while other do not work at all! Dr. Halleck, the genital lecturer, claimed to have researched "the whole of man's sexual nature"[11] and experimented with thousands of cases. The result, he claimed, was the formulation of a compound which, when employed, increased and maintained sexual power, or restored it when lost. This miraculous substance, contained *thirty-three* different ingredients (some of them not more than the hundreth part of a grain), including *musk* and *Haschisch.* The price for a package was only FIVE DOLLARS! which he said should be sufficient in all ordinary cases. When sent through the post, it was perfectly disguised *in flat gold boxes* so no one could ever guess the nature of its contents. A bit could even be kept in the vest pocket without detection, in case a sudden need for it occured. The difference between this substance and others, it was claimed, was that it acted as a *renovator*, producing *New Brain* and *new Seminal* fluid.

(The actual substance comprising the Nervous System was almost identical with that of the Seminal fluid in Man and the Ovae in Woman. The Brain and Sexual apparatus were placed at opposite extremities of the body, connected by the Spinal marrow. When one or the other was overcharged, the vital power of the other was undercharged. This explained why mental exertion was often followed by Sexual Impotence, and why sexual abuse destroyed intellect. Expelling too much semen thus caused a portion of the brain to die, and could lead to *Softening of the Brain*. This was the reason why a *renovator* must work at both poles of the body.)

"The only one of the kind which can possibly be obtained because Dr. Hollick alone possesses the *secret of its composition*, and has secured the ENTIRE MONOPOLY of the rarest and *most powerful* ingredients, "the Aphrodisiac was said to benefit the CHILDLESS, the IMPOTENT, the NERVOUS, the FEEBLE both in BODY and MIND, the HYPOCONDRIAC and those who find they are becoming OLD TOO SOON, of BOTH SEXES (from an advertisement of the times). For those who worried about confidentiality, all letters of inquiry were destroyed as soon as notes were taken for the casebook. Unfortunately, this "miracle drug" appears to have died with its creator.

Dr. Hollick claimed to have cured thousands of patients with his treatments directed at the derangement of the genitals. His observations of sperm under the microscope helped explain sterility (it wasn't, after all, always the woman's fault), and his series of Popular Lectures introduced many uninformed 19th Century citizens to the internal and external reproductive system. His pleas for sex education could as well be used today, and his suggestions from the Casebook were often useful, if not amusing. His limits of anatomy, physiology and psychology are apparent now, but one hundred and thirty years ago, he was paving a new road. He certainly did not starve for his efforts (his treatments were, by admission, for those blessed with the golden calf), but his was another step in the growth of knowledge.

FOOTNOTES

[1] Frederick Hollick, M.D., *The Diseases of Woman* (New York: T.W. Strong, 1855), p. 362.

[2] *Ibid*, p. 363.

[3] *Ibid*.

[4] *Ibid*, p. 367.

[5] *Ibid*, p. 290.

[6] *Ibid*, p. 292.

[7] *Ibid*, p. 305.

[8] *Ibid*, p. 397.

[9] S.G. Howe, M.D., "Report on the Subject of Idiocy," presented to the Massachusetts Senate, February, 1848.

[10] Hollick, *Diseases*. p. 433; see also Robert Hooper, M.D., *Lexicon Medicum* (New York: Harper & Brothers, 1842), p. 79.

[11] *Ibid*, p. 434.

WHIMSEY

A Bad Split.— It is said that Chang and Eng, the Siamese twins, differ in politics. Both are veteran democrats, but Chang is now for Breckinridge, and Eng for Douglas.
(**The New York Herald**, Monday, August 6, 1860.)

Breaking camp? Moving out? Troop transfer? Don't forget to notify Headquarters: communicate new location to THE KEPI.

NAME ______________________________

ADDRESS ______________________________

THE GENTLEMAN AGRONOMIST

by Karen Murphy

The "gentleman agronomist" in the mid-19th Century was much like the physician, the educator or the professional soldier: involved in all aspects of his business. Specialization was a practice that did not come into vogue until much later in the next century. The recurring theme in *any* occupation a hundred and twenty years ago was that a man had to be diverse to succeed. If he were a farmer, he had to be familiar with crops, seed, tilling of ground, management of a stable and a barnyard, veterinary medicine and animal husbandry.

In fact, it was alleged that a farmer was a *working chemist* and a *fighting soldier,* surrounded by enemies. One particular enemy of the "country gentleman" in the 1850's was fraudulant practices conducted by vendors and compounders of fertilizers. When the price of one well-known manure increased radically, agents immediately advertised and put on the market spurious "cheap," "economical" & c. fertilizers. Attached to them were certificates from farmers praising the extraordinary results obtained from these products. Research done by concerned cultivators revealed, however, that such statements were commonly expressed in very general terms and seldom reflected the true end product. Additionally, some unscrupulous manufacturers gave out samples of fertilizers for experimentation in exchange for a farmer's testimony of its worth. After good results were obtained, and the lauditory statements of the farmer put on record, the product was sold, using inferior blends.

There were no laws protecting the purchaser from such fraudulant schemes at the time of the Civil War. Just as there was no licensing of physicians, there were no fair practice standards for advertisers, and no punishments for those who made a living by cheating in this manner. The only defense a buyer had was to deal with people he knew and products he trusted. Legislation and laws would develop slowly from those who found out the hard way money often came before honor.

The care of livestock was critical to the economic well-being of the farm, and tips for the care of horses filled agronomy journals. It was suggested that it was cruel and injurious to go to supper and leave a team in the field without refreshment, and questioned whether colic took place from long fasting and the greedy stuffing which followed. Mules were highly recommended to the thrifty man as being superior to the horse as regarding farm work. The mule (a cross between an ass and a common horse) could do more work, stand the heat better and pull as much, if not more, than the horse. Being less liable to disease, and able to eat every kind of rough fodder, they cost less to keep and could be raised profitably for market, selling in 1855 for $70.00 a head.

Weeding a crop was encouraged, and even the rudiments of pest control were practiced. In order to eradicate the Curculio bug from fruit trees, some nurserymen recommended cotton be wrapped around the trunk of a tree to keep the insect from crawling up it. When it was discovered, the Curculio had wings, a process of *jarring* was suggested. This consisted of placing a sheet under the young tree and then shaking it violently. When the pests fell off onto the sheet, they were stomped on and killed. This proc-

(Continued on page 31)

MILITARY REVIEW

Fightin' Tom Rosser, C.S.A., by Millard K. Bushong and Dean M. Bushong, Beidel Printing House, Shippensburg, PA, 17257, 1983; available from the authors at: "Arcadia," Route 2, Berryville, VA, 22611; $19.00: 281 pps., maps and illustrations.

This biography of Confederate cavalry leader Thomas L. Rosser traces his life from boyhood through West Point, his Confederate service and post bellum years. Beginning his career as a lieutenant of artillery, Rosser fell under the eye of JEB Stuart, who had him appointed full colonel, 5th Va. Cavalry. He spent the rest of the war in this branch of the service, eventually rising to the rank of major-general. When asked what his greatest wartime accomplishment had been, he replied that, when Lee surrendered at Appomattox, he took his men, "rode over a brigade or more of the enemy, captured the Lynchburg road and left that fatal field *in triumph*, refusing to surrender either myself or my command to the enemy." He was finally captured outside his Virginia home.

After the war, Rosser took his family west, where he worked at various jobs, including engineering duty for the expanding railroads. In 1886, he returned to Virginia, where he dabbled in politics and quarreled with old soldiers about battlefield tactics. After training soldiers for the Spanish-American War as a brigadier-general, U.S.V., he preformed the duties of postmaster at Charlottesville until his death in 1910.

The husband and wife team of Millard and Dean Bushong wrote in their preface: "The historian rarely has access to so many unpublished facts concerning the subject he is studying." They interviewed members of the Rosser family (the general's granddaughter and great-grandson), drew material from an unpublished master's thesis, and referred frequently to the *Thomas L. Rosser Papers*, from the National Archives, and the University of Virginia. With the great volume of primary source material at hand, it is unfortunate that so little was actually presented to the reader in it's original form. Rosser was a man of strong emotions, and appears to have expressed himself freely in his correspondence. Unfortunately, the reader is all too frequently presented with only a summary of what is available. The student must choose either to accept the authors' brief synopses, or personally research the source material. As a case in point, when Rosser criticized Lee for his failures at Gettysburg, not one direct quotation is taken from the Rosser papers; rather, we are left with: "According to Rosser, Lee could have directed the details of the battle if he had so desired..." In the lull after Gettysburg, Rosser appeared to suffer from depression, and perhaps a touch of fear, when he wrote his wife that he rejoiced in the fact that he had missed a fight in which several of his fellow officers had been wounded and talks of resorting to drink, and resigning the army to go to South America. His dilemmas and his personality would be more easily understood if we could have divined Rosser's feelings from his own words, as was possible when full excerpts of letters to and from General Stuart were given.

The book contains an excellent appendix, containing the muster rolls for the 7th, 11th, 12th, 25th Virginia Cavalry, and Chew's Battery. Photographs of Rosser and period illustrations supplement the text well.

The dust jacket of "Fightin' Tom Rosser, C.S.A." states: "This book is the first attempt to acquaint the public with his remarkable career." As such, it succeeds, and should become the groundwork for scholars studying this Confederate cavalryman.

S.L.Kotar J.E. Gessler

Robert E. Lee by Jefferson Davis and Alexander Stephens, Edited and with an Introduction and Notes by Col. Harold B. Simpson, Hill Junior College Press, Hillsboro, TX, 76645, 1983; $10.00: 121 pps., Index, photographs.

Jefferson Davis and Alexander Stephens are the stated authors of this limited edition work (1000 copies), which is a bit misleading. The primary focus of the book is two short articles, one by the President, and one by the Vice-president of the Confederate States of America, but the introductions and notes by Col. Simpson comprise about two-thirds of the text.

Davis wrote a study of R.E. Lee for *The North American Review* in 1889, and it was later reprinted in the *Southern Historical Society Papers*. Stephen's article appeared in the *Southern Bivouac* after his death in 1883. Hill Junior College Press as combined both works, so they may be readily available for study. The thorough notes presented after each provide extensive background on the people, places and events in Lee's life.

Davis' essay traces Lee's career, touching briefly on his West Point career, the Mexican War and his civil engineering days, through the Civil War and Reconstruction. His comments are lauditory, a bit tenuous, and his memory more gracious than accurate. Written shortly before his death, the twelve-page text is as interesting a study of its author as it is of its subject.

The same comment is equally applicable to Stephens and his essay. Dictated to one of his biographers, then rejected by the author, the copy was revised by the editors of *Southern Bivouac* after Stephen's death. Stating that he had no knowledge of Lee until they were brought together by war, he describes his first encounters and how he came to admire him. More specific than Jefferson Davis' comments, he describes Lee's early place in the Virginian and Confederate armies, and includes two poems that he thought significant, which describe Gen. Lee.

Complementing the text are fifteen photographs of R.E. Lee, one of his death mask, and ones of Davis and Stephens. This is a very good addition to the work, and facilitates an understanding of the man whose face grows older and more careworn as the pages are turned.

Robert E. Lee is a very useful book which allows the reader to view its subject through the minds of two men he was closely associated with over four years of turmoil. The footnotes provide valuable information for the general reader, although they are so detailed, it might be better to read the two essays completely through first, then reread them with the footnotes.

S.L. Kotar

(Continued on page 31)

Articles of War

An Act For Establishing Rules and Articles For The Government Of The Armies Of The United States.

ART. 87.* No person shall be sentenced to suffer death but by the concurrence of two-thirds of the members of a general court-martial, nor except in the cases herein expressly mentioned; *nor shall more than fifty lashes be inflicted on any offender, at the discretion of a court-martial;* and no officer, non-commissioned officer, soldier, or follower of the army, shall be tried a second time for the same offense.

ART. 88. No person shall be liable to be tried and punished by a general court-martial for any offense which shall appear to have been committed more than two years before the issuing of the order for such trial, unless the person, by reason of having absented himself, or some other manifest impediment, shall not have been amenable to justice within that period.

ART. 89. Every officer authorized to order a general court-martial shall have power to pardon or mitigate any punishment ordered by such court, except the sentence of death, or of cashiering an officer; which, in the cases where he has authority (by Article 65) to carry them into execution, he may suspend, until the pleasure of the President of the United States can be known; which suspension, together with copies of the proceedings of the court-martial, the said officer shall immediately transmit to the President for his determination. And the colonel or commanding officer of the regiment or garrison where any regimental or garrison court-martial shall be held, may pardon or mitigate any punishment ordered by such court to be inflicted.

ART. 90. Every judge advocate, or person officiating as such, at any general court-martial, shall transmit, with as much expedition as the opportunity of time and distance of place can admit, the original proceedings and sentence of such court-martial to the Secretary of War; preserved in the office of said Secretary, to the end that the persons entitled thereto may be enabled, upon application to the said office, to obtain copies thereof.

The party tried by any general court-martial shall, upon demand thereof, made by himself, or by any person or persons in his behalf, be entitled to a copy of the sentence and proceedings of such court-martial.

ART. 91. In cases where the general, or commanding officer may order a court of inquiry to examine into the nature of any transaction, accusation, or imputation against any officer or soldier, the said court shall consist of one or more officers, not exceeding three, and a judge advocate, or other suitable person, as a recorder, to reduce the proceedings and evidence to writing; all of whom shall be sworn to the faithful performance of their duty. This court shall have the same power to summon witnesses as a court-martial, and to examine them on oath. But they shall not give their opinion on the merits of the case, excepting they shall be thereto specially required. The parties accused shall also be permitted to cross-examine and interrogate the witnesses, so as to investigate fully the circumstances in the question.

ART. 92. The proceedings of a court of inquiry must be authenticated by the signature of the recorder and the president, and delivered to the commanding officer, and the said proceedings may be admitted as evidence by a court-martial, in cases not capital, or extending to the dismission of an officer, provided that the circumstances are such that oral testimony cannot be obtained. But as court of inquiry may be perverted to dishonorable purposes, and may be considered as engines of destruction to military merit, in the hands of weak and envious commandants, they are hereby prohibited, unless directed by the President of the United States, or demanded by the accused.

From "Revised Regulations for the Army of the United States, 1861," Philadelphia, J.P. Lippincott & Co., pp. 512-514.

CONCURRENT EVENTS

WHAT IT COSTS — in 13 days the Allies rained on Sebastopol 780,000 balls, averaging 45 lbs. in weight, making a daily shower of 2,700,000 lbs of iron, or a total storm of 35,100,000 lbs., worth, at the rate English pig iron is selling, (let alone transportation and manufacture,) $313,380. If the cannon balls fired from the Allied lines, during the thirteen days, were rolled into rail bars weighing sixty pounds to the yard, the bars would extend three hundred and thirty-two miles; or if laid as a railroad, would suffice for a single track road from New York to Albany, with all the necessary turn-outs. It is estimated that 4,680,000 lbs. of powder was required to push the iron. At 15 cents per lb., this cost $702,000. Lives lost not counted. [From *The Country Gentleman*, Albany, New York, May 24, 1855.]

* So much of these rules and articles as authorizes the infliction of corporeal punishment by stripes or lashes, was specially repealed by Act of 16th May, 1812. By Act of 2d March, 1833, the repealing act was repealed, so far as it applied to the crime of desertion, which, of course, revived the punishment by lashes for that offense. Flogging was totally abolished by Sec. 3 of Chap. 49, 5 August, 1861.

UNCLASSIFIED ADS

THE STUART-MOSBY SOCIETY

The Stuart-Mosby Society celebrated the 150th birthday of Colonel John Singleton Mosby at Warrenton, Virginia, on December 4, 1983. Ceremonies included the placing of wreaths at Mosby's grave and at the Mosby monument located on the town square prior to a memorial service conducted in the auditorium of. Warrenton High School. Master of Ceremonies was U.S. Army Colonel J.E.B. Stuart IV, descendent of Confederate General J.E.B. Stuart. Admiral (ret.) Beverly Mosby Coleman, grandson of Colonel Mosby, represented the Mosby family and welcomed the descendents of the officers and men who served in Mosby's command during the Civil War. The memorial address was delivered by Dr. Daniel Jordan of Virginia Commonwealth University. Specifically recognized were author Virgil Carrington Jones for his many Mosby books, and Adele Mitchell, President of the Stuart-Mosby Society, for arranging the memorial service. Over 300 members and guests attended the activities.

Membership in the society is open to all interested in the Civil War. A newsletter, *The Southern Cavalry Journal,* is published for members. Dues are $10.00 annually payable to:

Adele Mitchell
216 Garland Drive
Carlisle, Pennsylvania, 17013

(Continued from page 28)

ess, repeated daily at noontime for two weeks, was promised to rid the tree of bugs and produce a good crop. If the striped bug appeared on tomatoes, egg plants or peppers, a timely application of tobacco dust after a shower or watering would effect a remedy. It had to be repeated, however, after every rain.

The mid-19th Century farmer witnessed the beginning of scientific experimentation and control. He benefited from the new techniques, and shared his practical wisdom and experience with the scientists of his day to make even greater advances. He was a versatile provider of foodstuffs for the growing American population, and his contributions enabled a young country to expand and survive, even amid the devastations of civil war.

(Continued from page 29)

Civil War Guitar, Campfire Memories, produced by Ray Herbeck, Jr., performed by Kirk Browne, Legend Records, P.O. Box 1941, Glendale, CA, 91209; $8.98. (Also available from Ray Herbeck, Jr., 15541 Buttram Street, Hacienda Heights, CA, 91745 at $6.98 plus $2.50 postage & handling.) 1983.

Civil War Guitar, produced by Ray Herbeck, Jr., features the artistry and interpretation of Kirk Browne, who plays all the instruments used in the album: guitar, banjo, fiddle, harmonica and mandolin. An insert entitled "The Songs & Their Origins" capsulizes the history and lyrics to each song included. The Model 1850 Martin 2 guitar (shown on the album cover) is featured in each of the 16 pieces chosen. This guitar was found in a Confederate camp by a Union officer, and was preserved and played by his family, until Kirk Browne acquired it in 1978. Not only the guitar but the period style of playing was imparted to Browne by the granddaughter of the officer. Especially touching are "When Johnny Comes Marching Home," "Dixie," "Somebody's Darling" and "The Star Spangled Banner." But each piece is searchingly reminiscent, stirring in the listener what seems to be the shadow of a memory.

These tunes represent the music of the soldier as heard beside the campfire, as opposed to the commanding, emphatic band music used for a march or parade. The *Producer's Note* states: "...a 'rustic flavor' has purposely been imparted to this recording. It is intended for listeners who love the sound of a guitar and the realities of the instrument...Additionally...The best complete performance for each song was used."

Side A is The Union, while Side B is reserved for The Confederacy, yet both sides shared many of the same tunes, with the same sorrows, vitality and tears of the heart. For appreciators of music, this album should not be overlooked; and for sheer listening pleasure, *Civil War Guitar* may be recommended to all.

J.E. Gessler

CERTIFICATE OF SERVICE.

To all Whom it May Concern:

This is to Certify, That John Jeffers, who was enrolled on the 21st day of June, one thousand eight hundred and Sixty one to serve three years, was discharged on the 23rd day of May, 1864 by reason of disability at Davids Island N.Y.H., while holding the grade of Private in Company G, 40th Regiment of New York Vols and serving honorably in the military service of the United States.

This certificate is given upon evidence that the original discharge has been lost or destroyed, and in all cases upon the condition imposed by the Act of Congress approved March 3, 1873, that it "shall not be accepted as a voucher for the payment of any claim against the United States for pay, bounty, or other allowance, or as evidence in any other case."

Given at the Adjutant General's Office, War Department, Washington, D. C., this 8th day of September, A. D. 1888

Assistant Adjutant General.

THE KEPI
P. O. Box 4542
North Hollywood, California 91607

BULK MAIL

THE KEPI

JUNE-JULY 1984 $3.50

"A New Concept in Civil War Reporting"

To Raise A Flag
by Samuel J. Martin

A Smuggler and a Guerilla
by Thomas J. Evans

Breaking the Code at Vicksburg
by Hector De Marco

THE KEPI

"A New Concept in Civil War Reporting"

JUNE-JULY 1984

VOLUME II, NUMBER 3

THE KEPI is published bi-monthly for $3.50 per issue, or $18.00 per year. Canadian single copy, $4.50; subscription $23.00 per year. United Kingdom single copy, $4.50; subscription $28.00 (Air Mail) per year. Correspondence to the editors may be directed to: **THE KEPI,** P.O. Box 4542, North Hollywood, CA, 91607. Unless expressly stated, **THE KEPI** neither endorses nor is responsible for the content of advertisements appearing in its pages. **THE KEPI** requests authors to query with specific topics before submitting articles, and assumes no responsibility for unsolicited manuscripts not accompanied by return postage. Work not fully documented and footnoted where applicable will not be considered. All inquiries should be directed to the above address. **THE KEPI** reserves the right to edit and copyright all letters. Please allow 6 weeks for delivery of first issue.

ILLUSTRATIONS

Cover: Gen. Richard Ewell from *Moore's Rebellion Record;* Gen. P.G.T. Beauregard from *Embattled Confederates;* Gen. Rodes from *Lee's Lieutenants;* Gen. Trimble from *Stonewall in the Valley.*

Illustrations from "To Raise A Flag" from *The Official Military Atlas of the Civil War.*

Illustrations and photographs from "A Smuggler and a Guerilla" courtesy of the author.

Photographs from "Shoes" courtesy of the author.

Illustrations from "Printing: The State-of-the Art" from *Harper's New Monthly Magazine* September, 1855.

Back cover: courtesy of Charles Lowe.

THE REGULARS

THE IRREGULARS

GENERALLY SPEAKING

To all Americans today, the flag has always meant one thing — the stars and stripes. The number of stars may increase, but the basic configuration remains the same. For the citizens of the newly formed Confederacy, change was desirable only so far as the flag's representation of the Union. Their ancestors had fought England in the 1770's to form a new nation, and these latter-day rebels felt as close to the red, white and blue as their estranged Yankee brothers. Creating a new national symbol to represent the CSA proved a difficult and often unrewarding task. The new standard had to retain as much of the old as possible, as well as be distinguished readily in smoke-filled battlefields. Dedicating the flag would prove equally difficult, as the officers of General Richard Ewell's command discovered. Author Samuel J. Martin, while researching the Southern leader, discovered new information concerning the raising of one such flag in 1863. This humorous incident underscored the tragedy that awaited these same men a short way down the road at Gettysburg, Pennsylvania.

Thomas J. Evans' "Echoes of War" tells the tale of a man, too young to enlist in the Confederate army, who spent his war years along the Virginia/Maryland border as an occasional smuggler and guerilla. His adventures with Mosby, John Mobberly and "Sink" Sinclair reveal a side of war the military histories often overlook. The occult dangers to soldiers and civilians alike was very real, and the bitterness that sprang up between Northern and Southern sympathizers smolders to this day. As Mr. Evans' field investigations prove, "hatred dies hard."

From 1861 to 1865, the constant need for shoes obsessed Confederates, from the private with bleeding feet, to the general, who could not afford to see his men an army of sorefooted stragglers. Charles A. Lowe writes of the Yankee invention that increased the efficiency of manufacturing shoes, thus significantly aiding the Union war effort by providing shoes for their soldiers.

"Breaking the Code at Vicksburg," by Professor Hector DeMarco, skillfully ties in the 19th Century with the Civil War, by demonstrating that a short story by Edgar Allan Poe helped Confederates learn the plans of their enemies before the surrender of that city. Mr. DeMarco has also prepared a cryptogram for KEPI readers, similar to the one Poe used in "The Gold-Bug."

Beginning with this issue, a new feature appears, entitled GENERAL THOUGHTS. This will serve as the editorial column for the Editors-in-General. This month's topic is the new book printed by the University of Illinois, "How The North Won," and some thoughts on book-reviewing and book-writing in general.

We would like to take this opportunity to wish everyone a happy 4th of July. On Independence Day, let us remember the past, and work toward the understanding that all people are created equal.

S.L. Kotar *J.E. Gessler*

Editors-in-General

Head-Quarters of THE KEPI
1st June 1984

GENERAL ORDERS
No. 7
by Command of the Editors-in-General

After the disaster at Bull Run in 1861, the Union army reorganized, and Gen. George B. McClellan placed at its head. The "Young Napoleon" proved an overly cautious leader, and before he had ever fought a major battle, army officers and civilian representatives called for his removal. The August/September issue of THE KEPI features the complete manuscript written in 1893 by Rush C. Hawkins, the commander of the 9th New York Volunteers, of his involvement in the plot to have McClellan replaced. TIMELINE chronicles the overall view of events as they transpired in this unsettled period of Union indecision.

Complementing "A Mission to President Lincoln" are three articles that deal with the nation's 16th president: a narrative account by Lincoln's boyhood friend, a visit to the White House in 1861 and the diary of a young Northern woman who witnessed the last stages of war and the great leader's funeral procession.

Rounding out the issue are the regular/irregulars, and Family Bible continues.

"To Raise A Flag"

By
Samuel J. Martin

First, there was a cloud of dust. As the marching men drew nearer, the color of their uniforms provided a clue. Gray was not Union, but dark blue could be either North or South. Even when the soldiers were close, Brigadier-General Pierre Gustave Toutant-Beauregard, commanding the Confederate forces at First Manassas, could not discern the flag — the symbol of friend or foe — that waved in front of the approaching columns.[1]

Beauregard could see the banner. The problem was the Stars and Bars looked too much like Old Glory.[2] The Rebel ensign had a field of blue, containing a circle of seven stars, tucked in the upper left corner of its rectangle. Three broad horizontal stripes — red at the top and bottom, and white in the middle — completed the colors.[3]

Confederate Stars and Bars.

The soldiers so difficult for Beauregard to identify that Sunday, July 21, 1861, proved to be Southern (Early's brigade).[4] Their timely arrival on the Rebels' left flank tipped the tide of battle in favor of the Confederacy.[5]

After the victory, Beauregard took steps to prevent future problems in recognition. He appealed to friends in Richmond to have the official flag, (which had been adopted in March, 1861)[6] changed.[7] A committee of five was selected by Congress to review his proposal, and they voted four to one against the Creole's request.[8] William Miles, the only member to support Beauregard's petition, wrote: "It's impossible to tear the people away from a desire to retain some reminiscence of the United States flag."[9]

Frustrated by the Legislature, Beauregard turned to the War Department. He asked them to approve a second ensign "made of red with two blue bars crossing each other diagonally"[10] for use in the field. The battleflag, in vogue today as the most recognizable emblem of the Confederacy, was accepted as a substitute for the still official Stars and Bars.

Confederate battle flag.

The first battleflag was created out of skirts, provided by the Cary's.[11] Connie and her cousins, Hettie and Jennie,[12] were visiting the front, and Beauregard convinced them and other ladies to donate their silks. When the new banner was presented to the troops on November 28, 1861,[13] it was described as sacred because mothers, wives, and sweethearts had sacrificed their garments for the making.[14] "Under its untarnished folds," Beauregard wrote in his orders, "beat back the invador, and find nationality, everlasting immunity from an atrocious despotism, and honor and renown for yourselves — or death."[15]

To celebrate the unveiling, a series of boisterous parties were held that night.[16] The officers behaved rather badly, perhaps a harbinger of things to come.

From that day on, either Beauregard's battleflag or a state's colors were carried by the columns in the field. It must have bothered the officials in Richmond that their army refused to wield the national ensign. But in typical governmental fashion, two years passed before a better banner replaced the Stars and Bars. The bill approved by the Congress on May 1, 1863,[17] provided that the third flag to represent the Southern States should be "white, the length double the width...with the union (now used as the battleflag) to be a square of two-thirds the width... having the ground red; thereon a broad saltier of blue...emblazoned with white...five-pointed stars, corresponding in numbers to...the Confederate States."[18] In plain words, the battleflag designed by Beauregard was installed in the upper left corner of an all-white field.

The first use of the new standard was to drape the coffin of General T.J. (Stonewall) Jackson. The great leader had died May 10, 1863, and his body was brought to Richmond the following day. It was laid in the center of the reception room of the Governor's Mansion, and covered with the new colors, which henceforth was known as the Jackson flag.

Although the people of Richmond were now aware of the new flag, the troops in the field remained ignorant of the change. When the Army of Northern Virginia crossed into enemy territory in June, 1863, Confederate officials wanted the correct colors flying at the head of their advancing columns. A plan was developed: there would be a flag-raising.

Lee chose Jackson's old corps, now led by Lieutenant-General Richard S. Ewell, to be the first to hoist the new flag. Ewell passed the honor on to Major-General Robert Rodes' Division; Rodes gave it to Brigadier-General Junius Daniel's Brigade; and Daniel designated the 32nd North Carolina Regiment to unfurl the new ensign.[19] The ceremony was scheduled for Sunday afternoon, June 28th, outside the U.S. Army Barracks (occupied by Ewell's men) at Carlisle, Pennsylvania.[20]

"It was a grand occasion," remembered Private Henry London of the 32nd North Carolina.[21] Jed Hotchkiss, one of Ewell's staff, described it as "an animating scene."[22] However it was recalled, the rites were *not* what the officials in the capital had envisioned.

The ladies of Richmond had sewn the new flag and it was sent north to Ewell.[23] When he saw the banner, he decided it was too small to impress the "ignorant citizens" of Carlisle.[24] Tailors were summoned from the ranks and ordered to make a bigger ensign. The full-sized battleflag of the 32nd North Carolina was stiched to a larger, all-white field.[25]

It was a hot, sultry summer's day. Given a respite while the tailors improvised the new banner, the officers sought shade and something cool to drink. A keg of lager beer was tapped.[26]

"It was the strongest I ever saw — probably mixed with whiskey," Campbell Brown, Ewell's A.A.G. offered as an excuse. "I never saw Rodes intoxicated before..."[27]

About 4:00 P.M. that Sunday, General Ewell rose on the balcony of his headquarters building before the whole of Rodes' Division — 8,000 men — gathered in the yard.[28] He was probably sober, as his dyspepsia would have kept him away from the contents of the dangerous keg.[29] The General said a few words as the flag was raised, then retired from the scene.[30] He claimed a headache, but more likely wanted no blame for what was apt to come.[31]

Rodes spoke next. Red-faced and weak-kneed, he managed to hold himself together. His remarks were appropriate, intelligible, and delivered with aplomb.[32] When Junius Daniel followed with a similar dry speech, the men probably groaned inwardly.[33] They expected a series of repetitive, dull appeals to their patriotism.

The "Jackson" flag.

Major-General Isaac Trimble's appearance shattered that illusion. Assigned command in the Shenandoah Valley,[34] he had journeyed north to serve as a volunteer member of Ewell's staff.[35] He had drunk his share (or more) of the beer. Campbell Brown described him as "quite jolly," and said his comments were "not too neat."[36] Responding with joyous shouts, the troops begged to hear more.[37]

Major Ben Greene, Ewell's A.C.S., seized the podium. He smiled, waved his hand, introduced himself and started to talk.[38] He was "utterly incoherent."[39] Amidst the jeers and laughs of the men, someone on the porch reached out to yank him by the coattails from view.[40]

With the soldiers calling for more speakers, consternation swept the stage. No one wanted to come forward. As the officers milled about, trying to decide what to do next, a volunteer surged to the front and began babbling: it was Greene again. He put on a "worse exhibition" than before.[41]

It took an act of God to end the ceremonies. At the height of the confusion, the heavens opened and a downpour, accompanied by thunder and lightening, scattered the troops.[42]

The new ensign had been raised but it still did not replace the familiar battleflag. A flaw in the design of the official banner was discovered. If there was no wind to unfurl the flag, it hung limp — the white field only was visible to the enemy.[43] The bearers seemed to be advancing to surrender.

Richmond made one last attempt at perfecting their national colors. Taking almost two more years to address the issue, the Congress passed a bill on March 8, 1865, to add a vertical red stripe to the right edge of the flag.[44] This provided a brilliant hue that could always be seen, whether the ensign was waving or at rest.

The fourth Confederate flag.

Major Ben Greene may have viewed this passage as his chance to complete the talk he began at Carlisle, Pennsylvania. If he looked forward to another raising, however, he was disappointed. The Civil War ended before the appropriate ceremonies for the fourth flag could be held.[45]

FOOTNOTES

[1] Douglas Southall Freeman, *Lee's Lieutenants* (New York, New York: Charles Scribner's Sons, 1942), Volume I, p. 71.

[2] T. Harry Williams, *Napoleon in Gray* (Baton Rouge, Louisiana: Louisiana State University Press, 1954), p. 109.

(Continued on page 26.)

Echoes

of

War

"A Smuggler And a Guerilla"

An anonymous interview by:
Thomas J. Evans

"Well, my Grandpa was a smuggler and a guerilla, and he rode with John Mobberly a few times. He rode with Mosby, too, and he scouted some for Lige White. He never liked for people to know he had been a smuggler and guerilla, and once after the war some of Mosby's men invited him to go to one of their reunions, and he wouldn't even consider it.

"Now, Grandpa's name doesn't appear on the roster of any unit, for he never got around to joining the Confederate Army. He was too young to join until way late in the war. A lot of the boys who rode with the guerillas like Mosby never were formally signed up. They just showed up for a raid, and once it was over, why they just went back to being farmers until they got word about the next one. You take that farm boy up at Front Royal, for instance. He got wind somehow Mosby was going to raid the Union Army supply wagons thereabout, and he got his father's gun and borrowed a neighbor's horse so he could go along. He was captured that very first time, and Union General George A. Custer hanged him and four or five Mosby Rangers from an oak tree. Well, when those Indians out West later killed and scalped old General Custer, there were a lot of folks around that didn't shed any tears.

Harper's Ferry

"Grandpa [was going on fourteen when he] started working full time on the farm, and in Autumn, after the first freeze, word came out from Harper's Ferry a storekeeper there needed some hams. They had finished killing some hogs, and it was decided Grandpa would take several hams, some potatoes, honey and a few other things over there to sell. He had never made the trip by himself before; he had only been there with his father. They would leave the farm at dawn driving a two horse team pulling a wagon, and they would arrive there in the afternoon in time to do their trading and get away before dark. About half-way home, they would camp overnight at a big spring that flowed into a good sized creek. There used to be gypsies camp there from time to time, years ago. Grandpa and his father stayed overnight rather than travel in the dark, for fear a horse would step in a hole and break a leg. The roads were pretty rough during those times, and horses were valuable. If one broke a leg all you could do was shoot him, and that was a hard loss for a farmer.

"One morning early, Grandpa loaded the wagon and made ready to go to town. His father came out and said he had something for him now he was doing a man's job. He handed him a new Colt's pistol, a little long barreled .31 caliber five-shooter, and a box of skin cartridges. He also put a double-barreled shotgun in the wagon, one of those old muzzle-loading guns with the mule-ear cocking hammers, and told Grandpa to shoot a squirrel for dinner that night when he was camping out. They used to do that on those trips and roast the meat over the camp fire. Grandpa was pleased with his new pistol and said there wasn't any reason to worry about his getting back safely with the money from selling those hams.

"Grandpa arrived at Harper's Ferry in the early afternoon and went right to the store. An old man ran it, and he had a couple of young clerks to help him. He was right pleased to see Grandpa, and he came right out and shook hands and inquired about great-grandfather's foot and how the family were. He was glad to get those hams and paid top dollar for them. Grandpa took out the list his mother had given him, and ordered a few things: sugar, meal, that kind of thing. Then he paid off what was already owed on credit and for the new purchases. The old storekeeper figured it all up and then gave the bill to Grandpa.

'You just check that when you got time, you being good at arithmetic, and if you find a mistake, why you know where you got it.'

"Grandpa stuck the bill in his pocket, for at no time did anyone ever know the storekeeper to make an error. Most folks who traded with him couldn't read or write and dealt with him because of his reputation. He was known to be an honest man. Well, Grandpa had some other errands to tend to, and when he was leaving the store he noticed two men, a big fellow and a small man, sitting on a nearby porch. They were petting one of those short-haired yellow dogs you don't see many of any more. They didn't have long hair on their tails at all. Folks called them 'Yaller Dogs' and every poor family had two or three around. Grandpa felt like they were watching him, so he made sure he didn't show any sign he suspected anything. They were dressed rough and looked like hard characters. He sure didn't want to get involved with them.

"It was after finishing some errands his mother had given him to do that Grandpa met John Mobberly. Now, John was the son of Polly Mobberly and a fellow by name of Sam Fine. They were never married. Fine was pretty much of a drifter, and headed west when it looked like a war was coming. 'Poor white trash' is how people would probably classify that family. John didn't get much schooling and was dirt poor to go with it. He wasn't afraid of work, though, and he was at the Ferry helping out a butcher, a free black man named Joe Hagen, when Grandpa met him. They fell to talking, and Grandpa showed him his new pistol and invited him down to the river to try it out. John was glad to go because he didn't have many friends, and he took a lot of teasing and all. They went down to the river and shot at turtles, and John proved to be an excellent shot. When they had finished and Grandpa was leaving, John warned him to look out for those two men up near the store, for he had heard some talk and didn't trust them.

John Mobberly, Confederate guerilla leader.

Yaller Dogs

"Grandpa headed on back to the store where he had left the wagon, and headed on out for home. He saw those two men again, and felt uncomfortable since he believed they were watching him and probably had guessed he was carrying some money. He drove slowly on out of town and checked to make sure he wasn't being followed, and everything seemed all right. He made good time towards home, shot a squirrel out of a beech tree for his supper, and he pulled into the camping place well before dark. He unhitched the horses, fed and watered them, and tied them where they could graze. He roasted the squirrel and had plenty else to eat with it, and was sitting back resting when he heard a slight noise. He looked up, and a yaller dog was going past on the road. The hair on Grandpa's neck raised up like a scared cat, for he recognized that dog as the one that belonged to the two men back at Harper's Ferry.

"Well, Grandpa never let on or did anything that would look unusual. He studied the surrounding ground, and figured if anyone was coming after him they would probably come down the mountain nearby and cross the spring branch to where he was camped. If they came any other way, they would be out in the open a good bit of the time, and he figured they were too smart to do that. He looked the mountain over closely, and saw it had been timbered and had grown back up in scrub trees, bushes and saw briars. It would be hard to get through the thick growth without making a lot of noise. That was in his favor. He put some flat rocks behind his campfire to reflect the light away from the wagon and towards the mountain. Then he put his blankets out near the fire when it began to get dark, just in case he was being watched. Soon as it was dark, Grandpa walked in front of the fire a few times to make it look like he was getting ready to go to sleep. He stuffed some brush under his blankets and put his hat at the end, and made a right good dummy that looked like someone laying there in the dark.

"After he was set up, Grandpa slipped underneath the wagon with his guns. He had bought some buckshot at the store, and he poured some down both barrels of the shotgun on top of the loads that were already in there. You couldn't blow up those old black powder guns if you loaded them heavy. Then he got behind the front wheels of the wagon where he could see the mountain, and he laid out his pistol, after making sure it was loaded and ready with some extra cartridges where he could get to them in a hurry. He didn't intend to be robbed, and he aimed to fight it out. Along after midnight, Grandpa could hear those fellows coming down the mountain just as he had expected. He hadn't worried about going to sleep because his heart had been beating so loud he was afraid those men would hear it. He aimed the shotgun where the noise was loudest, and when they got well in range, he cut loose with both barrels.

'Shotgun!' somebody yelled.

'I'm hit!' said another voice.

"Now, those fellows went to tearing back up the mountain through the briars like a couple of bears, stumbling, cursing, and making all kinds of a racket. Grandpa fired three quick shots from his pistol so they would know he was well-armed and meant business. Then he quickly reloaded both guns and waited to see if they planned to come back. He didn't believe they would, but he couldn't be sure. Come daylight, Grandpa eased around the mountain through the trees and brush, and he found where they had tied their horses. They were long gone, but left behind a bloody rag on the ground. He was sure at least one of those buckshot had hit the target. He drove on home and told his father what had happened, and he was sure proud he had brought the money back home safely. His father alerted the neighbors, and everybody kept an eye out for those two fellows, but they had left the county. And it was a good thing they did, too, for there were folks there who would have shot them on sight, my great-grandfather being one.

Smuggling

"Virginia was full of rumors of war, and before most farm folks realized it, the Civil War was under way. Grandpa thought he ought to go, but he was still only a boy, and his parents weren't in favor of it. Anyway, he rode to Waterford during the first winter, and stayed about a week at the Quaker Church, where Confederate Captain E.V. White, 'Lige' they called him, had his company camped. Lige told him he was not quite ready yet to join his company. Lige was always good about not hurting a person's feelings, but that he would use him as a scout from time to time. Grandpa got teased a bit by the boys on account of his age, but they liked him, and gave him a .36 caliber Colt Navy revolver. They told him they would have given him one of those big .44 caliber Army pistols, but it was so heavy, and Grandpa so skinny, his pants would fall down if he didn't wear two pair of suspenders all the time. Guerillas all used pistols, for they didn't have any use at all for swords. They might carry a carbine on their horse if they were going to ride through open country, but pistols were their main weapons. Every one of them had at least two. Something else that is not generally known is that each of them carried two or three spare loaded cylinders on their person. When a pistol was emptied during a fight, they would slip in a spare cylinder quick as a wink, and be ready to shoot again. They didn't have metal cartridges for pistols until way late in the war, and then they were of too small a caliber to be very effective.

"Grandpa went on back home, but he wasn't satisfied. John Mobberly was scouting for Lige White, and put in a good word for Grandpa with a smuggler so that soon he had steady work. Smugglers would buy up goods and medicines throughout the North, and ship them into the border state of Maryland. Then at night, they would slip them across the Potomac River into Northern Virginia and on down south into the Confederate lines, where they could sell them at a high price. They made big money, and paid their help in gold. It was risky business, for the Union Army was always looking for them on both sides of the river, and so were the Union guerillas, and even some of the Confederate guerillas, when they thought they could get away with it. It was a thriving business all throughout the war, even more so towards the end, when the Union Navy blockaded the Southern ports. The smugglers also would carry secret papers to the Confederacy from spies in the North, and they would sometimes bring a spy back with them and help him cross over into Maryland.

"Grandpa became a right talented smuggler, and his services were very much in demand. He knew how to handle horses and mules, and he could move them through the darkness and over the mountains as well as any grown man could, and better than most. He was well-paid, and used to give his parents money that they got to needing more and more as the war went on. It finally got too hot for him to live at home, for the Union Army was beginning to suspect him, and the Loudoun Rangers started looking for him.

The Loudoun Rangers

"I need to tell you about the Loudoun Rangers. Their Captain was a Quaker named Sam Means. He had a brother who fought in the Confederate Army, and I want to say I believe Sam was plain hounded into joining up with the Union. The Confederates had confiscated his flour mill and butchery business, so he moved over into Point of Rocks, Maryland, and set up business. It was a bad day for the Confederacy when Sam decided to go with the Union, and he recruited a company of cavalry from around Waterford where he had been living. Those boys were tough and dedicated, and were feared throughout Northern Virginia. Of course after the war when the former Confederates got in office, they tried to make out the Rangers were a bunch of thieving rascals. Well, they were wrong, for those Rangers were as good a bunch of fighters as ever came out of Virginia, and old 'Quaker Sam' sure made it rough on the smugglers and the guerillas.

"There was a Loudoun Ranger named Sammy Tritapoe, who just used to give the smugglers fits. He always seemed to show up when least expected. Once, Grandpa was the last of three smugglers crossing the Potomac one night, and he had three mules with packs under his care. As he came up out of the water, Sammy opened up on them from behind some rocks. Grandpa had expanded his arms to two Colt Navy revolvers and three spare cylinders, and he sat there and shot it out with Sammy until the other boys could get away. Sammy killed one of the mules, and Grandpa had to leave the pack behind when he made his run for it. That was one of only two mules that Grandpa ever lost. The other one was killed by a long range shot by a bushwacker, but Grandpa always figured it was Sammy who fired that shot. He didn't know anyone else who shot that well. Mosby's men were all out to get Sammy, but he outwitted or outshot them all, and killed off a number of them before the war was over.

"As I was saying, Grandpa had to leave home to get away from the Union Army and the Loudoun Rangers. He took up with a family back in the mountains for a time. Guerillas didn't live in camps like soldiers, although some folks seem to think they spent most of their time sitting around a campfire singing songs. They would stay with a family until they received word there was going to be a raid, then they would gather at some prearranged spot and start out from there. If any of Mosby's men missed answering two calls, they had to go to the regular Confederate Army. Of course, Grandpa was too young, and he didn't have to worry. When he wasn't busy smuggling, he would ride with Mosby, White, or Mobberly, as he pleased. The guerillas never imposed on those families they were staying with. It was the unwritten rule they paid twenty-five cents for each meal they ate, and they would help out around the farm when they could. Now, needles and pins were hard to get in the South after the war had been going on for a spell, and the smugglers would carry them through the lines once in awhile. Grandpa had all he wanted, so he always left a package or two where he was staying or where he ate several meals. The women appreciated his thoughtfulness very much, and were glad to see him, and the kids took to calling him 'Mister Needles.' That was a good nickname, and he was known as that by many folks who never did know his real name.

Bite Off It's Head

"Well, now, Grandpa was sitting on the front porch of the cabin where he was staying one morning, when he recognized one of Mosby's Rangers riding up the path towards him. It was a fellow known as 'Sink.' I'll recall his name in a minute. Sink was one of the best Rangers in Mosby's command, and he was full of fight. He rode on up to where Grandpa was sitting, wondering if Sink was bringing news that there was going to be a raid.

'Mister Needles,' Sink said, 'I have a powerful urge for a drink of good whiskey. In fact, I am in need of a drunk, and have taken a leave to throw a big one. I need a smuggler like you who knows everybody to take me over on Short Hill Mountain so I can buy a few jugs.'

'I'm your man,' said Grandpa. 'Just let me saddle my horse. The old man that lives here has been mighty good to me, and I have been wanting to get him a jug as a present.'

"Grandpa figured there was not much action going on, and he would just go ahead and guide Sink over to a place where he knew they could buy some whiskey. Now, Mosby didn't allow his men to drink on the job, and he was against distilleries. He would break up any stills he found because he said the grain was needed for the hungry people in the Confederacy. However, his men had ways of getting liquor, just as Sink was prepared to do. Let me tell you about Sink, because he had a habit that is hard to believe, but I will swear it is true on a stack of Bibles. When he decided to do some serious drinking, he would catch up a bag of small snakes — garter snakes, black snakes and the like. After he had a good drink, he would pull one out of the bag, bite off its head, and proceed to eat it. Just munched right along on them as he talked and drank. If the whiskey was good, he might eat seven or eight at a sitting. Well, Grandpa could see Sink had a bag tied to his daddle that seemed to hiss and move around ever so often, and he wanted to see if what he had heard about Sink's drinking habits was true. Grandpa mounted up, and took Sink over the mountains by some back trails, and in a few hours they rode out of a pine thicket overlooking a long, narrow cabin. Sink and Grandpa studied the countryside to make sure it was safe, then rode down and tied their horses to a rail in front of the shack next to two other horses. They crossed the porch and went into a long room in front of the building. There were a couple of tables with chairs in there, and a rough bar made from some planks laying across some old barrels. There were two men wearing odds and ends of Confederate uniforms sitting drinking at one of the tables.

"Sink walked up to the bar and said: 'You got drinking whiskey?'

'You drinking or jugging?' asked the fat man tending bar. What he meant was were they buying by the drink or did they aim to buy a jug. If you bought a jug the first drink was a free sample.

'Both,' said Sink throwing a twenty-dollar gold piece on the bar causing the bartender to grin from ear-to-ear.

'Payne's men,' the bartender said, nodding at the two guerillas sitting at the table.

'Mosby's men,' said Sink by way of introduction, raising his hand casually to the other guerillas.

"Sink turned to the bar and took a drink of whiskey. Grandpa did the same, and said it tasted like turpentine, and not very good turpentine at that. He noticed the two guerillas were looking at him kind of funny, and suddenly his heart came right up in his throat and liked to choked him. They were the two fellows that had tried to rob him at Harper's Ferry. He dropped his hand slowly down to his side and opened the flap on his holster, for if there was going to be any trouble, he wanted to be ready. Sink untied his bag of snakes, and had started to smile in anticipation of a good time, when one of the guerillas spoke.

'I know you,' he said real mean-like, looking at Grandpa, 'You are the fellow who shot me at the Ferry.'

"Before you could blink an eye, Sink threw his open bag of snakes upon their table and snakes went crawling in all directions.

'Rattlers!' Sink shouted, 'Don't get bit!'

"The men both struggled to get to their feet, slapping away at snakes and cursing. Sink had his .44 caliber revolver out, and shot the big man through the heart. He fell over backwards and never moved a muscle. Sink then shot the smaller man who was reaching for his gun by then, and his bullet spun him sideways. Grandpa had managed to get clear of Sink, and he shot the man twice real quick with his pistol. That fellow just slumped down with all that lead in him, and died right there. Sink next jumped around and shot the bartender through the arm when he noticed he was reaching for a gun back of the bar. It wasn't any of the bartender's business, and he shouldn't have mixed in.

"Sink stood there in the smoke-filled room and said to Grandpa sort of surprised-like: 'I sure don't remember shooting that fellow at Harper's Ferry. I must have been drunk. In fact, I would call that man a liar, but I don't want to speak ill of the dead.'

"Grandpa started to tell Sink it was him those guerillas were talking about, but didn't think right then was a good time, now that Sink had lost all of his snakes. Every fellow was expected to fight his own fights in those days. However, Grandpa hadn't asked for any help, so he didn't feel guilty about the matter. They patched up the bartender, and Sink gave him ten dollars for a doctor, and he told him to keep those fellows' horses for the trouble of burying them. Well, that bartender was crippled up for life in that arm, but he didn't farm, or anything where you had to work hard, so it wasn't so bad on him. He knew he shouldn't have gone for his gun, and it hadn't been any of his affair. Sink paid for four jugs, and Grandpa paid for one to take back. They strapped the jugs on their horses, and headed on back towards the place Grandpa was staying.

"Grandpa told Sink to follow him, and he turned his horse off the trail and cut through some low land along a creek. Grandpa was watching close, and he soon saw a small snake. He was down off his horse in a flash, and caught it, and put it in Sink's sack. Sink got in a better mood and pitched in, and in no time they had caught six or seven. Sink started tapping a jug, and soon hauled out a snake, bit off its head, and proceeded to eat the squirming body. He offered Grandpa a bite, but Grandpa said he wasn't feeling hungry just then. They finally got back to where Grandpa was staying, and Sink shook his hand, and thanked him for helping out during the shooting, and especially for helping catch those snakes. He also said he was going to put in a good word for Grandpa with Mosby, but Grandpa figured that might not be worth too much if Mosby got wind of what they had been up to. Sink rode off, and Grandpa gave that jug to the old man he was staying with, and the old fellow was well-pleased with it, and said it was about the best he ever tasted.

"Years later, after the war, Grandpa ran into Sink. Sink recollected who he was right off, and thanked him for helping catch that bunch of snakes. Grandpa felt he ought to confess, and said he was sorry about the shooting that day. Well, Sink looked puzzled, and asked what shooting he meant.

Grandpa could see Sink had forgotten the event, so he bought him a bottle, and considered the matter closed.

French Bill

"Now, I have said smuggling was dangerous work. Once, during a bad winter evening, Grandpa was helping smuggle a load of goods south, when it came up a snow storm. He said at first it got so quiet, it was almost scarey, no sound at all, and the horses and mules seemed to make a tremendous amount of noise. They sighted a group of riders, and were preparing for the worst, when they recognized 'Billy the Frenchman,' or 'French Bill,' as he was better known. He was a deserter from the Union Army who headed a gang of guerillas, and they were returning from a raid. They talked for a couple of minutes, and then they went their separate ways as the snow started to come down. It was good the smugglers moved on, because Sammy Tritapoe and some of the Loudoun Rangers were following French Bill. Sammy kept after French Bill's gang, tracking them through the snow, and apparently saw no sign of the smugglers' tracks that the snow must have covered up quickly. He caught up with French Bill, who was drinking, and had stopped at a house to get out of the bad weather, and he captured him without a fight. French Bill was taken to Harper's Ferry, to the Union Army headquarters, court-martialed, and hanged. Grandpa always said afterwards if French Bill hadn't of been drunk, he wouldn't have been so careless as to have been taken without a fight, or even so much as a shot having been fired. Bill just wasn't a match for Sammy that night.

John Mobberly

"Now, you are interested in John Mobberly, so let me get around to him. He deserted from Lige White's company, and started a guerilla band of his own. Lige didn't hold it against him, used him as a scout, and took any information John passed to him. It was on account of John that Grandpa nearly got killed or captured by the Sixth New York Cavalry, over near Short Hill Mountain. Lige White had been promoted to Colonel of the 35th Battalion of Virginia Cavalry by then, and he decided he could sneak in and capture a couple companies of the Sixth that were stationed at the foot of the mountain. He had John and one of his men, Lum Wenner, I believe it was Lum, scout out the camp. They reported back to Lige there was about two-hundred and fifty Union cavalrymen there. Lige was always ready for a fight, so he mustered his men, and sent out word to the guerillas for a raid. Some of Mosby's men responded, and some of John's, and Grandpa was relaxing between smuggling trips and figured he would go along. Altogether, Lige had about ninety men when he followed John in a thundering night charge into the camp, expecting to find only a couple hundred Union boys, and them ready to run. Well, there was six hundred cavalrymen in that camp, veterans they were, and as ready to fight as Lige. Bullets were flying around, singing like bees, Grandpa said, and the Confederates made a desperate retreat. I think that was the last skirmish of the war in Loudoun County. Anyway, Grandpa said relations between Lige and John were somewhat strained after that little affair.

Front view of John Mobberly's gravestone.

"John was doing some smuggling of his own accord, and he sent word to Grandpa he needed him for a trip. Grandpa joined John and two men, and they took twenty loaded horses all the way south, nearly to Waynesboro. Grandpa didn't comment on the fact they were using fast horses instead of mules, for he didn't figure it was his business. It was all right with him how John moved his goods, and he just kept his mouth shut, and waited to see what was going on. After the goods were sold, they went back up in the mountains and camped. Near morning, twenty black men, slaves they were, showed up, and were heartily greeted by John. He told them they were practically free men, and he was taking them North to help President Lincoln free all the slaves. Grandpa saw at once they were all young and healthy. As the slaves were mounting the horses bareback, John winked at Grandpa, and told him to make sure the five he was responsible for got to Maryland, and didn't have any second thoughts about leaving the South. They rode north through the mountains, and the night before they got to the border, they were joined by a man dressed in black, called the 'Preacher.' The Preacher gave a fiery sermon, and told the blacks they should join the Union Army to help free their brothers in bondage, and that he would help them. The next night, they crossed the border into Maryland, and, true to his promise, the Preacher helped each black enlist in the U.S. Colored Troops. He also collected the bounty for each one, and sent it back to John Mobberly. Well, John paid Grandpa double since he had taken a load both ways, and it was a right profitable trip. John was never stingy when it came to paying.

"The war changed John Mobberly, Grandpa said. John had nothing as a boy, and suddenly he had power and money. He dressed in the best of Confederate uniforms, and started referring to himself as 'Captain.' He attacked the Union Army regularly, and often dressed his men in captured Federal uniforms so he could surprise the Yankees, and get up close to them before they realized what was happening. He also got to taking what he wanted from the local farmers without paying. There was this teenage girl, Betsy Compher was her name, who wouldn't let John set fire to the family home when he decided they were Unionists, and he was going to burn them out. They got to struggling, and John pistol-whipped that girl about the head. She never fully recovered from it, and caught typhoid fever while she was weak, and died. Well, he was responsible for the death of that girl, just as much as if he had shot and killed her. John got to running after the girls, too. 'Sweet Johnny,' folks got to calling him. He was all over the county, telling every girl she was his only true love, and that he was going to marry her after the South won the war. There wasn't any truth in it, and he had a couple of those poor girls in a family way, before long.

"Oh, Lord, I'm gone"

"It was that mean streak in John that finished him. He was upset with the Loudoun Rangers chasing him all the time, and he happened to get into a skirmish with some of them over at Waterford. One of the Rangers, Sergeant Charles B. Stewart, was shot off his horse during the fight, and left wounded on the field. Mobberly rode over to where he was laying, and shot him in the face to finish him off, then he rode his horse over him a couple of times, before getting off and stealing Stewart's boots. It was a bad thing to do, particularly to a Loudoun Ranger, for they weren't too fond of John to start with. Now, Sergeant Stewart wasn't killed, nobody knows why, and he was taken back to Harper's Ferry. While he was recovering, he had time to think about John Mobberly. He knew John loved fast horses almost as much as he loved the girls. He also had heard that when one of John's band was away, John's horse could be seen tied in front of his girl friend's house. Stewart felt he just might have a way of dealing with old John.

"As soon as he had recovered, Sergeant Stewart made his way across the Potomac River with two hand-picked Rangers known to be crack shots. He went to the barn of Luther Potterfield, just outside of Lovettsville. Soon after, the angry boy friend of that girl John had been sneaking around seeing on the sly told John there was a real fast throughbred horse for sale out at the Potterfield farm.

"Early on an April morning in 1865, just as the war was drawing to a close, John Mobberly and his lieutenant, Jim Riley, rode out to the Potterfield farm to buy or steal that horse. Jim Riley stopped to water his horse in a little spring brach there in the barnyard, while John rode on up to the barn to see the horse. Just as he rode up to the barn door that was standing open, he recognized there was a man standing inside with a revolver in each hand. John's last words were: "Oh Lord, I'm gone." He was right, too, for Sergeant Stewart and those two Rangers who were hidden up in the barn loft shot him full of holes, and he died right there in the barnyard. Jim Riley heard the shooting, turned, and galloped away, but that night he and the rest of Mobberly's band returned, and burned the Potterfield barn to the ground.

"The Loudoun Rangers took John's body back to Harper's Ferry, and dumped it in front of the Union Army headquarters there on Shenandoah Street. Folks noticed right off John wasn't wearing any boots, and figured Sergeant Stewart had gotten his back, all right. People came from all over and cut the buttons off John's uniform, then cut pieces off the uniform, and finally cut off most of his hair and beard for souvenirs. The Union Army let John's body lay there for three days before they gave it to his mother. She took him over to Salem Church and buried him, and it wasn't long before all those girlfriends of his got together and put up a grave stone for him. One of them wrote a poem, and they had it carved on there, and on the front they had it say he was assassinated. Just like President Lincoln, you know. They used to put flowers and little Confederate flags on his grave over the years, but everyone interested in John had died off by now. If Grandpa ever went over there and paid his respects to John, I never heard mention of it.

"Well, that is the story of John Mobberly, and of my grandfather, who was a smuggler during the Civil War. There wasn't much difference between a smuggler and a guerilla, I suppose, for most of the boys along the border did a little of both. It sure didn't make any difference to the Union Army, for they shot them all, any chance they had. They wore civilian clothes inside Union territory, and were armed, and many a smuggler that was shot was counted as being a guerilla.

"Unlike most that came back from the war, Grandpa had money. He bought a farm, lived a quiet life, and never spoke much about the war until he was older. I remember him well, but he died when I was young. He had a beard and big mustache, and he looked like a grandfather was supposed to look. To see him, you wouldn't think he would kill a fly, but he had known them all, Mosby, White, Mobberly, and the rest, and he had ridden with them. Now, when you write about those days, don't make it sound all romantic, but you tell it like it really was. There are a lot of smugglers buried along the border in graves that no one knows who they were; and a lot of mothers waited for years for boys who never came home, and nobody knew what happened to them. Those were hard times, and a lot of ill will is still left over, even though all who fought then are all gone now."

EPILOGUE

The role of the smugglers along the Northern Virginia-Southern Maryland border during the Civil War is not widely recognized. Similar to those engaged in spying, smugglers neither desired, nor sought, publicity. Anonymity was their goal, and after the war had ended, they faded back into the civilian population, leaving little hint of their wartime escapades. Many were rich from their activities, as were numerous Northern merchants who profited from the illict trade.

Quick to recognize the impact of smuggling in prolonging the war was Union Army Captain Samuel C. Means of the Independent Loudoun (VA) Rangers. Writing to his Commander, General Franz Sigel, on April 6, 1864, Means stated: "I am just getting out of a severe spell of sickness & not able to write much but learning that you was about to raise the Blockade on the Potomac River from Harper's Ferry to Mouth of Monococy (the Head Quarters of all smugglers). Ask that you telegraph giving me permission to come to Cumberland to see you before you act."[1] However, General Franz Sigel was not disposed to listening to advice from anybody, particularly a mere captain, and Means' efforts were unsuccessful.

Captain Means' company was composed of hard-fighting Virginians, many of whom, like himself, had brothers or relatives fighting for the Confederacy. They were a dedicated group, ready to take on smugglers and guerillas at any time. Corporal Samuel E. Tritapoe was one such Ranger, and the Union Army Headquarters at Harper's Ferry was notified of the following incident on December 6, 1864:

"Corporal Tritapoe, who took French Bill, shot one of Mosby's men yesterday; he died this morning at Lovettsville. He is out again after more."[2] Fearless Corporal Tritapoe was a man to be reckoned with by smugglers and guerillas alike. He did not differentiate between them.

The fate of the luckless French Bill, who had the misfortune to be captured by the relentless Corporal Tritapoe, is chronicled in the following messages:

"HARPER'S FERRY
December 1, 1864.

Maj. Gen. Sheridan,
I caught French Bill yesterday, a notorious murderer and bushwacker, belonging to White's battalion, who was with the party that murdered the surgeon of the 6th Pennsylvania Cavalry. He is a deserter from the 61st New York Infantry.

Respectfully,
JOHN D. STEPHENSON,
Brigadier General Commanding.[3]

HEADQUARTERS MIDDLE MILITARY DIVISION,
December 1, 1864—1.15 p.m.

Gen. STEPHENSON,
Commanding District of Harper's Ferry.
As soon as you can have fully ascertained that you have French Bill as your prisoner, take him out and hang him. This will be your authority.

P.H. SHERIDAN
Major General Commanding.[4]

HARPER'S FERRY,
December 1, 1864.

Maj. Gen. SHERIDAN,
I have, undoubtedly, French Bill. He will be hanged at 2 p.m. tomorrow.

JOHN D STEPHENSON,
Brigadier General.[5]

HARPER'S FERRY
December 2, 1864.

Gen. SHERIDAN,
French Bill has been hanged in accordance with orders.

JOHN D. STEPHENSON
Brigadier General."[6]

Accordingly, White's Battalion lost one member more. There can be no mistaking the policy of the Union Army was that of "no prisoners" for those suspected of guerilla activity. Brigadier General E.B. Tyler formalized this position in the following extract from a message reporting the activities of guerillas along the Potomac River: "These men are supposed to belong to a Lieutenant Mobberly's gang of outlaws. I have instructed my command not to bring any of them to my headquarters except for interment."[7] Life was hard for a guerilla or smuggler along the Potomac.

Generally unrecognized as an outstanding partisan leader and army cavalry commander was Elijah "Lige" V. White. On leave at his home near Leesburg, Virginia, during October, 1861, he heard the opening gunfire at the Battle of Ball's Bluff. He rushed to the scene where he rendered such conspicuous service to the Confederate commanders he received mention in three different dispatches. As the result of his gallant actions, he was commissioned a captain and given authority to recruit a company of cavalry. He fought brilliantly throughout the war, and, always in the front of his command, was wounded seven times. He was promoted to the rank of lieutenant-colonel, and commanded the famed Laurel Brigade at the time of the final surrender at Appomattox.[8]

The notorious guerilla leader Payne was captured by the hard-riding troopers of Brigadier General Thomas C. Devin, who reported: "Payne should be tried by a military commission and shot. Since I have been here he has been robbing all around, and shot one man and nearly beat another to death for his money. I ordered that he should not be brought in alive, but he was not recognized in time. If he is tried I will furnish the evidence. I have him tied hand and foot."[9] The guerilla fighting was brutal, and little quarter was either given or expected.

The adventures of Colonel John Singleton Mosby, the "Gray Ghost," are too well known to elaborate upon. However, the "grapevine" of small distillers operating throughout the Short Hill-Round Hill-Snicker's Gap area of the Blue Ridge Mountains defeated his efforts to stop their making of whiskey to save the much needed grain for the Confederacy. The Union Army also had little success in stopping them. A raiding party of the 14th Pennsylvania attempted to close one still by drinking all of the captured spirits. However, the operators persevered, and had little trouble in selling their product to soldiers of both causes throughout the war.[10]

The interest displayed in the reptile world by one of Mosby's Rangers, "Sink" Sinclair (pronounced "Sinkler"), is worthy of further mention. He survived the war to meet often with his former comrades. His unique snacking habits could be observed in the town of Arcola, Virginia, even in his old age. Many former members of Mosby's command rose to prominence after the war as doctors, lawyers, ministers and similar respected occupations. Sink's occupation was drinking, and he labored at it successfully. He remains forever an interesting character of that period.[11]

John Mobberly's place in Civil War history will always be the subject of dispute. He never was widely accepted as a Southern hero. From the Union Army point of view, he was deceptive, and violated the rules of war by dressing his men in captured Federal uniforms so that he could surprise the unwary.[12] He proved impossible to catch, in spite of the concentrated efforts of Union forces, and "Mobberly was not found" was often the outcome of numerous attempts to end his career.[13] Of quick temper, he cruelly pistol-whipped fifteen-year-old Elizabeth Elmira Comper at her father's (Joseph Comper) farm near Lovettsville, Virginia, when she tried to stop him from burning the house. Her injuries resulted in failing health, and she soon contracted and died of typhoid fever.[14] That he was brutal is evidenced by his shooting of the helpless Sergeant Stewart as he lay wounded following a skirmish. Hearing the shooting, Dr. Bond, a Quaker physician from nearby Waterford, gathered several young ladies from the town, and rushed to the scene to aid the wounded. Witnessing Mobberly's wanton act, Dr. Bond shouted: "Surely no hell is too hot for a man that would do a thing like that."[15]

Following the attempted murder of Sergeant Stewart, Mobberly's days as a guerilla and smuggler were numbered. He was feared by the residents of Northern Virginia who wished him captured, and his continued operations were an embarrassment to the War Department. After the ambush and shooting at the Potterfield barn, the following messages were exchanged:

"HARPER'S FERRY
April 5, 1865
(received 6.20 p.m.)

E.M. STANTON,
I sent out, on Monday, a small party to wipe out the notorious guerilla, Mobberly, and his band. They returned today with the body of Mobberly, and in the fight mortally wounded his right-hand man, Riley.

Respectfully,
J.D. STEPHENSON
Brig.-Gen. Commanding."[16]

"WAR DEPARTMENT
Washington, April 5, 1865.
(9.48 p.m.)

Brig.-Gen. STEPHENSON,
Harper's Ferry.
Accept the thanks of the Department for your diligence, skill, and success in the achievement mentioned in your telegram of this date.

EDWIN M. STANTON,
Secretary of War."[17]

John Mobberly's prophetic words, "Oh Lord, I am gone!" had come true.[18] But he was not forgotten. His grave in the churchyard of the Old Salem Methodist Church near Hillsboro, Virginia, was marked with a plain white marble marker. On the front was inscribed:

JOHN W. MOBBERLY
Member of Company A, White's
Battalion of Va Cavl'y,
was born
June 1. 1844
was assassinated
April 5th 1865
Aged 20 years 10
mo's & 4 days.

On the back of the stone his female admirers had carved the following epitaph:

God bless thee, brave soldier,
Thy life dream is ore;
For country and freedom,
Thou will battle no more.
To the land of the blessed,
Thou has gone to depart;
With a smile on thy face,
And a joy in thy heart.
Thrice hallowed the green spot,
Where our hero is laid;
His deeds from our memory,
Shall never more fade.
The stranger will say,
As he lingers around;
'Tis the grave of a hero,
'Tis liberty's mound.

The war was ended for John Mobberly. His men accepted paroles and returned to their homes.[19]

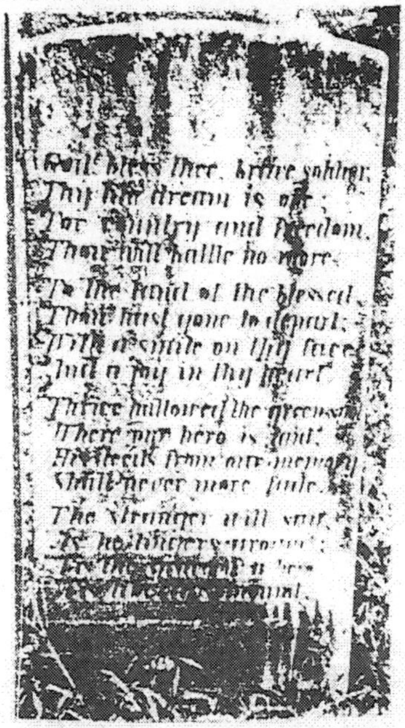

Back view of John Mobberly's gravestone.

Of the subject of this article, he remains anonymous, as he had lived, and as he would have wished. Only the readers of KEPI know of his adventures as one of the little-known smugglers who silently slipped through the mountains of Virginia all during the Civil War. "Mister Needles" rests in a churchyard in the countryside of his beloved Virginia, beneath a plain grave stone erected by his loving family. Altough never entered on the muster role of any unit, he was as much a part of the war as any man who fought in it. Peace to his ashes!

FOOTNOTES

[1] National Archives, Captain S.A. Means, Service Record.
[2] *OR*, Series I, Vol. XLIII, Part 2, p. 749.
[3] *Ibid.*, p. 721.
[4] *Ibid.*
[5] *Ibid.*
[6] *Ibid.*, p. 727.
[7] *Ibid.*, Part 1, p. 693.
[8] Captain William N. McDonald, *A History of the Laurel Brigade* (Baltimore, 1907), p. 381.
[9] *OR*, Series I, Vol. XLVI, Part 2, p. 365.
[10] Research from Allan Tischler, Historian-Researcher, Cross Junction, Virginia.
[11] Interview with Ted Ballard, U.S. Army Historian, Washington, D.C. Ted's stepgreat-grandfather served with Ranger Sinclair in Mosby's command. His stepgrandfather observed Sinclair consume a snake at Arcola, Virginia, after the war.
[12] *OR*, Series I, Vol. XLIII, Part 2, p. 604.
[13] *OR*, Series I, Vol. LI, Part 1, p. 177.
[14] David H. Brown, "John W. Mobberly-Renegade Hero," Manuscript, June 5, 1960.
[15] Interview with John Devine, Historian-Author, Leesburg, Virginia. John's grandmother was one of the young ladies Dr. Bond brought to the skirmish site to aid the wounded. John's grandfather served in the Loudoun Rangers.
[16] *OR*, Series I, Vol. XLVI, Part 3, p. 590.
[17] *Ibid.*
[18] Briscoe Goodhart, *History of the Independent Loudoun Rangers*, (Washington, D.C., 1896), p. 197.
[19] *OR*, *op. cit.*, p. 840.

"BREAKING THE CODE AT VICKSBURG"

by

Hector De Marco

"During the negotiations we noticed that General Grant and Admiral Porter were communicating with each other by signals from a tall tower on land and a mast-head on Porter's ship. Our signal-service men had long before worked out the Federal code on the principle of Poe's 'Gold Bug,' and translated the messages as soon as sent. We knew that General Grant was anxious to take us all as prisoners to the Northern prison-pens. We also knew that Porter said that he did not have sufficient transportation to carry us, and that in his judgement it would be far better to parole us and use the fleet in sending the Federal troops to Port Hudson and other points where they were needed. This helped to make General Pemberton more bold and persistent in his demands, and finally enabled him to obtain virtually the terms of his original proposition."[1]

"The Gold-Bug," written by Edgar Allan Poe, and first published in *The Philadelphia Dollar Newspaper* in 1843, was one of the author's most popular works. Prior to the Civil War, it had a wide reading, was often reprinted, and was even adapted into a play by Silas S. Steele.[2]

The story, set on Sullivan's Island, Charleston Harbor, South Carolina (where Poe had served a tour of duty while in the army), concerned William Legrand, a man of good family, but reduced to poverty. While living on the island, he had found a golden-colored *scarabaei* (a large black or nearly black dung beetle), and took it home for his natural science studies. In sketching it for his friend, the narrator of the story, Legrand realized a scrap of parchment his servant had found to wrap the treasure in contained a drawing of the insect, sketched so as to resemble a skull, or death's-head. By a series of clever techniques, Legrand was able to elicit from the paper a set of letters that appeared to be a cipher. He solved the cryptogram, deduced it to be a treasure map, and eventually led his friend to a huge tulip tree, where they found buried nearby a fortune in gold and jewels.

In unraveling the enigma for his skeptical companion, Legrand carefully explained how he had solved the seemingly impossible code. The explanation of one character to another in a fiction published eighteen years before the Civil War thus became the basis for helping Confederate officers decipher a Union cryptogram, which, in turn, was directly responsible for General Pemberton's strong position at the bargaining table.

When Poe first wrote his series of articles on Cryptography in 1841 (appearing in *Graham's Magazine*), he was exploring a field which was relatively new to this country.[3] By following his essays with a short story, he aroused the awareness and curiosity of those whose intellect drew them to puzzles of logic. In "The Gold-Bug," Poe's outline for decoding "the very simplest species of cryptography"[4] was so precise, that it was remembered and applied by a fellow Southerner fourteen years after the author's death.

Poe's character, Legrand, explained to his bewildered listener that the first step in cracking the series of enigmatic characters (or, in the case of Vicksburg, flag signals), was to determine which language was employed. In the story, it was discovered to be English. Next, the task was to ascertain the predominant letters, as well as the least frequent. Counting all the characters, the deducer could then construct a table, placing the symbol or letter opposite its number of occurances. Legrand then explained that the letter which appeared most frequently in the English language is *e*, and the succession afterwards, *a o i d h n r s t u y c f g l m w b k p q x z*. (Mr. Poe mysteriously omitted the letter *j*.) Taking the letter which appeared most frequently and substituting *e* for it, the next step was to check for a three letter repetition of letters which appear frequently. These being isolated, their probable meaning is *the*. This allows the *t* and the *h* to be inserted. With three letters filled in, words may be guessed at by trial-and-error, until sentences form. What is left is the addition of punctuation to make sense of the riddle.

Although born in Boston, Massachusetts, Edgar A. Poe was raised and educated in Richmond, Virginia. He spent much of his life in the South, and considered himself allied with the wealthy aristocrats. It would have suited his fancy to have known he aided, in some small way, the destiny of a Southern army.

FOOTNOTES

[1] Robert Johnson and Clarence Buel, Battles And Leaders of the Civil War (Castle Books: New York, 1956), p. 492.

[2] Philip Van Doren Stern, *The Portable Edgar Allan Poe* (Viking Press: New York, 1945), p. 332.

[3] *Ibid.*, p. 507.

[4] Edgar Allan Poe, "The Gold-Bug," from *The Portable Edgar Allan Poe*, *op. cit.*, p. 501.

ANOT TF X LUN JPR SLUE HLU X BTCTGXUK PXO. EMUT JY GRN PTDT SXU, GRTF GKON LH RNXEYNKU SXF SLUJ IK ILGR JLUGR XJE FLMGR, GRMF INPLBTJ X IT—OXUGTFXJ FKBILC LH GRN SXU.

GENERAL THOUGHTS

Analysis and Commentary:

"How The North Won"

How The North Won, by Herman Hattaway and Archer Jones, The University of Illinois Press, Box 5081, Station A, Champaign, IL 61802; $24.95. 1983; 763 pages, maps, diagrams.

No one likes to write a bad book review. But then, no one appreciates spending $25.00 (or more) for a work only to find the lauditory review they read was written by someone who had probably taken the summary from chapter titles and dust jackets. Who has ever seen a dust jacket with anything but the most lauditory comments on it? (For that matter, who's seen artwork on one that bears any resemblance to what the book's about?) I don't want someone else's praises/phrases running through my mind when I sit down to read the book. What I begin with is the Preface, or Introduction. This is the "author's corner" — their chance to talk directly to their readers and explain why they wrote the book, what influenced them and what they hoped to achieve. I always take this seriously, for these comments are what they're going to have to live up to. It also tells me a little about these impersonal people I've never met before — are they sincere, offhand, smug, superior? Did they write the book on grant money, or did they research on weekends, supported by their own efforts? Are they professors being printed by the home press, did they sell the book to a prestigious publishing house, or are they printing it themselves? It makes a difference. In a field as "small" as the Civil War, there are no best sellers; a biography of John Bell Hood doesn't make the New York Times best seller list. There isn't big money to be made, and few lucrative film contracts. (The historical works that do make it to the screen are usually the fictional ones — invariably, the end product is fictional). There are few ways the researcher can get his book in print, and it's important to assess that before getting into the text.

Usually, reviewing a book is clear-cut: it must be readable, accurate, well-footnoted and informative. It must be fair — I don't like to read books that "take sides" — no one is here to re-fight the war; we're here to understand and learn. I don't like hindsight — the 19th Century is not the 20th. Too much reliance on "today" can cloud the realities of the 1860's — the Vietnam War may be like the Civil War, but the Civil War cannot be like the Vietnam War. In reviewing a book with a specific topic, as *"The Lincoln Image,"* I look to see whether the authors have established a theme, and systematically presented/proved it. With a book such as *"How The North Won,"* the job is much more difficult. The text covers a very wide range, and requires a careful analysis. In this instance, therefore, I decided to let the writers speak for themselves, and I reread their Introduction several times. Authors Hattaway and Jones suggest that their book "offers something to a number of different groups of readers and purports to tell the military history of the war from the viewpoint of the higher commanders on both sides." They promise to emphasize strategy and logistics, rather than tactics, and, therefore, provide maps in place of "minutely extensive accounts of individual battles."

Since my specialty is the War in the East, I determined to concentrate on that part of the book first. If I found "new insights" applied here, then I would systematically read the rest of the text. I was immediately disappointed. Although discussing the various lines of operations adequately, the chapter on First Manassas was routine. True to their preface, "extensive accounts of individual battles" was avoided. However, this total lack of detail completely deemphasized the importance of the actual fighting, and eliminated any impact the study of turning movements and interior lines should have. When they did choose to add "details" they were out-of-keeping with the unemotional thread of the text: "Jackson, incidentally, waved his hand high in the air quite often — little wonder that it attracted a bullet!" (pg. 48). Their opinion that Jackson asked President Davis for 10,000 men to move on Washington City was "probably more in delirium than seriousness" (pg. 48) is highly speculative, and contrary to Dr. McGuire's own statements and Jackson's subsequent actions. It must also be noted that Jackson was a *brigadier-general* on July 21, 1861, not a colonel of a Virginia brigade, as stated on page 46, (Curiously, this same mistake appears in William C. Davis' first volume of *"The Image of War,"* pg. 199.)

Chapter 7, "Simultaneous Advances and the Southern Response" dealt, in part, with the Valley Campaign. Again, the entire series of maneuvers were traced out on overly simplistic maps, while the text treated only the basic strategies, adding no new information. The Seven Days Battles are likewise only briefly touched upon; presumably, the five maps were to speak louder than words.

After 204 pages, I still had not found what the Introduction promised. It is essential to study strategy, interior and exterior lines, transportation, communications and the organization of staffs, but the *impact* of all those factors on the war and the men who fought it was still unapparent. The battles and maneuvers still seemed like they were being played on a chessboard, the pieces being set up to demonstrate a particular tactic, then rearranged to show another. In the discussion of war, it is critical to convey not only the logistics of what actually happened, but the motives behind the action, and what it cost. Stressing "the influence of purely military factors" (itself an undesireable impossibility) is to make war a game played by lead soldiers. To truly understand it is to study mankind. Without the human element, the resulting facts and figures amount to nothing.

On page 223, I came across a description of the Battle of Cedar Mountain: "Jackson's men moved too slowly, however, largely because he failed to keep his division commanders informed of his plans and their subsequent modifications." Not only is this an oversimplification, its vagueness makes it an actual misrepresentation. To compound matters, it is footnoted: "Lee to Jackson, Aug. 7, 1862, O.R. XII, pt. 3, 925-26." To any serious student, this reference to the OR's need not be examined to know Lee never wrote any such thing to Jackson. The footnote actually belongs several lines above where it was placed. If it were a simple case of a typesetting error, it could be overlooked. Unfortunately, it is not. Throughout the work, whole paragraphs are left unfootnoted and then at the end of one, as in the case of the Cedar Mountain quote, a footnote number is placed. Usually, several works are then cited, which are presumed to refer to various parts of the preceeding paragraph, or paragraphs. This is a very serious flaw to *"How The North Won."* As a text book, its sole value is to teach. The way a person with a "mature interest" learns is by reading the work and judging it against his own knowledge, as well as checking the author's basis for his statements by cross-referencing with works cited in the footnotes. That is a primary reason footnotes are used. When they are used incorrectly, it is not only annoying to the student, it can also prove to be misleading. A footnote indicates a quote or an opinion originating from another source. By putting it after the authors' opinion, it *appears* to be substantiated by the Official Records. To a student not familiar with the campaign, or one who does not have access to the OR's, it would seem Lee was chastizing Jackson. In fact, it is nothing of the sort, but the damage has been done. Whether this was done to conserve space, or for some other reason, the choice was ill-advised.

At this point, I turned to the dust jacket. *"How The North Won,"* it suggested, reads like an epic narrative, as well as an analytical military history. Chapter 9, "The Maryland and Kentucky Raids Repulsed," begins with Gen. Halleck going to Washington: "He had done his 'best to avoid it,' because he had 'studied out' and could 'finish the campaign in the West' and felt he did not 'understand' and could not 'manage affairs in the East.' His friend Sherman agreed that he 'would prefer to finish' what he had 'so well begun' in the West." Not a particularly flowing narrative. It might have been clearer if the reader was given the entire quotation (taking short phrases out of context can often change their original meaning), and forgone the 'epic proportions' of seven quotes. This type of writing persists throughout the work.

In the Introduction, the authors state that they have offered new theories, and modified other standard interpretations. The promise is a bold one, but should have been more elucidative. If their specialty was Lincoln and Grant, and the primary focus of their new ideas based on them, it might have been wiser to have clearly delineated this in the beginning and omitted the logistics of the war that this did not pertain to. In this way, the reader would not have read one third of a book or more, to get into the heart of their philosophies.

Taking advantage of knowledge gained in Viet Nam and World War I (a potentially dangerous proposition of looking at history through hindsight), the book suggests that Lee's 1862 "invasion" of Maryland was, in reality, a "raid": since the ANV could not sustain itself in enemy territory, its occupation must be of short duration, and thus be catagorized a raid. The comparison of McClellan's line of supply covering his 1862 occupation, with the Confederates' movement north of the Potomac without supply lines is a good one. However, it seems more a question of semantics than military objectives, and deserves a more careful consideration than what was discussed in the text.

Another point stressed was that "Most Civil War officers seemed to understand the immense power of the defense," (pg. 47). This is in direct contrast to Grady McWhiney and Perry Jamieson's 1982 work *"Attack and Die.* In their preface (pg. xv), they state: "We contend that the Confederates bled themselves nearly to death in the first three years of the war by making costly attacks more often than did the Federals," and that many Southern leaders were unable to change their military tactics to adjust to the increased efficiency of the rifled musket. Also, the West Point training and the Mexican War/Indian War training many Civil War commanders received is a subject both books touched on. Both are of incalculable importance, and neither gave it the depth required.

"How The North Won" eventually develves on Grant and his strategies in the West, and with Sherman's campaigns. Lincoln's role as Commander-in-Chief, and the structure of his advisers, from Halleck to the Cabinet are analyzed and laid out. This is the strength of the book, and it clearly could have stood on its own. Involving the entire war makes the book muddled and detracts from the benefits obtained by the latter half of the work. It is a book that should not be undertaken by those unfamiliar with the war, but rather by those with a solid background which would enable them to judge it with deliberation and experience.

The book provides military logistics, but fails to convey a sense of reality: it gives the impression of judgement from hindsight, rather than from day-to-day living in the years 1861-65. Comments such as: "Ultimately the North's allegiance to its cause was adequete to the needs of achieving victory; that of the South was not" (pg. x) stand out as being judgements that do not fit well in a book purported to stress purely military factors. War is indeed not a science, but an art.

S.L. Kotar J.E. Gessler

• SIGNALS •

Dear Editors:

Thank you very much for the Feb/March issue of THE KEPI. It is a fantastic work of journalism and literature of the Civil War period. I especially liked the fact that you published a picture of Confederate general Simon Buckner that I have never seen. I enjoy seeing pictures of major figures of the Civil War age that are not published very often.

These past couple of years seem to be the age of new Civil War journals. Yours is an especially welcome magazine since it looks and reads like a contemporary paper of the Civil War. It amazes me how you people can research and publish all the articles that you do in the magazine. You have certainly grown since your first issue of last year.

I wish to receive your magazine, and I wonder could I pay for each issue before it comes out? Because I am going to college (as an historian), I cannot spend very much money at once. Thus, if I could pay for each issue individually, could you let me know what time of the month to send the money so as not to miss an issue? Also, do you have back issues? I wish to get the rest of your issues at all costs so that I can have a complete run of this wonderful magazine. If you do have back issues, please let me know the price so that I can get the money to you as fast as possible. Thank you again.

David Miles
Pine City, NY

[You may buy each issue as it comes out for $3.50. All back issues are available for the same price. For those of you who are interested, you can earn a free subscription by having five of your friends subscribe to THE KEPI for a year. Just have them mail their names, addresses and subscription fees together with your name, or the person you wish to receive a gift subscription.]

Dear Editors:

Just received my first two issues, and I think they are great and down-to-earth. The small little unknown things that happened in the war are just as interesting as the major events. I was so pleased that I have enclosed my check for the back issues that are available. I am in the process of writing a book about my two uncles who were in the war. Your magazine will help set the tone. After three-and-one-half years of research, it is now time to put the words on paper. Thanks again.

John Adams
Tewksbury, MA

Dear Editors:

Thank you for the current issue. We really enjoy it. Your publication is unique - researched, erudite, footnoted, and still very entertaining reading.

Janet Burgess
Davenport, IA

Dear Editors:

In the April/May '84 issue, in the regulars section called Family Bible, there was a mistake, confusing "California Joe" Moses Embree Milner (who was a scout for Custer), with Truman Head of Berdan's Sharpshooters.

Darrell Stridh
Whittier, CA

Paul Kallina replies:

Two of my sources, "The Dictionary of American Biography" and "Webster's American Military Biographies," identify "California Joe" and Moses Embree Milner as being the same man—a soldier in the 1st U. S. Sharpshooters (Berdan's), and after the Civil War, a scout with Custer. Dennis Mcloughlin, in his "Wild and Wooly, An Encyclopedia of the Old West," states that "such alternatives as Joseph Milmer and Joseph Hawkins cannot be disregarded as being just 'a load of bullrushes.'" However, neither the above books nor any of the several I consulted (I found "California Joe" to be a rather fascinating character), equates him with Truman Head.

MILITARY REVIEW

"*The Lincoln Image: Abraham Lincoln and the Popular Print*," by Harold Holzer, Gabor S. Boritt, and Mark E. Neely, Jr., Charles Scribner's Sons, 597 Fifth Avenue, New York, NY, 10017; 1984; price $35.00. 234 pages, Index, illustrations.

This book is the result of an exhibit of engravings and lithographs on Abraham Lincoln presented by Gettysburg College, PA, to commemorate the 175th anniversary of his birth. The object of the exhibit and the book is to present the image of America's sixteenth president as his contemporaries saw him, and to "explain how a primitive industry shaped its unappreciative subjects for a passionately political audience."

Before the Civil War, newspapers and magazines were unable to reproduce photographs on their pages. To represent likenesses of political or military figures, they had to use a process of engraving or lithography. This art form was also used to transfer pictures of American political leaders to mass-produced prints, and was, in effect, the primary means of 'getting one's face around.' When Lincoln became the Republican presidential nominee in 1860, most of the country did not know what he looked like. To quickly dispurse his likeness to the voting public, commercial print manufacturers sent artists to Springfield to paint him. These paintings were engraved on plates by another artist and run off by the thousands. They were then either distrubuted by political associates or individually sold by the publisher as a commercial venture.

Prints had several advantages over the newly developing field of photography: they could be mass produced quickly and inexpensively, they could depict entire scenes, whether real or imaginary (Lincoln reading the Emancipation Proclamation to his Cabinet; Lincoln being welcomed into heaven by George Washington) and they could be printed in color.

The emphasis of "*The Lincoln Image*" is to follow the beginnings of his public prominence in 1860, through his death and subsequent 'sainthood' by examining the many prints for him made over this period. In doing so, the authors also follow the developing print industry, and provide an excellent commentary on the different artists and publishers involved in the business. The text expertly takes the reader through Lincoln's involvement with the image-making process, in both the political and visual sense. They detail the men who came to paint him, and thoroughly illustrate the steps that the artist took from life sittings with Lincoln, to the photographs they copied, to the finished canvas, the subsequent prints and finally the pirated versions put out by rival companies.

The authors present numerous prints of Lincoln, as he emerged from a presidential nominee to president-elect, and through the war. His career unfolds and the modern reader is enabled to view his transformations, as a citizen of the 19th Century would. Through the prints and political cartoons we are introduced to a beardless Lincoln, a handsome one (features smoothed over), a Lincoln head etched over less saleable politicians' bodies; poses suddenly grow whiskers (vividly misrepresented by the engravers)and studies of various oil paintings made during his White House years, including ficticious family scenes. What emerges is a fascinating study of the man and his times.

"*The Lincoln Image*" is a living study of the mid-19th Century. It examines America's passion for politics and traces the use of visual images in shaping Lincoln's destiny. What is lacking from this book is any definition of lithography or the engraving process. It would have been very useful to have had a short description of both processes at the beginning of the text. There are also no color illustrations, although the oil paintings and many of the prints were hand-colored. One of the advantages of a print was its color, and not to have a single one reproduced as such is a loss for the researcher. The greatest flaw in this book is one of layout, however. The text is profusely illustrated, but too frequently the picture is on one page and the caption on the other. While reading the book and referring back and forth through many pages, it is difficult to locate the proper illustration. This problem could have been corrected by identifying the picture by a simple "Figure—" directly underneath it. As a final note, the page numbers are put close to the binding instead of on the outer edges of the page, and this is difficult to get used to.

Messers. Holzer, Boritt and Neely have done an excellent job with this book, and its use as a study of Abraham Lincoln, the political process and mid-19th Century America cannot be overstated.

S.L. Kotar

Grant and Lee, The Virginia Campaign, 1864-1865, by William A. Frassanito, Charles Scribner's Sons, New York, 1983. List price $24.95. 442 pages, Index, profuse illustrations.

Without a doubt, this book is one of the most important research tools to be published on the 1864-65 Virginia campaigns. In Chapter One, author Frassanito delineates the primary purpose of the text: to provide scholars with the first detailed, comprehensive examination of the hundreds of photographs taken during this period. Using primary sources, including original glass negative plates with negative numbers scratched on them, captions placed on the scenes by cameramen, original issue views and photographic catalogs, the attempt was made to accurately date, identify and describe the war images. What emerges from this investigation is a fascinating piece of detective work that is so skillfully written, it is impossible to put down.

The bulk of photographic work was done by the three competitive firms of Mathew Brady, Alexander Gardner and E. & H.T. Anthony Brothers. These men and their agents (most notably Timothy H. O'Sullivan), and Andrew J. Russell, official army photographer, were an elite set of individuals who viewed the campaigning as an opportunity to make money, while documenting the historically significant sights of war. Following the army, or at times hurrying to battle sites after the fact, these men captured the horrors and details no pen could describe. *Grant and Lee* traces these photographers' activities, painstakingly connecting their whereabouts with the images they created. In this fashion, as well as extensive research into the OR's, diaries and a collaboration of techniques (analyzing shadows, terrain and personalities), the author has been able to correctly date (frequently at variance with original issue dates) and group together series of photos. He has provided the reader with important new interpretations, as in the case of terrain images taken prior to the Wilderness. Frassanito concludes that, contrary to current belief, "the foliage was by no means in full bloom, and hence visibility was probably greater than has been assumed in the past." He has also challenged traditional caption descriptions of various buildings and scenes, and by comparing negative numbers, placed them in their proper locale.

The book explains why certain objects, such as "The Dictator," were extensively photographed, leading scholars to erroneously place more emphasis on them than is historically accurate; why gaps involving long periods of time occur between pictures, and why certain subjects were unfortunately beyond the camera's reach.

Capturing an image in 1864 required ten seconds of inactivity, thus preventing actual battles from being photographed. Photographers often traveled behind the army, not with it: they arrived after the danger was over, to capture the aftermath. Dead bodies (invariably Confederate), fortifications, pontoon bridges and personality studies were the primary targets of these men.

In addition to the unparalleled work on photo identification, an interesting portion of the text concerned Mathew Brady, the renowned photographer. At the time of the war, Brady was a middle-aged businessman who did little, if any, actual camerawork; all the "Brady photos" were the products of his operatives. This dispels the common misconception that Brady was the photographer, and finally gives credit to the artists (although they remain anonymous) who performed the creative, and sometimes dangerous, work.

(Continued on page 23.)

FAMILY BIBLE

Birthdays of Civil War Notables

by

Paul Kallina

(N = North. S = South.)

JUNE

June 1, 1806—Gen. John B. Floyd, Smithfield, Montgomery County, Virginia. (S)
June 1, 1814—Gen. Philip Kearny, New York City. (N)
June 1, 1825—Gen. John Hunt Morgan, Huntsville, Alabama. (S)
June 1, 1829—Gen. Cullen A. Battle, Powelton, Georgia. (S)
June 1, 1831—Gen. John B. Hood, Owingsville, Kentucky. (S)
June 1, 1833—Col. John M. Harlan, Boyle County, Kentucky. (N)
June 1, 1843—Col. Charles R. Ellet, Philadelphia, Pennsylvania. (N)
June 1, 1844—Gen. Galusha Pennypacker, near Valley Forge, Pennsylvania. (N)
June 2, 1831—Gen. Ben Hardin Helm, Bardstown, Kentucky. (S)
June 2, 1845—Col. Arthur MacArthur, Chicopee Falls, Massachusetts. (N)
June 3, 1808—President Jefferson Davis, Christian County, Kentucky. (S)
June 4, 1803—Gen. Gabriel J. Rains, Craven County, North Carolina. (S)
June 4, 1825—Rep. John DeWitt Clinton Atkins, Henry County, Tennessee. (S)
June 5, 1825—Col. Jabez L. M. Curry, near Double Branches, Lincoln County, Georgia. (S)
June 5, 1827—Gen. Beverly H. Robertson, The Oaks, Amelia County, Virginia. (S)
June 7, 1824—Gen. Alfred Pleasonton, Washington, D.C. (N)
June 8, 1806—Gen. Gideon J. Pillow, Williamson County, Tennessee. (S)
June 8, 1813—Adm. David D. Porter, Chester, Pennsylvania. (N)
June 8, 1824—Gen. William M. Gardner, Augusta, Georgia. (S)
June 9, 1785—Gen. Sylvanus Thayer, "the father of West Point," Braintree, Massachusetts. (N)
June 9, 1812—Rep. Thomas B. Hanly, Nicholsville, Kentucky. (S)
June 9, 1833—Gen. John R. Cooke, Jefferson Barracks, Missouri. (S)
June 10, 1816—Rep. William H. Tibbs, Appomattox, Virginia. (S)
June 10, 1833—Pauline Cushman, Union spy, New Orleans, Louisiana. (N)
June 11, 1816—Gen. Robert H. Milroy, Washington County, Indiana. (N)
June 11, 1822—Gen. Samuel D. Sturgis, Shippensburg, Pennsylvania. (N)
June 11, 1823—Gen. James L. Kemper, Madison County, Virginia. (S)
June 12, 1798—Gen. Samuel Cooper, Hackensack, New Jersey. (S)
June 12, 1799—Col. Josiah Harlan, Newlin Township, Pennsylvania. (N)
June 12, 1810—David L. Yulee, railroad promoter, St. Thomas, West Indies. (S)
June 13, 1786—Gen. Winfield Scott, Laurel Branch, near Petersburg, Virginia. (N)
June 13, 1809—Gen. Philip St. George Cooke, Leesburg, Virginia. (N)
June 13, 1821—Gustavus V. Fox, Assistant Secretary of the Navy (1861-1866), Saugus, Massachusetts. (N)

June 13, 1838—Col. Walter H. Taylor, *Four Years With General Lee*, Norfolk, Virginia. (S)
July 13, 1824—Lt. James I. Waddell, C.S.N., Pittsboro, North Carolina.
June 14, 1805—Gen. Robert Anderson, Soldier's Retreat, near Louisville, Kentucky. (N)
June 14, 1811—Harriet Beecher Stowe, *Uncle Tom's Cabin*, Litchfield, Connecticut. (N)
June 14, 1812—Fernando Wood, mayor of New York and congressman, Philadelphia, Pennsylvania. (N)
June 15, 1826—Maj. Charles H. "Bill Arp" Smith, humorist, Lawrenceville, Gwinnett County, Georgia. (S)
June 15, 1833—Gen. Edward M. McCook, Steubenville, Ohio. (N)
June 16, 1834—Gen. Wesley Merritt, New York City. (N)
June 17, 1830—Gen. Richard M. Gano, Bourbon County, Kentucky. (S)
June 18, 1840—Benjamin F. Stringfellow, scout and secret agent, Culpeper County, Virginia. (S)
June 19, 1844—Midshipman Bowman H. McCalla, Camden, New Jersey. (N)
June 20, 1823—Gen. Jesse L. Reno, Wheeling, Virginia. (N)
June 21, 1817—Gen. James B. Ricketts, New York City. (N)
July 21, 1831—Mary Anna Jackson, *nee* Morrison, near Charlotte, North Carolina. (S)
June 22, 1814—Sen. and Gen. James H. Lane, Lawerenceburg, Indiana. (N)
June 24, 1811—John A. Campbell, Assistant Secretary of War (1862-1865), Washington, Wilkes County, Georgia. (S)
June 24, 1813—Henry Ward Beecher, abolitionist, Litchfield, Connecticut. (N)
June 24, 1842—Maj. Ambrose Bierce, Horse Cave Creek, Meigs County, Ohio. (N)
June 25, 1823—James D. Bulloch, naval agent, near Savannah, Georgia. (S)
June 26, 1819—Gen. Abner Doubleday, Ballston Spa, New York. (N)
June 26, 1838—Lt. Col. John W. Barlow, Perry, New York. (N)
June 27, 1836—Maj. Robert Stiles, *Four Years Under Marse Robert*, Woodford County, Kentucky. (S)
June 28, 1823—Gen. William T. Wofford, Habersham County, Georgia. (S)
June 30, 1832—Rep. Samuel St. George Rogers, Pulaski, Tennessee. (S)

JULY

July 1, 1802—Gideon Welles, Secretary of the Navy (1861-1869), Glastonbury, Connecticut. (N)
July 1, 1810—James S. Gibbons, abolitionist, Wilmington, Delaware. (N)
July 1, 1818—Gen. Josiah Gorgas, Running Pumps, Dauphin County, Pennsylvania. (S)
July 1, 1825—Gen. John Adams, Nashville, Tennessee. (S)
July 1, 1828—Col. Thomas H. Herndon, Erie, Alabama. (S)
July 1, 1833—Gen. Alfred T. A. Torbert, Georgetown, Delaware. (N)
July 2, 1810—Gen. Robert Toombs, Secretary of State (1861), Wilkes County, Georgia. (S)
July 2, 1830—Gen. John B. Villepigue, Camden, South Carolina. (S)
July 2, 1836—Gen. Henry E. Davies, New York City. (N)
July 2, 1842—Capt. Charles Chaille-Long, Princess Anne, Somerset County, Maryland. (N)
July 3, 1828—Gen. John A. Wharton, near Nashville, Tennessee. (S)
July 4, 1804—Nathaniel Hawthorne, writer, Salem, Massachusetts. (N)
July 4, 1807—Giuseppe Garibaldi, Italian patriot and soldier, Nice, France.
July 4, 1812—Col. Thomas Johnson, Montgomery County, Kentucky. (S)

July 4, 1822—Rep. William P. Miles, Walterboro, Colleton District, South Carolina. (S)
July 4, 1826—Stephen C. Foster, songwriter, Pittsburgh, Pennsylvania. (N)
July 4, 1828—Gen. James J. Pettigrew, Bonarva, Tyrrell County, North Carolina. (S)
July 4, 1830—Maj. William B. Wright, Columbus, Georgia. (S)
July 5, 1801—Adm. David D. Farragut, Campbell's Station, near Knoxville, Tennessee. (N)
July 6, 1832—Archduke Maximilian, Emperor of Mexico (1864-1867), Schoenbrun Castle, near Vienna, Austria.
July 8, 1819—Gen. Alexander Hays, Franklin, Pennsylvania. (N)
July 8, 1826—Gen. Benjamin H. Grierson, Pittsburgh, Pennsylvania. (N)
July 8, 1839—Sgt. John H. Worsham, *One of Jackson's Foot Cavalry*, Richmond, Virginia. (S)
July 9, 1819—Pvt. Elias Howe, Spencer, Massachusetts. (N)
July 11, 1841—Gen. William P. Roberts, Gates County, North Carolina. (S)
July 12, 1807—Gen. Silas Casey, East Greenwich, Rhode Island. (N)
July 12, 1817—Henry David Thoreau, Concord, Massachusetts. (N)
July 12, 1821—Gen. Daniel H. Hill, York District, South Carolina. (S)
July 12, 1822— Cdr. Robert H. Wyman, Portsmouth, New Hampshire. (N)
July 13, 1815—James A. Seddon, Secretary of War (1862-1865), Fredricksburg, Virginia. (S)
July 13, 1821—Gen. Nathan Bedford Forrest, near Chapel Hill, Bedford County, Tennessee. (S)
July 13, 1824—Lt. James I. Waddell, C.S.N.,Pittsboro, North Carolina. (S)
July 13, 1839—William Travis, soldier and artist, Staunton, Virginia. (N)
July 14, 1818—Gen. Nathaniel Lyon, Ashford, Connecticut. (N)
July 15, 1838—Col. Loyd Wheaton, Pennfield, Michigan. (N)
July 16, 1806—Col. Jacob Zeilin, U.S.M.C., Philadelphia, Pennsylvania. (N)
July 16, 1829—Gen. Robert B. Potter, Schenectady, New York. (N)
July 18, 1843—Virgil Earp, 83rd Illinois Inf., Hartford, Kentucky. (N)
July 19, 1813—Sen. Williamson S. Oldham, Franklin County, Tennessee. (S)
July 19, 1814—Samuel Colt, arms manufacturer, Hartford, Connecticut. (N)
July 19, 1817—Mary Ann Ball "Mother" Bickerdyke, hospital worker, Knox County, Ohio. (N)
July 19, 1828—Gen. Roger A. Pryor, near Petersburg, Virginia. (S)
July 20, 1822—Col. John R. Baylor, Paris, Kentucky. (S)
July 21, 1802—Gen. David Hunter, Washington, D.C. (N)
July 21, 1817—Gen. Joseph K. Barnes, Surgeon-General of the Army (1863-1882), Philadelphia, Pennsylvania. (N)
July 21, 1821—Cdr. Jonathan M. Wainwright, New York City. (N)
July 21, 1826—Gen. James G. Blunt, Trenton, Maine. (N)
July 22, 1822—Gen. John G. Walker, Cale County, Missouri. (S)
July 23, 1822—Gen. Darius N. Couch, Southeast, Putnam County, New York. (N)
July 24, 1798—Gen. John A. Dix, Boscawen, New Hampshire. (N)
July 25, 1806—Maria W. Chapman, abolitionist, Weymouth, Massachusetts. (N)
July 25, 1821—Rep. Muscoe R.H. Garnett, Elmwood, Essex County, Virginia. (S)
July 27, 1812—Gen. Thomas L. Clingman, Huntsville, North Carolina. (S)
July 27, 1840—Gen. Ranald S. Mackenzie, New York City. (N)
July 27, 1849—Pvt. Luther S. Kelly, Geneva, New York. (N)

July 28, 1809—Gen. Ormsby M. Mitchel, Morganfield, Kentucky. (N)
July 29, 1794—Thomas Corwin, minister to Mexico (1861-1864), Bourbon County, Kentucky. (N)
July 29, 1820—Rep. Clement L. Vallandigham, New Lisbon, Ohio. (N)
July 30, 1815—Gen. Thomas J. Rodman, inventor, near Salem, Indiana. (N)
July 30, 1837—Gen. Elon J. Farnsworth, Green Oak, Livingston County, Michigan. (N)
July 30, 1838—Col. Henry A. DuPont, near Wilmington, Delaware. (N)
July 31, 1800—Judge Robert L. Caruthers, governor-elect of Tennessee (1863), Smith County, Tennessee. (S)
July 31, 1803—John Ericsson, inventor, Langbanshyttan, Varmland Province, Sweden. (N)
July 31, 1816—Gen. George H. Thomas, Southampton County, Virginia. (N)
July 31, 1817—Gen. Philip Cook, Twiggs County, Georgia. (S)
July 31, 1837—Capt. William C. Quantrill, Canal Dover, Ohio. (S)

BIBLIOGRAPHY

Biographical Directory of the United States Executive Branch, 1774-1971. Ed. Robert Sobel, Westport, Connecticut: Greenwood Publishing Company, 1971.
Dictionary of American Biography. 20 vols. New York: Charles Scribner's Sons, 1928-1936.
The National Cyclopaedia of American Biography. 57 vols. New York and Clifton, New Jersey: James T. White & Company, 1892-1977.
Wakelyn, John L. *Biographical Dictionary of the Confederacy.* Westport, Connecticut: Greenwood Press, 1977.
Webster's American Military Biographies. Ed. Robert McHenry. Springfield, Massachusetts: G. & C. Merriam Company, 1978.

PENNSYLVANIA CIVIL WAR FLAG PROGRAM

The Capitol Preservation Committee's "Save-The-Flag" program is now well underway. The Committee plans to do some badly-needed conservation work on the collection of approximately 360 Civil War flags and 23 Spanish-American War colors now housed in six glass display cases in the Capitol Rotunda. To help meet the $1,000 per-flag cost of the treatment, private contributions are being sought. For further information, please contact Capitol Preservation Committee, Honorable Joseph R. Pitts, Chairman, Room 144, Main Capitol, Harrisburg, PA 17120 (717) 783-6000.

As part of this project, the Committee plans to publish a book that will include a brief history of each Pennsylvania regiment and illustrations of its flags. In order to facilitate research, the military historian who is researching the history of the flags is seeking access to historical materials written by or about Pennsylvania soldiers in the Civil War. Such materials include letters, diaries, post-war reminiscences, newspaper clippings, and photographs. Also, he is interested in locating any flag-related memorabilia that could be photographed for possible inclusion in the book. For further information, contact: Capitol Preservation Committee, House of Representatives, Mr. Richard A. Sauers, Military Historian, House P.O. Box 231, Harrisburg, Pennsylvania 17120 (717) 783-6484.

(Continued from page 20.)

Grant and Lee is a classic text that is a must for all researchers of the Civil War, or for those interested in the mid-19th Century. William A. Frassanito has done an incredible job, and his efforts cannot be praised enough. His inclusion of modern day photos to compare with those of the period are an excellent addition to the book, and his integrity to his subject will set standards to be emulated for years to come.

S.L. Kotar

Articles of War

An Act For Establishing Rules and Articles For The Government Of The Armies Of The United States.

ART. 93. The judge advocate or recorder shall administer to the members the following oath:

"You shall well and truly examine and inquire, according to your evidence, into the matter now before you, without partiality, favor, affection, prejudice, or hope of reward So help you God."

After which the president shall administer to the judge advocate or recorder the following oath:

"You, A.B., do swear that you will, according to your best abilities, accurately and impartially record the proceedings of the court, and the evidence to be given in the case in hearing. So help you God."

The witnesses shall take the same oath as witnesses sworn before a court-martial.

ART. 94. When any commissioned officer shall die or be killed in the service of the United States, the major of the regiment, or the officer doing the major's duty in his absence, or in any post or garrison, the second officer in command, or the assistant military agent, shall immediately secure all his effects or equipage, then in camp or quarters, and shall make an inventory thereof, and forthwith transmit the same to the office of the Department of War, to the end that his executors or administrators may receive the same.

ART. 95. When any non-commissioned officer or soldier shall die, or be killed in the service of the United States, the then commanding officer of the troop or company shall, in the presence of two other commissioned officers, take an account of what effects he died possessed of, above his arms and accoutrements, and transmit the same to the office of the Department of War, which said effects are to be accounted for, and paid to the representatives of such deceased non-commissioned officer or soldier. And in case any of the officers, so authorized to take care of the effects of deceased officers and soldiers, should, before they have accounted to their representatives for the same, have occasion to leave the regiment or post, by preferment or otherwise, they shall, before they be permitted to quit the same, deposit in the hands of the commanding officer, or of the assistant military agent, all the effects of such deceased non-commissioned officers and soldiers, in order that the same may be secured for, and paid to, their respective representatives.

ART. 96. All officers, conductors, gunners, matrosses, drivers, or others persons whatsoever, receiving pay or hire in the service of the artillery, or corps of engineers of the United States, shall be governed by the aforesaid Rules and Articles, and shall be subject to be tried by courts-martial, in like manner with the officers and soldiers of the other troops in the service of the United States.

ART. 97. The officers and soldiers of any troops, whether militia or others, being mustered and in pay of the United States, shall, at all times and in all places, when joined, or acting in conjunction with the regular forces of the United States, be governed by these rules and articles of war, and shall be subject to be tried by courts-martial, in like manner with the officers and soldiers in the regular forces; save only that such courts-martial shall be composed entirely of militia officers.

ART. 98. All officers serving by commission from the authority of any particular State, shall, on all detachments, courts-martial, or other duty, wherein they may be employed in conjunction with the regular forces of the United States, take rank next after all officers of the like grade in said regular forces, notwithstanding the commissions of such militia or State officers may be elder than the commissions of the officers of the regular forces of the United States.

ART. 99. All crimes not capital, and all disorders and neglects which officers and soldiers may be guilty of, to the prejudice of good order and military discipline, though not mentioned in the foregoing articles of war, are to be taken cognizance of by a general or regimental court-martial, according to the nature and degree of the offense, and be punished at their discretion.

ART. 100. The President of the United States shall have power to prescribe the uniform of the army.

ART. 101. The foregoing articles are to be read and published, once in every six months, to every garrison, regiment, troop, or company, mustered, or to be mustered, in the service of the United States, and are to be duly observed and obeyed by all officers and soldiers who are, or shall be, in said service.

SEC. 2. *And be it further enacted,* That in time of war, all persons not citizens of, or owing allegiance to, the United States of America, who shall be found lurking as spies in or about the fortifications or encampments of the armies of the United States, or any of them, shall suffer death, according to the law and usage of nations, by sentence of a general court-martial.

SEC. 3. *And be it further enacted,* That the rules and regulations by which the armies of the United States have heretofore been governed, and the resolves of Congress thereunto annexed, and respecting the same, shall henceforth be void

(Continued on page 31.)

"SHOES"

By
Charles A. Lowe

Arnold Shoe factory, looking northwest.

Hundreds of thousands of shoes poured out of the 82 shoe factories of Little Abington, Massachusetts, for Union troops during the Civil War. Originally established in 1668 as the village of Manamooskegan (Land of Many Beavers), it became incorporated as the Town of Abington on June 10, 1712. In 1861, it encompassed 24 square miles, and had a population of 8,527. (Between 1874 and 1886, East Abington became Rockland, and South Abington became Whitham. All have a census of about 13,000 and are still rural in flavor.)

Napoleon said, "An army marches on its stomach." It is also true that an army needs to be well-shod to get anywhere and do anything. This was especially true in the Civil War, when many a Confederate soldier with worn-out shoes, or no shoes at all, would kill a Yankee to get his shoes, hoping they'd fit. Long, arduous marches were the norm in both nations, rather than the exception, and there was little tolerance (although much sympathy) for the bleeding, rag-wrapped feet of the infantryman.

Seth Bryant of East Bridgewater, Massachusetts (next to Abington), was a man who knew a good thing when he saw it. Armed with an invention patented in 1860 by Lyman Blake of Abington, he journeyed to Washington, D.C. He sought and received an audience with Secretary of War Edwin M. Stanton. Blake had perfected a machine that could sew shoe uppers to soles, making mass production of shoes easier, doing away with hand-stitching (expensive and time-consuming) and the system of using wooden pegs and tacks to hold shoes together. It also alleviated the problem of the tacks and pegs working through the soles, which caused blisters and discomforts.

Bryant convinced Stanton he could deliver shoes that were better and cheaper, and produce them quickly, which was essential. He was awarded the contract.

The inventor of the machine, Lyman Blake, was born in 1835 on East Market Street, Abington, Massachusetts. At the suggestion of Edmund Shaw of East Abington (who was the Singer Sewing Machine agent for the area), Blake worked on his own time at the small shoe factory of Gurney and Mears, 21 Vernon Street, until he perfected and patented the badly needed device. Lacking funds, he sold the rights of the machine in 1861 to Col. Gordon McKay, of Lawrence, Massachusetts, for $6,000 and promises of future monetary returns. Blake's invention became the McKay Stitcher, eventually the backbone of the United Shoe Machinery Company of Beverly, Massachusetts, which was the giant of the shoe industry, world-wide. Before his death, Blake realized $175,000.

Many owners in Abington, unable to pay for the McKay Stitchers, took them on a "royalty" basis: as each pair of shoes was completed, part of the sales price went to McKay, Bryant and Blake. Thus, leasing was born.

Another advent of the Industrial Revolution happened at the same time. The techniques of harnessing steam power, fueled by wood-burning furnaces, to the machines, gave the North a tremendous advantage in producing shoes, and all the other necessities for waging war.

Bryant, realizing he had assumed a large responsibility when he accepted the War Department contract, sublet over half of his contract to shoe shops in East Bridgewater and Abington, and was able, by this method, to fulfill his obligation to Stanton. He eventually supplied over 300,000 pair of shoes to the Union troops, and another 600,000 were supplied by his subcontractors. Ironically, Bryant never became wealthy, although others did. He was an unsung hero.

Blanche Evans Sprague, in her dissertation, "The Organization of the Boot and Shoe Industry in Massachusetts Before 1875," said: "It was interesting as it was natural that Abington, the home of Blake, the inventor of the McKay machine, should have been the boot and shoe center which made on McKay machines more than half of the army shoes provided Northern soldiers during the Civil War." The late Elroy Thompson, in his "History of Plymouth, Norfolk and Barnstable Counties of 1928," wrote: "more than half of the army shoes provided for Union troops were made in Abington."

For the Union, it was fortuitous that Blake had the ingenuity to invent the shoe-stitcher, steam power was har-

Gurney and Mears Shoe factory, 21 Vernon Street, Abington (now Whitman), where Lyman Blake worked on and perfected his machine to stitch uppers to soles for shoes.

nessed to it, and Col. McKay and Bryant had the know-how and financing to insure its success.

There are no shoe factories in Abington, now. Brockton (bordering Abington), in later years, became the Shoe Capital, and also the home of Rocky Marciano and Marvelous Marvin Hagler, world champion boxers. Brockton still has shoe factories in operation, but from 1861-1865, Abington was Shoe City.

BIBLIOGRAPHY

Town Clerk, Abington, Massachusetts.
Massachusetts Department of Labor.
Mrs. Jean Lothrop, Head Librarian, Abington, Massachusetts.
Rockland Memorial Library, Research Reporter Files, Mrs. Martha Campbell.
Elroy Thompson, History of Plymouth, Norfolk and Barnstable Counties, 1928.
John W. Barber, Historical Collections, 1837.
Mrs. Blanche Evans, Sprague, The Organization of the Boot and Shoe Industry in Massachusetts, 1921, Harvard University Press.
Captain James Bates, Plymouth County Commissioner and Surveyor, 1830.

(Continued from page 6.)

[3] *The Official Military Atlas of the Civil War* (Washington, D.C.: U.S. Government Printing Office, 1891), Plate CLXXV.
[4] Freeman, *Lee's Lieutenants*, Volume I, p. 72.
[5] *Ibid*, p. 72.
[6] *Atlas*, Plate CLXXV.
[7] Williams, *Napoleon in Gray*, p. 109.
[8] *War of the Rebellion: A Compilation of the Official Records of the Union and Confederate Armies* (Washington, D.C.: U.S. Government Printing Office, 1880-1901, Volume LI, Part 2, p. 272.
[9] Williams, *Napoleon in Gray*, p. 109.
[10] *O.R.*, Volume LI, Part 2, p. 272.
[11] Williams, *Napoleon in Gray*, p. 109.
[12] Campbell Brown, *Military Reminiscences* Nashville, Tennessee: unpublished, Tennessee State Library and Archives), p. 16.
[13] Williams, *Napoleon in Gray*, p. 110.
[14] *Ibid*, p. 110.
[15] *Ibid*, p. 110.
[16] *Ibid*, p. 110.
[17] *O.R.*, Series IV, Volume II, p. 534.
[18] *Ibid*, p. 534.
[19] Glenn Tucker, *High Tide at Gettysburg* (Indianapolis, Indiana: Bobbs-Merrill Company, 1958), p. 63.
[20] Archie P. McDonald, *Make Me a Map of the Valley* (Dallas, Texas: Southern Methodist University Press, 1973), p. 155.
[21] Tucker, *High Tide at Gettysburg*, p. 63.
[22] McDonald, *Make Me a Map of the Valley*, p. 155.
[23] Tucker, *High Tide at Gettysburg*, p. 63.
[24] Campbell Brown, *My Confederate Experience* (Nashville, Tennessee: unpublished, Tennessee State Library and Archives), p. 47.
[25] *Ibid*, p. 47.
[26] *Ibid*, p. 47.
[27] *Ibid*, p. 47.
[28] Tucker, *High Tide at Gettysburg*, p. 63.
[29] Robert G. Tanner, *Stonewall in the Valley* (Garden City, New York: Doubleday and Company, 1976), p. 182.
[30] Brown, *My Confederate Experience*, p. 47.
[31] *Ibid*, p. 47.
[32] Brown, *My Confederate Experience*, p. 47.
[33] *Ibid*, p. 47.
[34] O.R., Volume XXVII, Part 3, p. 923.
[35] Brown, *My Confederate Experience*, p. 47.
[36] *Ibid*, p. 47.
[37] *Ibid*, p. 47.
[38] *Ibid*, p. 47.
[39] *Ibid*, p. 47.
[40] *Ibid*, p. 47.
[41] *Ibid*, p. 47.
[42] Wilber Sturtevant Nye, *Here Come the Rebels* (Baton Rouge, Louisiana: Louisiana State University Press, 1965), p. 310.
[43] Spokesman, Confederate Museum.
[44] *Atlas*, Plate CLXXV.
[45] Spokesman, Confederate Museum.

"Printing: The State-of-the-Art"

By Fritz Steinner

The initial Q, used as a chapter heading, designed in Rome in 1475.

By the dawn of the Civil War, the art of printing and the printing press were routine to most Americans. Everything, from daily newspapers to literary magazines, antislavery standards, religious tracts, advertisements and every conceivable type of book made their way into homes and libraries. As the country grew and people migrated west, schools sprang up to teach reading and writing, creating a literate and expanding population. While it was illegal in some parts of the country to teach slaves to read, their incessant desire to become educated drove them to find people who would teach them. All this increased the demand for the printed word; to supply this need, inexpensive publications flooded the market.

The technique of printing had come a long way from its roots in the 15th Century. Prior to the invention of the printing press, all books and manuscripts were copied by hand. This laborious process was time-consuming and expensive: often it would take a solitary monk half his life to copy one religious work. The transcription was done by pen and ink onto parchment (the skin of a sheep or goat). This writing surface had been developed in the second century B.C., after the ruler of Egypt forbade the exportation of papyrus in an attempt to stop the production of books for his rival's library. This early parchment was called Carta Pergamena or Pergamenta.

Woodcutting was the intermediate step between the handcopied manuscript and printing. By the 14th Century, it was practiced in Europe, primarily in the reproduction of illustration, rather than words or sentences.

Between 1450-60, John Faust, John Guttemberg and Peter Schoeffer established an office for the production of books in Mentz, Germany. Their new process consisted of arranging movable metallic letters in the form of words, sorting these into sentences, putting ink over them, then placing paper on the letters and pressing on it. In this manner the famous Guttemberg Bible was printed. This Latin version consisted of 637 leaves of vellum, designed to imitate manuscript. In the next twenty years, this process produced more Bibles than had been made in all the centuries before.

It was said the Guttemberg process was based in part on the technique of Laurentius Coster, who, before his death in 1440, had used a series of wooden blocks strung together with string to print small tracts. Guttemberg, however, directed his attention to the development of metallic type by casting from molds. His experimenting took eleven years, after which time he

Guttemberg's first proof of his famous printing of The Bible.

Portrait of Laurentius Coster.

ran out of money. Entering into partnership with the wealthy Faust, they perfected the type and revolutionized the world. The partnership dissolved in 1455; Faust and his son-in-law, Schoeffer, declared themselves to be the inventors of metallic type, as did Guttemberg's assistant, Mentelius. Guttemberg himself died in 1467, in obscurity. The art of printing was brought to England by William Caxton, in 1474, and he subsequently printed the first English book, "The Game of Chess," that same year.

The new process was far from the science it had become by 1860. Books in the early years were issued on ponderous folios, published at immense cost, hand-colored and gilded. It wasn't until the early part of the 16th Century when Hans Holbein and others substituted initial letters that the massive work performed by hand went out of style. These large decorative initial letters, as well as the illustration, were cut from wood and used to beautify the work.

The manuscripts and early books did not contain any paragraphs to break up the heavy columns of print. In some books, the use of red or blue paint on the first letter of a sentence gave variety to the page and broke up the monotony of solid text. The comma was not in use; printers employed only the full point and the double point, or colon. The character / was used to separate phrases, and this eventually became the comma. Capital letters at the beginning of sentences did not come into use until much later, and early book readers had to bear with the frequent loss of a letter (where a character had fallen out of the press), imperfect type faces and frequent abbreviations, such as *xpus* for Christ.

Caxton's portrait.

Caxton's device.

The binding of early books was often massive and costly. Frequently, heavy oaken boards were covered with hog skin or leather, and decorated with inlaid gold. At the time of the mid-nineteenth Century, it was already profitable to collect these old books, and their sale at auction brought the liveliest bids from speculators, as well as book lovers.

In 1860, the state-of-the-art had reached the point where Guttemberg's wooden press had been replaced by steamdriven iron ones, which employed electrotyped matter, with each page a solid piece of copper, presenting every letter, illustration, line and dot on its surface, requiring approximately 600 persons to transform the author's copy into a printed page. It was a system that mass-produced news and information, and kept Americans on both sides of the Mason-Dixon line up-to-date on world events and civil war.

Solution to cryptogram from page 15.

KEPI IS A FRENCH WORD FOR A MILITARY CAP. DURING THE CIVIL WAR, THIS TYPE OF HEADGEAR WAS WORN BY BOTH NORTH AND SOUTH, THUS BECOMING A BI-PARTISAN SYMBOL OF THE WAR.

Tracts

"All Equal in the Eyes of God"

by

Rev. Allen Kelley

While religion was a dominating aspect of 19th Century life, it wasn't all serious. Reverends, deacons, parishoners, church rites and even biblical passages were a source of humor to the American of the 1800's. Yankee, Southerner, Westerner and immigrant, all found release from day-to-day living by having an occasional (albeit a respectful) laugh at the religious quarter. The following 1855 offerings are from *Harper's New Monthly Magazine.*

A new reading of the marriage service is mentioned by Silk Buckingham, in his forthcoming autobiography. He says that some students gained access to the Clarendon printing-offices in Oxford, and roguishly substituted the letter *k* for the letter *v* in the word live; and thus the vow "To love, honor, comfort, etc., so long as ye both shall live," was made to read "so long as ye both shall *like.*" The change was not discovered until the whole of the sheets were printed off. Not a few of our modern reformers would be willing to have the alteration stand as the text.

In Georgia, a preacher of the violent order - that is, one who thinks to take the kingdom of heaven by violence - was praying with great vehemence of lungs - fairly shouting in the ear of Infinite Mercy - when a little girl in the house drew her mother's ear close to her lips, and said, 'Ma', don't you think, if he lived near to God, he could make him hear without speaking so loud?"

It was near New Haven, Connecticut, the City of Elms, as our correspondent writes, that the Rev. Mr. Smitkins was describing the peaceful departure of an aged saint, on whose last hours it had been his recent privilege to attend. Mr. Smitkins was one of the unlearned clergy, who despise grammar, and spake as they were moved. He said: "When I *arrove* at the house of my *diseased* friend, he was *perspiring* his last. I went and stood by his bedside. He was too far gone to talk, but I said, 'Brother, if you feel happy now, jist *sque-eze* my hand,' and he *squo-oze* it."

We shall not mention the name of the distinguished divine who has often told the following story, but it is genuine, and on his credibility we become responsible for its truth. A good man, but ill-instructed Out West, had a call to preach. Being unable to read, he employed a friend to read the selection from the Scriptures. On one occasion the chapter was the twenty-second of Genesis, which contains the word - "These eight did Milcha bear to Nahor, Abraham's brother." From these words he proceeded to discourse as follows: "Brethern and sisters, let us consider our blessings. We have all the comforts of life. We have flocks and herds, and our hearts are filled with food and gladness. Morning and evening our wives and daughters milk the cows, and our wants are all supplied. In the days of good old Abraham, the case was different; for then, as you have heard, it *took eight to milk a bear,* and they did not get much at that."

Truth and Falsehood traveling one warm day, met at a river, and both went to bathe at the same place. Falsehood, coming first out of the water, took his companion's clothes, and left his own vile raiment and then went on his way. Truth, coming

out of the water, sought in vain for his own proper dress, disdaining to wear the garb of falsehood. Truth started all naked in pursuit of the thief, but not being so swift of foot, has never overtaken the fugitive. Ever since he has been known as "Naked Truth."

Oliver Millikin is as much of a wag as a musician, and to make his waggery the more successful, he affects the dress and manner of a clergyman. With his white cravat and sober black coat he makes not a bad imitation. He was at one of the hotels in Springfield, a few weeks ago, and his decidedly clerical appearance attracting the attention of the company as they sat down to dinner, he was called upon to ask a blessing. He did not refuse, and got on very well for a sentence or two; but when he came toward the conclusion, he found as it was his first attempt, that he was utterly unable to remember the usual manner of closing such an exercise. After hesitating a moment, he recollected the usual subscription of his letters, and he brought his petitions to a close by adding, with great solemnity, "Yours truly, Oliver Millikin!"

To play upon Scripture is neither proper nor witty, and the following, sent to us by a Virginia friend, is to be excused only for its having happened to be a veritable hit, and made by a Quaker, too.

"During the last political canvass in Virginia, at the time of Mr. Wise's visit to Charleston, Jefferson County, quite a large crowd was attracted to see and hear him. Among them were two of our country people, a Yankee and a Quaker, the former of whom was given to asking questions, and the other was more of a wag than Quakers are wont to be. As they were walking up street, they passed by two of our well-known citizens, both of whom are remarkable for their extensive learning and sound judgement, and with them was a stranger of somewhat distinguished appearance. The Yankee wondered aloud,

" 'Who can that be with Mr. L. and Mr. S.?'

" 'That is Mr. Wise,' promptly replied the Quaker.

" 'How on earth do you know him, when I heard you say a little while ago you had never seen him,' said the Yankee.

" 'Easily enough,' the Quaker answered. 'Are we not told that he who walketh with wise men *shall be wise*?' "

There is not only a good degree of characteristic oddity and humor in the subjoined anecdote of the celebrated Rev. Rowland Hill, but it contains besides a *lesson in health,* which is worthy of the consideration of all persons of sanitary habits or associations at least:

"Rowland Hill used to ride a good deal, and *by exercise,* he always preserved vigorous health. On one occasion, when asked by a medical friend what physician and apothecary he employed, he replied:

" 'My physician has always been a *horse,* and my apothecary an *ass!*' "

Not *personally* complimentary, perhaps, to either profession, but at the same time conveying a "practice" which each would undoubtedly recommend to a friend who was a patient and a customer.

Clergymen sometimes unbend from their general seriousness of demeanor, and in the way of humor or harmless satire make as good a "point" as they could in a sermon, and perhaps be quite as effective for good. In strong illustration of this, look at Sydney Smith. But it is not of him that the following anecdote is told, but of Parson D———, an orthodox minister of Marblehead, Massachusetts:

"Parson D——— liked a joke amazingly; and so, for that matter, did Parson A———, who was a Baptist. The latter being near the house of the former when a shower came

UNCLASSIFIED ADS

25¢ per word, 20 word minimum ($5.00). Name and address run free. All advertising must be accompanied by check or Money Order made payable to THE KEPI, P.O. Box 4542, N. Hollywood, CA 91607. Tear sheet sent to all unclassified advertisers upon publication.

AMERICAN MILITARY BELTS AND RELATED EQUIPMENTS - *New Book.* A wealth of photographs and information on accoutrements of the 1833-1910 period. *See ad, page 15.*

THE PALMETTO LIGHT ARTILLERY is recruiting serious re-enactors in the western Carolinas and northern Georgia. The battery has one of the largest collections of privately owned cannon in the nation. Safety and authenticity are stressed. *Contact:* JIM CEVASCO, 308 RAINWOOD DR., SIMPSONVILLE, SC 29681.

WANTED: O.R.'s, odd volumes, especially Vols. 2, 25, Hamlin's *Old Bald Head;* White's *Robert E. Lee and the Southern Confederacy.* A. ZIMMERMAN, 316 ISINGLASS RD., HUNTINGTON, CT 06484.

2ND ANNUAL GETTYSBURG CIVIL WAR HERITAGE DAYS. June 30th-July 8th, 1984; Living History Encampments, Church Services, Band Concerts, Civil War Exhibits, "History of the American Soldier" Pageant, Lecture Series, Civil War Collectors' Show. "You've Got A Friend in Pennsylvania." *For Information:* GETTYSBURG TRAVEL COUNCIL, DEPT. 445, 35 CARLISLE STREET, GETTYSBURG, PENNSYLVANIA 17325.

CIVIL WAR ARTIFACTS, Swords, Arms, Accoutrements, Relics, Documents. Buy & Sell. Subscription, $3.00/year, 7 issues of our list. THE CIVIL WAR SHOP, P.O. BOX 2091, FT. OGLETHORPE, GA 30742.

BATTLES, Confederate & Union Portraits, & other scenes by famous Civil War artists, framed & unframed, send $1 for full color brochure. THEME PRINTS, LTD., P.O. BOX 123-L, BAYSIDE, N.Y. 11361. *(See our display ad page* 26*)*

CIVIL WAR BOOKS, NEWSPAPERS, DOCUMENTS, IMAGES. PLUS MUCH MORE. COMPREHENSIVE CATALOG $1. GORDON TOTTY, 576 MASSACHUSETTS AVE., LUNENBURG, MA 01462.

OLD MAPS — OLD CIVIL WAR MAP KITS. Also old State, Railroad and County maps. 70 to 120 years old. All States. Also have Revolutionary War Map Kit. Stamp for catalog. NORTHERN MAP CO., DEPT. TK, DUNELLON, FL 32630.

R. STEPHEN DORSEY ANTIQUE MILITARIA - Quality arms and accoutrements, carefully selected by a senior collector. Subscription to lists, $4 per year. R. STEPHEN DORSEY, P.O. BOX 263, EUGENE, OR 97440.

BOOKS FOR SALE: McDonald's *Make Me A Map of the Valley* — $25.00; Dowdey's *The Wartime Papers of R.E. Lee* — $12.00; Taylor's *Destruction and Reconstruction* (ed. by Harwell) — $12.00; Sorrel's *Recollections of a Confederate Staff Officer* (Neale Pub. Co., 1905, *First Edition)* — $50.00. All excellent. Prices postpaid. CHONA RABAYA, 11805 EAST CRESTA VERDE DR., WHITTIER, CA 90601.

KERR .45 cal. Cap and Ball Patent revolver. Anchor and initials on front grip. Serial number in CONFEDERATE series. Checkered grip sharp. Needs adjustment. Best offer. MATT GILLMAN, C/O GENERAL DELIVERY, UPPER JAY, NY 12987.

---

up, called on Parson D———, and requested the loan of an umbrella:

" 'I thought,' said Parson D———, 'that you *liked* water?'

" 'So I *do,*' responded the Baptist; 'But I wish to avoid *sprinkling!*' "

A capital story is told of a spirited young lady on her way to the church to be married. At the door her intended husband paused, and said very unexpectedly, "My dear girl, there are three things of which I ought to have spoken to you before, and if you do not like them, it is not too late to retire even now from our engagement. To tell you the whole truth, I must insist, first, to sleep alone, secondly, to eat alone, and thirdly, to find fault where there is no occasion."

"Is that all?" she said. "I agree at once. If you sleep alone, I shall not; if you eat alone, I shall eat first; and as to your finding fault, I shall take care that you never want occasion."

The conditions thus being adjusted, they entered the church, the ceremony was performed, and at the latest accounts both parties were doing very well, the husband having fairly given in to the high spirit and good-humor of his wife.

(Continued from page 24.)

and of no effect, except so far as may relate to any transactions under them prior to the promulgation of this act, at the several posts and garrisons respectively, occupied by any part of the army of the United States. [APPROVED, April 10, 1806.]

From: *Revised Regulations for the Army of the United States, 1861,* Philadelphia, J.P. Lippincott & Co., pp. 514-516.

Remember: Flag Day is June 14. Display your colors proudly!

Arnold Shoe factory, Wales Street, corner of Railroad Avenue, Abington, Mass., looking west, showing railroad siding for factory. Arnold was one of the major shoe suppliers for the Union Army in the Civil War. The building currently houses light industry.

THE KEPI

P.O. Box 4542
North Hollywood, California 91607

BULK MAIL

THE KEPI

AUGUST-SEPTEMBER 1984 $3.50

"A New Concept in Civil War Reporting"

A Mission To President Lincoln
by Peter Harrington
Lincoln's Playmate
by Thomas J. Evans
Every Thing Is New To My Eye
by William F. Howard

THE KEPI

"A New Concept in Civil War Reporting"

AUGUST-SEPTEMBER 1984 **VOLUME II, NUMBER 4**

THE KEPI is published bi-monthly for $3.50 per issue, or $18.00 per year. Canadian single copy, $4.50; subscription $23.00 per year. United Kingdom single copy, $4.50; subscription $28.00 (Air Mail) per year. Correspondence to the editors may be directed to: **THE KEPI**, P.O. Box 699, Baldwin Park, CA, 91706. Unless expressly stated, **THE KEPI** neither endorses nor is responsible for the content of advertisements appearing in its pages. **THE KEPI** requests authors to query with specific topics before submitting articles, and assumes no responsibility for unsolicited manuscripts not accompanied by return postage. Work not fully documented and footnoted where applicable will not be considered. All rights reserved. It is prohibited to reproduce this issue in any form without permission in writing from the editors. All inquiries should be directed to the above address. **THE KEPI** reserves the right to edit and copyright all letters. Please allow 6 to 8 weeks for delivery of first issue.

ILLUSTRATIONS

Cover: Col. Rush C. Hawkins, from the Anne S.K. Brown Military Collection, Brown University.
Col. Rush C. Hawkins, p. 5, from *Battles and Leaders of the Civil War*, Vol. I, p. 662.
Gen. John E. Wool, p. 9, from the Anne S.K. Brown Military Collection, Brown University.
Gen. George B. McClellan, p. 13, from the Anne S.K. Brown Military Collection, Brown University.
Austin Gollaher, p. 16, courtest of Beatrice Crady, Hodgenville, Kentucky.
Back cover: Typical steamboat landing from *Gleason's Pictorial Drawing-Room Companion*. April 1, 1854.

THE REGULARS

THE IRREGULARS

• GENERALLY SPEAKING •

In a continuing policy to present our readers with original, unpublished material on the Civil War, we are pleased to introduce our cover story, "A Mission to President Lincoln." This article, actually written by Rush C. Hawkins about his participation in the military policies of 1861, reveals the temper of the times, and gives insights into the workings of the "Grand Army of the Republic." The editor, Peter Harrington, is a native of England, and is currently Assistant Curator of the Anne S.K. Brown Military Collection at Brown University. His interests include 19th Century British and American military history.

As editors of THE KEPI, we are proud of the magazine's role as a disseminator of unknown facts and original interpretation, and it is our pledge to continue to supply our readers with this type of material whenever possible.

In conjunction with the cover story, Tom Evans' ECHOES OF WAR provides an unusual look at Abraham Lincoln as a boy. In an interview given by a childhood friend, "Father Abe" emerges as a very normal, very human boy, growing up on America's frontier. The second part of the article details a visit to President Lincoln in 1861, revealing a careworn man with the future of the country on his shoulders, still able to take a moment and share a little humor with his visitors.

Through the diary of a young woman, the world of 1865 comes alive, in "Every Thing is New to My Eye." Her experiences chronicle the events on a ship, as it travels up and down the coast from Washington to points in Virginia, and records the tragic procession of Lincoln's funeral. In keeping with the format of THE KEPI, wherein the 19th Century is presented as it was lived, the story's editor, William F. Howard, has maintained spelling and grammar true to the original text.

Paul Kallina's chronicle of birthdays and birthplaces of Civil War notables continues, and MILITARY REVIEW presents a scholarly look at newly published books.

Of additional import, please note THE KEPI has a new address: THE KEPI, Post Office Box 699, Baldwin Park, CA, 91706.

S.L. Kotar J.E. Gessler
Editors-in-General

Head-Quarters of THE KEPI
1st August 1984

GENERAL ORDERS
No. 8
by Command of the Editors-in-General

It was described as one of the greatest moments in the history of the Confederate armies. Those who participated in it never forgot it; those who were left out always regretted it. It was an act of intense mortification for the Union. It was "Christmas in August," Gen. T.J. Jackson's brilliant raid around John Pope, that netted him the fabulously supplied Federal base at Manassas Junction. This article, and the accompanying TIMELINE trace "Stonewall's" route as he brought his army behind enemy lines, captured and subsequently treated his troops to a feast that would become folklore before the day was past.

"Atrocity at Hunter's Mill, Virginia," describes the Rev. John D. Read and his spying activities for the Union army. Very little has come to light about the life and personalities of the people who risked their lives transmitting information. This article details one man's fight for survival, and the enemy's activities against him.

Premiering in this issue will be a new feature, PRIVATE THOUGHTS. Individual soldiers, their lives and participation in the war will be capsulized, and supplemented with their pictures, whenever available.

The regular/irregulars continue, as does FAMILY BIBLE, with birthdays from October and November.

"A Mission To President Lincoln"

Edited by: Peter Harrington

Brevet-Brigadier General Rush C. Hawkins.

Rush C. Hawkins was the flamboyant founder and commanding officer of the Ninth New York Volunteer Infantry, or "Hawkins' Zouaves" as they were more commonly known. The regiment served on the Carolina coast, at South Mountain, Antietam and Frederickshurg before being mustered out of service in May, 1863.[1] In later life, Hawkins established a Memorial to his late wife Annmary Brown, which is now part of Brown University; his collection of early printed books is now housed in the building and his grave lies next to hers at the rear of the building.

According to his librarian,[2] perhaps the most vital service which he rendered his country was the attack upon his commander-in-chief, McClellan. As Hawkins explained: "I was absolutely full of hatred for McClellan, his incompetance and his pomp, and I was very firm in my denunciation of the man. I went for him without gloves."[3]

Among the records at the Annmary Brown Memorial is a manuscript notebook written by Hawkins in 1893, and containing the original letters quoted in the text. It gives a very circumstantial account of his personal campaign against McClellan. The text, which is titled "A Mission to President Lincoln" is given here in full.

"Among the Regiments stationed at Newport News, Virginia during the summer of 1861, were the First and Ninth New York Volunteers. Before the incident about to be related took place, the latter command had been rather successful in eliminating from its rolls, a long list of incompetant and unfit men who, at the time of its organization had been elected Company Officers. The effect of the weeding out was instantaneous. The Regiment took on a new aspect and laid a solid foundation for future usefulness. The other Regiment named was not so fortunate. The weeding process had been neglected and several unfit men remained upon its Roll of Officers.

The evening of the first of August not long after retreat, Colonel Phelps[4] of the First Vermont Volunteers, who commanded the first, sent an urgent message for me to report to his headquarters; when there, he informed me that there was an emeute in the First New York Regiment that might end in an open revolt, and that I was to hold my Regiment ready to suppress a possible disturbance.

The cause of the trouble was a very aggravating one. Captain Leon Barnard of Co. K., without cause had shot and seriously wounded a member of his Company, who was greatly respected by his fellow-soldiers. That officer, as soon as he had committed his dastardly act, took shelter in the tent of the Colonel, where he had the protection of his own Lieutenants, and several more of the Regimental Officers, against the uprising of the members of his Company, who had been joined by a large number of their companions. The Officers took a very decided stand in favor of law and order and succeeded in sending the men to their quarters, but not before my Regiment had been ordered to hold itself in readiness for a possible emergency.

This sad incident was the camp talk of the day, and nothing more, at that time, was heard of the wounded soldier and his would-be murderer.

I remained at Newport News until the last week of August, when with three companies of my Regiment, I joined the expedition for the capture of the Rebel fortifications at Hatteras Inlet, North Carolina. What was accomplished by that expedition has passed into history and does not need repeating here; we were left at Hatteras Inlet and the three Companies were soon joined by others of the Regiment, and the newly captured Rebel stronghold was made a permanent Army post. In the meantime, Captain Barnard had exchanged his commision with a Captain of my Regiment, whose courage had been seriously questioned by his brother Officers, and in consequence his position in the Corps had become very uncomfortable. Both of these worthies were glad to get out of their old and into new Regiments, where they were not partiularly well known.

The whole matter had been kept secret and nothing was known of the exchange until Captain Barnard reported to me for duty at the Camp at Hatteras. To say that I was surprised would be putting it very mildly. I was indignant almost beyond expression and quite beyond control. I had spent months of most vexatious labor in ridding my Regiment of unfit Officers, and had finally succeeded; and now, for my honest exertions in the interests of the service, was rewarded by having the Governor of my State[5] appoint, over my head, without my leave, knowledge, or consent, one of the worst characters he could have selected as a Captain in my command. Under these circumstances I felt warrented in refusing to assign him to duty and he returned to Fortress Monroe.

Soon after Captain Barnard had returned to Fortress Monroe, Brigadier-General Thomas Williams,[6] a senseless martinet, reported and assumed command at the Inlet, and so occasioned a new opportunity for the obnoxious Captain, which he soon acted upon. He again reported for duty and again was refused admission to the Regiment. He communicated the refusal to Williams, was again ordered to report to me, which he did, and with no more success than formerly. Then came a letter from Williams asking for an explanation of the reasons for my un-officer-like conduct, and a preemptory demand to know whether I would or not assign Captain Barnard to duty.

The entire correspondence relating to this, for me, unfortunate incident, is now given for the purpose of showing my exact position, but is not quoted for vindication.

Fort Clark
Hatteras Inlet N C
Oct 8 1861

Genl John E. Wool.[7]
Com'g Dept. of Va.
Fortress Monroe, Va.

Sir,

I do not acknowledge the right of Gov. Morgan to appoint and transfer officers over my head, into my regiment without my consent, and I will resist the appointment and transfer of Capt. Barnard even though it cost me my Commission.

If Capt. B. in the end should succeed in obtaining a company in my regiment, I shall then if not dismissed beforehead, resign my commission.

With my great respect, I remain
Your faithful servant,
Rush C. Hawkins.
Col 9th Reg. N Y V

Head Qrs. Fort Hatteras. N. C.
October 22nd 1861

Colonel,

I forward to you at the same time with this, a copy of the order (Special Orders. No. 5) furnished to Captain Leon Barnard. The General desires me to ask you if you refuse to obey said order. Answer immediately.

Very respectfully,
Your obt. Serv.
C. Hook.
1st Lt. 1st Art. A.A.A.G.

Colonel R. C. Hawkins
9th N.Y. Volunteers

Special Orders No. 5

Captain Leon Barnard, 9th regiment N. Y. Volunteers, having

TIMELINE

July 26, 1861
George B. McClellan is appointed to command the city of Washington, D.C., and the troops around it. He replaces General Irwin McDowell.

August 8
McClellan writes General Scott that he believes the enemy has at least 100,000 men in front of Washington or in the vicinity, and the capitol is in imminent danger.

August 9
Scott writes to the Secretary of War, advising him of McClellan's letter of the 8th. He states he is of the opposite view and declares his age and infirmities compel him to resign.

August 10
At Lincoln's request, McClellan gives his authority to withdraw his letter of the 8th. The President then requests Scott to withdraw his reply.

August 12
Scott feels he cannot withdraw his letter because McClellan, the "ambitious junior," has grown indifferent and bypassed his authority, as have some Cabinet members.

August 26
Combined forces of infantry and navy set sail for Hatteral Inlet, North Carolina, with Flag-officer Silas H. Stringham in command of the fleet and Major-General B.F. Butler of the land forces.

August 27
Fleet begins bombarding Forts Clark and Hatteras. Troops are landed, but due to adverse weather conditions, the main force cannot join them.

August 28
Confederates surrender Fort Hatteras, giving the Union possession of the best sea entrance to the inland waters of North Carolina.

Union losses: 1 wounded. Confederate losses: 670 prisoners, 1000 stand of arms, 35 cannon. Rush C. Hawkins is left to command the captured forts with detachments of the 9th and 20th N.Y. Vols.

September 6
Hawkins writes to General John E. Wool, suggesting Roanoke Island be occupied, as well as Beaufort and Beacon Island for the security of the Federal forces in the area. Wool indorses these suggestions.

September 10
Provisions arrive to supply the troops on the Hatteras expedition, and 6 more companies of the 9th N.Y. under Lieut.-Colonel George F. Betts join those already there.

September 11
Hawkins urges acceptance of his September 6th proposals; Wool again approves them, and forwards them to the War Department. Union forces defeat General R.E. Lee at Cheat Mountain, securing the western portion of Virginia for the U.S.

September 16
An expedition of land and sea forces under Lieut. James G. Maxwell are sent to destroy the forts of Beacon Island and Portsmouth. They find them abandoned by the Confederate forces.

September 27
At a Cabinet meeting, Lincoln and McClellan discuss plans for a Virginia offensive and the need for Union victories in the state to offset a growing public concern.

September 29
Hawkins embarks the 20th Indiana (recently arrived with the 7 remaining companies of the 9th N.Y.) on gun-boats, and accompanies them to a point opposite Chicamacomico, with the intent to protect loyal natives and prevent a surprise landing by the enemy.

September 30
Lack of military action in Virginia

reported at these Head-Quarters in compliance with Special Orders No. 77, dated, Head-Quarters, Dept of Virginia, Fort Monroe, Oct 18th 1861, he is assigned to duty with his regiment, and will report for duty to Colonel R. C. Hawkins, comd'g 9th regiment N. Y. Volunteers.

By Order of
Brigdr Genl Thos Williams
(signed:) C. Hook
1st lieut. 1st Artillery A.A.A.G.

Camp Wool, October 22nd 1861

Colonel R. C. Hawkins of the Ninth Regiment N. Y. Vols, through the A. A. A. General C. Hook, 1st Lt. 1st Art. U.S.A., would inform General Williams that Special Order No. 5 is not sufficiently explicit to admit of either an affirmative or a negative reply to his note. But Colonel Hawkins would now state that he does refuse to allow Captain Leon Barnard to enter upon any duty in any way connected with his Regiment.

Head Qrs, Fort Hatteras, NC
Oct. 22nd 1861

Sir,

Having received your note of this date in which you refuse to allow Capt Leon Barnard to enter upon any duty in any way connected with his regiment, and which is in direct disobedience of Special Orders No. 5, from these Head Qrs., you are hereby placed in arrest. You will confine yourself to the limits of your Camp.

By order of Brigdr Genl Thos Williams
(signed:) C. Hook
1st Lieut. 1st Arty. A.A.A.G.

Colonel R. C. Hawkins
9th N. Y. Volunteers.

Fort Hatteras, N. C.
Oct. 23, 1861

My dear Colonel:

Don't put yourself in the wrong by disobeying orders: - so *gravely* in the wrong! If the man is not fit to be an officer, try the prescribed course by charges & a ct. martial to get rid of him. The good of the service requires you should be in command of your regiment - to discipline, instruct, control & lead it - but the paramount good of the service requires that orders be obeyed. No necessity can dispense with this.

Try regular means; don't put yourself in the wrong - don't reduce yourself to inaction at the moment your services may be most needed for your own reputation & the general success.

Ask, officially, to recall your note; give the man his company, & I'll be most happy to release you.

The Revd. Sidney Smith was not only a wit but a wise man & of manifold experience thro' a long life, & he used to say that the easiest way to be *right* was the best way: - that, hardly any success would compensate for the annoyances, pains, disgusts, & uncertainties of contention.

This is not the time for contention among ourselves, except such contention as shall show who is most zealous for his country's cause, forgetting his own.

Yours very truly, & with much esteem
T. Williams.

Colonel Rush C. Hawkins.

Head Qrs. Fort Hatteras, N. C.
Oct 29th 1861

Colonel,

Under the obligation of law, regulations and the custom of service, to enforce all legal orders by all legal means, to disregard a case of positive and persistent disobedience of orders might properly be ascribed to culpable indifference or more unworthy evasion. The commanding officer regrets he cannot accede to your request.

By order of Brigdr. Genl. T. Williams
(signed:) C. Hook,
Lt U. S. A.
A. A. A. G.

Colonel R. C. Hawkins,
9th N. Y. Vols.

Head Quarters, Fort Hatteras
Hatteras Inlet, NC
Oct 31st 1861

Special Order no 11

Colonel R. C. Hawkins, 9th N. Y. Vols, will proceed to Fort Monroe by the first steamer, and report himself to Maj Gen'l John E. Wool, Comd'g Dep't of Virginia, &c.

By order of Brigdr Gen'l Thos Williams
(signed:) C. Hook
1st Lt. 1st. Arty. A. A. A. G.

Head-quarters Deparment of Va.
Fort Monroe Va.
November 5th 1861.

Colonel.

The Major General Commanding the Department, directs me to inform you that the charges preferred against you by Brig. General Thomas Williams are dismissed and you are hereby released from arrest.

I am sir very resepctfully Your obdt. servt.
Wm. D. Whipple.[8]
Asst. Adjt Genl.

Col. R. C. Hawkins
9th Regt N. Y. Vols.

General Wool took a common sense view of the arrest and the causes that led to it. He said that while I had committed a grave offense, in a wilful disobedience of orders, that I was clearly in the right in so doing. That being solely responsible for the character and efficiency of my Regiment, it was my right to choose my Officers, and also my duty to see that unfit men did not secure commissions. But the matter was of such importance that he would not, upon his own responsibility, take the liberty of deciding it. So he gave me a special Steamer and sent me to Washington, where with other matters, I was to present my case to the President for his decision.

In the meantime, the men of the Regiment settled the Barnard question in a way quite their own. Soon after my leaving Camp for Fortress Monroe, it came the Captain's turn for a tour of duty as Officer of the

continues to bring criticisms from Northern supporters.

October 1

The *Fanny*, after delivering supplies to the Union troops at Chicamacomico, is captured by Confederate gun-boats. In Washington, Scott, McClellan and Lincoln discuss plans for a November coastal expedition.

October 4

Confederate troops under Col. A.R. Wright land at Chicamacomico, causing the Federals, under Col. Brown, to retreat.

October 5

Hawkins meets Brown at the light-house, and in view of these reinforcements, Wright retreats.

October

Col. Hawkins is superseded by Brig.-General J.K.F. Mansfield after having been severely censured for dividing his small force, and because of the excitement over the loss of the *Fanny*.

October 13

Brig.-General Thomas Williams relieves General Mansfield and assumes command of the post.

October 18

McClellan is reluctant to provide General Sherman with sufficient troops for his proposed coastal expedition, and Lincoln is called on to pacify both parties.

October 21

Acting under McClellan's orders, General Charles Stone and Col. Baker move troops across the Potomac from Ball's Bluff and Edward's Ferry to Leesburg. When Confederates succeed in pushing them back, Baker is killed and the Federals retreat in disorder. Men are drowned and shot during the panic to escape across the river. Union losses: 49 killed, 158 wounded, 714 missing and presumed drowned; Confederate losses: 36 killed, 117 wounded, 12 missing. A public cry against Stone is raised.

October 31
The month ends with no major battles or change in territory for either side.

November 1
Upon the retirement of General Winfield Scott, George B. McClellan assumes command of all the armies, except the Department of Virginia.

November 2
Scott leaves Washington, and McClellan begins his new job.

November 5
Col. Hawkins is sent by General Wool to urge President Lincoln to abandon Hatteras Inlet or erect suitable accomodations for the troops.

November 6
Hawkins reports to Lincoln, presents his letter from Wool, and is asked to appear before the Cabinet.

November 7
Hawkins appears before Lincoln's Cabinet, and argues that Hatteras Inlet be held. General McClellan questions him about operations in Virginia, with the necessity of making Fort Monroe a base of operations. The council decides to hold Hatteras, and Hawkins is instructed to wait until necessary orders are prepared before returning.

November 11
Washington celebrates the new General-in-Chief with a torchlight parade.

November 13
When Lincoln calls on McClellan at his home, McClellan refuses to see him.

November 20
McClellan reviews some 60,000 troops in Washington.

November 30
Another month has passed without any significant movement of the Army of the Potomac commanded by McClellan.

Major General John E. Wool.

Guard. The tent of the Officer of the Guard was very near the great Atlantic breakers. He had turned in for the night, within it, and probably was asleep, when the guys were cut, the Captain securely rolled within the folds of the tent, and he was being carried towards the water with great rapidity, when another Officer passing by, seeing something mysterious going on, came forward, and rescued a very badly frightened Captain from a watery grave. The next day he left the Regiment and was heard of no more.

At this time, thirty-two years and more after the event, I find it quite impossible to remember exactly when I arrived in Washington, whether on the morning of the sixth or seventh of November but am almost certain it was the latter date.

Soon after my arrival at the Capital, I proceeded to the White House, and sent to the President the personal letter provided by General Wool, which in brief set forth the cause of my having been ordered to Washington. In a few minutes, Mr. Lincoln sent word he would see me and I was shown into the Executive Chamber where I found him alone sitting by his desk (with his long ill-shapen legs crossed, body bent forward and looking down at the floor - *line deleted by Hawkins*). Soon as

I came near, he looked up with those sincere sad eyes of his, into my face, extended towards me one of those great honest hands, said he was glad to see me, and wanted me to tell him all about how matters were progressing in General Wool's department. In that department at that time, the point of the greatest interest was Hatteras Inlet. To make it more comfortable for the Troops stationed there, to hold it for future operations or to abandon it, were the questions which General Wool had instructed me to present for consideration to the President and the military authorities at Washington. After going over the whole matter, explaining the advantages of holding the Inlet as a base for future operations, he said the matter was too important for him to decide, and he would present it to the next regular Cabinet meeting, which was to take place within two or three days. He suggested that I ought to be present at the meeting to explain the situation as I had to him. He said he could not consider the matters in which I was personally interested at that time, but if I would come in at half past seven the next evening, and take a cup of tea and a piece of toast with him we would be alone and have plenty of time to talk the matter over. I promised to keep the engagement, and after a cordial shake of the land, left the room.

It is quite impossible for me to describe my feelings upon the occasion of that interview. I had never before had my sympathies so excited by any human being. Never before had I witnessed such an abject expression of sadness and depression. The cruelties and mortality incident to war seemed to be ever uppermost in the mind of our good President, and several times during the interview this absorbing thought found expression in speech, when entirely out of order and not at all germane to the matters under consideration. I passed the rest of the day, the whole of the evening and a part of the next day by myself; the sad, sad eyes I had looked into haunting me the whole time. I had but one feeling and that was to have the war ended as soon as possible, so that the great load might be lifted from the shoulders of the man, who in a few minutes, I had learned to love and respect, as I had never loved and respected before. "Dear old Uncle Abe" was a revelation to me, so sadly quaint, so melancholy, so simple in his manners, humane, sympathetic, and generally so different from anybody I had ever known. The great impression was made and I would willingly have given my life to serve him.

The next evening at the appointed hour, I was in the room of the private secretary and soon after sent my name to the President, who was ready to receive me. After the greeting, which was most simple and cordial, and the advent of the tea and toast, I proceeded to state my case, elaborating, as I went along, with remarks about the effect of Governor Morgan's act upon the discipline of New York Regiments, if he should continue as he was then doing. After listening most attentively, without interruption, to my statement, Mr. Lincoln straightened himself back in his chair, and in substance said: — "You are clearly in the right and I sympathise with you in your desire for a clean reputation for your Regiment. But we are in the midst of a great war and have a terrible business before us, and I cannot afford to enter into a controversy with the Governor of a State that I rely upon more than any other to assist in putting down this terrible rebellion, and you must say as much to General Wool, and tell him that I say he must fix it up with Governor Morgan, so that justice will be done to you, and in such a way as to avoid irritation in the future."

At this point a messenger announced Colonel Charles Devens,[9] of the Fifteenth Mass. Infantry, and George Ashmun; the former with his command had participated in the disaster of Ball's Bluff for which McClellan was responsible, but the scapegoat was General Stone,[10] who spent six months in a military prison in the expiation of an offense which he did not commit.

As soon as these gentlemen had been shown in and Colonel Devens

December 1
Lincoln questions McClellan about how long it would take to get the army in motion.

December 2
Congress convenes, and the legislators express concern that no major offensive move has yet been made; they recommend a forward movement into Virginia before winter.

December 18
McClellan and Lincoln confer about the General's plans for the advancement of the army.

December 20
McClellan does not report to headquarters as he is ill with typhoid fever.

December 31
Lincoln writes General Halleck, asking if he and General Buell are in concert; again, the month has passed and there has been no general movement of the Army of the Potomac.

January 6, 1862
A senatorial group approaches Lincoln with the suggestion that McClellan be replaced. Along the Virginia/Maryland border, General "Stonewall" Jackson is conducting a winter campaign — McClellan seems unable to emulate this. Lincoln rejects the proposal.

January 13
McClellan resumes the duties of his office.

January 27
Lincoln issues "General War Order Number One," indicating February 22 as the day for a general land and sea movement against the Confederacy. He hopes this will force McClellan to move.

January 31
Lincoln supplements General War Order Number One, hoping to move the Federal troops into a confrontation with the enemy at Manassas Junction.

February 3

Lincoln writes McClellan, who has publicly disagreed with him about army operations.

Lincoln was unsuccessful in his attempts to prod McClellan into a major overland move. Eventually, "Little Mac" directed the campaign that took the Union army by water to Virginia's shore and set up the protracted siege of Richmond. It was not until Gen. Jackson's brilliant Valley Campaign forced Lincoln to reorganize the command structure that McClellan lost some of his authority. When the Federals were unable to defeat the combined Confederate armies, and were pushed back in a "change of base," the President finally gave up on the commander-in-chief. John Pope was brought from the West to achieve victories and the man who felt he had a divine directive to save the Union was left to the task of withdrawing his troops northward. McClellan was again called to reorganize a beleaguered army after Second Bull Run, commanding it through Antietam. When he was subsequently removed, McClellan never again led the men in blue into battle.

S.L. Kotar

CONCURRENT EVENTS

The Chesapeake Bay Line of Steamers from Baltimore, connected with the James River boats at Norfolk, to City point and Richmond, are in operation. Passengers to and from Charleston and the intermediate places will find this by far the most comfortable route.

[*The Albion*, Oct. 31, 1846.]

introduced, the President asked him to give an account of the disaster to our arms. While the narration was being given, the President was more cast down that I had seen him before, and several times involuntarily exclaimed: — "When will this terrible war be over? Is there no way of stopping this shedding of blood?" Colonel Devens' description was vivid and pathetic, and every sentence carried with it the conviction to my mind, that the General at headquarters had been responsible for a most unwarrentable and cruel blunder.

The whole interview lasted fully an hour, and my tongue, during the latter part of it, had, to say the least, performed its duty, and the President showed unmistakeable signs of weariness; so after being reminded of the coming Cabinet meeting, and expressing my gratitude for kind words and the favorable decision, with the others I took my departure, more impressed if possible, than before, with the nobility and simplicity of the finest character I had ever known.

Between the interview and the Cabinet meeting I spent the better part of my time in observing men and things, and came to the conclusion that pygmies in great places were trying to make the world believe that they were full grown men.

At the Cabinet meeting there were present: the President, Attorney General Bates, Postmaster General Blair, Secretary of the Interior Smith, Assistant Secretary of the Navy Fox and General McClellan representing the Secretary of War. The President read the letter from General Wool and then called upon me for a verbal statement, which I made as briefly as possible, and at General McClellan's request, took a seat near him at the Cabinet table, and while the others were consulting about the continued occupation of Hatteras Inlet, we discussed a plan which I had proposed, of operations against Richmond with Fortress Monroe for a base. At McClellan's suggestion, I made a rough drawing of the water and land approaches to Richmond, which he folded and put in his pocket. The President then asked the General's opinion about keeping possession of the Inlet, and received an answer to the effect that the position was an important one to be used as a base for future operations along the Coast and in the inland waters of North Carolina. It was then decided to continue the occupation and General McClellan was instructed to prepare a special order of instructions, which I was to wait for and take to General Wool. I then took leave of the President and bowed myself out of the room followed by Secretary Smith, who stopped me in the ante-chamber of the private secretaries, and made an engagement for a meeting with him next day.

At the interview with Secretary Smith, he told me frankly that there was a growing dissatisfaction with the conduct of General McClellan; his seeming want of earnestness; the mystery surrounding his intentions of future action of the army; his delay in moving toward Manasses; his incapacity to command a large army, and that a few men in high position had begun to suspect his loyalty. The Secretary was particularly anxious to know the opinion of General Wool about present conditions and the future outlook. I told him that the morning of my leaving Fortress Monroe, General Wool had said to me that he could not account for General McClellan's inactivity. He felt certain that a good opportunity had been lost, i. e. the getting into Richmond early in the Autumn before the Confederates had had time to equip, organize and fortify. The Secretary said this was clearly his opinion, and he was glad to know it was also General Wool's, and he wished the old General would speak out.

My week spent in Washington waiting for the order, was devoted to observation. I made the acquaintance of several of McClellan's staff Officers, and a few of the Generals commanding in the field. Only two or three were enthusiastic about the General commanding the army. Blenker[11] was particularly outspoken in his condemnation of the unexplainable inactivity.

Early one morning — about half

past seven o'clock, I met in front of Willard's Hotel, General Wadsworth,[12] who had just ridden in from his camp in front of Munson's Hill. He had heard that I was not a very strong McClellanite and was very glad to see me, and after asking what I thought about present conditions, unburdened himself in a very vigorous manner. Pointing to the Forts on Munson's Hill armed with "Quaker" (wooden) guns, he said: — "I have offered repeatedly with my division, to clear the rebels out of that position, and have been answered every time, that such a movement would bring on a general engagement and that the army was not ready for that yet. The damned coward and traitor who made those answers was not ready, but the army has been for months, and he knows it — the damned traitor!" He gave a glowing description of the tooting up of "Little Mac," the "Young Napoleon" etc. etc., *ad nauseam*, and condemned the newspapers of the large Cities for bestowing unstinted praise upon an untried man who had fought no battles nor showed a disposition to do so. I told him that I intended to induce, if possible, General Wool to write a letter to the President, criticising the waste of time, of soldiers' lives by disease, and of the Nation's money. He said it was a good move and it might induce Lincoln to prod McClellan; but nothing else would, as Lincoln had been stuffed to the full so that he now believed McClellan was the greatest military genius "that ever lived." The interview with General Wadsworth lasted more than an hour; he did nearly all the talking, which was devoted to tremendous denunciation of present conditions and the men who had brought them about.

My period of observation in Washington convinced me that the high military Officers were only playing at war; that there was no serious intention of hurting the enemy. They were drawing good pay, had but little to do, and were enjoying themselves greatly, and the McClellan clique permitted no one to question the supreme greatness of their Chief. In this respect there was a little reign of terror; no one dared to say ought of him but praise. Senator "Zack Chandler" of Michigan, while in the barber shop of Willard's Hotel, dared to question his ability, and was instantly met with a threat, from Colonel A. V. Colburn of McClellan's staff, of personal violence for daring to express a query even in relation to the greatness of his chief. The Colonel afterwards being a little in his cups, in my presence, boasted of having shut up "old Zack Chandler" when speaking against McClellan. I could not stand it, my hot blood mounted to the brain, and I informed the belligerent Colonel that if he had undertaken the job of thrashing everybody who did not believe in the greatness of his "young Napoleon," he would have his hands full and he might commence his contract with me. He never spoke to me afterwards and the entire staff gave me the cold cut. I managed however to preserve my appetite, and to get on without their society. But I became *famous* at once.

As a sample of Headquarter's humbug with which fools, in abundance, were hoodwinked, I will give one incident.

I was walking up Pennsylvania Avenue one afternoon about two o'clock and looking up the Avenue towards the Treasury Department, saw a great commotion and heard a great banging of hoofs against the pavement, and clanging of sabre scabbards. McClellan at full gallop riding at the head of thirty staff Officers and a squadron of Cavalry, was on his way to the Artillery camp east of the Capital, to inspect a newly arrived battery of Massachusetts Artillery! Such was the pomp and circumstance of his style of warfare.

At length the order came, and I had the last of my series of interviews with Secretary Smith. I submitted a plan of campaign against McClellan to which he fully assented, wished me God-speed and promised me all the assistance in his power, and soon after I left in a special boat for Fortress Monroe where I arrived about one o'clock in the morning and went directly to the bedroom of General Wool.

From time to time while in Washington, I had written to him giving my impressions of men and affairs, and I found my letters had excited considerable anxiety and a desire on his part to hear more of Washington conditions. Without the least reservation and in the most vivid language at my command, I gave him a full account of what I had seen and heard. I gave him a clear picture of McClellan's posings, his demeanour showing a poor imitation of Napoleon; his seeming want of plan for future operations; the blind idolatry of his enormous cabal of well-paid followers; the unwarrentable confidence in his capacity on the part of the President and others high in Office; the dangerous praise of the whole press of the nation of a man who had accomplished nothing; the great mortality among the troops from Camp fever; and I gave in full the conversations with Generals Wadsworth and Blenker, and with Secretary Smith, and ended with stating that I believed that the army would never achieve success while commanded by McClellan, and on behalf of Secretary Smith and myself, urged him, as the senior General of the Army, to use his influence with the President, in favor of some sort of army action or the removal of McClellan.

During my description, the patriotic old soldier was affected to tears, and when I had finished, said he had the worst fears for the fate of his Country, and that it was perfectly clear something ought to be done. But what and how?

I suggested that the troops at Fortress Monroe should be increased to 25,000 men and that a simultaneous move of the main army upon Manasses, and of his command, by the Peninsula, in the direction of Richmond, should be made as soon as possible. I also suggested to General Wool that he should state it as his belief that the Fortress Monroe column would probably get into Richmond within two weeks after receiving the order to move. The weather was then fine, and up to that time there had been very little mud in Virginia. I asked that this plan should be embodied in a letter for me to take to the President. This he would not consent to do, but might write it in a personal letter to Erastus Corning, an intimate friend, then a member of Congress, who might show it to the President. He wanted time to think the matter over and gave me permission to remain at Fortress Monroe until further orders. The second of December he sent for me, and read for my information a letter to Mr. Corning, which embodied the plan of operations I had suggested. He instructed me to be ready to leave with the letter for Washington that evening.

I arrived in Washington the next morning and went directly to the office of Secretary Smith. He read and approved of the letter and said he would make an early engagement for a meeting with the President. I then delivered the letter to Mr. Corning, who was pleased with the conto the Executive Chamber. The President was alone, looking quite as promised co-operation in any movement which would compel action. He gave me a personal letter to the President recommending my mission to his favorable consideration, and for his information inclosing the letter of General Wool.

The second day after this interview I received a letter from Secretary Smith informing me that the President would see me the morning of the sixth of December.

The time between my arrival in Washington and the interview with the President was occupied in interviewing Senators and Members of Congress. Senator Collamer who was from my native State, I knew to be honest and patriotic, and those well acquainted with him esteemed him wise. One of my first interviews was with him. I made bold to present my case against McClellan in rather vivid terms. The aged Statesman was fairly frightened, and I am sure regarded me as a curiosity, a sort of a cross half crank and half traitor. He gave the same answer made to me by nearly all the wise men I had consulted: — "McClellan was the President's choice for the head of the Army, and the people and their representatives must support him." This was blind faith without the slightest foundation to stand upon. It was saying practically that the power that furnished the men and paid expenses could not question the acts of their servants.

Major General George B. McClellan.

My call upon Senator Sumner was equally fruitless. I had from him a two hour talk about Fabius and McClellan's resemblance to that General. The senator was sure that a great batch of strategy was being hatched, and that sometime we would see very great results. He did however admit from his civilian point of view, that it seemed as though precious time was being wasted, and great opportunities lost. But like the others he was tongue and hand tied. McClellan must not be interfered with, and the duty of the people and their representatives was to support him. I left the Chamber of the Senator feeling that I had spent a very pleasant evening, but entirely without profit to the cause I had so much at heart.

Half past eleven o'clock of the morning of December sixth, 1861, found me in the room of the Private Secretary of the President, and within five minutes I was admitted to the Executive Chamber. The President was alone, looking quite as sad and dejected as when I saw him before. The old feeling came back, my heart warmed towards him beyond my control, and before he had time to rise I ran forward, seized one of his hands in both of mine, and without speaking, turned aside to conceal the tears which a fast beating and pitying heart had sent to the surface. The silent greeting over and, after answering many questions about affairs at Fortress Monroe and along the Coast, we came to the object of my interview.

I commenced my argument against McClellan by giving the impressions made upon my mind during the previous visit. I dwelt particularly upon the want of earnestness, frivolity and ostentatious show which characterized the daily military life in Washington and the bluster of the well-paid McClellan Cabal. This part of the conversation the President did not like, but he made no attempt to refute a single allegation. The next part pleased him better. The views of General Wool were given in full. His surprise at the seemingly unwarrantable delay in not moving the army; the practical betrayal of the cause of the loyal people who had so bountifully supplied men and money; and upon my own account I urged the proposition that the Rebels were not as well organized, armed and equipped as ourselves, and that every hour's delay on our part was an important addition to their opportunity.

At this point in the conversation and without any word of introduction, I handed to the President the letter of General Wool to Mr. Corning. He read it in silence and after he had finished looked through the window towards the Potomac for a short time, then turned towards me and said: 'I am glad General Wool wrote that letter.'

The President took the floor in defense of the action or rather inaction of the Government and expressed himself with great earnestness. He said the Generals in command at different points were urging the importance of moves in their departments; these importunities were embarassing to the Government and he intimated that they were annoying. I suggested that these importuning Generals might be patriotic, that they saw the necessity of doing something, and hence their importunities. He said he had not taken that view and possibly I might be right. He cited difficulties in organizing a great army out of raw materials; the time needed for drilling, the seasoning of soldiers in the field, etc., etc., and closed with about these words: "But General Wool knows more about army matters than we do, and it has seemed to me that McClellan ought to do something, and I wish you would see him and talk to him as you have to me." I answered I would do anything he could ask of me, but told him he must give me a note of introduction to insure an interview. He then wrote on a card: 'Gen. McClellan, please see Col. Hawkins, who comes from Gen. Wool. A. Lincoln. Dec. 6, 1861,' — handed it to me and suggested that I should call upon the General at about nine o'clock, at his house the next morning. After the interview he wanted me to report at the White House for the purpose of telling him the result.

The next morning at the hour suggested I was at the door of the commanding General, sent up the card of the President to his room, received back word he could not see me, but if I wanted to make any suggestions about the distribution of troops that they must be made to Adjutant General Thomas. I informed him that I would do so, asked for a note of introduction and for the return of the President's card. Both these requests were complied with and I besought the Adjutant General, had easy access to him and without my mentioning the object of my call, was blandly informed that he was only the mouthpiece of the commanding General, and had nothing to do with the moving of Troops. I informed him that I knew the functions of his office and bowed myself out. The object of my mission had been found out, talked over, and my repulse agreed upon. Of this there can be no possible doubt.

From the interview with the Adjutant General I went directly to the office of Secretary Smith and told him what had occurred. He said that he was not astonished at the General's not complying with the President's request. It was not an uncommon occurrence.

During my previous visit to Washington I had heard a Staff Officer of McClellan's boast of the latter's incivility to the President. His exact words were these: 'When old Lincoln and Seward want to see the General at the White House, he sends word that he is engaged and they have to go to him and wait their turn.' The frequency of this incident was cited as evidence of McClellan's greatness. The argument was: 'He must be a great man. If he were not, the President of the United States and the Secretary of State would not go after him as they do.' Upon this occasion I turned away as soon as possible with both hands in my pockets, and my tongue well secured between my teeth. The want of respect ascribed to the man, as well as to the high office he filled, made me indignant beyond expression. My Failure upon that occasion, to explode, has been a puzzle to me ever since.

The evening after the defeats recounted, I called upon the President and reported results. I asked him if nothing could be done. He answered: "I think not. I have placed General McClellan at the head of the army, and I must rely upon him to put down this terrible rebellion, by giving him the unqualified support of the Government.' I said: 'My President, your position precludes all enquiry into the capacity of a General who has never fought a battle?' His response was: 'I suppose so, for the present at any rate.' Seeing he was wearied and troubled, I prepared to withdraw, the President took my hand at the same time saying: "If all the Officers in the army had your spirit this rebellion would not last six weeks.' Thanking him for all his kindness to me I bade him a very sad and feeling good-bye, and left him with a broken spirit, but not quite discouraged, and went directly to the rooms of Mr. Corning at Williard's Hotel, found him alone and told him all that had occurred. He said my efforts had ended about as he had supposed; that everybody in Washington, but Wadsworth, was wedded to an idol, and none of them could tell why. The next morning, by appointment, I saw Mr. Corning again and proposed he should call a meeting of the New York Congressional delegation, for the purpose of considering a proposition to call upon the President and ask the removal of McClellan. This was rather startling, but did not frighten the old gentleman in the least. He gave me the names of three or four members of the delegation to consult and arranged for another meeting that evening.

During the day I succeeded in seeing two of the parties indicated, and to my great astonishment, found they were in favor of the proposed action, but they said the delegation would not, as a delegation, join in it, but that such a meeting would serve a good purpose. I was encouraged, and reported to Mr. Corning in the evening full of hope. He said the move was very grave, and wanted time to think it over, would give me his answer the next afternoon, and in the meantime I was to continue to feel the pulse of the members of the New York delegation, which I did not do, but occupied myself with certain affairs at the War Department, where I had been invited to give information about the Coast and inland waters of North Carolina.

The afternoon appointed, Mr. Corning informed me that he had about made up his mind to call a meeting for consultation, to which I would be invited for the purpose of stating my views; that a general discussion would probably follow and possibly something might grow out of it; he would consult with one or two of his confreres with reference to time, etc. This was the end. The next day, Mr. Corning received a telegram announcing the serious illness of his wife at Albany, wrote me a hasty note of regrets, went home and did not return for several months. At the same time a letter reached me from General Wool describing coming events along the Atlantic Coast in which my command was to be interested, and advising my immediate return to Fortress Monroe.

Thus ended my campaign against McClellan. It was at once a victory and a defeat. As events proved I was right in my estimate of that officer, but my foresight cost me a world of opprobrium, and I was pointed out as the D—d Volunteer Colonel who ought to have been cashiered for having spoken disrespectfully of General McClellan.' This ardent desire to be of great service to my country, also prevented my promotion.[13] After I was mustered out of the service, Simon Draper asked Secretary of War Stanton, 'Why was Hawkins not promoted?' The answer, as I have it in a letter from Mr. Draper was: 'I remember him well. He had the faculty of discovering fools and traitors in advance of the rest of us, and was not discreet enough to keep his discoveries to himself. Such an officer could not be popular, consequently undesirable for promotion.' This was literally true. For President Lincoln told Senator Conness, of California, that Burnside said I was one of the most efficient and reliable Officers in his command, but he could not recommend me for promotion because I was so unpopular with his general Officers.

At Washington, one beautiful Sunday in the month of May, 1864, I had my last interview with the good President. It was a social morning call with Hugh McCulloch and Simon Draper. The former, not knowing that I had ever met the President before, presented me to him. Mr. Lincoln came towards me with an extended hand, and said, 'We don't need an introduction,' and then, turning to the others, said to them, 'This is the Colonel who in the fall of 1861, tried to induce me to remove McClellan from the command of the Army. At one time I thought I would have to arrest him for a traitorous conspiracy,' and then after a short pause he turned to me and said, 'Poor George, you knew him better than any of us. I did all I could for him, but he could do nothing for himself.' Later on, in the general conversation, he told us he had kept McClellan in command fully six weeks after he was convinced of his unfitness to command the Army.[14] 'But what could I do? Everybody and all the newspapers said he was the only man that could command it.'

This interview lasted fully an hour and all through it the President was in the minor key. His was the saddest expression I have ever seen; once in a while he would brighten a little and tell one of his inimitable jokes, and at once go back again to a condition of absolute dejection. In me, he excited, as no other human being ever has, admiration, affection, and pity.

The parting with him upon this occasion was the last. I never saw him again. Undoubtedly he was one of the most remarkable characters that ever lived, and the World can never begin to know how much he meant when he gave utterance to the immortal sentence — "With malice towards none, and charity for all." His last inaugural was the death song of the martyr, and one of the most pathetic and beautiful compositions in any language. More than any other, it reminds us of the sayings and teachings of the most beautiful character known to sacred or profane history. As with Him, the whole life of the martyred Lincoln, was filled with Love and Charity for the whole human race.

The foregoing is neither a vindication or an explanation. It was written for the purpose of giving permanence to an exceptional series of incidents which were somewhat connected with important public interests. And also, to show that those who stand, uncompromisingly, for what they know to be right usually fail, while those who do wrong, compromise with dishonesty, and violate every principle of natural justice, succeed. The former have but little sympathy or support, are unpopular, languish hopelessly through life and end their existence without friends. The latter, after having, by nefarious methods, obtained wealth, can if they desire, achieve the so-called highest positions in church, state and society, and when they die a venal press sounds peans of praise in honor of virtues they never possessed. Such conditions, I know, do not obtain in several other countries, but in mine they are the acknowledged rule. In the United States, at this time, successful wrong is accepted right.

New York, March 31, 1893.
Rush C. Hawkins."

(See page 20 for footnotes.)

Echoes of War

"Lincoln's Playmate"

by
Thomas J. Evans

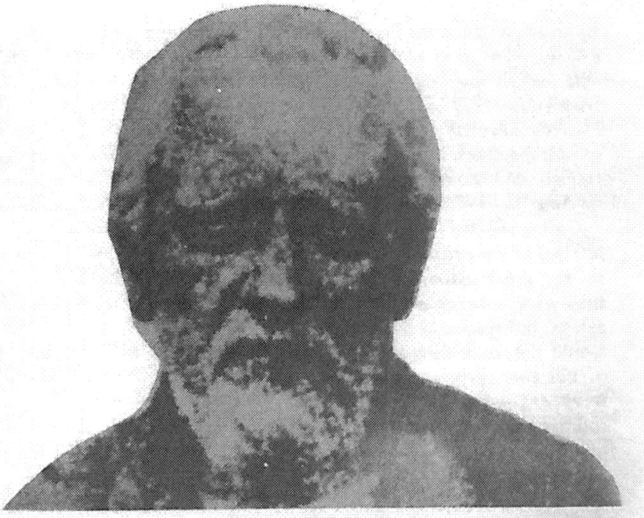

Lincoln's playmate — Austin Gollaher. This is the only known photograph of him.

During May, 1891, Austin Gollaher, a childhood playmate of Abraham Lincoln, was invited to address the students of the Buffalo, Kentucky, school at their picnic ending the school year. H.J. Graham, a student at the school, copied the speech as follows:

"The Lincoln Family moved from Abe's birthplace to their new home in Knob Creek in 1811. The new home was near my home and our parents visited each other often and Abe and I soon became acquainted, played together in the fields close to our homes and attended the school of Zachariah Riney at what is now Athertonville, Kentucky, and near our homes. Mr. Riney was Abe's first school teacher. Occasionally we went with our fathers to the grist mill of John Hodgen, which was at the present site of the town of Hodgenville and for whom the town takes its name. John Hodgen lived with his widowed mother, Sarah Hodgen, near the mill. She was a kindly disposed woman and would often take us to her house and make us an 'ash cake.' She did this by raking coals of fire and ashes out of the fireplace and kneading the wheat dough and spreading it on the coals and ashes and it was soon ready to be eaten and it was good, too.

"I can see Abe now in my memory, as he sat and ate his cake, smiling and wiggling his toes in delight, while Mrs. Hodgen looked on graciously and kindly, in her motherly way.

"As stated, Abe and I would go out into the fields near our homes and play and throw rocks at rabbits and birds, wrestle with each other and occasionally tear our clothes or soil them badly, and our mothers did not like that at all and grumbled about it at times.

"On one occasion after a hard rain, and a rise of the water in the Creek, they both looked right at us and said: 'Now, boys, run along and play but don't cross that creek. (Knob Creek.) If you cross it, we will whip you. Remember that and run along and play.' We played along, getting closer and closer to the creek all the time. I never saw it look so pretty on the other side of the creek before in my life. We finally reached the creek, at a point where a foot-log lay across it and I said: 'I don't suppose our mothers would care if we crossed if we did not fall in or get our clothes wet, would they?' Abe replied: 'I don't know, Austin, they might and they might not.' I could not stand it any longer and bolted across the log and turned and said: 'Now, come on, Abe.' He said: 'I cannot walk that log; the water whirling around under it will make my head swim and I will fall in.' I said to him: 'Yes, you can walk it - don't look at the water - look at me.' He started and got about half way across, looked down at the water and off he went. I was scared stiff and did not know what to do, so I grabbed a pole and threw it to him and then jumped into the water and pulled and tugged at him and finally got him out but I thought he was dead - his eyes were half closed and his body was limber. I took him by his heels and slung him around several times and the water just poured out of his mouth and he soon revived. Then we knew that the whipping was sure to follow but a bright idea came to us -the sun was shining bright, the weather was warm and the bushes on the bank of the creek were thick; we would pull off our clothes, wring the water out of them, and hang them on the bushes to dry and when dry, we would put them on, go home, and no one would ever know about it at all, but they did not dry as fast as we expected and when it came time to go home, we had to put them on wet, face our mothers and confess to them that we crossed the creek and that Abe fell in and came near drowning. They were frantic and almost speechless.

Question: "Mr. Gollaher, did you get the whipping?"

Answer: "No, I think they were scared so bad that they forgot to whip us."

"We played along during the years until the Lincolns decided to go to Indiana in 1816 - that was a sad thing for us and the Hodgens, and Abe's mother much preferred not to go, but Thomas Lincoln had had a lot of trouble in Kentucky with his land titles, and resolved to leave the state and go where he hoped it would be better for him.

"A few days before they started for Indiana, my mother, Abe's mother, Abe, and I went to the cemetery where near the home of the Lincolns - he lived only three days after his birth.
ed only three days after his birth.

When we arrived at the cemetery, we all knelt down, my mother prayed, Abe's mother sobbed bitterly and wiped the tears from her eyes. We started back home and she looked back at the grave of her baby and bade him a long farewell as he slept.

"When the day came to leave for Indiana, the Lincolns came by Hodgen's mill to say 'Goodbye' to them - father and I were with them. Mrs. Hodgen patted little Abe on his back and hugged him 'Goodbye,' as her eyes glistened with tears, and we started on - father and Thomas Lincoln, walking in front of the wagon, Abe's mother and his sister Sarah Lincoln, rode in it with their household goods, and Abe and I walked behing it to look after an unruly cow that had a rope around her horns at one end, and the other end of the rope tied to the wagon. Father and I went as far as Middle Creek, which is about five miles from Hodgenville and toward Elizabethtown. We all said 'Goodbye' again, Abe and I hugged each other, he pulled his cap down and they started on again, while father and I stood and watched the wagon slowly climb the long slant and finally drop over the top out of sight. I was so overcome that I could not eat anything for a day or two. That was the last time I ever saw Abe.

"Many years later, it was announced in the community that one Abraham Lincoln had been nominated for President of the United States! People were amazed and anxiously said to one another: 'Do you suppose it could possibly be Tom Lincoln's boy, Abe?' When it became an assured fact that he was the man, I resolved to go to the election early and vote for him, and I went, but a bunch of rough fellows took charge of the election and wouldn't let me vote and I was greived and mad about it.

"Dr. Jesse Rodman and David T. Brownfield, of Hodgenville, went to Washington to see him the next day after the first Battle of Bull Run. They sent in their cards and he admitted them to see him, although the city was in a state of great confusion. They said he asked about me and others of the community and told them to tell me that he wanted to see me and that if I did not have the money to make the trip, he would send it to me. They told me what he said but I did not go.

Question: "Why did you not go?"
Answer: "I was afraid to ride on the cars."

"When I heard that a fellow had shot and killed Abe, I was nearly struck down with sorrow and I have never got over it to this day. It has been many, many years since we played together on Knob Creek but I will never forget it. Sometimes now I dream that we are boys again, playing as we did in the years gone by, but when I awake, I realize it is only a dream and that I am an old man with rheumatism and other aches and pains of old age, and that my playmate has gone on to the other world, never to return, yet, I cannot help hoping that the Good Lord will in some way, or somehow, let us meet again in the world to come."[1]

Austin Gollaher's reference in his speech to the trip to Washington, D.C., by Dr. Jesse Rodman and David T. Brownfield to see Abraham Lincoln was well-rememberea by Mr. Brownfield who often related the following story:

"Dr. Jesse Rodman and I were in Washington the next day after the first Battle of Bull Run, and the city was in a high state of excitement and anxiety — the question uppermost in every mind was, 'What will the Confederates do next? Will they follow up their great victory and burn the city and commit the crimes that attend such occasions or can it be prevented in some way? What would Mr. Lincoln do about it?' Everyone, high and low, wanted to see him and hear what he had to say. General McDowell's army had crept leisurely and slowly to meet the Confederates at Bull Run, and fought all day July 21, 1861, with promise of victory to first one side and then the other, but late in the afternoon Confederate General Jackson arrived with fresh reinforcements, and the Federals broke and fled back to Washington — McDowell said it was a complete rout of his forces. The soldiers threw away their hats, shoes, and whatever hindered their flight and ran back to Washington. Mr. Lincoln and all were exasperated, but, as usual, he did not become panicky under the trying circumstances.

"In the midst of all this confusion and anxiety, we sent in our cards to see Mr. Lincoln, but with little hope whatever of seeing him, but we were told he would see us and to have seats. In a short time we saw Mr. Lincoln coming, swinging his long arms and legs, walking as leisurely as if nothing unusual confronted him. He held out his hand and said: 'Good morning, gentlemen. You say you are from Hodgenville, Kentucky?' 'Yes, Sir, we are,' I told him. 'What has become of my friend Austin Gollaher?' I told him Mr. Gollaher was still out on Knob Creek, where he had always lived. 'Tell him when you go home that I want to see him and if he does not have the money to make the trip to Washington, I will send it to him,' said Mr. Lincoln.

"We had been standing and talking, but now took seats and Mr. Lincoln looked at us rather quizzically and said: 'I suspect that you fellows are Southern sympathizers, are you not?' I replied that we were, and we all laughed heartily. Then Mr. Lincoln's face took on a very serious look and he said: 'Gentlemen, I will put down this rebellion or I will break the government in the attempt.' He looked like he meant business. Then a twinkle came in his eyes and he said: 'I do not understand this war — it is a great puzzle to me.' I asked him why it was a puzzle and he said: 'Well, for instance, it took General McDowell three weeks to go from Washington to Bull Run and only three hours to come back yesterday!' This remark broke up the meeting in laughter; he arose, shook hands with us once more, and bowed himself out on his way to his office, shaking his sides laughing as he went. That was the only time I ever saw him"[2]

RESEARCH NOTES

Mr. Graham observed Mr. Gollaher closely at the time of his speech and remembered: "He was a rather striking figure — six feet in height, florid complexion, grey eyes, broad shoulders, long shaggy eye-brows, snow-white long hair, a friendly smile and 85 years of age — he was born in 1806, on Knob Creek, in what is now LaRue County, Kentucky, some seven miles north of Hodgenville, Kentucky, and lived in the same community all his life and devoted much of his time to the best interests of the community. He taught school for two years and then turned his attention to farming but taught Sunday School as long as his health permitted it. When the hour arrived for him to speak to us, the old fellow tottered up the steps to the platform with the aid of two canes. In giving this account of the speech of Mr. Gollaher, it has been the aim to give it as nearly as possible in his words, without adding to or taking anything from it.''[3]

An 1884 interview with Austin Gollaher contained in J.E. Gollaher's *Best Lincoln Stories* provides a slightly different version of the drowning incident. In this interview Mr. Gollaher supposedly indicated it was Abe Lincoln's idea to cross Knob Creek. He stated, "While we were wandering up and down the little stream called Knob Creek, Abe said: 'Right up there' — pointing to the east — 'we saw a covey of partridges yesterday. Let's go over.' The stream was too wide for us to jump across. Finally, we saw a footlong, and we concluded to try it. It was narrow, but Abe said, 'Let's coon it.' '' It was further noted in the interview that Mr. Gollaher was then nearly a hundred years old, (he was 78), but his mind was bright and clear.[4]

Upon Mr. Gollaher's death in 1898, he was buried in the cemetery of Pleasant Grove Baptist Church located in White City, Kentucky. This historic church served as a Union Army hospital following the bloody Battle of Perryville. A Kentucky Historical Highway Marker with the following text is located at the entrance to the church: "LINCOLN'S PLAYMATE. To the west, in Pleasant Grove Baptist Church Cemetery, is the grave of Austin Gollaher, 1806-98. Lincoln, while president, once said, 'I would rather see (him) than any man living.' They were schoolmates and playmates when the Lincoln family lived in this area, 1813 to 1816. Gollaher is credited with rescuing Lincoln from flooded waters of Knob Creek.''[5] Mr. Gollaher's grave is marked by a tombstone bearing the inscription, "Lincoln's Playmate.''

Although land deed problems are given as the primary reason Thomas Lincoln decided to move to Indiana, Mr. Gollaher apparently did not remember 1816 as being "The Cold Year" — the year without a summer. In Kentucky frost occurred in every month of 1816, with ice a half-inch thick in May and a skim of ice on water on July 5th. In New York a 3 inch snow was recorded in May and most of the country experienced cold weather. The reason for this unusual season was the eruption of the volcano "Tambora" in Indonesia. Airborne derbis interfered with the sun's rays resulting in an extremely cold summer and widespread crop failure. The Lincolns lived in the valley of Knob Creek surrounded by high hills where the sun was late to rise and early to set. This, coupled with the cold summer, may well have hastened Thomas Lincoln's decision to move to Indiana.[6]

The log cabin owned by Austin Gollaher has been restored and relocated to the Lincoln Knob Creek Farm where it now serves as a tourist attraction and is open to public inspection. A Kentucky Historical Highway Marker containing the following text is locates at the site: "LINCOLN KNOB CREEK FARM. Abraham Lincoln, (1809-1865) lived on this 228 acre farm, 1811-1816. He wrote in 1860 'My earliest recollection is of the Knob Creek place.' A younger brother was born here.''[7]

The gravestone of Thomas Lincoln Jr., was badly damaged over the years by the chipping of souvenir seekers and was removed for its protection. The pioneer cemetery where he is buried lies on the hiking route of the Boy Scouts of America's Lincoln Heritage Trail, and in its place the Scouts have erected a concrete replacement inscribed "Tommy Lincoln.''[8]

Mr. Brownfield told his story of the visit to see Lincoln over and over again as long as he lived and would laugh until the tears ran down his cheeks into his long gray beard at Mr. Lincoln's jest at General McDowell's change of speed on his return to Washington. He would usually say in conclusion: "He was certainly a great man, cool and collected, and even joked with us while the city was in extreme anxiety and excitement. This feeling was not because he did not realize the mountain of responsibility that rested on his shoulders — he was just a big man and did not lose his head.''[9]

Austin Gollaher acquired neither wealth or fame during his lifetime, however he was respected by all who knew him. One can only speculate on the outcome of the Civil War if he had failed to save Abraham Lincoln's life that fateful day on Knob Creek.

FOOTNOTES

[1] From the files of Mrs. Bessie Merrifield, now deceased, granddaughter of Austin Gollaher. Her extensive knowledge of Lincoln lore and her collection were the source of information for this article.

[2] *Ibid.* Dr. Jesse Rodman remembered Lincoln as saying also: "I would rather see (him) than any man living.''

[3] *Ibid.*

[4] *The Works of Abraham Lincoln*, Volume I, The University Society, Inc., 1908, pp 16-18.

[5] Kentucky Historical Highway Marker, #827. (Pleasant Grove Baptist Church, Ky. 84, Larue Co.)

[6] "Richard Burks Award goes to Jim LaRue at Lincoln Luncheon,'' *The Herald News*, Hodgenville, Ky., Feb. 19, 1981, page 1. The article above references *The Herald News* for May 23, 1901 and *The National Geographic* magazine issue of Nov., 1980.

[7] Kentucky Historical Highway Marker, #120. (6 mi. NE Hodgenville, US 31E, Larue Co.)

[8] The old Redmon Cemetery is not accessible by automobile.

[9] *Op. cit.*, Merrifield files.

"Every Thing Is New To My Eye"

by

William F. Howard

Emma Levenia Gray was twenty years old in April, 1865, when she sailed on the *U.S.S. Cossak* and penned her thoughts in the little leather-covered diary. Emma's sister, Josephine, was married to Daniel Sheppard Thompson, the steamship captain who sailed the Chesapeake Bay and adjacent rivers carrying troops, supplies, horses and prisoners of war for the Union navy. Captain Thompson often took his wife Josephine and her unmarried sister Emma on trips with him in the Bay. The following heretofore unpublished account covers a period in April, 1865, during the time Emma was in Washington and witnessed the funeral procession for President Abraham Lincoln.

We are indebted to Woodland G. Shockley, a great-nephew of Emma Gray for providing us with the following excerpt from Miss Gray's delightful Civil War-period diary.

● ● ●

U.S.S. Cossack
April 1st 1865

We left Washington last night a 5 o clock for Eiscburough point. stopt the over-night. We took on board 100 horses for City point. left this morning at 10 clock. We past Fort Washington at 11 o clock. it tis a beautiful place. We were stoped by the Guard Ship at Alexandria to show our pass. we are now on the Potomac River. the secenery is beautiful to me for every thing is new to my eye. We are now opsite Mount Vernon the place were Washington lived and died. We are most to far form it to see the Tomb and each Hills as there is on this river I never before beheld, and there is some very handsome residents also on the banks of the river. We passed Port Tobaco at 3 o clock. We anchored at New Port News a[t] 1 o clock. We had splendid time April Fooling each other to day but I was quite fortuanote for I was not fooled once during the day.

James River
April 2nd 1865

I was quite sea sick last night for about 2 hr but I fell but little effects of it this morning. about 8 0 clock this morn we passed James - town one of the Oldest settlements in the U.S. I saw the walls of the oldest Church in the U.S. its a Catholic and it lies on the East side of the James River. the seanery is not quite so grand hear as on the Potomac R. the land is more level. 1/2 9 o clock we are now passing Wilson landing. I can see the House were president Tyler lived. it tis a long white frame house situated on a large Hill on the East side of James River. crossed Harris's Landing at 20 minutes to 11 oclock. Major Gen. Fritze Johnson H.R. 1862. & the Birth place of President Harris [son]. I have an resplendant view of of all the winters quarters of the soldiers. 12 oclock we are now at City Point. I have had Glorious news. our troops are burning Peattersburg and have got Richmond. I was at P_______ Point H_______. We stoped at Point Lookout April 17th for the first time.

Apr. 19th

It has bin some time since I have wrote in my jornel. I have bin quite neglectful. Well to day has bin a day long to remember. our Presidents remains were carried into the Capitol to day. I went to see the processions had an excleant view stood on the railings in the Capitol park. it was very warm for the month of April. I thought I should melt almost one time. the procession was 4½ miles long. it taken 2½ hr for it to pass. it was 6 o oclock when we get to the ship again and I was almost gave out for I had not bin on land for two weeks. after supper I promenaded the decks a while. after that I came in looked at them playing cards and conversed till 10 oclock and now I have bin writing for a half an hour & I think its time I had bid adieu to for the day or night.

(Continued from page 15.)

FOOTNOTES

[1] For a detailed history of the regiment, see Mathew J. Graham, *The Ninth Regiment New York Volunteers (Hawkins' Zouaves)*. New York, 1900.

[2] Margaret Bingham Stillwell, *General Hawkins as he revealed himself to his Librarian*. Providence, 1923, p. 19. (Reprinted from the Papers of The Bibliographical Society of America).

[3] *Ibid*, p. 19.

[4] John Wolcott Phelps became colonel of the 1st Vermont in May 1861. The regiment saw action at Newport News and later in the Lower Mississippi. In 1880, Phelps ran for President on the Antimason ticket. (Ezra J. Warner, *Generals in Blue*, Louisiana State University Press, 1964, p. 368).

[5] Governor Edwin D. Morgan of New York.

[6] Thomas Williams became Brigadier-General of Volunteers in September 1861. He took part in Burnside's North Carolina expedition and was in command at Fort Hatteras until March 1862. Five months later, while defending Baton Rouge, he was mortally wounded. (*Warner*, p. 563).

[7] General John Ellis Wool was the oldest officer to exercise command on either side having served in the army since the War of 1812. In 1861, he was Commander of the Department of the East and was responsible for the relief of Fort Monroe. He retired from the army in August 1865. (*Warner*, p. 573).

[8] William Denison Whipple served with Hunter's division at First Manasses as assistant adjutant general. In November 1863, he became assistant adjutant general of the army. (*Warner*, p. 555).

[9] At Ball's Bluff, Devens was saved when a rifle bullet hit a button on his uniform. He commanded a brigade at Seven Pines and saw action at Fredericksburg and Chancellorsville, where he was wounded for a third time. (*Warner*, p.122).

[10] Charles Pomeroy Stone was colonel of the 14th Regular Infantry Brigade. He served at First Bull Run and later on the Upper Potomac, where he was made to bear the burden of the Union disaster at Bull's Bluff for which he was arrested and confined for 189 days. In 1863 he was assigned to the Department of the Gulf and saw action at Port Hudson and the Red River Campaign. (*Warner*, p. 480).

[11] Louis Blenker was appointed Brigadier-General of volunteers in August 1861. He commanded a brigade at First Manasses, and later served in the Shenandoah Valley. (*Warner*, p. 37).

[12] General James Samuel Wadsworth served under McDowell at First Manasses and later fought at Gettysburg. He was killed at the Wilderness in May 1864. (*Warner*, p. 532).

[13] The promotion, however, came in 1866 when Hawkins was raised to the rank of

(Continued on page 26.)

The Order of Presidents procession April 19th 1865 Washington

10th Veteran R.C. with regimental Band
9th " Reserved Corpes. Colored Gile with regimental Band.
Battalion of marine Corpes & band.
2 Batteries 89th U.S. battery
16th N.Y. Cavalry Colonel Sweltzer
8th Illinois " " Clendenin.
13th N.Y. mounted band.
General Ketchum & Staff.
" Slough " "
Commander of Escort and Staff mounted.
Navy Officers on foot
Army " " "
Surgeons & Assistant — surgeons on foot.
Cavalry & Infantry officers mounted
Clergy
General Grant & Officers in carriages.
Catafalco, containg Corpse
and guarded by 1st Virginia Artillery on foot.
Presidents Horse gray with boots of presidents in stirrups
Presidents Guard.
6 Carriages single file containing the relatives of deceased.
The delegations of the State of Illinois & Kentucky as mourners.
President Johnson.
The Cabinet Ministers.
Foreign Legation.
Senators & Congressmen.
The Chief Justice & Associate Justice of Supreme & City court. the Assistants Secretaries of the differants departments.
Knights Templar & Band.
Phila city councils.
Citizens of Washington.
4th U.S. battery & band.
Delegations representing Ohio, New Jersey, & California.
Clerks of the Treasury department
" " " Ordnance "
" " Post Office "
" " War "
" " " Interior "
Army Surgeons mounted.
Battalion of Q.M. General's Office and band.
Baltimore City councils and Custom house officers.
Quartermasters band.
1st Regiment Meigs home guard.
2nd " " " "
Employes Q.M. general's office
8th Illinois Cavalry band.
U.S. Military R.R. employes.
National republican Union association of the 7th ward.
Citizens of Alexandria with band
Friendship No 1. & S[?] No 2. five companies of the same city.
Potomac hose company of Georgetown.
M. Vernon Association of Alexandria Va.
Lincoln hospital soldieres.
Finley hospital band and soldiers of hospital.

Arsenal Employes.
Roman Catholic schools led by Father Wiget.
Gonzaga & Georgetown Carriages.
Merrity's band.
East Baltimore Union league.
Hebrew congregation.
Turners.
Baltimore city cornet band.
Order of Ancient good fellows.
Federal city division No 2. Sons of Temperance followed by
Carver Hospital band.
Good Samaritan No. 1.
Ansora division No 9. Equal division No 4. Lincoln D. No 17.
Central D. No 12. Empire — D. No 19. Hope Division No 18.
Everett D. No 25. Mount Vernon Division.
Grand division: 1,500 in all.
Mount pleasant band
Columbia Typographical society.
Emory hospital band and soldiers of Hospital.
Washington United benevolent association - - Colored.
Harmony Lodge No 18. C.U.A. of D.F. - - " "
King Heizehiah's pasture No 3. F.A.A.M. = " "
Union Grand Lodge — F.A.A.M. - - " "
Citizens & Strangers..

Apr 20th Jones R.

We left Washington this morning for Geasaburough and taken in 100 horses for City Point. it has bin raining all day and its has bin so disagreable on deck that I was forced to stay in the saloon. but I have enjoyed myself this day stealing the mens hats and hideing them and I romped so much that I am near about give out to night. Mr. Reiley tied my hands 3 times behind me but I still have one of his hats witch I will keep for a while. the officers are now singing in the after saloon and I will have to stop to listen.

Postscript

Emma Levenia Gray was born in Philadelphia, Pennsylvania, January 8, 1845, the second of seven children born to William and Emeline Gray. On January 8, 1868, she married Alanson H. Chittenden in Milford, Delaware. The couple produced four children, two of whom died shortly after birth, and a son, Alanson, who drowned in a boating accident on the Mispillion River near Milford in 1874.

After the death of her husband, Emma Gray Chittenden later married Alfred E. Scott on October 20, 1887. They had one child, which apparently did not live to maturity. Emma died on December 30, 1913, in Milford, and is buried in the Odd Fellows Cemetery there.

ADDENDUM

Daniel Sheppard Thompson (1826-1897) was appointed as an acting ensign on temporary duty with the U.S. Navy on October 9, 1862. He was stationed first at Brooklyn Navy Yard, and then shipped out for instruction on the *Savannah* from October 17 to November 22, 1862. He served on the *Young Rover* from November 22, 1862, to April 23, 1864, and on the *Victoria* from April 24 to May 27, 1864. While serving on the *Victoria* of Wilmington, North Carolina, he contracted rheumatism and bronchitis from exposure, and was sent to Norfolk Hospital, Virginia, where he remained from May 28 to June 12, 1864. He resigned from the Navy on June 3, 1864, in consideration of his illness, and was discharged for disability June 10, 1864.

Nine months later, however, Thompson was given command of the U.S. Transport *Cossack*, and carried Confederate prisoners from Richmond to Baltimore. After the war, Thompson reportedly served as captain of the *Florida* from 1866-1867. From: *Pension Records*, National Archives and family records of Marion Shockley, Milford, Delaware.

CONCURRENT EVENTS

The Cherokees — In the Senate on Friday, Mr. White of Tennessee, from the Committee on Indian Affairs, in answer to the petition of John Ross and others, of the Cherokee Tribe of Indians, reported the following resolution: Resolved, That the President of the United States be authorized and requested to cause to be ascertained upon what terms the claims of the State of Georgia and its citizens, to the lands of the Cherokees east of the Mississippi, can be extinguished, and communicate the same to Congress at the next session — and that the further consideration of the memorial be postponed to that time. [*American Railroad Journal and Advocate of Internal Improvements*, June 28, 1834.]

"Abraham Stoefield: Black American Soldier"

Edited by: Daniel Johnson

Slavery was an issue that had perplexed the people of North America before there was an independent nation called the United States. The Founding Fathers themselves were divided on the question, and often personal wishes could not be carried into law or acted out in their lives. George Washington freed his slaves upon his death, but Thomas Jefferson was unable to. Virginia law dictated that if a slave-owner desired to free his servants, he must provide for them all their lives so they would never be destitute or a burden to society. The money the first president set aside covered the needs of his former slaves, until the last one died, while Jefferson, who died in literal poverty, could not provide an income for his slaves, and thus they were not liberated.

The feeling that Negroes were inferior to the white race was widespread, as was an almost universal fear of blacks individually and collectively. Harsh laws were enacted to prevent servile insurrection, and religion was encouraged in the hope that the black man would seek his rewards in Heaven rather than on earth.

As men became more enlightened and the abolitionist movement more accepted, the gap between those who supported slavery and those who did not widened. Economic factors, interpretation of the Bible and basic prejudices worked against the black race, while the humanitarians labored for their basic rights of life, liberty and pursuit of happiness. As differences between North and South intensified, the question of slavery took on more and more significance, until the nation was brought to Civil War. Hundreds of thousands of men and boys went to war, in what would become irrevocably the war over slavery, on January 1st, 1863. That the Emancipation Proclamation was a political weapon is indisputable. It changed the complexion of the war and kept England and France from interceding in the conflict. Also, the way was opened for the Union army to recruit, mobilize, and finally use black troops in combat. What it could not do was erase the uncomfortable or hostile feelings the Union soldiers had, or assure commanding officers that these black men could fight. That, only time and experience would prove. With every duty done, every opportunity and battle willingly faced, the Colored Troops made believers out of their white companions.

The story of one black man, first slave, then soldier, is outlined below. Written by a wounded soldier and printed in *Hospital Register*, a newspaper for the convalescing Union men, September 5, 1863, this article represents a significant positive attitude in the acceptance of the black soldier. It was won with pride, courage and determination, and paved the way for others to follow. Not an end, but a beginning, Abraham Stoefield was an *American* soldier.

THE CAMP STOOL TO THE CANVAS-BACK CHAIR.

Will these blacks fight? In answer I will tell you the record of one; let him be the type of all.

Abraham Stoefield was of the despised and degraded race; his skin was dark, his soul uninformed; the life of a slave had been his portion, and that of his forefathers for three generations back; the iron had entered his soul, but he felt it not; he was callous and indifferent, without knowledge of the past, enjoyment of the present, or hope for the future. He was sullen and slow about his work, and his back was scarred by the lash, inflicted as a punishment for his shortcomings. — All this he bore as the ox does its burden — without thought.

And if he met with his fellows in secret down in the swamp at night, and with vigilant watchers set out against surprise, talked together and prayed in their poor way, that the day of the Lord might come, bringing with it deliverance, still he knew little of what he said, and the next day went to his unwilling labor on the plantation, hardly remembering the bitter words and prayers of the night before.

And so he might have done, he and his descendants, for all time, just as his people have been doing for so many years, but for one thing. The power which held him down there, unpaid and hopeless, essayed to do a mightier thing; even to lay hold of the nation and enslave it likewise. To make the strong North bow down to the weak South, even as labor had been held to be the property of sloth.

When the army of the North was slowly but surely forcing its way northward from the conquered city of New Orleans to meet the armies of the West in their triumphant southward march, it passed over, and encamped for a time on the plantation where Stoefield labored. His master and overseer had fled, there were none left but those of his own race; so with mingled curiosity and awe, he followed in the train of this force, which appeared to him to be a host numberless and omnipotent.

He had then no thought of taking the musket to avenge his wrongs, and if he had fled from oppression, it was but a dim gleam of light that entered his dark soul, when he felt that the day of

(Continued on page 30.)

FAMILY BIBLE

Birthdays of Civil War Notables

by

Paul Kallina

(N = North. S = South)

AUGUST

Aug. 1, 1813—Capt. Daniel B. Ridgeley, U.S.N., near Lexington, Kentucky. (N)
Aug. 1, 1814— Gen. Maxcy Gregg, Charleston, South Carolina. (S)
Aug. 1, 1819—Herman Melville, *Battle-Pieces and Aspects of the War*, New York City. (N)
Aug. 1, 1835—Rep. Elias C. Boudinot, near Rome. Georgia. (S)
Aug. 1, 1843—Capt. Robert Todd Lincoln, Springfield, Illinois. (N)
Aug. 2, 1839—Col. Scott Ship, Warrenton, Virginia. (S)
Aug. 3, 1808—Hamilton Fish, commissioner for prisoner exchange (1862-1865), New York City. (N)
Aug. 3, 1820—Gen. William Miller, Ithaca, New York. (S)
Aug. 3, 1823—Gen. Thomas F. Meagher, Waterford, Ireland. (N)
Aug. 4, 1823—Oliver P. Morton, governor of Indiana (1861-1867), Salisbury, Wayne County, Indiana. (N)
Aug. 5, 1818—Gen. Carnot Posey, Wilkinson County, Mississippi. (S)
Aug. 5, 1821—Commander Foxhall A. Parker, New York City. (N)
Aug. 6, 1811—Judah P. Benjamin, cabinet official (1861-1865), St. Croix, British West Indies. (S)
Aug. 6, 1819—Rep. Preston S. "Bully" Brooks, Edgefield, South Carolina. (S)
Aug. 6, 1819—Gen. Samuel B. Carter, Elizabethton, Tennessee. (N)
Aug. 8, 1812—Commodore John Rodgers, near Havre de Grace, Maryland. (N)
Aug. 8, 1819—Charles A. Dana, Assistant Secretary of War (1863-1864), Hindsdale, New Hampshire. (N)
Aug. 8, 1820—Gen. Winfield S. Featherston, near Murfreesboro, Tennessee. (S)
Aug. 8, 1822—Gen. George Stoneman, Busti, New York. (N)

WHIMSEY

Railway Smoking - It is announced that the smoking salons are only for the fast trains of the Eastern Counties Railway. We presume that the fast trains on the line are those which are stuck fast in consequence of stoppages. It is certainly desirable to allow the passengers, under such circumstances, the opportunity of smoking, instead of leaving them to fume, as they have done hitherto.
[*Punch*, 1846.]

Aug. 8, 1839—Gen. Nelson A. Miles, near Westminster, Massachusetts. (N)
Aug. 9, 1823—Gen. Daniel M. Frost, Schenectady County, New York. (S)
Aug. 10, 1801— Sen. Robert W. Barnwell, near Beaufort, South Carolina. (S)
Aug. 10, 1814—Gen. John C. Pemberton, Philadelphia, Pennsylvania. (S)
Aug. 10, 1814—Sen. William L. Yancey, Warren County, Georgia. (S)
Aug. 10, 1821—Jay Cooke, financier, Sandusky, Ohio. (N)
Aug. 10, 1841—Lt. George C.Remey, U.S.N., Burlington, Iowa. (N)
Aug. 10, 1843—Ens. Charles E. Clark, Bradford, Vermont. (N)
Aug. 12, 1819—Gen. Daniel D. Bidwell, Black Rock, New York. (N)
Aug. 12, 1829—Gen. John H. Forney, Lincolnton, North Carolina. (S)
Aug. 13, 1830—Capt. John T. Wood, C.S.N., Fort Snelling, Michigan Territory. (S)
Aug. 13, 1831—Gen. Henry L. Abbott, Beverly, Massachusetts. (N)
Aug. 13, 1851—Sgt. John L. Clem, the drummer boy of Shiloh, Newark, Ohio. (N)
Aug. 14, 1821—Gen. Clement H. Stevens, Norwich, Connecticut. (S)
Aug. 14, 1824—Gen. William Terry, Amherst County, Virginia. (S)
Aug. 16, 1817—Rep. Henry W. Davis, Annapolis, Maryland. (N)
Aug. 16, 1828—Gen. Joseph B. Carr, Albany, New York.(N)
Aug. 18, 1792—Lord Russell (John Russell), British Foreign secretary (1859-1865), London, England.
Aug. 18, 1807—Charles Francis Adams, U.S. minister to England (1861-1868), Boston, Massachusetts. (N)
Aug. 18, 1818— Gen. William F. Barry, New York City, New York. (N)
Aug. 18, 1824— Gen. John W. Davidson, Fairfax County, Virginia. (N)
Aug. 18, 1846—Ens. Robley D. Evans, Floyd Court House, Virginia. (N)
Aug. 19, 1815—Gen. John P. McCown, near Sevierville, Tennessee. (S)
Aug. 19, 1831—Gen. Stephen G. Burbridge, Scott County, Kentucky. (N)
Aug. 20, 1802—Commodore Cadwalader Ringgold, Washington County, Maryland. (N)
Aug. 20, 1824—Gen. Absalom Baird, Washington, Pennsylvania. (N)
Aug. 20, 1832—Thaddeus S. C. Lowe, aeronaut, Jefferson Mills, New Hampshire. (N)
Aug. 20, 1833—Gen. Benjamin Harrison, North Bend, Ohio. (N)
Aug. 20, 1834—Gen. Francis R. T. Nicholls, Donaldsonville, Louisiana. (S)
Aug. 21, 1789— Thomas Garrett, abolitionist, Upper Darby, Pennsylvania. (N)

• SIGNALS •

Dear Editors:

I have enclosed a check for $18.00 to cover the cost of a year's subscription to THE KEPI. I just received the April-May issue which I purchased singly, and I found it informative and entertaining. Please begin my subscription with the June-July issue.

Gerald R. Gannon
Hurst, TX

Dear Editors:

I particularly enjoyed reading "To Raise A Flag." As a student of the Gettysburg period, I appreciated the new insights this article afforded me.

Tim Doran
Houston, Texas

Dear Editors:

Tom Evans' work is excellent! I had despaired of ever discovering *new* information on Mosby, but after reading about his Rangers in your June/July issue, I know nothing is impossible. Keep up the good work. Your maga-

Aug. 21, 1821—Gen. William Barksdale, Smyrna, Rutherford County, Tennessee. (S)
Aug. 22, 1827—Gen. Joseph A. Mower, Woodstock, Vermont. (N)
Aug. 22, 1834—Gen. Nathaniel H. Harris, Natchez, Mississippi. (S)
Aug. 23, 1788—Gen. Joseph G. Totten, New Haven, Connecticut. (N)
Aug. 23, 1824—Sen. William T. Dortch, near Rocky Mount, North Carolina. (S)
Aug. 24, 1810—Theodore Parker, abolitionist, Lexington, Massachusetts. (N)
Aug. 24, 1827—Gen. Walter H. Stevens, Penn Yan, New York. (S)
Aug. 24, 1828—Gen. George H. "Maryland" Steuart, Baltimore, Maryland. (S)
Aug. 25, 1819—Allan Pinkerton, secret service agent, Glasgow, Scotland. (N)
Aug. 26, 1811—Gen. Danville Leadbetter, Leeds, Maine. (S)
Aug. 26, 1819—Albert, Prince Consort of Great Britain (1840-1861), Rosenau, Germany.
Aug. 26, 1840—Lt. John W. Philip, U. S. N., Kinderhook, New York. (N)
Aug. 27, 1800—Gen. William S. Harney, Haysboro, Tennessee. (N)
Aug. 27, 1809—Vice President Hannibal Hamlin, Paris Hill, Maine. (N)
Aug. 27, 1824—Gen. Hiram G. Berry, Rockland, Maine. (N)
Aug. 27, 1824—Gen. Max Weber, Achern, Grand Duchy of Baden, Germany. (N)
Aug. 27, 1827—Gen. Roger W. Hanson, Clark County, Kentucky. (S)
Aug. 27, 1832—Gen. James A. Walker, near Mt. Sidney, Augusta County, Georgia. (S)
Aug. 27, 1839—Gen. Emory Upton, Batavia, New York. (N)
Aug. 28, 1822—Frederick W. Howe, Arms manufacturer, Danvers, Massachusetts. (N)
Aug. 28, 1828—Gen. William A. Hammond, Surgeon-General of the Army (1862-1864), Annapolis, Maryland. (N)
Aug. 30, 1820—George F. Root, song writer, Sheffield, Massachusetts. (N)
Aug. 31, 1811—Gen. Goode Bryan, Hancock County, Georgia. (S)
Aug. 31, 1822—Gen. Fitz John Porter, Portsmouth, New Hampshire. (N)

SEPTEMBER

Sep. 1, 1795—James G. Bennett, editor *New York Herald*, Banffshire, Scotland. (N)
Sep. 1, 1798—Gen. Richard Delafield, New York City. (N)

zine and Tom Evans are outstanding.

Max Bailey
New Haven, Conn.

[*Another excellent article by Tom Evans on Col. Mosby entitled "The Affair At Fairfax Court-House" appeared in the April/May issue. Mr. Evans diligent research and dedication to the preservation of oral history is a valuable edition to THE KEPI and we are proud to present his work to our readers. The Editors.*]

Dear Editors:

"*A new concept in Civil War Reporting*" accurately describes an excellent magazine even when I hold a different point of view. Such things being subjective I have to satisfy myself with being picky over a side comment in your review of *How The North Won* in the June-July 1984 issue. My copy of William Davis' first volume of *The Image War*, p. 199, calls Jackson

Concurrent Events

Slave Population

1790 - 697,897
1800 - 893,041
1810 - 1,191,364
1820 - 1,538, 064
1830 - 2,009,031
1840 - (est) 2,700,000

[*The Emancipator*, April 2, 1840.]

a general at First Bull Run.

William F. Banner
Greensboro, NC

[*The first edition of* "Shadows of the Storm" *(Images of War), 1981, page 199, reads:* "...'*There stands Jackson like a stone wall.' It was the brigade of Colonel Thomas J. Jackson, shown here in a 1855 daguerreotype. A man of inordinate peculiarities, he would become the greatest legend of the War.*" *Immediately after receiving our copy, we wrote to Mr. Davis, informing him about Jackson's rank. He was, of course, a brigadier-general in July, 1861.*

We are extremely pleased to know that subsequent editions have corrected this gross error. Incidentally, the description "*inordinate pecularities*" *is a* "*cheap shot,*" *being both inaccurate and out-of-keeping with the tone of the book. The Editors.*]

CONCURRENT EVENTS

Important from Harper's Ferry - There is no truth in the report that employees of the Baltimore and Ohio Railroad Company have been imprisoned there for refusing to serve in the Virginia army, or that the company intend to close their machine shops here. The Virginia authorities give as a reason for detaining the (coal) trains that it was understood that the coal was for the use of the government. Assurances were sent by the railroad authorities that such was not the fact, and it is supposed that the trains will soon be permitted to run as usual.

[*The New York Herald*, May 26, 1861.]

(Continued from page 20 *.)*

Brigadier General by brevet.

[14] On November 7, 1862, McClellan was directed to turn the army over to Burnside. Among other things, this was due to his indecisiveness and failure to move against Lee after Antietam in September.

Sep. 1, 1824—Gen. Isaac H. Duval, Wellsburg, Virginia. (N)
Sep. 1, 1829—Gen. James Conner, Charleston, South Carolina. (S)
Sep. 2, 1798—Sen. Thomas H. Hicks, governor of Maryland (1857-1862), Dorchester Copunty, Maryland. (N)
Sep. 2, 1833—Henry Hotze, diplomat and propagandist, Zurich Switzerland. (S)
Sep. 2, 1837—Gen. James H. Wilson, near Shawneetown, Illinois. (N)
Sep. 3, 1825—Gen. Armistead L. Long, Campbell County, Virginia. (S)
Sep. 3, 1831—Gen. States Rights Gist, Union District, South Carolina. (S)
Sep. 4, 1793—Edward Bates, Attorney General (1861-1865), Belmont, Virginia. (N)
Sep. 5, 1793—John L. Burns, hero of Gettysburg, Burlington, New Jersey. (N)
Sep. 5, 1804—Sen. William A. Graham, Lincoln County, North Carolina. (S)
Sep. 5, 1847—Jesse W. James, guerilla, Kearney, Clay County, Missouri. (S)
Sep. 6, 1797—Gen. William "Extra Billy" Smith, governor of Virginia (1864-1865), Marengo, King George County, Virginia. (S)
Sep. 6, 1815—Gen. St. John R. Liddell, near Woodville, Mississippi. (S)
Sep. 6, 1816—Col. Francis S. Bartow, Savannah, Georgia. (S)
Sep. 6, 1819—Gen. William S. Rosecrans, Little Taylor Run, Delaware County, Ohio. (N)
Sep. 7, 1811—Gen. William H. Emory, Poplar Grove, Queen Annes County, Maryland. (N)
Sep. 7, 1815—Gen. Howell Cobb, Jefferson County, Georgia. (S)
Sep. 8, 1811—Gen. Lucius B. Northrop, Charleston, South Carolina. (S)
Sep. 8, 1821—Gen. Henry Baxter, Sidney Plains, Delaware County, New York. (N)
Sep. 8, 1828—Gen. Joshua L. Chamberlain, Brewer, Maine. (N)
Sep. 8, 1828—Gen. George Crook, near Dayton, Ohio. (N)
Sep. 9, 1809—Commodore William Radford, Fincastle, Virginia. (N)
Sep. 9, 1819—Gen. Martin L. Smith, Danby, New York. (S)
Sep. 9, 1829—Rep. Julian Hartridge, Daufuskie Island, Beaufort District, South Carolina. (S)
Sep. 9, 1834—Gen. William MacRae, Wilmington, North Carolina. (S)
Sep. 10, 1787—Rep. John J. Crittenden, Versailles, Kentucky. (N)
Sep. 10, 1832—Gen. Randall Lee Gibson, Woodford County, Kentucky. (S)
Sep. 10, 1836—Gen. Joseph Wheeler, near Augusta, Georgia. (S)
Sep. 11, 1821—Erastus Beadle, dime novel publisher, Pierstown, New York. (N)
Sep. 11, 1835—Gen. William W. Allen, New York City, New York. (S)
Sep. 12, 1806—Adm. Andrew H. Foote, New Haven, Connecticut. (N)
Sep. 12, 1818—Richard J. Gatling, inventor, Hertford County, North Carolina. (N)
Sep. 13, 1813—Gen. John Sedgwick, Cornwall Hollow, Connecticut. (N)
Sep. 13, 1836—Gen. John McCausland, St. Louis, Missouri. (S)
Sep. 14, 1833—Col. Robert J. Breckinridge, Jr., Baltimore, Maryland. (S)
Sep. 14, 1835—Col. Jacob F. Kent, Philadelphia, Pennsylvania. (N)
Sep. 14, 1838—Maj. John Pelham, Benton County, Alabama. (S)
Sep. 16, 1817—Sen. Waldo P. Johnson, Bridgeport, Harrison County, Virginia. (S)
Sep. 16, 1832—Gen. George Washington Custis Lee, Fortress Monroe, Virginia. (S)
Sep. 17, 1800—Adm. Franklin Buchanan, Baltimore, Maryland. (S)
Sep. 17, 1811—John Brough, governor of Ohio (1863-1865), Marietta, Ohio. (N)

Sep. 17, 1820—Gen. Earl Van Dorn, near Port Gibson, Mississippi. (S)
Sep. 17, 1825—Col. Lucius Q. C. Lamar, Eatonton, Putnam County, Georgia. (S)
Sep. 18, 1812—Sen. Herschel V. Johnson, Burke County, Georgia. (S)
Sep. 18, 1818—Gen. Marcellus A. Stovall, Sparta, Georgia. (S)
Sep. 19, 1813—John Luke Porter, naval architect, Portsmouth, Virginia. (S)
Sep. 20, 1799—Commodore George N. Hollins, Baltimore, Maryland. (S)
Sep. 20, 1809—Gen. Sterling Price, Prince Edward County, Virginia. (S)
Sep. 20, 1820—James P. Holcombe, C.S. commissioner to Canada (1864-1865), Powhatan County, Virginia. (S)
Sep. 20, 1820—Gen. John F. Reynolds, Lancaster, Pennsylvania. (N)
Sep. 20, 1829—Gen. Albert J. Myer, Newburgh, New York. (N)
Sep. 21, 1817—Gen. Carter L. Stevenson, near Fredricksburg, Virginia. (S)
Sep. 21, 1827—Gen. Michael Corcoran, Carrowkeel, County Donegal, Ireland. (N)
Sep. 21, 1832—Gen. Samuel S. Carroll, Takoma Park, Maryland. (N)
Sep. 22, 1796—Commodore Cornelius K. Stribling, Pendleton, South Carolina. (N)
Sep. 22, 1821—Gen. Patrick T. Moore, Galway, Ireland. (S)
Sep. 22, 1822—Gen. Eppa Hunton, Fauquier County, Virginia. (S)
Sep. 22, 1829—Gen. William W. Belknap, Newburgh, New York. (N)
Sep. 22, 1833—Gen. Stephen D. Lee, Charleston, South Carolina. (S)
Sep. 24, 1816—Henri Mercier, French minister to the U. S. (1860-1863), Baltimore, Maryland.
Sep. 25, 1806—Commander James H. Ward, Hartford, Connecticut. (N)
Sep. 25, 1847—Vinnie Ream, Sculptress, Madison, Wisconsin. (N)
Sep. 27, 1803—Adm. Samuel F. Dupont, Bergen Point, New Jersey. (N)
Sep. 27, 1809—Adm. Raphael Semmes, Charles County, Maryland. (S)
Sep. 27, 1809—Capt. David S. McDougal, U.S.N., Chillicothe, Ohio. (N)
Sep. 27, 1830—Gen. William B. Hazen, West Hartford, Vermont. (N)
Sep. 27, 1838—Gen. Lawrence S. Ross, Bentonsport, Iowa. (S)
Sep. 27, 1840—Lt. Alfred Thayer Mahan, U. S. N., West Point, New York. (N)
Sep. 27, 1840—Thomas Nast, political cartoonist, Landau, Bavaria. (N)
Sep. 28, 1828—Gen. John Gregg, Lawrence County, Alabama. (S)
Sep. 28, 1828—Maj. Theodore Winthrop, New Haven, Connecticut. (N)
Sep. 28, 1833—Gen. James Fessenden, Westbrook, Maine. (N)
Sep. 28, 1841—Georges Clemenceau, correspondent for *Le Temps*, Mouilleron-en Pareds, Vendée, France.
Sep. 29, 1831—Gen. John M. Schofield, Gerry, New York. (N)
Sep. 29, 1840—Col. Henry Kyd Douglas, *I Rode With Stonewall*, Shepherdstown, Virginia. (S)
Sep. 30, 1805—Col. Samuel P. Heintzelman, Manheim, Pennsylvania. (N)
Sep. 30, 1819—Gen. Thomas Jordan, Luray, Virginia. (S)
Sep. 30, 1824—Gen. Charles P. Stone, Greenfield, Massachusetts. (N)

BIBLIOGRAPHY

Biographical Directory of the United States Executive Branch, 1774-1971. Ed. Robert Sobel, Westport, Connecticut: Greenwood Publishing Company, 1971.
Dictionary of American Biography. 20 vols. New York: Charles Scribner's Sons, 1928-1936.
The National Cyclopaedia of American Biography. 57 vols. New York and Clifton, New Jersey: James T. White & Company, 1892-1977.
Wakelyn, John L. *Biographical Dictionary of the Confederacy*. Westport, Connecticut: Greenwood Press, 1977.
Webster's American Military Biographies. Ed. Robert McHenry. Springfield, Massachusetts: G. & C. Merriam Company, 1978.

"The Father And The Savior"

by
James P. Kelley

The image of George Washington, America's "patron saint" was very much a part of the 19th Century. Woodcuts, engravings and paintings of the first president adorned homes all across the United States, up to and during the Civil War. A reverence of Washington was professed by both North and South, and frequent accusations were made, as landmarks associated with the Father of His Country, including Arlington and the White House, found themselves in the path of war.

It was natural that Abraham Lincoln would try to associate himself with Washington. As a child, he avidly read Mason Locke "Parson" Weems *Life of Washington*.[1] The good parson, a medical student turned Anglican priest, made a living selling books, especially the Bible. A "bibliopolist" (one who links God with gain),[2] he worked the states in the South, preaching against dueling, slavery, gambling, tobacco, alcohol and advocated religious tolerance, public virtue, civic duty and education.[3] His books caught the imagination of America's youth and inspired them with tales of virtuous heroes. Like the young Lincoln, a boy named Thomas Jackson grew up on Weems' subtle but influential pages. Among his favorites was *The Life of Francis Marion*,[4] the legendary "Swamp Fox."

As a politician, an association with Washington helped promote a positive image. In 1842, on the anniversary of Washington's birthday, Lincoln gave a speech declaring "Washington... is the mightiest name on earth." By 1850, he was declaring "The Republicans are walking in the 'old paths' " of Washington and his contemporaries.[5] Running for president he promised Kentuckians on the issue of slavery that "we won't hurt you.

(Continued on page 31.)

MILITARY REVIEW

Five Tragic Hours: The Battle of Franklin by James Lee McDonough and Thomas L. Connelly, The University of Tennessee Press, Knoxville, 1983; $16.95; 185 pps., maps and illustrations, index and organizational tables.

One result of the historical bias that focuses attention on the Eastern Theatre of the Civil War while largely ignoring the Western campaigns is that many important Civil War battles have not been considered in detail by professional historians. *Five Tragic Hours: The Battle of Franklin* by James Lee McDonough and Thomas L. Connelly is an insightful volume that fills a long-neglected void in Civil War scholarship. McDonough and Connelly, two accomplished scholars eminently qualified to pursue the intricacies inherent to any study of a particular battle, tell the sad story of the Franklin battle with both compassion and understanding. Like the best Civil War monographs, *Five Tragic Hours* is more than a mere tactical history, it is also a well-researched personal story — a memorial to those who fought and died in the terrible battle that once raged on the rolling fields around Franklin, Tennessee.

Five Tragic Hours, however, succeeds on other levels as well. McDonough and Connelly explore the Franklin battle in all its dimensions. Most interesting in their approach is the authors' concentration on the personalities of Union and Confederate commanders. A well-developed psychological portrait of General John B. Hood, who commanded the Southern Army of Tennessee is advanced and reaps the benefits of Richard McMurry's recent biography, *John Bell Hood and the War for Southern Independence*. Contextual supporting evidence is also gleaned from Grady McWhiney and Perry Jamieson's strategic study *Attack and Die*. There is little doubt that the authors' portrayal of General Hood as a shell-shocked soldier whose grand delusions shattered the once invincible Army of Tennessee will arouse great controversy. McDonough and Connelly argue most convincingly that Hood lived in a world of broken dreams, and thought a great victory in Tennessee (if Franklin fell, so too would Nashville) could salvage Confederate hopes, as well as Hood's own tarnished military reputation.

McDonough and Connelly propose that Hood's own psychological problems had much to do with the ultimate Confederate slaughter at Franklin. Hood's tangled strategy, the absence of any real "plan" aside from the campaign as a diversionary move to pull Sherman out of Atlanta, was delusory. Hood seriously overestimated his own skills while virtually ignoring enemy strength and the condition of his army. Hood tended to blame his personal failings on the soldiers he commanded (frequently calling them cowards) and sought to impose discipline and raise troop morale by adopting aggressive attack tactics. But if he misunderstood his army, Hood also failed to appreciate the way war had changed since his Gettysburg wound had taken him out of service in 1863. Back in command in late 1864, Hood tried to emulate the grand offensive posture exemplified by Lee and Jackson in the early years of the war. "The Tennessee Campaign," write McDonough and Connelly, "was one last irrational act of attempting to duplicate the strategic movements of Hood's idols, Lee and Jackson."

Hood wanted to fulfill his own early promise in the tradition his military idols had already established. The problem was Hood lacked their brilliance. He was no Lee or Jackson, and lacked the intellectual capacity to turn bold maneuvers into military gains. Hood sought to fight a style of offensive war that was already archaic. By 1864, the Civil War was largely a contest of industrial might and available manpower reserves. Even Lee now fought his battles from the trenches around Petersburg. The nature of war had changed: by 1864 Hood was a warrior strangely out of keeping with his times.

The result of Hood's command failings was a confused battle that raged furiously for some five hours, and pitted the Army of Tennessee against the Union army of General John M. Schofield south of Franklin, Tennessee, on November 30, 1864. Although a relatively brief affair, Hood ordered his army to make a number of suicidal charges headlong into the strongly fortified Union line, literally sacrificing the flower of the Southern army for his hopeless dream of personal glory and Confederate triumph. In five hours, approximately 1,750 Confederates were killed, more than were lost by General McClellan during the entire Seven Days' Battles, or by General Hooker's massive army at Chancellorsville. The Franklin battle annihilated the Southern officer corps. Six Confederate generals were killed, six others wounded or captured. The authors theorize that had Schofield not ordered his Union army to fall back toward the impregnable defenses at Nashville held by General Thomas' Union army during the night of November 30, 1864, Hood would have renewed the assault the next day. McDonough and Connelly describe the Franklin battle as the "act of a tormented man driven by visions and dreams."

Although Hood considered Franklin a victory, the battle did nothing to advance the Southern cause. Far from helping to liberate Tennessee and Kentucky, the Franklin battle emphasized the hopelessness of the Southern cause and took the Army of Tennessee out of the war as an effective military force. Just two months after the battle, when its full impact had been tested in a series of battles around Nashville, Hood led his crippled ghost army back to Franklin and requested to be relieved of command. Shortly thereafter, Richmond authorized Richard Taylor to replace Hood. Until his last years, Hood acidly defended his actions at Franklin. His popular memoir, *Advance and Retreat*, admitted no apology for the frontal assaults he ordered that took so many Southern lives so needlessly.

Five Tragic Hours is the first scholarly work to critically interpret the Franklin fight, and while the authors' indictment of General Hood may discomfort some, the monograph has a solid foundation and presents arguments that are well-supported. Nonetheless, this fine study is sure to cause a great deal of controversy once its findings are synthesized in other volumes. McDonough and Connelly, in the fine tradition of their other influential Civil War volumes, have performed a great service. Immensely readable, filled with the echoing sounds of cannon and musketry that come alive on its pages, *Five Tragic Hours* is a book for all those who thought they were not interested in the Western Civil War, as well as those who already appreciate the significance of the "war in the West." It should not be overlooked.

William F. Howard

The Children of Pride, A New, Abridged Edition, edited by Robert Manson Myers, Yale University Press, 92 A Yale Station, New Haven, CT, 06520; 1984; price $25.00. 671 pages, illustrations.

This book is entirely the compilation of letters written to and from the various members of the Rev. Dr. Charles Colcock Jones family. This influential, wealthy set of individuals resided in Georgia, and owned several extensive plantations and a large number of slaves. The older son, Charles Colcock Jones, Jr., was mayor of Savannah in 1861, and later became an officer in the Chatham Artillery; the other son, Joseph Jones, was a doctor and spent time in the Confederate army as a surgeon; the daughter, Mary Sharpe Jones Mallard had married a Presbyterian clergyman and also lived in Georgia at the beginning of the Civil War. The parents, Rev. Charles Colcock and Mrs. Mary Jones shared time between their homesteads at Maybank, Montevideo and Arcadia.

The five hundred letters reproduced in this book, from October 1860, to January 1868, provide a fascinating look inside the family circle of an elite Southern aristocracy. The early letters, containing avowals of Southern supremacy and fighting to the death, if need be, gradually give way to ones of hesitant speculation, and finally, despair, as the Confederacy collapsed. Throughout the correspondence, religion, health and personal finances played predominently in their thoughts. There were worries the slaves would escape, worries about diseases and stock dividends, leaves of absence, and having portraits painted in miniature. What emerges is indeed a history as readable as a good "epistolary" novel. However, a great deal has been sacrificed to make it such. Author Myers writes in the introduction, "Seeking to make *The Children of Pride* a happy blend of accuracy and readability, I have resisted all temptation to intrude upon the text with explanatory notes." While the total lack of footnotes may have achieved his purpose of preventing the epistolary reader from becoming unknowingly wearied, it has had the opposite effect on me. I found that reading letters referring to people, situations or printed matter not previously mentioned in the text extremely aggrevating. For example, Dr. C.C. Jones wrote a review of Dr. Hodge's "The State of the Country" in 1861. Jones was considered by his fellow reverends to be the one person qualified for the task because his whole life had been "devoted to the study of the 'slavery' subject"; yet barely a hint is given as to the original article or the actual rebuttal. If there was not room to reprint both articles, a short synopsis is clearly indicated, and its absence is keenly felt.

As a research tool, its worth is severely impaired because of the lack of editorial notes, but an even more serious problem is that the book lacks an index. The researcher has to comb the text meticulously, and take copious notes for future reference, or else (as I did) spend a great deal of time underlining significant passages, then trying to make some organizational notes.

The Children of Pride is an extremely valuable book for those studying Georgia, or religion in mid-19th Century America. The letters superbly represent how the "other half," or landed gentry passed the war years, but the greatest worth of the book is found in the journal entries that describe the Yankee occupation toward the end of the war, and the Southerners' views of emancipation after peace was declared. How the Jones' attempted to cope with the freedmen and their new roles in society is an excellent, although tragic study of the plight of the newly freed black population. For this alone, the work is well worth reading.

It is unfortunate that the new trend in publishing historical works is to appeal to the lover of historical novels. History is fascinating and certainly doesn't have to be dry or boring, but no explanatory notes or even an index makes the balance tip unfavorably away from the historian. In such a multifacted work as this, it is certainly to be regretted.

S.L. Kotar

(*Continued from page 20.*)

the Lord had indeed come to him; for as yet he knew not what it meant to be free, his faculties had been dulled if not obliterated by the corrosion of more than a century of wrong. Is it difficult to comprehend that it should be so?

There was in this army, which Stoefield followed, a force of two or three regiments of negro troops, recruited from the immediate neighborhood of New Orleans. They presented a soldierly appearance in their new uniforms, and bearing their new bright rifles, but they were regarded as an experiment; as yet they had not been tried by the ordeal of battle, and further than in a few unimportant and trifling skirmishes, they had not even been under fire. They had been mustered in for some time, and had done the heaviest labors of the army; for such work they were justly appreciated, for they bore, uncomplaining, the hottest beams of a tropical sun, and there was less sickness among them than among the white troops. The question which was so warmly argued throughout the camps, — will these men fight? — remained undecided up to the day and hour of their fierce and fiery trial.

But Stoefield saw in this dark-faced soldiery, the answer to all his prayers and cries to the Lord, and at last, in a dim way to be sure, but still irresistibly, to his benighted heart, came this unanswerable fact: these strange northern men, coming, for all he knew, from the moon, had beaten his oppressors, and crushed a power which he had for so many years believed to have triumphed over his God. Is God dead? he used to ask of himself; he saw now that He lived, and that His day had almost come. These Yankees, murderers, thieves and destroyers, as his master had called them, had lifted from off his neck the wearisome yoke and offered him a fair chance to strike a blow for his freedom, and for the freedom of his race; and when he stood for the first time with his weapon, it was natural that his heart should be filled only with one dark, earnest longing for vengence upon the power that had been his bane for so long. But would he fight when his time came?

The white men fought, not alone and uncheered, as he would have to do. Their past, rich in the history of an heroic ancestry, their minds stored with precedents of past greatness and hopes of future renown; a hundred battle fields to add the names of departed heroes, who, though long ago gone from the scenes of their victories, still, in spirit, moved in the van of charging columns and waved towards the foe the time-honored battle flags of the past.

Sentiment and honor, romance and pride, would lead these Northern warriors down even unto the gates of hell; but had Stoefield and his comrads such incentives? The isolated cases of the heroism of his people had never reached his ears; of Denmark Vesey, of Nat. Turner, and of the few other names that make up the list of those who had died for the freedom of their race, he had never heard. Even though their names had been silently handed down as strange traditions among his people, throughout the regions wherein they lived, plotted, and died, and though it was comparatively near to him, he knew of them not, so how could he have heard of those loftier names, of Ogi, Chavannes, Buckman, Praloto, Maurepas, and of Toussaint, the heroes of San Domingo?

No such stimulant fired his blood, or the blood of his brothers, but still they eagerly looked forward to the hour of their triumph; — and it came.

The story of the assault of the colored troops upon Port Hudson, is now written in the history of the nation; how although defeated, they fought as few men ever fought; how of one regiment alone six hundred dead and wounded laying close up under the guns and upon the slopes of the rebel batteries, spoke eloquent and unanswerable argument for the freedom and manhood of the race for which they bled and died.

Abraham Stoefield, in the turmoil and fury of this his first and last fight, bore himself with the same stubborn and undaunted courage that had sustained him through his hard, joyless life.

Fear he felt not; hope, glory, renown, he cared nothing for, knew them not even; he felt but one wish, but one purpose; to get over into the batteries, in amongst the enslavers of his people, bayonet in hand.

He fought for an hour or two unhurt, and had reached the ditch, before he received his first wound; it was severe, and the pain and blood maddened him; he heard not, he heeded not the recall, he struggled on across the ditch and up the face of the battery, felt not the dozen balls that struck him as he passed on, but climbed up with feeble step upon the parapet, and fell there dead, while charging alone against thousands.

The record of this people during the past year has been plain, the lesson cannot be gainsayed.

Milliken Bend, Vicksburg, Port Husdon, and Morris Island, will be forgotten before the heroic deeds of this enslaved and long degraded people, shall be obliterated from the pages of history. If they fall short of the high standard of the white races, they have still proved themselves worthy of a higher sphere than that our countrymen have heretofore awarded them; and the act of justice is even now being accomplished. The crushing load which we northern men helped to pile upon them, toppled over upon ourselves, and in freeing us from the incubus, we must lift them up also. But not all shall be done by us; in the events of the future, with all its coming trials and sorrows, they will not be found wanting when their hour comes. Little can we expect from them indeed, for in what school have they been tutored for the great day; but as their opportunities are developed, and their manhood tried, the record of their deeds will be such, that we shall not be called upon to blush when we reflect into whose hands we have confided the battle-flags of the republic.

UNCLASSIFIED ADS

25¢ per word, 20 word minimum ($5.00). Name and address run free. All advertising must be accompanied by check or Money Order made payable to THE KEPI.
Tear sheet sent to all unclassified advertisers upon publication.

A U.S. GRANT, *or* W.T. SHERMAN MATTED PORTRAIT, with solid OAK FRAME and GLASS, *only $29.95 each*, plus $3.50 shipping. *(Illinois residents add $1.90 tax.)* SATISFACTION GUARANTEED! Send check or Money Order to: MIDWEST PRINT AND FRAME, P.O. BOX 33355, JOLIET, IL 60434.

OLD CIVIL WAR MAP KITS. Also old State, Railroad and County maps. 70 to 120 years old. All States. Also have Revolutionary War Map Kit. Stamp for catalog. NORTHERN MAP CO., 9343 S.W. 183rd COURT, DEPT. TK, DUNNELLON, FL 32630.

"GONE WITH THE WIND" — MEMORABILIA AND GIFT TYPE ITEMS FROM THE GREATEST CIVIL WAR PERIOD MOVIE EVER MADE, FOR LIST, SEND $1.00 TO: TARA ENTERPRISES, P.O. BOX 1013, WILLOW GROVE, PA 19090.

CIVIL WAR ART by eyewitness soldier/artists. Portraits of Confederate & Union Generals, Battle Scenes, and Army Life. Framed & unframed. Free Brochure. THEME PRINTS, LTD., P.O. BOX 123-K, BAYSIDE, NY 11361.

AMERICAN MILITARY BELTS AND RELATED EQUIPMENTS - *New Book.* A wealth of photographs and information on accoutrements of the 1833-1910 period. *See ad, page 15.*

UNIFORMS: LOW PRICED, PRECISE, AUTHENTIC, historical uniforms. Civil War uniforms our specialty. Other uniforms made upon request. WOOL YARDAGE AVAILABLE! FREE CATALOG: BROOKLYN ARSENAL UNIFORMERS, P.O. BOX 248, CEDARVILLE, NJ 08311, [609] 327-5017.

R. STEPHEN DORSEY ANTIQUE MILITARIA - Quality arms and accoutrements, carefully selected by a senior collector. Subscription to lists, $4 per year. R. STEPHEN DORSEY, P.O. BOX 263, EUGENE, OR 97440.

CIVIL WAR ARTIFACTS, Swords, Arms, Accoutrements, Relics, Documents. Buy & Sell. Subscription, $3.00/year, 7 issues of our list. THE CIVIL WAR SHOP, P.O. BOX 2091, FT. OGLETHORPE, GA 30742.

KERR .45 cal. Cap and Ball Patent revolver. Anchor and initials on front grip. Serial number in CONFEDERATE series. Checkered grip sharp. Needs adjustment. Best offer. MATT GILLMAN, C/O GENERAL DELIVERY, UPPER JAY, NY 12987.

CIVIL WAR BOOKS: Quality books, fast service, discount prices. Large descriptive list, $1.00. JOHNNY-YANK BOOKSTORE, 220 N. JEFF DAVIS JR., FAYETTEVILLE, GA 30214.

CIVIL WAR Books, Paper & Photos. Current catalog for two stamps, MacDONALD'S MILITARY, EUSTIS, ME 04936.

CIVIL WAR BOOKS, NEWSPAPERS, DOCUMENTS, IMAGES. PLUS MUCH MORE. COMPREHENSIVE CATALOG $1. GORDON TOTTY, 576 MASSACHUSETTS AVE., LUNENBURG, MA 01462.

BOOK FOR SALE: *Tarheel Cossack W.P. Roberts: Youngest Confederate General.* Paperback, 1983. Footnotes, bibliography, 103 pages. $6.00. Robert's military career. THOMAS B. KEYS, 1721 FLAMINGO DR., ORLANDO, FL., 32803.

BOOKS FOR SALE: McDonald's *Make Me A Map of the Valley* — $25.00; Dowdey's *The Wartime Papers of R.E. Lee* — $12.00; Taylor's *Destruction and Resconstruction* (ed. by Harwell) — $12.00. All excellent. Prices postpaid. CHONA RABAYA, 11805 EAST CRESTA VERDE DR., WHITTIER, CA 90601.

WANTED: O.R.'s, odd volumes, especially Vols. 2, 25, Hamlin's *Old Bald Head;* White's *Robert E. Lee and the Southern Confederacy.* A. ZIMMERMAN, 316 ISINGLASS RD., HUNTINGTON, CT 06484.

(Continued from page 27.)

We will treat you as Washington, Jefferson, and Madison treated you, and will leave slavery alone where it already exists among you."[6] In 1860, he spoke against his rivals who were "imploring men to unsay what Washington said, and undo what Washington did."[7] Leaving his home in Springfield, he told friends and well-wishers that he had a task before him "greater than that which rested upon Washington."[8]

After the assination in 1865, Washington, the Father of His Country, and Lincoln, the Great Emancipator, were inexorably linked together. Although Washington had not died in the service of his country, he and Lincoln were the martyrs who had sacrificed everything for the Union. Lithographs and engravings of the first president welcoming the 16th into heaven, became an instant commercial success. Washington's apotheosis prints turned into Lincoln's, and many typical lithographs portrayed the two as "Columbia's noblest sons."[9]

Today there is not a schoolchild who does not know the story of Washington and the cherry tree (a Weems creation), or that Lincoln freed the slaves. Even on citizenship exams, the two names are always mentioned and their relative places in history examined. Carved together on Mount Rushmore, honored by monuments in the city named for the first president, these two men stood for and held the Union.

FOOTNOTES

[1] Stephen B. Oates, *With Malice Toward None* (New York: New American Library, 1977) p. 12.

[2] Garry Wills, *Cincinnatus; George Washington and the Enlightenment (Garden City, N.Y.: Doubleday & Co., 1984), p. 42.*

[3] *IBID*, pp. 47-50.

[4] Frank E. Vandiver, *Mighty Stonewall* (New York: McGraw-Hill Book Co., Inc., 1957), p. 10.

[5] Harold Holzer, Gabor S. Boritt and Mark E. Neely, Jr., *The Lincoln Image* (New York: Charles Scribner's Sons, 1984), p. 191.

[6] *Malice*, p. 179.

[7] *Image*, p. 191.

[8] *IBID*, p. 188.

[9] *IBID*, p. 195.

Typical steamboat landing, such as the type Emma Levenia Gray would have seen on her trip.

THE KEPI
P. O. Box 699
Baldwin Park, CA 91706

BULK MAIL

THE KEPI

OCTOBER-NOVEMBER 1984 $3.50

"A New Concept in Civil War Reporting"

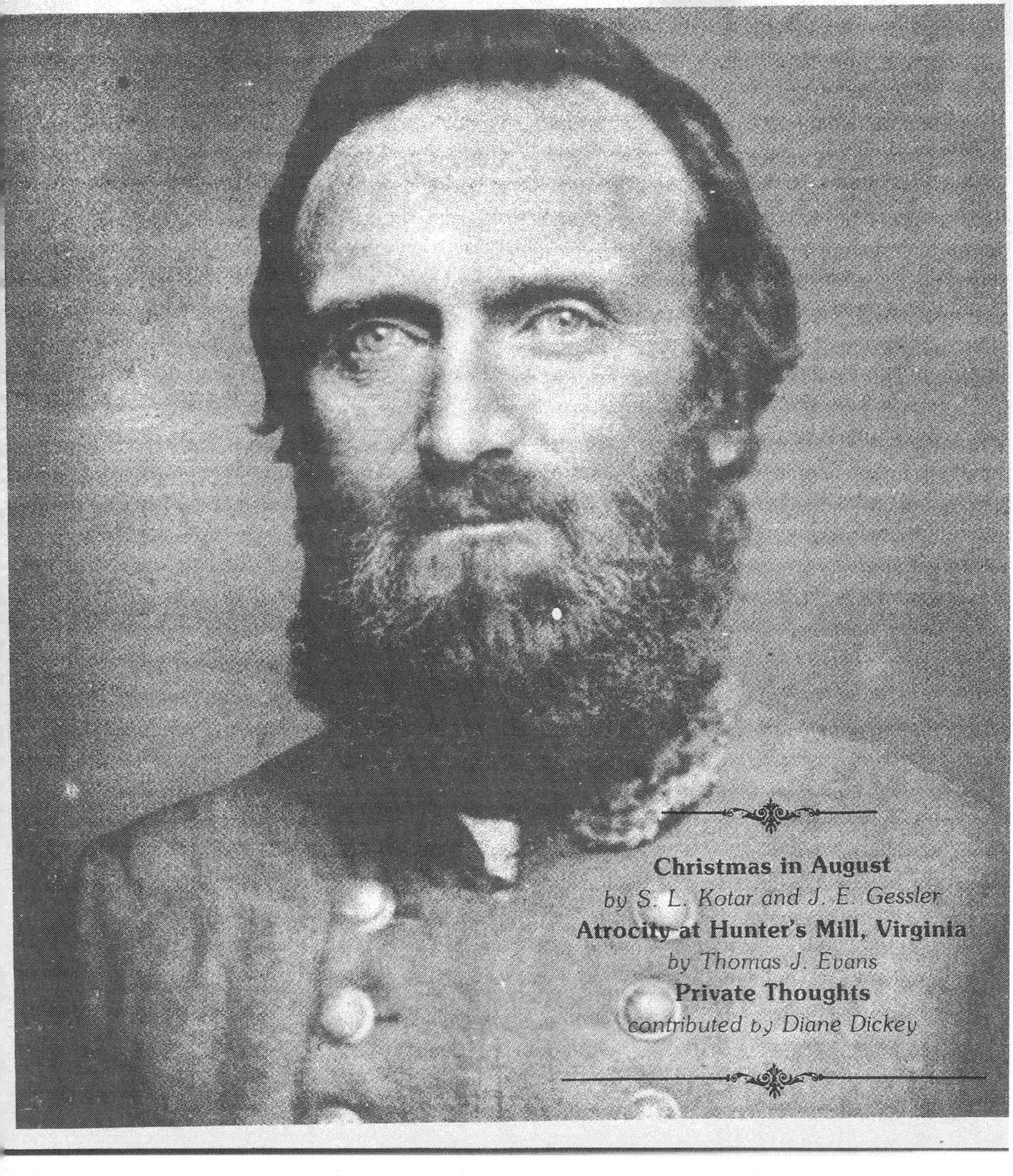

Christmas in August

by S. L. Kotar and J. E. Gessler

Atrocity at Hunter's Mill, Virginia

by Thomas J. Evans

Private Thoughts

contributed by Diane Dickey

PUBLIC NOTICE

To All Subscribers:

This is to inform you that should a renewal fee be received late, your sequence of issues *will not* be broken. However, due to the high cost of postage, that edition will be mailed to you at the next bulk mailing.

This is THE KEPI's guarantee that your collection will remain complete. (Please note THE KEPI's address has changed: THE KEPI, P. O. BOX 699, BALDWIN PARK, CA 91706. The return envelope accompanying your renewal form should already have this pre-printed on it. Thank You—*Editors-in-General.)*

BACK ISSUES OF THE KEPI — Vol. I, Nos. 1-5; Vol. II, Nos. 1-5, $3.50 each, bulk mail. Add 85¢ per issue for 1st Class postage.

This Christmas, treat yourself by adding research material to your Civil War library!

THE KEPI

"A New Concept in Civil War Reporting"

OCTOBER-NOVEMBER 1984 — VOLUME II, NUMBER 5

THE KEPI is published bi-monthly for $3.50 per issue, or $18.00 per year. Canadian single copy, $4.50; subscription $23.00 per year. United Kingdom single copy, $4.50; subscription $28.00 (Air Mail) per year. Correspondence to the editors may be directed to: **THE KEPI**, P.O. Box 699, Baldwin Park, CA, 91706. Unless expressly stated, **THE KEPI** neither endorses nor is responsible for the content of advertisements appearing in its pages. **THE KEPI** requests authors to query with specific topics before submitting articles, and assumes no responsibility for unsolicited manuscripts not accompanied by return postage. Work not fully documented and footnoted where applicable will not be considered.

ILLUSTRATIONS

Cover: Gen. T. J. Jackson, vintage image by Meiley, of Lexington, Va. From the authors' private collection. "The Rebel Stonewall Jackson," p. 5 from *Harper's Weekly*, August 30, 1862.
R. S. Garnett, p. 8 from *Stonewall In The Valley*.
Francis H. Smith, p. 17 from *100 Years at V.M.I.*
William Gilham, p. 19 from *100 Years at V.M.I.*
R. E. Colston, p. 20 from *100 Years at V.M.I.*
Rev. John D. Read, p. 29, Courtesy of Dr. Clarence R. Hartman, Falls Church Historical Collection by H. Douglas.
Rev. Hiram W. Read, Courtesy of Betty Danielson.
J. D. Read gravestone, p. 32. Courtesy of the author.
Back Cover: "The Situation In Virginia," from *The New York Tribune*, August 11, 1862.

THE REGULARS

THE IRREGULARS

GENERALLY SPEAKING

Three years ago, when we first began putting ideas together for THE KEPI, it was our intention to write several major articles a year for it. Once we had the first several issues out, however, we realized this was going to be impossible. There was always something more urgent demanding our time, or a new article we just received that had to get in the next issue.... Writing that feature was pushed further and further back on the schedule, until more than a year had passed, and we finally realized it was time to "make time." We didn't take as much as we wanted, but we did get "Christmas in August" done, along with two accompanying pieces, TIMELINE and "Missing in Action." These three articles were something we had promised ourselves, and we are proud to present them to you.

"Christmas In August" was originally planned to cover Major General T.J. "Stonewall" Jackson's daring raid around the enemy's rear, in August 1862. Once we started it, though, it became obvious that the story could not be picked up in the middle: explaining how it came about was as important as describing the raid itself. This explanation necessarily centered around Jackson and his relationship with R.E. Lee.

It was alleged that Jackson had not done well during the Seven Days: the period that followed would see "Stonewall" on trial. Everyone, it seemed, was keeping an eye on that quiet man from the Valley, about whom, few really knew. The correspondence from Lee to Jackson during this period was anywhere from blunt, to accusatory, to degrading. On August 27, Lee wrote: "Do not let your troops run down if it can possibly be avoided by attention to their wants, and comforts & c., by their respective commanders. This will require your personal attention...." Lenoir Chambers, in his two volume biography, *Stonewall Jackson*, wrote on page 101: "What Lee had in mind is not clear. Jackson...had no name for mistreatment or hardship in normal circumstances." Lee's *meaning* is not unclear — the reason *why* he wrote that to Jackson is. This is one of the several points explored in the article, that we feel will shed new light on both men and the circumstances surrounding August 27th — one of the most unique days in Confederate history.

TIMELINE chronologically follows Jackson's advance on Cedar Run, the Confederate concentration, and their aborted attempts to catch the Union army in a vulnerable position.

"Missing in Action" is a study of Francis Henney Smith and William Gilham, two of Jackson's colleagues at the Virginia Military Institute prior on the war. Both men were summoned to Richmond to assume important roles as the state prepared for civil war. Bright futures seemed to be before them — but their early promise faded into near obscurity by the end of the war. Their careers are traced in an unusual look at VMI's other officers.

Tom Evans continues his excellent series for THE KEPI with "Atrocity at Hunter's Mill, Virginia," a look at the dangers and consequences of spying along the Virginia/Maryland border. It isn't often a minister is accused of being a spy, and there is little information on the men and women who risked their lives intelligence-gathering for "their cause." This article explores both, in a fascinating look at the life of the Reverend Read.

We are introducing a new KEPI feature in this issue: PRIVATE THOUGHTS. In our renewal survey, many expressed an interest in reading more about the common soldier in the Civil War. In response (Continued on page 25.)

GENERAL ORDERS
No. 9
by Command of the Editors-in-General

Headquarters of THE KEPI
1st October 1984

They strongly opposed secession, and when their state broke away from the Union, they threw their lot in with the Federal cause. They secretively planned an audacious campaign to burn bridges and were promised an army to help them. They wanted to rid their land of Confederates and form a new state. They were not Virginians, but Tennessians, and unlike their fellow dissidents, they were unsuccessful in creating the state of East Tennessee. "A Fire Bell at Night" vividly depicts the desperate situation these pro-Unionists found themselves in, as promises went unkept, and they were left "hanging."

Sue Mundy was a Confederate guerrilla whose name might have been lost in the annals of history, except for one thing: no one knew whether Sue Mundy was a man or a woman. If a man, then he was just another rebel who happened to be related to Col. John S. Mosby's wife; but if a woman, she was not only a thorn in the Federal's side, she was a political embarrassment. Find out the secret, and see Mundy's photograph for yourself in the December/January issue of THE KEPI.

Read how a Congressional Medal of Honor winner lost his rifle to Col. Mosby — while bathing; search for Pvt. Thompkins and follow the life of John Householder of the 2nd Ohio in PRIVATE THOUGHTS. FAMILY BIBLE concludes with birthdays from December and January; included is an additional bibliography.

Christmas in August

by
S. L. Kotar and J. E. Gessler

"If the army is to move with efficiency, rigorous measures must reduce this luxury of transportation. Can anyone believe that Jackson's troops regularly receive 20 carloads of oats per day, or that they move with 23 wagons to each regiment of 600 strong, or that they have two sets of tents to a company?"

M.C. Meigs
Quartermaster-General,
Union Army
August 18, 1862

"The Rebel Stonewall Jackson"

It was obvious to everyone, even the high command of the Union army, that General Thomas Jonathan "Stonewall" Jackson was something special. After his promotion in October, 1861, to a major generalcy and his assignment to the Valley District of Virginia, Jackson left the main body of Confederates under Gen. Johnston and started forming what would become the immortal Valley Army. Bringing with him only one aide, Lt. Alexander (Sandie) Pendleton, he forged militia and recruits into solid fighting units. He managed to pry from Johnston his old brigade, and with these troops, he planned and carried out an audacious winter campaign. In the spring of 1862, pressed back from his winter headquarters in Winchester, Va., by the superior number of enemy, he withdrew up the Shenandoah Valley. Constantly alert to the possibilities of surprising the Union forces, he seized an opportunity on March 23, and attacked them at a nondescript place called Kearnstown. Although retiring from the field after a very strenuous contest, Jackson had long range goals in mind when he declared the contest a victory for the Confederates. He had achieved his purpose directed to him by his superiors - keep Federal forces from leaving the Valley, so they could not reinforce those ready to make a move on Richmond. In the subsequent weeks, he would more than succeed in this difficult assignment. He would march his men up and down the Valley in a series of innovative tactics that would mystify his opponents.

After the Seven Days Battles (June 26-July 2), architected by the new Confederate commander, Gen. Robert E. Lee, there was a lull in the bloody head-on clashes between his troops and those of Gen. George B. McClellan. Jackson's beleagured but triumphant Confederates were pulled back around the capital on July 2, amid rumors that Gen. John Pope had taken command of the Federal troops around Washington, under the ominous title "Army of Virginia." The speculation in Richmond was that Pope would reorganize and concentrate his troops for a united march on Gordonsville. From there, the path would be open to the Virginia Central Railroad, control of which could effectively cut communications with the Shenandoah

Valley, placing Richmond in a vulnerable position.

On July 13, the little army that had seen constant activity since evacuating Winchester, received orders to proceed "to Louisa Court House, and if practicable, Gordonsville, there to oppose the reported advance of the enemy from the direction of Orange Court-House."[1] Gen. Richard S. Ewell's Division, which had fought side by side with Jackson's troops, were to accompany them. Thus, the 11,000[2] "foot cavalry" marched out to face the 49,500[3] troops of the man Abraham Lincoln had brought east to achieve victories.

As was his custom, Jackson rode in advance of his troops, seeking information on the enemy's position and looking for an opportunity to strike. At Charlottesville, wild rumors circulated everywhere about the advance of Federal troops.[4] It was learned, however, that the enemy had not yet advanced to Gordonsville, so the troops were rushed forward, arriving there July 19.

By late July, Jackson had determined Pope's numbers were far superior to his, and wrote to Lee requesting reinforcements. Lee's answer of the 23rd states his belief that McClellan was "Being reinforced to the extent of the means of his government,"[5] and that he presumed Pope's responsibility was to guard the approaches to Washington. By the 25th, however, he offered the troops of Gen. A.P. Hill, if Jackson would consent to take them, along with their major general, who was currently under arrest.[6] Jackson notified Lee the next day he would take them if they could be spared, and promised their addition would enable him to take aggressive measures.[7]

He was as good as his word. On August 7, Jackson received news that Pope's advance had become separated from the main body during its move toward Culpeper Court-House. Seizing the opportunity to attack these troops while strung out, orders to move were quickly given.

"Banks is in front of me and he is always ready to fight - and he generally gets whipped."

After crossing the Rapidan, the army heard many stories from civilians about the atrocities committed by Pope's army under the auspices of his order to subsist off the land. Many fine homes were destroyed and the people made to suffer from the loss of their livestock and stores.[8]

Excessive heat and entanglement of troops slowed progress, but by evening, Jackson was in anticipation of a battle next day, even taking a moments release from the tension by remarking that "Banks is in front of me and he is always ready to fight - and he generally gets whipped."

His prediction proved accurate, for the next day, August 9, his forces and those of Gen. Nathaniel P. Banks clashed at Cedar Mountain, resulting in a Union withdrawal after dark to Culpeper. Because of the possibility that Banks had been heavily reinforced there, pursuit was inadvisable, so Jackson initiated the next best thing - he ordered his cartographer, Jedadiah Hotchkiss, to "at once make as many maps as he could of the region from where we are on to the Potomac."[9] All contingencies had to be planned for.

By August 13, Lee's attention had been drawn from McClellan on the coast toward Pope at Culpeper Court-House. Most indications were that Pope would make the next major offensive against the Confederate capital, so Gen. James Longstreet was sent to reinforce Jackson. They were transported by the Virginia Central Railroad to Gordonsville, while Hood and Whiting's brigades were ordered to proceed by the Fredericksburg railroad to Hanover Junction. These troops, it was hoped, could counter any move Burnside might make from Fredericksburg to the vital supply line of the VCRR. Lee himself remained behind.

Longstreet arrived at Gordonsville the same day, and Jackson immediately reported to him, tendering him command, as Longstreet was the senior officer. Henry Kyd Douglas, "Stonewall's" ADC, wrote in later years that Longstreet declined the overall command because he expected Lee to arrive shortly, and because Jackson knew the immediate field better than he.[10] In any case, the two generals discussed the military possibilities, and by 6 PM that evening, Longstreet wrote to their commander, suggesting two possible movements for his approval. He also stressed the point that Lee should come himself: Jackson did not agree with his projected moves, and "Old Pete" wished to press his case in person.

The two armies that met in Gordonsville that fine warm day had not yet come to think of themselves as a united body. Jackson's troops were the Valley Army, proud victors from the Shenandoah Campaign. They were Stonewall's men, and proud of it, invincible and ready to follow him anywhere. Longstreet's troops were the veterans of the Battle of Seven Pines and the Seven Days. They were openly bitter and resentful against Jackson, called him rude, arrogant, beastly and predicted "a month uncontrolled and he would destroy himself and all under him." They felt they had done more than their share of the fighting around Richmond, and believed Jackson and his troops had never faced the best the Union had to offer: their amazing triumphs in the Valley were achieved against inferior soldiers and commanders.

To some extent, their feelings of mistrust and suspicion were shared by the Confederate high command. In August of 1862, Jackson was still the unknown factor. Despite popular opinion and the increasing amount of lauditory comments that found their way into the press, "Old Jack's" position was not clear. Neither Davis nor Lee knew him personally. When Jackson had been introduced to the president at a strategy conference in July of 1862, he chose to salute him rather than shake his hand, as the others had done.[11] He was not interested in cordiality, and saw his own role as a military commander only. His first known experiences with Lee were hardly less formal on either man's part. In 1854, while a professor at VMI, Jackson had written to Lee, then Superintendent of West Point, for a recommendation, as he was seeking a new teaching position. It is not known whether the two met briefly during the Mexican War, but by 1861 they were virtual strangers to one another. War took them in opposite directions, and although they corresponded during the Valley Campaign, it could not be said they knew each other's minds. The first time they met during the Civil War was at a strategy conference immediately preceding the Seven Days Battles. Since Jackson had ridden 14 hours to attend this meeting, it is likely his hot and dusty appearance did not favorably influence his commander.

He didn't even share the comradeship of brother officers, such as Longstreet and Ewell, who had just resigned from the U.S. Army at the outbreak of hostilities. He was, even in August, 1862, an outsider with nagging questions surrounding him: he had performed well in the Valley as an independent, but could he act in concert with others? He praised God for his victories, but behind it, was he too ambitious? Was he brilliant, or was it true that he was just a second rater from the Valley? The Seven Days Battles around Richmond only accentuated these questions. The Confederate army had not worked well together, and most of the unofficial blame for it had fallen on Jackson. Officially, nothing was said, but it was evident Lee was keeping an eye on him, and had preconceived suspicions. After receiving complaints from Jackson in late July about his cavalry commander, Gen. Beverly Robertson, Lee had forwarded the letter to Davis for his "private perusal,"[12] and remarked to "Stonewall" on July 31 that Jackson expected too much from Robertson; in fact, Lee suggested that Robertson may have been readying his men for service, and preparing undisciplined men was no trifling undertaking. He did inquire, however, if there was anyone Jackson would prefer to lead his cavalry.[13] A week later, when Jackson met with Robertson and asked where the enemy was, the latter replied: "I really do not know." This caused Jackson to turn as "black as a thundercloud,"[14] prompting him to take Lee up on his offer to suggest a replacement. On August 7, he wrote, recommending "Grumble" Jones replace Robertson.[15] Lee responded on the 8th that he would see the Secretary of War, but "without knowing any of the circumstances attending it except as related by you, I fear the judgement passed upon him may be hasty. Neither am I sufficiently informed of the qualities of Col. W.E. Jones, though having for him a high esteem, to say whether he is better qualified."[16] (On July 31, Lee asked for Jackson's opinion, and on August 8, a mere eight days later, he told him his opinion was not sufficient. D. S. Freeman, Lee's biographer, calls this "frankness";[17] to a major general, it must have seemed insulting and frustrating to have his judgement so completely and suddenly disregarded.) Of additional importance pertaining to Lee's frame of mind was his communication of July 27 to Jackson, in which he told Jackson he would be reinforced by A.P. Hill. Lee added: "Do not let your troops run down it if can possibly be avoided by attention to their wants, comforts & c., by their respective commanders. This will require your personal attention; also consideration and preparation in your movements...A.P. Hill, you will, I think, find a good officer, with whom you can consult, and by advising with your division commanders as to your personal movements much trouble will be saved in arranging details, as they can act more intelligently."[18] The first part of advice Lee gave is somewhat startling in light of the fact that of all the high-ranking officers during the war, there were few, if any, who paid more "personal attention" to the needs of his men than Jackson. These admonitions may have been put in Lee's mind by Brig. Gen. R.S. Garnett. Garnett, who was a personal acquaintance of Lee's, had been in Richmond since July 11,[19] attempting to have his court-martial case brought to trial. Having been relieved of duty by Jackson for his actions at Kearnstown, Garnett had written a lengthy defense and submitted it to the War Department, where it had been received June 30.[20] Since Lee was already involved in Garnett's case, having had material previously forwarded to him by the Adjutant and Inspector General, Samuel Cooper,[21] it is likely he received Garnett's defense, if not from the War Dept., than from Garnett's

In August of 1862, Jackson was still the unknown factor.

Gen. R. S. Garnett

wealthy and influential relative (also a personal acquaintance of Lee's), R.M.T. Hunter. Since Garnett was in the capital, he would have found a sympathetic ear.[22] The brigadier's military record also contained a complaint lodged against him by Jackson for his disobedience of orders in January 1862. Garnett had deliberately countermanded his superior's orders to move his men out because he wanted to give them additional time to cook their rations. From this record, or from Garnett himself, Lee could have heard firsthand stories about Jackson's "running his troops down," and his "lack of personal attention" to their wants and needs. Garnett was a man with a grudge, and he blamed Jackson for making "covert attacks" against him. These, he wrote bitterly "are inconsistent with honor and justice, and should arouse grave doubts as to the motives and truthfulness of these secret allegations."[23] Here is a basis for Lee's sudden "frankness" when discussing Robertson, his denegrating order concerning his troops and his further suggestion to consult with his division commanders. Garnett had repeatedly complained his problems at Kearnstown stemmed from the fact Jackson had not consulted with him;[24] with this fresh in his mind, Lee, who was in accord with Garnett, and who had never heard Jackson's side, wished to see the situation was not repeated.* Whatever the reason, it is evident there were unanswered questions about Jackson that caused Lee to use care in dealing with him, and according to Gen. Jeb Stuart, "Gen. Lee came to us at Gordonsville with rather a low estimate of Jackson's ability."

It is no wonder then that Jackson considered the Battle of Cedar Run his greatest victory.[26] It proved to him that he could fight whatever Lincoln had to throw at him. He had beaten a portion of John Pope's new army, and he had done it by maneuvering the largest command he had ever taken into battle. It was enough of a clear-cut victory that he could put behind him the Seven Days Battles and proclaim that God had "blessed our arms with another victory."[27] In a personal sense, it was his bravery and powers of leadership that turned the tide of battle. He had ridden into the confusion on the east, drew his saber and literally turned men on the verge of rout into triumphant victors.[28]

When Longstreet arrived at Gordonsville on August 13, he was well aware of Jackson's new laurels. His immediate formation of battle plans and repeated requests for Lee to join him suggest that "Old Pete" wanted to be the main influence on the commanding general in any subsequent strategies. That he had Lee's ear is evidenced by the reply he received the following day: "I had arranged to leave in the cars to-morrow at 4 o'clock to join you. Let me know where I shall find you. I should like if convenient to see Jackson, too."[29] If Lee was contemplating a council of war, he was notifying Longstreet Jackson's presence was not mandatory.

Lee arrived August 15 and assumed command of the army. Jackson reported to him immediately and presented his intelligence reports, estimating Pope had between 45,000-50,000 men, and also provided him with accurate maps of the area. It was necessary during this high level session to decide when and how to attack Pope. The Union commander had placed himself in a vulnerable position between the Rapidan and Rappahannock rivers. If the cavalry, under Gen. Jeb Stuart, were able to destroy the Rappahannock Station Bridge, Pope would lose his avenue of retreat over the river and would be cut off. A rapid move on the part of the Confederates across the Rapidan would trap the Federals in this unfavorable position. Speed was of the essence, however, because at any moment Pope might realize the precarious position he was in and pull back across the Rappahannock.

Jackson advocated crossing the Rapidan on the 16th and giving battle on the 17th. Longstreet protested that his commissary was disorganized, and needed time to supply his troops. Jackson quickly offered enough hard bread for the march, but Longstreet remained skeptical that he could be ready to move by the morrow. Lee sided with Longstreet, and brought up the fact that the cavalry, who were essential to the plan, were not concentrated. The final decision was to approach the fords of the Rapidan on the 17th and attack the enemy on the 18th. In the evening, Jackson's troops moved out along the southern side of Clark's Mountain, and the afternoon of the following day, Confederates witnessed the enemy marching away from Crooked Run church. Later that night, Stuart left for Verdiersville, where he expected to unite with Fitz Lee and proceed to

* For further details on the Battle of Kearnstown, the charges Jackson brought against Garnett, and his subsequent trial, see THE KEPI, Vol. I, No. 1, "The Battle Behind Kearnstown," by S. L. Kotar and J. E. Gessler. Lee eventually dismissed the charges against Garnett, and he was reinstated to command.

destroy the bridge in Pope's rear. Instead, August 18 saw Fitz Lee telegraph R.E. Lee that his horses would not be in serviceable condition until the 19th. This, added to the fact that Longstreet's commissary had not provided enough bread for his troops, and Anderson's Division, arriving from Richmond, was not in place, caused Lee to postpone the general advance, first until the 19th, and then the 20th. The myriad delays cost the Confederates their chance, for on the 19th, Pope withdrew his army behind the Rappahannock. Lee attempted to follow, but found all fords to the river skillfully blocked by Pope's men. This forced the Confederates, with Jackson in the lead, to move from ford to ford, hoping to find an unguarded crossing. Futile attempts to force a passing of the river occupied Lee until Sunday, August 24. At that point, a strategy conference was called between the commanding general, Jackson, Stuart and Longstreet.

These four notables met in a tiny place called Jeffersonton. Set back from the river so as to be out of range of the enemy's guns, the town was serving as Jackson's headquarters when Lee rode up to confer with him. Jackson began the day out of humor because his quartermasters and commissaries had not gotten provisions up, and his men were hungry.[30] His mood changed, however, as a council of war was arranged. A table was placed almost in the middle of an empty field. Lee sat at the table with a map spread; Longstreet sat at his right, Stuart on his left, and Jackson stood opposite him. A group of staff officers waited on an adjacent knoll — only the highest-ranking of the Confederacy's officers would attend.[31]

Dr. Hunter McGuire, Jackson's Medical Director, reported on what he witnessed: "Jackson was so reticent that it was only by accident that we ever found out what he proposed to do... The day before we started to march round Pope's army I saw Lee and Jackson conferring together. Jackson — for him — was very much excited, drawing with the toe of his boot a map in the sand, and gesticulating in a more earnest way than he was in the habit of doing. Gen. Lee was simply listening, and after Jackson had got through, he nodded his head as if acceding to some proposal."[32] The plan was a daring one, similar to the strategies "Stonewall" had planned and perfected in the Valley: he would take his command, cross the Rappahannock above Waterloo Bridge and march in rear of Pope's army. He would then attempt to cut the Federal communications by destroying the Orange and Alexandria Railroad and the bridge at Bristoe Station, 4 miles below Manassas Junction. Destruction of this critical span would make quick repair and reopening of the line impossible. Papers Gen. Stuart had captured from Pope revealed that two of McClellan's Corps, the 3rd and 5th, along with Reynold's Pennsylvania Reserves, were within two days march from a union with Pope. Other troops, including 7,000 from the Kanawha region of Virginia were within a five-day march. Combined, these 130,000 men would be all but unbeatable. But if Pope behaved like his predecessors in the Shenandoah Valley -retreating to protect Washington once he discovered Confederates in his rear - he would be moving away from McClellan's forces at Fredericksburg. This would then allow Lee and Longstreet time to rejoin Jackson and attack the large Federal armies separately. The potential was great, but so were the risks.

"Jackson - for him - was very much excited..."

Traditional military strategies would not allow for the division of a numerically inferior army in the face of an enemy. But Jackson was not a traditionalist. He believed that the only way a weaker power could overcome a stronger was by avoiding the tremendous bloodlettings of head-on battles. Attack had to be swift, unexpected and to advantage, with alacrity and audacity the watchwords. Never do what the enemy expects, defeat him in small skirmishes, and by doing so, create a sense of superiority and faith in your own soldiers that will carry them beyond endurance, if need be. "Old Jack's" men had supreme confidence in him and in themselves -that was his ultimate weapon.

To be successful, the plan had to be carried out rapidly and secretly. If word that the army had divided reached the Union commander, he would be in a position to move between them, and turn the tables. While the Confederates had no reserves to summon in an emergency, Pope could attack one, then turn on the other, or wait for McClellan. Each Federal army could then attack half of Lee's. The chances were great.

The details of the plan were left open. Everything would have to depend on the moment. The key to getting behind the enemy was Thoroughfare Gap. If this was heavily guarded, progress could be delayed and the plan discovered. If it was undefended, then the Confederates could hope for success.

Jackson was eager to start at once, and, in fact, informed his staff officer, H.K. Douglas, after the meeting, "I will be moving within an hour."[33] Delays had already cost the Confederacy much - he wanted no more wasted time. Wagons were ordered up and his troops were concentrated around Jeffersonton. Rations for three days were issued until late at night, and men were forced to cook until early hours.

If Jackson was eager to move out, there were those who had doubts about the advisability of so daring a move. In Hotchkiss' diary, he included Jeb Stuart's account of the strategic conference, in which "He said Gen. Jackson was entitled to all the credit for the movement round the enemy and Gen. Lee had, very reluctantly, consented to it. He spoke of the great results it had..."[34] Hotchkiss added that Lee had inquired of Stuart "if he did not think it was very hazardous for Gen. Jackson to attempt to go round the enemy when we crossed the Rappahannock in August." Apparently, he did not, for the plan was put in motion and orders issued. Jackson was to move at dawn, "with the utmost promptitude, without knapsacks,"[35] carrying with him only ordnance trains, ambulances and a herd of cattle. A.P. Hill would continue his artillery duel with the Federals on the other side of the river until dark, then Longstreet's men would replace him, freeing Hill to join Jackson. Stuart would guard Waterloo Bridge until relieved, then follow and support Jackson.

Amid the sound of heavy cannonading that afternoon which kept the men "in an uncomfortable frame of mind,"[36] preparations hurried on. Hotchkiss sketched a map of the area for Longstreet,[37] while Jackson called for his engineer, Capt. J.K. Boswell, and ordered him to select "the most direct and covered route to Manassas."[38] Later that night, the General tried to sleep, but was unable because of the disturbances made by the troops getting ready to march.[39]

Monday, August 25, began cool, but soon turned very warm.[40] The troops were on the move early, some leaving behind half-baked bread.[41] They marched beyond Amissville eastward to a private ford called Henson's. Amid the lush vegetation of this valley, the "foot cavalry" crossed the tributary of the Rappahannock, Hedgeman's River, unmolested by the enemy. Some waded thigh deep in the cold water,[42] others reported crossing on a "poor but adequate bridge."[43] The horse cavalry was sent out in advance to scout for bluecoats and picket every converging road until the army had passed. As the morning progressed, orders were issued, and in Jackson-like fashion, guides were posted at each intersection to mark the route.[44] Boswell went with Ewell's van to guide the way. And everywhere was the General, riding from one unit to the next, keeping the column closed with his customary "Close up, men, close up." Rations were soon eaten and soldiers slipped out of ranks to raid the outlying fields for apples. To make the march easier for them, Boswell had fence rails taken down, making short cuts through fields, as well as keeping the men under cover.

In a moment, they forgot their weariness and cheered him.

The fact that a significant portion of Confederate troops had moved out that morning was known to the Yankees. As early as 7:15 AM, a Federal signal station at Waterloo Bridge spotted men with regimental colors moving along the river. Details were sent at 8:45, 9:30, 10:30 and 11:00, indicating a well closed-up column from Jeffersonton to the vicinity of Flint Hill, some 15 miles. By noon, John Pope had informed Gen. Halleck in Washington that 20,000 men were on the march. Pope interpreted this to mean the van of Lee's army was moving into the Shenandoah Valley. Since that movement appeared to be no immediate threat to him, Pope did not act on the information, further than alerting Gen. McDowell.[45] The rebels were allowed to continue unmolested.

While on the march, Jackson sent two of his officers ahead with a squad of cavalry to alert the people of their coming, and have them send provisions to his men. They responded by sending flour and meat, which Jackson acknowledged by saying the citizens had done good service that day.

The Black Horse cavalry, under Capt. Robert Randolph, and a company of 2nd Va. Cavalry, under Capt. W.W. Tobbs, went ahead through Orlean, Thumb Run Church and Salem. No enemy was seen, and the march continued unopposed.

Late in the afternoon, a mile from Salem, Gen. Jackson dismounted and climbed atop a large stone by the roadside. He removed his cap and, bareheaded, paused to watch the sun set behind the mountains. Weary of body, indefatigable of spirit, the man who carried the fate of the Confederate army on his shoulders was seen standing alone, cap in hand, by his men. In a moment, they forgot their weariness, and cheered him. "Old Jack," the general who symbolized their pride, courage and determination, made a swift, friendly gesture to silence them - not for him must they alert the enemy. Word went quickly through the ranks: obediently they passed in silence. But they would not let the moment go. They raised their caps in silent praise, creating a sublime moment for all who witnessed it. Jackson, proud and pleased, remarked with depth of feeling, "Who could not conquer with troops such as those?"[46]

Ewell's Division was stopped a mile south of Salem, and got a chance to rest that night. The remainder of the men, however, had to be brought up and did not get into bivouac until late. Those who had eaten their rations and were not fortunate enough to have received handouts along the way went to sleep hungry. The Confederates had covered twenty-five miles by hard marching, and Jackson was pleased.

When the citizens came to complain that "Stonewall" was abandoning them to the enemy, that concern (unnecessary as it was) may have pleased him, too. It could do no harm if that report reached Pope's ears.[47] By this time, manipulating information later received by the enemy was his forte, and this scene was reminiscent of one several months ago, when he had citizens and soldiers alike believing they were off for Richmond, when they actually headed the opposite way.

The order of March for Tuesday, August 26, was the same as the preceeding day — Ewell, Hill, Taliaferro. Jackson sent the 2nd Va. Cavalry ahead to scout Thoroughfare Gap. This critical passageway, through which the Manassas Gap Railroad passed to the Shenandoah Valley, was the crux of the move around Pope. Tense moments were spent that morning waiting for word from the cavalry. When it came, "Stonewall" must have sent a prayer heavenward - the pass was clear! Pope was paying no attention to them; prospects for success increased.

Deviating from his usual march regime of ten minute halts per hour, the commanding general allowed his troops to march at will. They straggled along the roadsides and the graded railroad beds, yet maintained a steady pace and kept good time. Their own eagerness grew with each step, for they soon surmised they were maneuvering around the back of the enemy. It was history in the making and Jackson was at the helm.

By midafternoon, Gen. Stuart and the cavalry joined them as the troops approached Gainsville. Jeb's watchful eyes now gave Jackson the security he needed; he allowed the soldiers to march slower, with greater intervals between units. At this time, he also had to make a decision on the route to take. The main road from Warrenton to Alexandria crossed Bull Run at Stone Bridge, but did not lead directly to the railroad bridge. To reach that, the shortest route was from Gainesville to Bristoe Station, 4 miles below Manassas Junction. Orders were given and the men were turned toward this objective. Footsore, hungry and tired, these veterans sensed something important was happening, and their primary concern was that they would "be up" when the excitement started.[48]

As the gray army advanced, citizens came up and informed the officers only a small contingent of Federals held Bristoe. Col. Munford's cavalry, supported by Forno's Brigade, were given the honor of sending the Yankees running. At about sunset, a charge was made upon the small station, and the lounging Union cavalrymen were chased off. This brought the rest of the Southern army up, and as they approached, the sound of cars was heard from the direction of Warrenton Junction. Hay's Brigade of Louisianans and North Carolinans smelled "booty," and frantically threw wooden sills across the track in hopes of derailing the train. As the iron horse came on, the infantry fired a volley of shots, but neither that nor the obstructions could stop it. It steamed on, out of the grasp of the excited rebels. Their disappointed spirits were lifted, however, when they discovered they had arrived at the hour when Federal supply trains ran from Manassas or Alexandria to Pope's front and back. In the hope of having another chance, the men found a derailing switch, and the 21st North Carolina lined up to give the crew a volley as soon as the switch was hit. The plan worked perfectly, and the excited troops had the intense satisfaction of seeing an enemy train plunge down an embankment. Sparks flew high into the darkness, lighting the way down, as they ran toward the wreck below. What met their eyes struck the humor of these grizzled soldiers - the locomotive was named "President," and through the picture of Abraham Lincoln on the steam dome, a Confederate bullet had passed.

They had barely enough time to savor this victory when the sound of another approaching train caught their attention. This one hit the preceding train: the result was a tremendous crash. Metal and wood twisted out of shape and the curious soldiers found the interiors to be fitted with wooden benches for troop transport. Inside, they also found a civilian, who had sustained a broken ankle in the crash. When informed his captors were "Stonewall" Jackson's boys, he expressed an eagerness to see the famous general. Jackson was busy talking to an engineer when he was pointed out. The Unionist took one look at his dusty uniform and cap pulled low over his face, and exclaimed, "Oh, my God! Put me down!"[49] "Old Jack's" men roared with laughter. They knew pomp and circumstance did not make a general.

What Jackson was trying to determine was whether the cars and locomotives could be salvaged for Confederate use. He summoned Boswell and ordered him to move the trains across Bull Run. Gen. Trimble assigned the 15th Alabama to assist in this task, but it proved impossible.[50] The engines were too badly damaged to be worth the effort to secure them. The General was regretful at the loss, well aware of how valuable the trains he had brought down the Valley turnpike in the spring had become, but there was no hope for it now. He quickly ordered the bridge over the Run burned, then turned his attention to a gathering of local residents. These excited patriots quickly told him of huge Federal stockpiles 7 miles up the railroad at Manassas Junction. There, they explained, Pope kept the supplies for his entire army. Warehouses and carloads of freight were ripe for the taking. They reported very few soldiers were left to guard the treasure, Pope having no suspicions the supplies could be in any possible danger. It was a challenge "the wagon hunter" could not resist. But there were factors that had to be weighed. A fourth train had come upon the wreck of the two derailed by the rebels, but had escaped by backing up along the track the way it had come. The alarm might already be given that the Federal army was in danger from the rear, and soldiers put on the march to counter the threat. Since their presence was known, time became a critical factor. There was also the question of which troops to send the extra miles to Manassas. Many of the men were not up yet, and those that were, were tired, having made a grueling two-day march. Some were already asleep where they had fallen, and the Stonewall Brigade, traditionally the one Jackson sought out for difficult assignments, was the rear unit.

The problem was solved when Gen. Trimble heard of the proposed move. He immediately volunteered his men and Jackson accepted the offer. After briefly explaining the situation as reported to him, as well as the risks involved, Jackson watched the 21st Georgia, under Major Thomas G. Glover, and the 21st North Carolina, under Lt. Col. Saunders Fulton (about 500 men),[51] march out. After these men had gone, Gen. Stuart rejoined the main body at Bristoe. Jackson, feeling Trimble would have greater success with the aid of the cavalry, ordered Stuart to move on Manassas and assist the infantry in its capture. The cavalry, sensing the excitement, rode quickly to the Federal supply depot, where the contest for possession was taking place.

A regiment had been put on either side of the tracks, advancing through the darkness in secrecy. Then, as they reached the long lines of boxcars, the ingenious rebels began shouting, resulting in such a din that the defenders thought the enemy was charging in great hordes. Firing their batteries wild in the ensuing panic, they allowed at least one gun to be taken by the rushing attackers before it could be reloaded.

In the dark of the night, the Confederates lost touch with each other as they shifted from one position to another; their banter revealed as much pride as accomplishment: "Halloo, Georgia, where are you?" "Here, North Carolina. We have taken a battery!" "So have we!"[52]

Both regiments did capture Federal batteries, each one containing four field pieces, horses, equipment and ammunition, as well as 300 prisoners. To Jackson, the general and artillerist, this was a great addition for his army, but for the foot soldier, the fun was just beginning.

For their reward in marching the additional 7 miles from Bristoe to Manassas, Trimble's men became the first to explore the Federal treasure trove. To be sure, guards were immediately posted around the warehouses, but they didn't keep out the cavalry or their companions. For his part, Gen. Trimble sent Jackson a message, briefly describing his success at taking Manassas and the guns he had captured. Jackson, in turn, relayed this information to Lee, adding that he had broken up the Orange and Alexandria Railroad and captured three trains of cars and prisoners. This reached him at Orleans, where Longstreet's men, having left their encampment on the Rappahannock that afternoon, had marched to.

Wednesday, August 27, 1862, Jackson was up early. No time was to be wasted, and he was eager to see for himself what the reports had only hinted at. Orders were given for the Stonewall Brigade to advance in support of Trimble; Poague's battery was also sent. Following them were A.P. Hill's troops and the remainder of Taliaferro's men. Gen. Ewell was left with three brigades to hold back the expected Federal advance until salvage work was concluded at Manassas. He was then to fall back on the main body, which would be all the troops Jackson had at his command should the enemy press him in force.

Jackson rode ahead of the column and reached the site of Trimble's victory by daylight. There before him lay what must have seemed like the stockpiles for a million men. Warehouses were filled with luxurious foodstuffs and liquors. Along the train sidings were row upon row of cars, each packed with something to fire the imagination of a Confederate soldier: immense quantities of ammunition, upward of 200 new issue tents, medical supplies of morphine and chloroform, brandy and military equipment, including blue Union uniforms. Trimble had placed guards around buildings and trains, but they could not stop the masses. Passing through town to a position further out, troops from the Stonewall Brigade broke ranks and pushed their way through a circle of guards to grab anything they could get their hands on. Finally driven off, they went in droves to the outlying buildings looking for booty. Barefoot, half-starved and in tatters, these long-suffering Confederates had found a dream come true. But for the moment, it would have to wait. "Old Jack" knew there was business to attend to first. A pesky Federal battery had been firing at them, so Poague was sent to silence it. Behind him went the rest of the troops, leaving only Gregg's men. Jackson himself was soon in advance of the army, following the road that ran along the tracks from Manassas toward Bull Run. This was familiar territory to him - the area had witnessed the first great battle of the war. He and his men had become immortal here; but in his mind now were the consuming responsibilities of a new campaign and new dangers.

About a mile ahead of him, he saw what he took to be the gleam of bayonets in the sun. A prolonged look couldn't confirm his suspicion, so members of his staff were summoned. It didn't take long, however, for the moving object of their attention to clarify itself: marching straight toward them, in perfect order, completely unaware of the waiting Confederates, was a column of Union soldiers. Carpenter's battery was rapidly brought up and Hill's men positioned so as to command the Federal left flank. As the untested New Jersey brigade, under Gen. George W. Taylor, approached, Carpenter's battery fired - prematurely, the infantry thought - bringing on a general artillery sweep of the field. The rookies kept marching, unknowingly advancing where no veteran unit would, making a perfect target for the rebel batteries. Sweeping the front and right, the Confederates then moved to better ground and shorter range, changing to canister for better effect. Still, the undaunted New Jersey men came on into the face of certain death, in a brilliant display of bravery. Gen. Taylor cheered them on, begging his men not to let this be a second Bull Run disaster for the Union forces.

"Stonewall" waved a handkerchief at them and shouted for their surrender.

He realized all too clearly the danger his raw troops were in, but he was determined to face the enemy bravely. That he did, at the cost of his own life and many of his soldiers. Even Jackson, the great warrior who had encouraged his men on other battlefields to shoot the brave ones so the others would run, could not stand by and watch an enemy be decimated. In a sublime act of mercy and personal courage, he rode to the front and halted his artillery fire. In front of those guns, as if to shield the enemy with his own body, "Stonewall" waved a handkerchief at them and shouted for their surrender. On a day that his army had captured John Pope's entire base of supplies, there need by no additional loss of honor. In reply, however, a Federal marksman took aim and fired. The bullet missed Jackson by inches, and was a good enough answer. He turned and gave the order to resume firing. They responded with vigor, and in a short time the ranks of the New Jersey brigade were completely broken. The men turned to run in panic and were pursued by the cheering Confederates back over the railroad bridge to the train they had just debarked from. The bridge was destroyed, the cars burned and 300 prisoners taken. One hundred thirty-five Union men were killed or wounded, among them Gen. Taylor, of whom Jackson wrote respectfully in his official report, by saying his army fought "with great spirit and determination and under a leader worthy of a better cause."[53]

By 11:00 AM, the unequal contest was over and Hill's men were sent back to Manassas. The infantry felt the entire Union brigade would have been wiped out had not the artillery under Carpenter, fired prematurely. The feeling prevailing in the ranks was that no serious attempt had been made to fly the black flag that day. Jackson, who was reported to have wondered about the benefits of a "no quarter" policy[54] demonstrated his personal beliefs this day in actions which spoke louder than words. In a situation where mercy to the enemy would not affect the outcome of battle, it was practiced. It was an acknowledgement of respect, both for bravery and for life that caused "Stonewall" to stay the sword of death; an action that would be remembered by both the blue and the gray when they wept at his graveside.

At Manassas, pandemonium and paradise awaited the returning rebels. Trimble, who had worked so hard during the night and early morning to protect the supplies, was furious at the conduct of the arriving troops. Nothing was left untouched in the quest for liquor and comestibles. Gen. Taliaferro, assuming command while Jackson was seeing to the enemy, had taken charge of supplies, and was furiously stuffing any vehicle on wheels with the irreplaceable items. A discovery here, a new find there, and more room was sought to pack his treasures. At the same time, he liberally dispensed items to the troops so that they could consume what he couldn't take. Like small children turned loose under a Christmas tree, the seasoned veterans tore covers from boxes, ripped open bags, and stove in barrel heads. Every sort of luxury dreamed of by a starving soldier was there for the taking. Cakes, candy, caramels, French mustard and all types of sutlers' stores were sampled, passed around and stuffed into pockets for future use. Staples such as flour, salt port, bacon and corned beef were passed over by the men, but not by their thrifty commander. Jackson had commandeered the large Federal bakery and put his prisoners to work baking fresh bread for his troops. In the midst of plenty, the lean days ahead could not be overlooked. Sutlers' stores and essential items for the well-dressed Union soldier, such as fine linen handkerchiefs, were gleefully swept up by the men. New drawers were converted into sacks to carry away coffee and sugar, while saddles and cavalry boots were treated with disdain.

What caught "Stonewall's" eye were two long rows of freight cars, each half a mile in length. Some of those valuable trains were filled with ordnance, and a find that must have made his heart catch - two newly developed 3" bronze rifled guns. These must not be overlooked, nor their companions, either, the long rows of pyramid-like piles of shot and shell carefully laid out along the track.

Over other railroad cars men scrambled, looking for that one very special item - spirits. Morphine and chloroform were thrown away in the mad frenzy for whiskey or brandy.[55] Distraught surgeons begged, threatened and finally sought help from the one person who could save them. They got what they wanted - a guard to disperse the men and orders from Jackson that all liquor not needed by the medical department be dumped on the ground. Men dropped to their stomachs to lap up the precious fluid, while others, raiding officers' store rooms, discovered cases of Rhine wine. Corks were popped and everyone's health was toasted, especially, it may be presumed, John Pope's, the Union leader who had earned for himself this day the sobriquet "Commissary General." Pope supplanted another Union general, Nathaniel P. Banks, who had been the first Federal so honored for his donations (albeit unwilling) to the Confederate diet.

John Pope earned for himself this day the sobriquet "Commissary General."

Fires sprang up everywhere, as delicacies were cooked, sampled and traded. So much was consumed or wasted that when Jackson summoned the artillerists back from their advanced position, all they got was canned goods. There was much hard feeling over that, and a general resentment toward the cavalry, who, it was claimed, carried great quantities away on their horses.

Jackson had intended to transport all the supplies he could away with him, but when he received word that Ewell was being hard pressed by Gen. Hooker, it was clear time was running out. All attempts at holding the men away from the foodstuffs were abandoned and "Old Jack" watched his boys enjoy themselves. The only stipulations he issued were that they should fill their haversacks with four days rations, they should not dress themselves in the plentiful array of Union uniforms and they should not interfere with the bakery that the prisoners were working. With these reservations, every man could take what he wanted. For a short time before they would be called back to the realities of war, it was indeed Christmas in August, good will to all, as "Stonewall" played Saint Nicholas.

Generals Stuart and Early skillfully withdrew from Bristoe, where they had been left as a rear-guard to hold back the advance of the Union Army. Pope had suddenly realized the troops in their rear were far more than Stuart's pesky cavalry, and Hooker had been sent to stop them. He could make no dent in the strong front the retreating Confederates put up, however. It was past dusk when Ewell arrived at Manassas and the men turned loose. The party was almost over, but they were issued meat, fresh bread, coffee and whatever else they could pick up. Orders were soon given to move out, and a unique and magnificent day drew to a close as a torch was put to all that was left behind. The fire lit up the heavens for miles around, a fiery testimonial to a quiet man's victory.

Within twenty-four hours, Jackson's troops were fighting again, this time Gen. McDowell's Corps of Hatch, Gibbons, Doubleday and Patrick's brigades. The successes of the day before were put aside for more immediate concerns, but they would never be forgotten. Those who were there that glorious summer day would, over many a campfire, recount the stories of "Old Jack" and his handkerchief, of sutlers' supplies and Rhine wine. For one short day, there had been plenty of everything; and through the bitter cold of winters, to the bitter road of defeat, these men would keep those memories alive, and perhaps the memories kept them alive. If that were so, it would have made "Stonewall" very proud.

FOOTNOTES

[1] *War of the Rebellion: A Compilation of the Official Records of the Union and Confederate Armies (OR)* (Washington: Government Printing Office, 1885), Series 1, Vol. XII, Part 3, p. 915.

[2] William Allen, *The Army of Northern Virginia in 1861* (Boston: Houghton, Mifflin and Co., 1892), p. 159.

[3] *IBID, p. 161.*

[4] Archie P. McDonald, Editor, *Make Me A Map Of the Valley* (Dallas: Southern Methodist University Press, 1973), p. 62.

[5] *OR*, XII, Part 3, p. 916.

[6] *IBID*, p. 917.

[7] *IBID*, p. 918.

[8] *Make Me A Map*, p. 66.

[9] *IBID, p. 68.*

[10] Henry Kyd Douglas, *I Rode With Stonewall* (Chapel Hill: University of North Carolina Press, 1940), p. 129.

[11] *Southern Historical Society Papers* (Millwood, N.Y.: Kraus Reprint Co., 1977), pp. 304-5.

[12] Douglas S. Freeman, *Lee's Dispatches* (New York: G.P. Putnam's Sons, 1915), pp. 42-3.

[13] *IBID.*

[14] Lenoir Chambers, *Stonewall Jackson* (New York: William Morrow & Co., 1959), Vol. II, p. 106.

[15] *OR*, XII, Part 3, p. 926.

[16] *IBID.*

[17] Douglas S. Freeman, *Lee's Lieutenants* (New York: Charles Scribner's Sons, 1942), Vol. II, p. 6.

[18] *OR*, XII, Part 3, pp. 918-9.

[19] *R.B. Garnett MSS*, Confederate Museum, Richmond; July 12 letter Garnett to Gen. S. Cooper.

[20] *IBID*, Garnett's defense, written June 20, 1862.

[21] *IBID*, letter of Cooper to Lee, May 6, 1862. Cooper enclosed a letter from Garnett and a copy of the charges.

[22] *IBID*, reply of Lee to Cooper, May 6, 1862.

[23] *IBID*, Garnett's defense.

[24] *IBID.*

[25] *Make Me A Map*, p. 118.

[26] R.L. Dabney, *Life And Campaigns of Lieut.-Gen. Thomas J. Jackson* (New York: Blelock & Co., 1866), p. 505.

[27] *IBID.*

(Continued on page 25.)

TIMELINE

The summer of 1862 had been a bloody one. The Seven Days Battles around Richmond cost thousands of lives and finalized nothing. McClellan had withdrawn to the James with his huge army, and in the northern part of Virginia, a new Union commander, Gen. John Pope, was preparing his Army of Virginia for a new move "on to Richmond." To counter this threat, Gen. Robert E. Lee sent the Army of the Valley, under Major-General T.J. Jackson, northward. Advancing on Gordonsville on July 19, Jackson, who also commanded A.P. Hill's Light Division, immediately looked for an opportunity to meet this new threat. His chance came on August 7, when he learned from spies that only part of Pope's army had concentrated at Culpeper. Moving rapidly, Jackson's Corps clashed with Pope's army on August 9, resulting in a victory for the Confederates at Cedar Mountain.

With the threat of Pope's possible reinforcement by Burnside and the rumors that McClellan would be withdrawn from the James to eventually unite with Pope, the immediate danger to the Confederates appeared to be on the Rappahannock. With the prospect of concentrating Lee's army with that of Jackson's and attacking Pope before he could be further strengthed, Lee gave orders for Longstreet's Corps to move north.

August 11, 1862

Confederate

Gen. Jackson issues a proclamation to his troops praising God for their victory near Cedar Run, and mourns the loss of Brig. Gen. Winder. Generals Stuart and Early arrange a truce with Generals Roberts, Hartsuff and Crawford so the Federals can remove their wounded from the field and bury their dead.

Union

King's Division of 10,000 men join Pope. They march from Fredericksburg, where Pope had called for them on the 8th after hearing of a Confederate forward movement. Included are two of Burnside's best batteries and the Harris Light Cavalry. His troops at Falmouth are prepared to advance at six hours notice. Troops of Gen. Cox are summoned from West Virginia to reinforce Pope.

August 12

Confederate

Gen. R. E. Lee writes Jackson: "The country owes you and your braves officers and soldiers a deep debt of gratitude."

Union

Pope informs Halleck that the enemy has retreated under cover of night toward Orange Court-House. Halleck, in turn, warns him to "beware of a snare. Feigned retreats are secesh tactics." Burnside is ordered to leave a minimum number of troops to protect Aquina and send the remainder to Pope.

August 13

Confederate

Lee orders Gen. Longstreet and his command to proceed by cars on the Virginia Central Railroad to Gordonsville. Longstreet arrives, and by 6 PM notifies Lee of two possible advance routes. Rumors abound that Burnside has either left, or is preparing to leave Fredericksburg to attack Jackson or go after the supply line of the VCRR.

Union

Burnside reports Reno's command of 12 regiments and four batteries near Rappahannock Station; troops arriving at Aquina Creek. Pope is ordered not to advance his force across the Rapidan. Cox prepares to move his troops toward Gen. Pope.

August 14

Confederate

Lee writes Longstreet that he has directed Stuart to prepare to sweep around the enemy's rear, cut his communications, and burn the bridge over the Rappahannock so Pope cannot use it for retreat. Whatever moves they plan should be executed as quickly as possible. Deserters from Burnside report that he has 12,000 troops, and he's received 21 regiments after his arrival from Aquina Creek.

Union

Pope reports the enemy has retreated beyond Gordonsville and complains about the "wretched and ineffecient" management of the railroad from Washington to Culpeper. Burnside is ordered to report to Halleck.

August 15

Confederate

Gen. Lee arrives at Gordonsville and assumes command of the army. In conference, Longstreet advocates making a move on their left. Jackson suggests crossing the Rapidan on the 16th and giving battle on the 17th, before Pope realizes the dangerous position he is in, and withdraws behind the Rappahannock. When Longstreet protests that his commissary is disorganized and he will need more time, Jackson offers to supply bread so there will be no delay. Lee determines the cavalry is not concentrated, so he puts off the prepared attack until the 18th.

Two citizens escape from Culpeper Court-House, and report Federal strength at 92,000 men. They report only supplies, not troops, are arriving by rail.

Union

Reno and McDowell are ordered to communicate with each other, their Corps being only 2½ miles apart. Reports indicate Jackson is 5 miles beyond Orange Court-House, on the road to Gordonsville.

August 16

Confederate

Jackson continues the move begun yesterday by marching his Corps along the southern slope of

Clark's Mountain, six miles from Orange Court-House, in anticipation of a Confederate offensive. Cavalry moves in the vicinity of Davenport's Bridge on the North Anna.

Union

108 vessels move along the James in less than 24 hours as McClellan withdraws his entire army northward. It thus becomes a race to see whether the Confederates can attack the Union forces at Culpeper before they are heavily reinforced. Halleck warns Pope again not to cross the Rapidan and thinks it would be better if the army were in rear of the Rappahannock. Cox reports he will be at Parkersburg on the 20th with 5,000 infantry. Negroes report large reinforcements have been sent to Jackson.

August 17

Confederate

After conferring with Lee, Stuart and a handful of his staff ride to Verdiersville, where they await the arrival of the cavalry under Fitz Lee.

Union

Pope reports a large force moving to reinforce Jackson, and anticipates the enemy will form a junction near the Rapidan at Germanna Mills or Ely's Ford, and interpose itself between his forces and those landing at Aquina Creek. McClellan's forces will all have crossed the Chickahominy by nightfall.

August 18

Confederate

Due to a misunderstanding, Fitz Lee's men do not join Stuart, resulting in a lost opportunity to assail the Federals, as well as endangering Stuart's life when Federal cavalry enter the town in the early morning. Toward evening, lookouts report Pope is breaking camp.

Union

Halleck warns that the enemy has intentions of attacking Pope before McClellan's troops are up. Pope acknowledges an enemy advance toward him and begins a withdrawal toward Culpeper and out of the Confederate trap that was being set for him. Halleck orders the fords well secured. McClellan reports his troops at Williamsburg.

August 19

Confederate

Lee issues orders for the army to cross the Rapidan, Jackson's forces being on the left and Longstreet's on the right wing. At a signal station, Lee witnesses the end phase of the Federal withdrawal out of their camp. Orders are re-issued for the troops to move on the 20th instead of the 19th, due in part to the poor condition of cavalry horses.

Union

Pope completes his withdrawal over the bridge of the Rappahannock, resulting in a lost opportunity for the Confederates. The withdrawal is efficient and well-planned. McClellan promises to hurry his troops to Pope's aid.

August 20

Confederate

The two Corps under Jackson and Longstreet move along the Rappahannock River, attempting to locate a crossable ford. The cavalry skirmishes with the enemy at Brandy Station.

Union

Troops are skillfull placed at all the bridges and fords, making it impossible for the Confederates to cross the river. This forces Lee to go further and further north, thus giving Pope a clear path to the virtually unguarded city of Richmond. Pope informs Halleck his force is 45,000, independent of Burnside, and states his plan to hold the Rappahannock until McClellan joins him.

August 21

Confederate

The day is spent attempting to cross the river. Heavy cannonading on both sides as the Federals defend their positions and hold back the Confederates. Lee informs Davis that it appears Burnside, Stevens and King have gone toward Fredericksburg, while Pope, Banks and Sigel have moved toward Winchester.

Union

The troops defending the crossways shadow the movements of the enemy, sucessfully holding them back. Burnside reports 6,000 troops landed at Aquina yesterday, and double may land this day.

August 22

Confederate

Stuart takes 1,500 men and 2 pieces of artillery, crosses the river and reaches Warrenton, where there is no sign of the enemy. He pushes on to Catlett Station, where he hopes to destroy the railroad bridge. His men are unable to blow the bridge, but find Pope's headquarters, where they capture a large number of officers, public property and Pope's personal belongings. Jackson pushes across Early's brigade at Warrenton Sulphur Springs, but a heavy storm breaks out, stranding them on the northern side.

Union

As Trimble's men move up river, Federal cavalry succeed in getting between them and attack his wagon train. When the head of Longstreet's column arrives, the cavalry are driven off, losing over 100 men and a brigadier general; Gen. Sigel's artillery fire has little effect.

Halleck reports Cox's troops from W. Virginia arriving in Washington, and will be sent to Pope.

August 23

Confederate

Jackson succeeds in bringing Early's men back across the river after constructing a rickety bridge. Stuart proudly displays Pope's uniform coat, and proposes to exchange it for his hat and plume, which he lost to the Federals at Verdiersville.

(Continued on page 18.)

"MISSING IN ACTION"

by

S. L. Kotar and J. E. Gessler

On April 21, 1861, Major T. J. Jackson of the Virginia Military Institute, marched the Corps of Cadets from their barracks at Lexington, Virginia, past the Mess Hall, down to the bridge and across the North branch of the James River.[1] As senior officer present, "Old Jack" was to lead the group of approximately 180 cadets to Richmond, where they would be housed in barracks at the Fair Grounds and employed in training recruits for service in the Confederate army. The march was successfully completed the night of April 22, as Jackson brought them safely to their new quarters, and turned them over to other VMI officers. The following day, under the command of Major William Gilham, the natty Corps of Cadets were reviewed by Governor John Letcher. Jackson, whose association with the Institute went back ten years to early 1851, when he had been recommended for a professorship by D. H. Hill of Washington College, had no duties assigned him by any ranking VMI personnel. Aside from answering an occasional question asked by an eager citizen-turned-soldier, Major Jackson had nothing to do in the capital but apply for a military position in the service of Virginia. His own peers, those professor/officers from VMI who had arrived ahead of or with him, were all assigned to tasks with the cadets and/or recruits. Surprisingly, they found no use for the man who had been a hero in the Mexican War. Major Robert E. Withers describes his first meeting with Jackson during this period of uncertainty.

"He was not then the recipient of any special attention, was very quiet and reticent, having little to say to any one. Being without any regular quarters, I invited him one night at bed time to share my blankets, as he appeared to have none of his own, an invitation which he promptly accepted. I spread a blanket on the floor, we took our saddles for pillows, covered with another blanket and passed a rather uncomfortable night, at least I found it so, and since that time I have never willingly slept on a plank floor — I always found the ground much more comfortable...."[2]

Francis H. Smith

It would not be long before this self-reliant man would make his mark. But what of his peers, the self-important VMI professors who had no place for him? The two prominent officers who outranked him were Colonel Smith and Major Gilham, both of whom had gone to Richmond before either Jackson or the Corps of Cadets. (In fact, the order calling out the Corps made no mention of Jackson; he assumed their command by right of seniority.[3]) In April 1861, it appeared they were well on their way to becoming important men in the Virginia service, and then, presumably, the Confederate military. Who were these individuals, and what happened to them while Jackson was making history?

FRANCIS HENNEY SMITH

Francis Henney Smith had been Superintendent of VMI since the first cadets arrived in 1839. He was a graduate of West Point, ranking 5th in the class of 1833. He had served as Assistant Professor of Geography, History and Ethics at the USMA, and resigned May 1, 1836, with the rank of second lieutenant. He then became professor of mathmatics at Hampden-Sidney College, Virginia, and left that position after two years to join VMI.[4] The author of several text books, "Old Specs" was considered so rusty in military drill that, in its first year of operation, a VMI cadet was appointed to handle that aspect of education at the Institute.[5] The rank of "Colonel" was bestowed on Smith, which provided him with a commission, but conferred no rank in the militia.[6]

In 1859, Smith was assigned the superintendence of John Brown's execution, which brought him statewide recognition, and in January 1860, with the passage of the Munitions Act, he was assigned by Gov. Letcher to a commission to upgrade the Virginia military. Throughout this year, the commission reviewed the militia, their arms and armaments, traveled to various foundries and purchased various types of munitions. On April 18, 1862, he left for Richmond, where Letcher appointed him to his "Council of Three," to aid and advise him as war approached. Two months later, on June 19, the Council was dissolved: Virginia had entered the Confederacy. The government was moved to Richmond, and the volunteers now came under the supervision of the CSA. There was no immediate offer tendered Smith, so the next day he submitted a proposition through another VMI professor, William Gilham, under which the Institute professors (presumably excluding Jackson, who was already a colonel in the Confederate service and in command of a brigade) would command a Cadet Battalion, adding, "if I cannot find a place in which I can render better service, I am willing to go into the field in command of the battalion of cadets."[7] The Corps of Cadets promptly rejected the offer, preferring to "select from their own body, officers, and attach themselves to what regiment we deem proper...."[8] "Old Specs" was not their idea of a field officer. Smith returned to Lexington, where VMI had suspended operations until fall. After meeting with the Board of Visitors, and winding up affairs (which included the marriage of his daughter Fanny), he left July 20 for Staunton. He was now a colonel in the 9th Virginia Regiment, Heavy Artillery, and assigned to Craney Island, commanding the defenses of Norfolk harbor. His duty in the field did not last long. By November 1861, he agreed to return to the comforts of wartime Lexington so VMI could be reopened on January 1 (the school had been closed since July). On December 18, he was relieved from Craney Island and given the brevet commission of major general "in order that the Institute might be reopened under his direction...Jackson had been promoted to the grade of major general, and this may have had something to do with the promotion of the superintendent."[9] (Smith claimed that Letcher had tendered him the rank of major general on April 19, but this was not recognized as the date of his promotion.)[10]

Aside from accompanying the Corps of Cadets on brief excursions, Smith was occupied with the operations of VMI. He was with the cadets near Balcony Falls when General Hunter entered Lexington and destroyed the Institute in June of 1864.[11] Upon returning, the cadets lived in buildings of Washington College, and those that could go home were ordered to do so on June 27.[12]

The fall term was scheduled to resulme that September, but as accomodations had not been prepared, the date was repeatedly changed. In October, cadets began organizing in Richmond to man the trenches. Smith, under authority from the governor, fought Confederate authorities who wished to conscript the cadets for field-service, but the need for men was greater than the need for education. During their wartime experiences, the cadets were primarily under the direction of Colonel Scott Ship, a VMI alumnus.

In December 1864, attempts were made to resume military instruction at the Alms House, City of Richmond, and those on field-service were allowed to resume their studies. Smith returned to the duties of Superintendent, but his service here would be of short duration. Calls for

Union

Pope feels Jackson has crossed the Rappahannock in force, and after Early's withdrawal, believes the Confederate general has given up hope of turning his flank.

August 24

Confederate

Generals Jackson, Stuart, Longstreet and Lee hold a conference to determine their course of action. It is determined that Jackson will take his command far around Pope's right, circle behind, and cut Federal communications with Washington. It is hoped this will cause Pope to retreat on Washington, thus moving him further away from reinforcements at Fredericksburg. Then Lee could bring up the rest of the army and attack Pope from the rear.

Union

Two of McClellan's Corps, the 3rd and 5th, and Reynold's Pennsylvania Reserves, are reported to be within two days marching. The 2nd, 4th and 6th Corps, including 7,000 men from the Western Virginia-Kanawha region, could unite with Pope in five more days, swelling the ranks from 50,000 to 130,000. Confederate strength is currently 54,500, including cavalry and artillery. Reinforcements of D.H. Hill and McLaws will add 17,000.

McClellan arrives at Aquina Creek. Halleck wires him that he does not know where either Pope or the enemy force is.

August 25

Confederate

Jackson's troops move north, successfully crossing the Rappahannock at Hedgemen's River (Henson's Mill). Waterloo Bridge. The troops march 25 miles and encamp near Salem.

Union

A signal station near Waterloo Bridge reports that by 7:15 AM troops are moving along the river. Reports continue past 11 AM that a large body of men and wagons are passing. By noon, Pope informs Halleck in Washington 20,000 enemy are on the march and that McDowell has been alerted. Pope believes this indicates the van of Lee's army is going to the Shenandoah Valley.

August 26

Confederate

Troops pass unmolested through critically important Thoroughfare Gap and pass Hay Market and Gainesville. Stuart and his cavalry join the army here. The main objective being the railroad bridge over the Run, troops march on Bristoe Station, 4 miles below Manassas Junction. At the station, cavalry and infantry cooperate, and spectacularly derail the oncoming trains. Local residents tell of huge piles of stores 7 miles up the tracks at Manassas Junction. Jackson sends Trimble, then Stuart, and they defeat the small force there. Immense supplies are captured.

Union

Pope believes the Southern movement behind him is only Stuart's cavalry. His own cavalry is overworked, at one period spending more than 83 consecutive hours in the saddle.

August 27

Confederate

Jackson marches all but Ewell's brigades to Manassas Junction. There they find undreamed of stockpiles of foodstufs, luxuries, medicine, liquor and military equipment. When Jackson realizes he cannot carry away this treasure, he lets his men loose, and they celebrate Christmas in August. When they are fired on by a lone battery, Jackson leads his troops toward Bull Run, where they encounter Gen. George Taylor's New Jersey Brigade, just debarked from the cars. The Federals are soon routed, and the Confederates, joined by Ewell's men, complete the evacuation of anything edible from Pope's supply base. Having destroyed the railroad bridge over Bull Run,

the cadets came as early as January 31, 1865, when General Ewell summoned them to help repel the enemy. They were held in readiness, but the governor felt they ought not be sent unless there was an extreme emergency.[13] Illness compelled Smith to return to Lexington in February, while acting Superintendent Preston advocated that the cadets be used as drillmasters for the organization of a Negro Corps. The war ended before any action on this was taken, however, and the remaining cadets entered the ranks as soldiers.

When VMI eventually reopened in Lexington October 16, 1865, Gen. Smith presided as Superintendent.

WILLIAM GILHAM

William Gilham was born in Indiana, but laid claim to being a Virginian when it was to his advantage. His father had been one before moving his family west. Gilham was appointed to West Point from Indiana, and graduated in the class of 1840, ranking 5th.[14] He spent a brief period in Florida under Braxton Bragg, then was recalled to the USMA to act as an assistant professor of Natural and Experimental Philosophy. He saw brief service in North Carolina and Texas, and in 1846, traveled with the 3rd Artillery, Army of Occupation, to Mexico, where he spent much time trying to obtain a professorship in any college. On July 3, 1846, he was appointed to VMI as a progessor of Physical Sciences, commandant of cadets and instructor of tactics. By October 17, he had resigned from the army with the rank of first lieutenant,[15] and arrived in Lexington. In 1850, he took the chair of Chemistry, Geology and Mineralogy. During the pre-war years, he spent a great deal of time analyzing soils and manures. His work brought him statewide recognition, and in 1859, he was elected to the new Chair of Agriculture at VMI. This position never came about, however, and he continued as the professor of chemistry.

Col. William Gilham

Gilham was considered "the disciplinarian and the champion of military ideals of the Institute," and was "regarded by many as a veritable martinet."[16] The "individual interest was never allowed under the regime of Gilham,"[17] and he was "disliked by those who shirked their duty and felt the iron of his hand...."[18]

He participated in the execution of John Brown, commanding the VMI infantry detachment and later that year was "detailed on special duty to prepare a synoptical work for the instruction of the militia service." By December 1860, the book went to press, and an early review accurately predicted its future:

> "The work is too elaborate and too costly and neither the State nor the officers will buy it."[19]

It was also early on "disregarded for cadet use" at VMI.[20]

Virginia passed the Ordnance of Secession April 17, 1861, and on April 20, Major Gilham was designated as officer in charge of the camp of instruction in the Central Fair Grounds, Richmond. He was ordered from Lexington to the capital, and left immediately. On April 23, it was he who led the Corps in review for the governor and in every way made himself conspicious.

Col. R. E. Colston

Along with Smith and Colston (another VMI professor), Gilham drafted a list of recommendations, with rank designation for VMI graduates, and he himself was promoted to colonel. He undoubtedly felt this would be just the beginning of his military achievements, but as spring turned into summer and the recruits were assigned into the field, no further opportunities offered themselves. When Smith proposed to gallantly lead the cadets into battle with the VMI staff as his immediate subordinates, Gilham presumably went along with the suggestion. After their rejection, however, it appeared he would have to make it on his own.

Gilham received an appointment as colonel, 21st Virginia Infantry; Major Scott Ship and Adjudant William H. Morgan (Gilham's colleagues at VMI) were also appointed officers.[21] On July 17, the 21st received orders to take the cars to Staunton, leaving the following day. They reached that destination, then were ordered to move west. They marched in that direction, opposite the one which would have taken them to the first major battle of the war. General William Loring, the commanding officer, commented on the 21st that they were "a fine body of men, but not soldiers.[22]"

General Robert E. Lee joined the troops in Pocahontas County, and assumed command of the department there. When Lee left for operations on Sewell's Mountain, he took all the troops with him, except the 21st Virginia, the Irish Battalion, a battery of artillery and a company of cavalry, which were left under Gilham's command. They soon fell back to Middle Mountain and left there September 28.

Nothing of significance transpired, and in October 1861, when the VMI Board of Visitors polled the Institute's professors on their willingness to return to the sedentary life of a teacher, Gilham expressed his intent to give up the glory of a soldier's life and return to Lexington.[23] The new session was scheduled to begin January 1, 1862.[24]

Gilham and his regiment, part of Gen. Loring's command, were ordered from Staunton to join Jackson's Valley Army, and take part in his winter campaign. They reached Winchester on Christmas day, having covered the short distance in a series of slow marches. Once there, the Irish Battalion, the 21st and 42nd Va. regiments were formed into the Second Brigade, Jackson's Division, with Col. Gilham the ranking officer, having approximately 2200 men.[25]

The brigades that moved out January 1 were vastly different from each other. The First Brigade was composed of veterans who had experienced the horrors of war firsthand. Loring's men had seen little actual fighting, their war service being less than arduous. They led the advance and confronted a Union picket outside of Bath on January 3. The following day, the advance crawled along, their lack of discipline and experience showing. To hurry the advance, Colonel Baylor, of Jackson's staff, was sent to lead the cavalry into town with the commanding general himself close behind. He found Gilham's brigade going down the wrong road, redirected it, then later criticized the VMI professor on his pursuit of the retreating enemy. Jackson wrote:

> "...he neither attacked them nor notified me of

Jackson directs his troops to a defensible position at Groveton to await Lee's arrival.

Union

Pope discovers it is not just Stuart's cavalry on a raid, but Jackson's entire army behind him. He breaks from the Rappahannock and moves north, planning to move on Gordonsville, thus putting himself between Jackson and his reinforcements. Toward evening, however, he determines Jackson will not evacuate Manassas Junction, but will take advantage of the fortifications there. Obsessed with bagging the whole crowd, he redirects his whole army to concentrate on Manassas.

August 28

Confederate

Jackson unites his troops at Groveton, and toward sunset determines to attack a large body of Federal infantry under Hatch, Gibbon, Doubleday and Patrick, who were marching down the Warrenton road toward Centerville. At point-blank range, both sides fight gallantly until 9 PM, when the Union forces withdraw. Gen. Ewell is wounded in the leg while leading a charge, which requires amputation and incapacitates him, keeping him from further combat until the following spring.

Union

Pope has three options: retreat to unite with McClellan and resume the offensive; withdraw into Washington, or move on Jackson. He determines on the latter course of action, and marches on Manassas. When he reaches it at noon, he discovers a smoldering ruin, and Jackson gone. Another column, marching straight for the Stone Bridge rather than southeast toward Manassas, are the ones attacked by the enemy. King's Division, Burnside's Corps, fight as stubbornly as their attackers, only withdrawing under cover of darkness. During the night, troops are interposed between the enemy and Washington.

August 29

Confederate/Union

Jackson directs his 18,000 troops to a defensive position behind an unfinished railroad line, waiting for Longstreet to come up. The Federals attack fiercely, pressing A.P. Hill on the left. Longstreet and Lee arrive, and form on the right of the Confederate line. They assess the situation from this point while the fight roars on the left. Toward evening, Jackson's men succeed in pushing the Union army back from the railroad cut to their original position, and the days fighting ends.

August 30

Confederate

At a headquarters conference, Lee suggests putting up a front, slipping across Bull Run and getting in Pope's rear; but Jackson, having concluded the enemy would offer a fight in the afternoon, returns to his command, where between 3 and 4 PM the enemy attacks his position. Artillery set up in the center of the line breaks up the Federal formation. They fall back on themselves, precipitating a general Confederate advance.

Union

Pope reports losing not less than 8,000 men killed and wounded, but feels the enemy lost 2:1. Halleck writes 30,000 men are marching to Pope's aid. Banks reports Jackson was recognized by a spy at Gainsville with 30,000 men. Porter writes McClellan the whole army has been badly whipped. He adds that the chief of corps were disappointed they had to stay and fight and the troops would not fight with heart when they learned the wounded would be left to the enemy, as they cannot retain the field.

August 31

Confederate

After their victory at Second Manassas, the Confederates bury their dead and collect arms. Jackson's command, being nearest the exposed flank of the retreating

> the cause of not doing so, nor even of his having overtaken the Federal forces, their artillery and infantry were permitted to escape "[26]

He would not forget this. In his reports, Jackson would criticize Loring harshly, and when time permitted, he preferred charges of "Neglect of Duty" against Gilham for his tardiness in his January 4 movement on Bath.

On January 9, Gilham and Ship had had enough of active campaigning and left for Lexington and the comforts of home. The troops presented the former with a fine horse,[27] which he would no doubt find useful, and they arrived at VMI on January 16. These "seasoned teachers" soon got the new session underway, unfortunately without the help of Colonel Preston, who had seen fit to stay with Gen. Jackson until the Romney expedition was completed. (Preston reported to VMI on February 1.[28]) Since Gilham resigned his position in the army, Jackson dropped the charges against him.

When Jackson called for the Corps of Cadets in April 1862, Gilham was left behind to "remain in charge of the Institution."[29] He remained at VMI during the war years, and was sent "to aid in providing every comfort for the cadets"[30] in the May 1864 campaign. He caught up with the Corps May 12, and followed them through the Battle of New Market, but "was usually not associated in the records of this campaign."[31] On May 20, Gilham was ordered to Lexington to collect clothes for the cadets,[32] and in October 1864, he made a "minute examination"[33] of the Alms House, Richmond, in anticipation of its use as cadet quarters.

After the war, when attempts were being made to reopen VMI, Gilham resigned his Chair of Chemistry, but maintained his claim to the inactive Chair of Agriculture "to see how things would develop." He had already decided, though, to move to New York City and work as an analytical and agricultural chemist.[34] The "late unpleasantness" was quickly forgotten when the need for money became acute. He wrote to Smith, October 18, 1865:

> "I had every reason to suppose that the proceeds of my tactics would be something handsome, and thus enable me to pay off my debts, but as I am getting nothing from that quarter, I am, I may say, practically without means."[35]

He never returned to teach at the Virginia Military Institute, although his honored name was carried on the faculty roster.

THE FORTUNES OF TIME

Francis Henney Smith and William Gilham, both graduates of West Point Military Academy, were called to the service of Virginia at the dawn of civil war. They had every reason to suspect their careers would be glorious. Neither one could have suspected that the quiet man they smugly left behind would receive the laurels they would not. The name Thomas Jonathan "Stonewall" Jackson has become legendary: the names of Smith and Gilham all but obscure. Jackson suffered privations both physical and mental, and died for his cause at the age of 39, having spent only a few days with his infant daughter before his death. Smith and Gilham spent the majority of the war in relative comfort and ease, in Lexington with their families. Smith's children were grown, and Gilham's eight children went east with him in 1865. Neither man was as much as wounded, and neither lived up to his expectations. All three started the war as colonels in the service of Virginia — one made history — two were lost in it. Some men march to a different drummer.

(Footnotes on page 35.)

enemy, is to follow, putting himself in their line of retreat. His proud troops move out and begin the chase over muddy roads.

Union

Pope's army destroys Stone Bridge, forcing his pursuers to follow by roundabout roads. 7,000 prisoners, 2,000 wounded and 20,000 small arms are left behind in the evacuation.

Toward afternoon, a shout warning of Federal cavalry startles Lee's horse, causing the General to fall and sprain his wrists. Halleck, believing Second Bull Run a Federal victory, writes Pope, blessing his noble army.

September 1

Confederate

Jackson's troops, starving and exhausted from continuous fighting at Second Manassas, march slowly after the retreating enemy. Toward afternoon, near the old Chantilly mansion, they encounter a defiant Federal rear guard. The column is deployed and a fierce contest ensues in the pouring rain. As darkness gathers, the Federals break off and continue on their way.

Union

Covering the rear guard of the army's retreat to Washington, Gen. I.I. Stevens' troops admirably meet the advancing enemy and hold them at bay until Stevens and Gen. Phil Kearny are killed while bravely leading their men. Loss of leadership and the fall of night dictate a withdrawal from the field. This is done, and they march unmolested. Kearny's body is brought to Jackson's headquarters, where many Confederates pass by to view their gallant enemy.

S.L. Kotar J.E. Gessler

FAMILY BIBLE

Birthdays of Civil War Notables

by

Paul Kallina

(N = North. S = South.)

OCTOBER

Oct. 1, 1808—Rep. Robert M. Jones, Choctaw Territory, Mississippi. (S)
Oct. 1, 1826—Benjamin B. Hotchkiss, ordnance inventor and manufacturer, Watertown, Connecticut. (N)
Oct. 1, 1832—Henry Clay Work, song writer, Middletown, Connecticut. (N)
Oct. 2, 1798—Gazaway, B. Lamar, banker and Blockade-runner, Richmond County, Georgia. (S)
Oct. 2, 1821—Gen. Alexander P. Stewart, Rogersville, Tennessee. (S)
Oct. 2, 1833—Fr. William Corby, Irish Brigade chaplain, Detroit, Michigan. (N)
Oct. 3, 1790—John Ross (Kooweskoowe), Cherokee chief, near Lookout Mountain, Tennessee. (S)
Oct. 3, 1823—Maj. Benjamin F. Stephenson, Wayne County, Illinois. (N)
Oct. 3, 1830—Col. Charles Marshall, Warrenton, Virginia. (S)
Oct. 3, 1839—Col. Charles Heywood, U.S.M.C., Waterville, Maine. (N)
Oct. 4, 1796—Capt. French Forrest, C.S.N., St. Mary's County, Maryland. (S)
Oct. 4, 1809—Gen. and Rep. Robert C. Schenck, Franklin, Ohio. (N)
Oct. 4, 1822—Gen. Rutherford B. Hayes, Delaware, Ohio. (N)
Oct. 4, 1824—Col. Lawrence M. Keitt, Orangeburg District, South Carolina. (S)
Oct. 5, 1804—Robert C. Parrott, ordnance inventor, Lee, New Hampshire. (N)
Oct. 5, 1830—Gen. Chester A. Arthur, Fairfield, Vermont. (N)
Oct. 6, 1809—John W. Griffiths, naval architect, New York City. (N)
Oct. 6, 1822—Commodore Benjamin F. Isherwood, engineer-in-chief of the navy, New York City. (N)
Oct. 6, 1846—Eng. George Westinghouse, U.S.N., Central Bridge, New York. (N)
Oct. 7, 1817—Gen. Bushrod R. Johnson, Belmont County, Ohio. (S)
Oct. 7, 1817—William Still, black leader, Shamony, New Jersey. (N)
Oct. 7, 1821—Gen. Richard H. Anderson, Hill Crest, Sumter County, South Carolina. (S)
Oct. 8, 1818—John H. Reagan, Postmaster-General (1861-1865), Sevier County, Tennessee. (S)
Oct. 8, 1826—Capt. William H. Parker, superintendent of Confederate Naval Academy (1864-1865), New York City. (S)
Oct. 8, 1838—Col. John Hay, Lincoln's private secretary, Salem, Indiana. (N)
Oct. 9, 1811—Rep. Warren Akin, Elbert County, Georgia. (S)
Oct. 9, 1813—John J. Pettus, governor of Mississippi (1859-1863), Wilson County, Tennessee. (S)
Oct. 11, 1821—Thomas C. Reynolds, governor of Missouri (1862-1865), Charleston, South Carolina. (S)

Oct. 11, 1835—Hunter H. McGuire, Stonewall Jackson's Medical Director, Winchester, Virginia. (S)
Oct. 11, 1841—Lt. Louis Kempff, U.S.N., Belleville, Illinois. (N)
Oct. 12, 1813—Sen. Lyman Trumbull, Colchester, Connecticut. (N)
Oct. 12, 1815—Gen. William J. Hardee, Camden County, Georgia. (S)
Oct. 13, 1826—Gen. La Fayette C. Baker, Secret Service agent, Stafford, New York. (N)
Oct. 14, 1814—Rep. Otho R. Singleton, Jessamine County, Kentucky. (S)
Oct. 15, 1818—Gen. Irvin McDowell, Columbus, Ohio. (N)
Oct. 15, 1836—Gen. Thomas L. Rosser, Campbell County, Virginia. (S)
Oct. 16, 1806—William P. Fessenden, Secretary of the Treasury (1864-1865), Boscawen, New Hampshire. (N)
Oct. 16, 1815—Col. Francis R. Lubbock, governor of Texas (1861-1863), Beaufort, South Carolina. (S)
Oct. 16, 1816—Gen. William Preston, Louisville, Kentucky. (S)
Oct. 16, 1819—Rep. Samuel A. Miller, Shenandoah County, Virginia. (S)
Oct. 16, 1835—Gen. William R. Shafter, Galesburg, Michigan. (N)
Oct. 17, 1818—Elizabeth L. "Crazy Bet" Van Lew, Union spy, Richmond, Virginia. (N)
Oct. 18, 1818—Gen. Edward O.C. Ord, Cumberland, Maryland. (N)
Oct. 18, 1822—Rep. Daniel C. de Jarnette, near Bowling Green, Virginia. (S)
Oct. 19, 1810—Gen. Cassius M. Clay, Madison County, Kentucky. (N)
Oct. 19, 1834—Gen. Francis C. Barlow, Brooklyn, New York. (N)
Oct. 20, 1816—Sen. James W. Grimes, Deering, New Hampshire. (N)
Oct. 20, 1820—Gen. Benjamin F. Cheatham, Nashville, Tennessee. (S)
Oct. 20, 1822—Gen. Mansfield Lovell, Washington, D.C. (S)
Oct. 20, 1825—Gen. Daniel E. Sickles, New York City. (N)
Oct. 21, 1819—Commander Robert Townsend, Albany, New York. (N)
Oct. 22, 1812—Colin J. McRae, financial agent in Europe, Sneedsboro, North Carolina. (S)
Oct. 23, 1828—Gen. Turner Ashby, Rose Bank, Fauquier County, Virginia. (S)
Oct. 23, 1831—Pvt. Basil L. Gildersleeve, Charleston, South Carolina. (S)
Oct. 24, 1818—Rep. William W. Boyce, Charleston, South Carolina. (S)
Oct. 25, 1837—Major Innes Randolph, "O, I'm a Good Old Rebel," Winchester, Virginia. (S)
Oct. 26, 1804—Gen. Lorenzo Thomas, New Castle, Delaware. (N)
Oct. 27, 1793—Eliphalet Remington, arms manufacturer, Suffield, Connecticut. (N)
Oct. 27, 1800—Sen. Benjamin F. Wade, Feeding Hills, Massachusetts. (N)
Oct. 27, 1823—Rep. William D. Simpson, Belfast, South Carolina. (S)
Oct. 27, 1828—Gen. Jacob D. Cox, Montreal, Canada. (N)
Oct. 27, 1837—Whitelaw Reid, war correspondent, Xenia, Ohio. (N)
Oct. 29, 1815—Daniel D. Emmett, "Dixie's Land," Mt. Vernon, Ohio. (N)
Oct. 29, 1824—Gen. Joseph H. Lewis, Glasgow, Kentucky. (S)
Oct. 29, 1845—Lt. Charles F. Crisp, C.S. Army, Sheffield, England. (S)
Oct. 30, 1822—Commander George Washington Rodgers, Brooklyn, New York. (N)
Oct. 31, 1812—John Forsyth, mayor of Mobile (1860-1866) and newspaper editor, Augusta, Georgia. (S)
Oct. 31, 1821—Commander Alexander C. Rhind, New York City. (N)

• SIGNALS •

Dear Editors:

Thank you for sending me the sample issue I ordered. I find THE KEPI to be quite different from any other magazine I have seen in the past, and I wish to subscribe. Enclosed is $18.00 for a one year subscription.

I am considering writing an article for your magazine; I would like to see the tone of your previous articles. Please send me the back issues as follows: Vol. I, #1-5; Vol. II, #1-3. (I enclose a check to cover the amount owed.)

Thank you, and I hope to be sending a piece soon.

C. Winfield
Anaheim, CA

Dear Editors:

The feature article, "A Mission To President Lincoln," [August-September 1984] by Peter Harrington, was particularly enjoyable and enlightening. I am a McClellan scholar, and have been on the look-out for material covering his command in 1861. What especially pleased me about this article is that it was presented in its entirety. I prefer to see all of such articles, correspondence, etc., rather than have someone else take lines out of context, and so interpret history for me. I also found TIMELINE to be of interest in sorting out events as they happened.

I look forward to receiving my next issue of your fine publication.

Jon Lorentz
Minneapolis, MN

Dear Editors:

I read your last issue [Vol. II, No. 4] cover to cover, and I was fascinated by every article. I felt as though I were back in time, reading these incidents as they were happening. Of special interest to me was the piece on the young Lincoln ["Lincoln's Playmate"], because I had visited the boyhood home of Lincoln, and now, having read Mr. Evans' story, I feel much closer to Lincoln. I feel as if I had watched a moment of his life—his private life, not public—and he has become much more real to me, no longer a distant legend.

Thank you for publishing a magazine which truly does make the 19th Century "come alive."

L. Cho
Portland, OR

Dear Editors:

Enclosed is my check for $18.00 to cover a years subscription to THE KEPI. I enjoyed the ½ year subscription I purchased from a special offer. It is most helpful to my research, since you include footnotes with each major article. The Concurrent Events also caught my fancy. To read what were late-breaking news events in the 1800's provides me with an insight I hadn't expected to find; comparing them with todays events, I realize how much, yet how little, times have changed. These little inserts are invaluable, especially to those who do not have a collection of period newspapers.

Much luck on your magazine.

William K. Long
Paramus, NJ

Dear Editors:

Please send me the following back issues of THE KEPI: Volume I, Number 5, Volume II, Number 1. I am enclosing $7.00 for the cost.

Thank You.

I am starting to really enjoy your magazine. Keep up the good work.

Stewart C. Yeakel
Kanawha, IA

Oct. 31, 1831—Gen. Daniel Butterfield, Utica, New York. (N)
Oct. 31, 1835—Gen. Adelbert Ames, Rockland, Maine. (N)

NOVEMBER

Nov. 1, 1809—Capt. Theodore P. Greene, U.S.N., Montreal, Canada. (N)
Nov. 1, 1835—Gen. Godfrey Weitzel, Cincinnati, Ohio. (N)
Nov. 2, 1810—Gen. Andrew A. Humphreys, Philadelphia, Pennsylvania. (N)
Nov. 3, 1794—William Cullen Bryant, editor, *New York Evening Post*, Cummington, Massachusetts. (N)
Nov. 3, 1798—James M. Mason, commissioner to England (1862-1865), Georgetown, District of Columbia. (S)
Nov. 3, 1810—Charles C. Burleigh, abolitionist, Plainfield, Connecticut. (N)
Nov. 3, 1816—Gen, Jubal A. Early, Franklin County, Virginia. (S)
Nov. 3, 1817—Col. Harris Flanagin, governor of Arkansas (1862-1865), Roadstown, New Jersey. (S)
Nov. 3, 1830—Capt. John E. Cooke, C.S. Army, *Surry of Eagle's Nest*, Winchester, Virginia. (S)
Nov. 3, 1845—Pvt. Edward D. White, LaFourche Parish, Louisiana. (S)
Nov. 4, 1805—Rep. John A. Gilmer, Guilford County, North Carolina. (S)
Nov. 4, 1835—Gen. Lunsford L. Lomax, Newport, Rhode Island. (S)
Nov. 4, 1842—Commander William B. Cushing, Delafield, Wisconsin. (N)
Nov. 5, 1818—Gen. Benjamin F. Butler, Deerfield, New Hampshire. (N)
Nov. 5, 1826—Col. Edward W. Serrell, London, England. (N)
Nov. 5, 1832—Gen. William W. Averell, Cameron, New York. (N)
Nov. 6, 1821—Lt. John Wilkinson, C.S.N., Norfolk, Virginia. (S)
Nov. 7, 1797—Adm. Silas H. Stringham, Middletown, New York. (N)
Nov. 7, 1828—Leonard W. Volk, sculptor, Wellstown, New York. (N)
Nov. 7, 1837—Gen. George A. Forsyth, Muncy, Pennsylvania. (N)
Nov. 8, 1817—Gen. Claudius W. Sears, Peru, Massachusetts. (S)
Nov. 8, 1830—Gen. Oliver O. Howard, Leeds, Maine. (N)
Nov. 9, 1795—Capt. Josiah Tattnall, C.S.N., near Savannah, Georgia. (S)
Nov. 9, 1810—Thomas Bragg, Attorney General (1861-1862), Warrentown, North Carolina. (S)
Nov. 9, 1825—Gen. Ambrose P. Hill, Culpeper, Virginia. (S)
Nov. 9, 1833—Capt. Sally L. Tompkins, hospital administrator, Poplar Grove, Mathews County, Virginia. (S)
Nov. 10, 1827—Gen. Alfred H. Terry, Hartford, Connecticut. (N)
Nov. 10, 1830—Gen. Albert G. Jenkins, Cabell County, Virginia. (S)
Nov. 11, 1811—Gen. Ben McCulloch, Rutherford County, Tennessee. (S)
Nov. 12, 1797—Hannah P. Cox, abolitionist, Longwood, Chester County, Pennsylvania. (N)
Nov. 13, 1804—Gen. Theophilus H. Holmes, Sampson County, North Carolina. (S)
Nov. 13, 1809—Adm. John A.B. Dahlgren, Philadelphia, Pennsylvania. (N)
Nov. 13, 1812—Sen. Edwin G. Reade, Mount Tirzah, North Carolina. (S)
Nov. 13, 1814—Gen. Joseph Hooker, Hadley, Massachusetts. (N)
Nov. 13, 1837—Col. Henry C. Merriam, Houlton, Maine. (N)
Nov. 14, 1819—Commander Christopher R.P. Rodgers, Brooklyn, New

York. (N)
Nov. 14, 1828—Gen. James B. McPherson, Sandusky County, Ohio. (N)
Nov. 15, 1797—Thurlow Weed, Republican politician, Greene County, New York. (N)
Nov. 15, 1836—Gen. Pierce M.B. Young, Spartansburg, South Carolina.
Nov. 17, 1809—Stephen S. Foster, abolitionist, Canterbury, New Hampshire. (N)
Nov. 18, 1824—Gen. Franz Sigel, Sinsheim, Grand Duchy of Baden, Germany. (N)
Nov. 19, 1802—Sen. Solomon Foot, Cornwall, Vermont. (N)
Nov. 19, 1811—Commodore John A. Winslow, Wilmington, North Carolina. (N)
Nov. 19, 1827—Gen. Isaac M. St. John, Augusta, Georgia. (S)
Nov. 19, 1831—Gen. James A. Garfield, Cuyahoga County, Ohio. (N)
Nov. 19, 1834—Col. Alexander R. Chisolm, Beaufort, South Carolina. (S)
Nov. 19, 1835—Gen. Fitzhugh Lee, Clermont, Fairfax County, Virginia. (S)
Nov. 19, 1839—Atticus G. Haygood, army chaplain, Watkinsville, Georgia. (S)
Nov. 20, 1802—Commodore James L. Lardner, Philadelphia, Pennsylvania. (N)
Nov. 22, 1805—Gen. Benjamin Huger, Charleston, South Carolina. (S)
Nov. 22, 1806—Sen. Lafayette S. Foster, Franklin, Connecticut. (N)
Nov. 22, 1818—Gen. Samuel G. French, Gloucester County, New Jersey. (S)
Nov. 23, 1815—William Dennison, governor of Ohio (1860-1862) and Postmaster-General (1864-1866), Cincinnati, Ohio. (N)
Nov. 23, 1819—Gen. Benjamin M. Prentiss, Belleville, Virginia. (N)
Nov. 24, 1844—Ens. Charles V. Gridley, Logansport, Indiana. (N)
Nov. 25, 1823—Henry Wirz, superintendent of Andersonville Prison (1864-1865), Zurich, Switzerland. (S)
Nov. 26, 1792—Sarah Grimke, abolitionist, Charleston, South Carolina. (N)
Nov. 26, 1816—Gen. William H.T. Walker, Augusta, Georgia. (S)
Nov. 26, 1832—Surgeon Mary E. Walker, near Oswego, New York. (N)
Nov. 27, 1808—Gen. Hugh W. Mercer, Fredericksburg, Virginia. (S)
Nov. 27, 1815—Commander Enoch G. Parrott, Portsmouth, New Hampshire. (N)
Nov. 27, 1843—Col. William Ludlow, Islip, Long Island, New York. (N)
Nov. 28, 1809—Capt. Samuel Barron, C.S.N., Hampton, Virginia. (S)
Nov. 28, 1819—Rep. Robert R. Bridgers, Edgecombe County, North Carolina. (S)
Nov. 29, 1798—Hamilton R. Gamble, governor of Missouri (1861-1864), Winchester, Virginia. (N)
Nov. 29, 1811—Wendell Phillips, abolitionist, Boston, Massachusetts. (N)
Nov. 29, 1815—Gen. Stephen A. Hurlbut, Charleston, South Carolina. (N)
Nov. 29, 1824—Gen. Thomas E.G. Ransom, Norwich, Vermont. (N)
Nov. 29, 1832—Louisa May Alcott, *Hospital Sketches*, Germantown, Pennsylvania. (N)
Nov. 30, 1828—Maj. Jedediah Hotchkiss, cartographer, member of "Stonewall" Jackson's staff, Windsor, Broome County, New York. (S)

Dear Editors:

I have received my last issue of THE KEPI, and I am enclosing a check for $18.00 for another years subscription, as well as the questionnare and this note. I felt I must congratulate you on bringing "fresh air" into the Civil War field by your scholarly input via THE KEPI magazine. I paricularly enjoy Military Review, TIMELINE, and those short articles about lesser known aspects of the war. I look forward to the article, "Christmas in August," which you said will appear in your next issue [October-November 1984], as stated in General Orders. I do not find that other magazines carry enough articles on Gen. Jackson, and I trust that your coverage of that period will be as enlightening as the rest of your articles.

I look forward to enjoying my second year with THE KEPI as much as my first. Good luck and much success!

Harold Lewis
Bennington, VT

(Continued from page 4.)

to this, PRIVATE THOUGHTS will be just that — a look at the men who fought in the ranks, through their letters and diaries. This installment concerns John A. Maxwell, 26th Indiana Volunteers. His three letters, the two photographs and this subsequent history were graciously supplied to THE KEPI by Diane Dickey, his great-granddaughter. *The Editors-in-General.*

(Continued from page 14.)

[28] William Poague, *Gunner With Stonewall Jackson*, Tenn.: McCowat - Mercer Press, 1957), p. 33.
[29] *OR*, VII, Part 3, pp. 929-30.
[30] *Make Me A Map*, p. 71.
[31] *I Rode With Stonewall*, pp. 132-3.
[32] Lieut.-Col. G.F.R. Henderson, *Stonewall Jackson and the American Civil War* (New York: Longmans, Green, and Co., 1911), Vol. II, p. 128; letter of Dr. McGuire to author.
[33] *I Rode With Stonewall*, p. 133.
[34] *Make Me A Map*, p. 118.
[35] *Hotchkiss MS Diary*, pp. 74-5; *OR*, XII, Part 2, p. 678.
[36] Letter of Hugh White to his family.
[37] *Make Me A Map*, p. 71.

(Continued on page 35.)

ATROCITY AT HUNTER'S MILL, VIRGINIA

Edited by:
Thomas J. Evans

The Reverend John D. Read, pastor of the Columbia Baptist Church, Falls Church, Virginia, was a Republican and an Abolitionist. These were two dangerous characteristics for anyone, even a minister, to possess while living in Virginia prior to and during the Civil War. Within his church, there was much dissension in the congregation and a bitterness against the Union and its sympathizers. When the war began, the Union Army invaded and occupied Northern Virginia, resulting in the extensive confiscation of crops, timber and livestock. Times were hazardous and difficult for the Reverend Read, his wife and six children: they were forced to flee to Washington, D.C.

John D. Read had succeeded his brother, the Reverend Hiram W. Read, as minister at the Columbia Baptist Church. Hiram served as an army chaplain in the West prior to the war, and early recognized a conflict was forthcoming. He resigned his pastorate at Falls Church to accept a clerkship in the first Comptroller's Office of the Treasury Department in nearby Washington, D.C. This position offered him the opportunity to secure liberal leaves of absence to serve the Union Army soldiers in the field and in hospitals.

Concerned with the problems suffered by his brother, he wrote the following letter to Salmon P. Chase (Secretary of the Treasury under President Abraham Lincoln), with whom he was acquainted:

(April 8/62 - Asks work for his brother)*

H.W. READ

*Note added to letter at Treasury Department.

Washington City, D.C.
April 8, 1862

Hon. S.P. Chase
Dear Sir:

Will you do me the favor to consider the following brief statement of facts. My Brother, Mr. John D. Read, formerly of New York - has during the last twelve years resided near Falls Church, Fairfax Co., Va. but last Spring was driven thence by the Seccessionists because he was a Republican & Voted for Mr. Lincoln. By fleeing from his home to this city in the night he barely escaped assassination. His wife & Six children immediately followed - being obliged to leave almost everything, even their family Bible. In their distress, sickness soon visited them & suffering and destitution have been their constant attendents ever since. My Bro. had been sick nearly a year from the effects of a Sun-stroke before being driven from Va. During the past summer & autumn, he labored in the Navy Yard where he won the esteem of all who knew him, but the labor there proved too severe. He met with an accident while in the Yard & was taken up for dead. A long and dangerous illness followed, from which he has now so far recovered as to be able to do only light work.

His little farm in Va. is ruined, not a vestage of his buildings or fences remain - all having been used for fuel by our troops, and therefore he cannot occupy his former home. His crops of grain & hay were also used by the troops & he has never been able to obtain a cent for them.

The total loss of everything — & his continuous sickness & that of his family, have prevented him from sending his children to School — even Sunday School — their clothes were not suitable & some of them have been obliged to go without even shoes during the winter. And within the last few days, I have learned that they suffered for both food & clothing. I have tried to do what I could from my scanty means for them, but having others beside my own family to assist, could not do half what My Brother needed. And now, Sir, I make this statement to you with reluctance, but with confidence — inasmuch as I remember your personal kindness to myself last Summer when you gave me a place in your Department. In view of the circumstances of my Brother, I beseech you to pity him & relieve him by giving him either a birth as Doorkeeper, Watchman or Messenger or would you direct his eldest daughter — a young woman of unexceptionable habits to be employed in cutting Treasury Notes -- or both. My brother and his wife are exemplary Christians.

He is a man of character, sober, honest, industrious, capable & every way trustworthy. He can bring to you the most satisfactory testimonials. His present necessities are very imperative. By relieving My Brother, by furnishing him employment for which he has sought in vain elsewhere, you will again demonstrate that you are a benefactor to the deserving, and God in Mercy will bless you —

With profound respect
I am Your Obt. Sevt.
Rev. H.W. Read — late Pastor
Baptist Church at Falls Church.

P.S. Both My Brother & myself have several times sought a personal interview with you but could not gain admittance to your office — hence the necessity of writing.

Please grant me the favor of a reply soon —

Yours etc.
H.W.R.[1]

Unfortunately, no work resulted from this letter for the Reverend John D. Read or for his oldest daughter, Betsy. The Read family returned to Virginia, settling in the village of Falls Church as the Union Army further garrisoned the towns and established camps surrounding Washington, D.C. The bitter "border war" between those with allegiance to the Union or loyalty to the Confederacy began to take its toll throughout Northern Virginia. Falls Church is located approximately ten miles from the nation's capital; however the Union Army did not conquer this area of Virginia until the final days of the Civil War. They constructed numerous forts and picket posts, but Colonel John Singleton Mosby, the "Grey Ghost" of the Confederacy and his Partisan Rangers moved freely over the countryside, particularly during the hours of darkness. They continually harassed the Federal forces, raiding supply wagons, cutting communications lines and savagely attacking military outposts. "Mosby's Confederacy" was born, and he controlled much of Northern Virginia.

The Reverend John D. Read was a man of strong convictions. He was positive the cause of the Union was just. He was proud but worried when his oldest son, Henry, joined the Union Army, and was heartbroken when his youngest daughter, Emma, was completely blinded from the accidental discharge of a gun. Read was aware he had access to important military information gained from wounded Confederate soldiers he visited in hospitals, and from the "grapevine" of the black people living throughout Fairfax County. Always interested in the plight of the blacks, he was held in great respect by the former slaves and received immediate news of any occurrence of interest in the area. Compassionate, worried and anxious to end the war, the Reverend John D. Read became a spy for the Union Army.

Mosby's Rangers soon observed an elderly black man making frequent trips to local Union Army encampments. He was halted one day, searched, and found to be carrying information relating to the movements of Confederate troops. Under threat of death, he revealed he was hired by the Reverend John D. Read to carry the message. Colonel Mosby was less than pleased with the incident, and sent word for the Reverend Read to confine his future activities to church business. No more blacks were intercepted by the Rangers. However, it was noted the Reverend Read became a frequent visitor to the various Union Army headquarters in the area. He ignored Mosby's warning, as evidenced by the following message: "HDQRS. DEFENSE OF WASHINGTON, SOUTH OF POTOMAC.

August 10, 1863.

Lieut. Col. J.H. Taylor, Asst. Adjt. Gen., Hdqrs., Dept. of Washington:

It is reported that one of our cavalry pickets was fired on last night while on duty near Falls Church. A Mr. Reed, resident of that place, says he knows that Mosby's headquarters are only about 5 miles from Falls Church, where he had about 40 men. My cavalry force is not sufficient to send out a large enough party to verify this information and keep up the regular nightly patrols.

Very respectfully,
G.A. De Russy
Brigadier-General
Commanding."[2]

This unfortunate dispatch identified the Reverend Read, also supplying his location. He had already been detected spying and was under surveillance by Mosby's men. He no longer had anonymity, the basic requirement for a successful spy.

Betsy Read, the oldest daughter in the family, shared her father's concern with the problems of former slaves, and established a school in Falls Church to educate them. She sought aid from Miss Emily Howland, a member of a well-to-do Quaker family from New York, who had once taught at a school for black girls in the District of Columbia.[3] Genuinely concerned with the problems encountered in running the school, the Reverend Read wrote:

August 3, 1864, Falls Church Va.

Miss Howland

As my daughter is writing, perhaps it would not be out of place for me to write you a few lines. I am very glad to see some one take interest in the colored folks in this place. The color'd people here have never called on the government for anything and are industrious well behaved set as can be found in any community. It appears to be their misfortune that my daughter started a sabbath school among them (Now near one year since) as it appears to cut them from all the charitable societies. I have made a number of applications but have not been able to get any thing towards sustaining a school among them. One says apply to the government, another say you do not come under our society, another say we can not do anything as your daughter does not belong to any christian church, and yet another say you are a baptist, you had better apply to that denomination. All began to make excuse as in olden times.

When the school became to large for Betsy I then felt it my duty to try and help her, and another daughter has joined us, and had we room to accommodate more schollars, more would come in. We have been obliged to turn away on account of not having room. Between 50 and 60 have been present and you your self no that is more than could be accommodated comfortable. We have been obliged to have nearly one half of them stand up and shift then round. I believe that had we had room the school would now number double that it does. Our efforts has been and still will be to learn them to read, spell, and write cipher and so forth. It is true I have tride to explain the scripture that is read. But sectional matters have been left entirely alone. The school is making some progress as could be expected. Some of them write a verry good hand. They use slates. But paper, I think, would be preferrable, but as we have no room for desks, have to use slates. We have about twenty that read in the testament.

The day school was started by the earnest solicitation of Dr. Johnson. I told him that we were not able to do it, as we had been three times strip'd of every thing by this unholy rebellion, and that it would be necessary for betsy to earn something. He said that he would se that she was paid for one month as he had funds in his hands. At the expiration of a month he wished her to keep on. I told that she could not do it unless she was paid for it as I was taxd nearly every day by some of the col'd people to a greater or less degree — I think that the principal of the school from house to house was the best that could have been done at the time, as many of them were but Just started in house keeping, and the mode of house keeping was taught some what with the studdies. Still some could not be taught as they lived in localities that I thought was not prudent for my daughter to go. Here is a work for somebodddy to do. The door is open. Men, wimmen, and children are anxious to learn. But as they are cut of from all help from the government and the charitable societies, it depends upon individual alone if any thing is done. The sabbath school we shall try to keep up untill some one else will take it up. Then we will willingly give it up. The day school, if kept up, must be Done by others. I hope you or some boddy else will take it as I believe these people should not be forgotten or neglected. But at the sum namd in your letter (if we understand it) ten dollars per month would not pay bord, say nothing about clothes. Shall have to decline the offer. But if a reasonable price, and she should be the choice, she can go in to it.

J.D. Read

My anxiety for this people must be my excuse for this long letter. Let us hear from you soon. And such things are just as necessary to them, in fact, it is part of the education that they need. As this general education must soon give out, they want instruction in cutting and making their own clothes. It is not necessary for me to mention as you are probably more acquainted with them and their wants than I am. I am Willing to put shoulder to the wheel and keep things rolling so far as I can. But as a direct tax, can do nothing, as it is as much as I can do to get Along these times with my own. It is as much as a bargain to get Along as we are over run with soldiers.

J.D. Read

PS Mrs Read had been verry sick but is now better so as to be about the house some. There is considerable sickness among the colord people and a number of deaths. The colord woman that livd in the house where we had the school dide of fever. We feel that in hur the colord people have met with a great loss. She was really a pious woman, and verry exemplary in all things, a patern of neatness. She was a field hand befor she came here, But being apt to learn, Became a gud house keeper. We feel hur loss deeply, but Gods will be done.

J.D. Read[4]

In addition to working with the school for former slaves, the Reverend Read was instrumental in establishing a Home Guard to protect Falls Church from guerilla raids, integrating local black men into the organization. He also developed a system of horn signals to alert the Home Guard members and a nearby Federal picket post of any surprise attacks. On October 18, 1864, a company of Mosby's Rangers conducted a surprise raid on Falls Church, and the following Union Army report relates the fate of the Reverend Read:

HEADQUARTERS CAVALRY BRIGADE

Near Fort Buffalo, Va., October 19, 1864

Lieut. Col. J.H. Taylor
Chief of Staff, and Assistant Adjutant-General.

COLONEL: I have the honor to inform you that as next senior officer to Col. H.M. Lazelle, the command of the force of cavalry here, has, on the receipt of his resignation devolved on me. The force of 400 men, mentioned in the report of October 17, left Annandale the same evening without making any demonstration at that point. One man who was on picketpost was taken prisoner while attempting to escape. About 2 a.m. on the morning of October 18, a force of Mosby's men estimated at 75, entered Falls Church village, halted at the church (brigade hospital), and after breaking open the barn of Mr. Sines, a citizen who lives opposite, and taking therefrom 5 valuable horses, passed up the Alexandria and Lewinsville pike towards Vienna. The post at the junction of the Lewinsville road with the pike, consisting of one Corporal and three men of the Sixteenth New York Cavalry, was captured, with one horse. A negro named Frank Brooks, belonging to the citizens home guard of the village, was shot dead while attempting to assist the picket in making defense. Mr. J.B. Reed, a citizen and a member of the same guard, with one of his negro employees, were taken prisoner at the same time. Mr. Reed was afterwards brutally murdered by the party who captured him, in a dense pine wood near Hunter's Mill, and his body has been found and brought into his house. An attempt to kill the negro taken with Mr. Reed was also made, and the rebels, supposing him dead, left him in the woods. He escaped afterwards, however, and has but a slight wound in the head, with the loss of an ear, blown off by a pistol shot. There is no doubt concerning the murder of Mr. Reed, as the surgeon, who has made an examination of the body, states that the skull at the base of the brain is blown to atoms, and the flesh about the wound is filled with powder, as if the pistol had been placed close to the head. The negro who escaped brings information that at or near Vienna, the force which visited Falls Church was joined by a reserve party of 100 or more men. The officer commanding at Annandale states that the party which visited him numbered 600 men. Everything has been quiet here since the attack.

Very respectfully, your obedient servent,
JOHN BIRDSALL
Major Thirteenth New York
Volunteer Cavalry[5]

The Rev. John D. Read, shot at Hunter's Mill, Va., by Mosby's Rangers

Mrs. Read was grief-stricken upon notification her captured husband had been taken twelve miles away and shot near Hunter's saw mill. She could not get a guard of either Union Army soldiers or Home Guard members to go with her to recover her husband's body, due to their fear of encountering Mosby's forces. While she was searching for help, a note arrived from Colonel Mosby giving Mrs. Read his personal guarantee of her safety while recovering the remains of the Reverend Read. Hiring an open wagon, the only vehicle available, she and her daughter, Lottie, drove to Fort Head at Tyson's Corner (where Read's body had recently been moved), and claimed her husband's remains. In later years, Lottie spoke of the long sad journey home, her mother sitting determinedly on the wagon seat, driving the horse, while she sat crying in the back, attending to the blanket covering her father's body. The war had ended for the Reverend Read. He was buried in the churchyard of the

Columbia Baptist Church.

The tragic death of the Reverend Read greatly discouraged his daughter, Betsy, who wrote Emily Howland:

Falls Church Oct. 23, 1864

Miss E. Howland - -

Dear Friend

I have before me three letters from you addressed to My Father. Who is no more in this world. Miss Howland, it is sad news to tell but My Dear Father has been shot, yes cruely murdered by the Rebels. As to the school I had concluded to take, I do not now feel as though I wanted to have any thing to do with it. And as you say you have too Teachers allready engaged, one of them had better take the school for I do not want it now. Though I do not think it would be advisable at this present time for any one to commence a school that is a colord School, for at this time the Rebs are looking down on us. And I should not be surprised if they should come down here again tonight. As to the School at Halls Hill, I do not know any thing about it. I feal that the Colord folks have cost us enough for it seems to me that they have been the innocent Cause of this our great trouble.

Your friend in deep affliction

B.J. Read

PS I shall be glad to hear from you at any time.[6]

The strain that Northern sympathizers living in Virginia endured is presented in this letter written by a Falls Church neighbor of the Read family, relating the incident:

Home Hill Nov. 1, 1864

Dear Mother Bartlett,

Are you looking for a letter. If so here it is. I have been thinking of writing to you so much of late, that if has farely haunted me, and now it is past. Gee, but I am determaind not to sleep untill, you are in a fair way to here from us. If you have read the papers you no doubt think the guerillas have got us, but they have not yet but I can't say they won't for they came pretty near the other night. I did not know one while but they would have the whole village, they came strong, they took Mr. Read a near neighbor and Colord man 7 miles from home and shot them, the most cold blooded affair that we have had here. As it happened the Colord man was not killed after the ruffins left he cralled off and got back to tell the horrid tale, Mr. R. Sister and daughter a girl of 14 years went alone for the body, found it and brought it home. They could not find but one man there that would help them put it into the waggon. It was a sorry sight I assure you may my eyes never witness another. But it is hard to tell what is in store for us. They took 4 horses from a barn across the road. It was a lovely night. I could see them as plain as though they had stood under my window. Johnnie took our horses from the barn and stood ready to excape with them should they come here. But fireing the alarm and blowing a horn, started them off. We had 250 Cavl. in half a mile of us but they were so long in getting here that they were of no use. There were to many Copperheads, among them, afraid they should hurt there friends if they came in time to fire into them. They threaten to come again and burn us out and shoot every man in Falls Church. There are good many moveing away, but all we have is here and we shall stick by the Ship, as long as there is a stick left. John has not undressed and gone to bed for 6 weeks, he takes his blanket and curles down by some bush near where he can here what is astir. I wonder he stands it so well as he does. But it wears on him. He looks older some way 10 years since the war. I feel as though it would be the end of us if we don't get butchered if it lasts much longer. I had calculated to send for you this fall but think it is well I didn't for I fear you could never bear the excitement which I have become enured to. I was down to Elizas Sunday. They are quite well now. Freedie has been very sick. They are doing very well and are in a safer place then we are although they were so frightened that they packed up ready to start. Sister Emilys family are afflicted, her daughters have been sick since last July, neither one is able yet to go from there rooms. I sometimes fear they never will. Johnnie went north a week ago yesterday to go to school, so I feel safe about him as to the Rebs getting him. I have just heard from him. He is on the Hudson a little way from N. York City.

Address

Hudson River Institute

Clarrack Columbia Co. N.Y.

Love to Mr. Sandersons family and all other friends, with love to yourself.

Emma G. Bartlett

P.S. Father & Mother are in Barre, somewhere. Ma. has not been home over a year. Pa came home and staid too or 3

days last spring. Last fall & winter they were west & in Byron last summer east.
One colord man they killed on the spot. I dare not let such a letter stay in the house overnight for fear they might come and find it. It is now 11 o.c. — if they come tonight it will be between now and 3 o.c. in the morning. Enclose five dollars. Let me no it you get it and I will send more.[7]

The Rev. Hiram W. Read; photographed by Alexander Gardner. (There is a picture of Abraham Lincoln sitting in the same chair.)

A minister was captured, taken into the dark pine woods of Virginia, made to kneel, and a pistol was fired into the back of his head killing him instantly. This atrocity at Hunter's Mill was only one of the many that happened in Northern Virginia. Both sides were guilty of similar incidents, and the bitterness of those stormy days is slow to heal. Captured spies could expect little mercy. It was a tragic period in the nation's history.

EPILOGUE

Following the war, information concerning the death of the Reverend Read was provided by former Confederates. During the early 1900's, a member of Mosby's Rangers wrote:

"(October 1864) Mont joy, with Company D, and with Bush Underwood as guide, moved off towards Falls Church and at night prepared to attack the camp. The pickets were captured, and some of our men were leading horses out of the stables, when the camp was aroused. The blowing of a horn, which at first was thought by our men to denote the assembling of a party of coon hunters, was discovered to be a signal given by a citizen named Reed to alarm the camp. Reed was shot by one of our men. The enemy, now thoroughly aroused, opened fire, which in the darkness did no damage. Three or four negro infantry were killed; 6 prisoners and 7 horses were brought out."[3]

A historian, writing about early Falls Church, recorded:

"There is an interesting sequel to the 'Reed Affair.' A number of years ago, before the First World War, Mack Crossman was visiting in Knoxville, Tennessee, and while there met a gentleman who told an interesting story. Learning that Mr. Crossman was from Falls Church, the gentleman told how he had been among the Group of Mosby's men who shot J.B. Reed. He overlooked this as an incident of war, but said it was a constant source of regret to him that they had 'killed the Negro' with Reed, since the Negro was innocent. The gentleman was delighted to learn that the Negro had survived, (with the loss of an ear), and was content as a servant of the Crossman family in Falls Church!"[9]

The widow, Charlotte E. Read, continued to live in Falls Church following the war. In 1873, the estate of the Reverend Read was awarded a War Warrent in the amount of $2,250 for damages to buildings and stores during the war.[10] Undoubtedly this was a great help to her. She died in 1886, and is buried in the family plot, her name appearing on the tombstone with her husband's.

Henry Read, the oldest son, served in Company K, 189th New York Volunteers during the war. Upon discharge, he became a harness-maker in Oswego, New York, preferring not to return to Virginia due to the bitterness existing there. He died in 1888, and is buried in Riverside Cemetery.[11]

Charlotte (Lottie) Read became a florist in Falls Church, and lived with her blind sister until her death in 1891. She is buried in the family plot.

"Miss Emma" Read, blinded since an early age by an accidental gun discharge, was active in the Baptist Church and won several blue ribbons for her crochet work at the Virginia State Fair. She died in 1938, and is buried in an unmarked grave in what is today the Falls Church Episcopal Cemetery.[12]

The fates of Betsy, John R., and the sixth child mentioned in the Hiram W. Read letter are unknown. The Minute Book (1863-67) in the Fairfax County Courthouse indicates the Reverend Hiram W. Read was appointed guardian of Betsy and Lottie, and that Charlotte E. Read was appointed guardian of her two younger children, Emma and John R., on Janueary 28, 1865. Henry did not require a guardian due to his age, and there is no mention of another child in the Minute Book. Possibly this child died during the time the Read family resided in the District of Columbia.

In keeping with President Lincoln's policy regarding ministers, the Reverend Hiram W. Read was granted liberal leave from his position at the Treasury to attend to the needs of the soldiers. He was so occupied when the Army of the Potomac under General George G. Meade had to abandon 2,500 wounded Union soldiers at Savage's Railroad Station on June 30, 1862, during the Seven Days Battles. Reverend Read volunteered to stay behind as doctor and chaplin to the wounded, and was captured. When word of his praying for the Union at burials reached Confederate President Jefferson Davis, he was court-martialed. On July 5, 1862, he was sentenced to be hanged for treason as "a Yankee, a Union Man, and an Abolitionist." Read replied to the sentence with, "Tell Jeff Davis he has not hemp enough in his bogus Confederacy to hang me."[13]

Read was saved from hanging when General Ambrose E. Burnside captured the noted Southern minister William F. Broaddus at Fredricksburg, Virginia. On September 21, 1862, Read and Broaddus were officially exchanged.[14]

During his confinement in Libby Prison at Richmond, Read contracted typhoid fever: his days with the army were finished. He traveled to the West on a mission for President Lincoln, stayed, and ultimately became New

The J. D. Read gravestone. Note the "apparition" of an angel with outstretched wings. Minerals leaching from the stone formed the phenomenon.

Mexico's first Baptist minister. He died at age 75 on February 6, 1895, and is buried at El Pasco.[15] For the next 54 years, he lay in an unmarked grave, until 1949, when a drive to erect a tombstone in his memory was successful. Unfortunately, on his stone his name is misspelled as "Reed."

The grave of the Reverend John D. Read was marked with a plain tombstone. The stone contained minerals that leached out over the years and formed an "angel." This was regarded as an omen by the black community, who felt Read was truly a martyr. Though the angel can still be seen, time has eroded the memory of the Reverend Read and the circumstances surrounding his death at Hunter's Mill.

FOOTNOTES

[1] Salmon P. Chase papers, Library of Congress, Washington, D.C.
[2] *OR,* Series I, Vol. XXIX, Part 2, p. 26.
[3] Josephine Pacheco, "A Civil War Freedmen's School at Falls Church," *Northern Virginia Heritage,* III-2:9, June 1981.
[4] Emily Howland papers, reproduced courtesy the Department of Manuscripts and University Archives, Cornell University, Ithaca, N.Y. (Punctuation added for clarity-grammar and spelling unchanged.)
[5] *OR,* Series I, Vol. XLIII, Part 2, p. 414.
[6] Howland papers, *op. cit.*
[7] Reproduced courtesty Henry Douglas, Historian, Falls Church, VA.
[8] James J. Williamson, *Mosby's Rangers* (Sturgis & Walton, New York: 1909), p. 271.
[9] Melvin Lee Steadman, *Falls Church by Fence and Fireside* (Falls Church, 1964), p. 65.
[10] Will Book, (1873), Fairfax County Courthouse, Fairfax, VA.
[11] Courtesy Michael Cahall, Director, Oswego County Historical Society, Oswego, N.Y.
[12] "Gravesites in the Churchyard of the Falls Church," *Falls Church, Va., Historical News and Notes,* 1970-1972, p. 29.
[13] Betty and Dale Danielson, "New Mexico's First Protestant Preacher," *Alburquerque Journal Magazine,* 6-37:8-11, June 30, 1981.
[14] *IBID.* For the viewpoint of the Southern hostage, see Archibald Thomas Robertson, *Life and Letters of John Albert Broaddus,* American Baptist Publication Society, 1910.
[15] Death Certificate, Bureau of Vital Statistics, El Pasco, Texas. The *El Pasco Times,* of February 8, 1895, ends the death announcement with: "Peace to his aged ashes."

PRIVATE THOUGHTS

In August, 1861, after the disastrous Federal defeat at Bull Run, Virginia, Lincoln called urgently for more troops. A twenty-year-old Indianian named John A. Maxwell was one of those bright, idealistic young men who answered the cry of his country. Joining Company K. 26th Indiana Volunteer Infantry that August, he and his companions went to war looking for a chance to show their mettle. Like most soldiers, he wrote home of camp life, weather, reviews, rumors and endless marching. Like many, Christmas was for him a day to do the washing; his thoughts wandered to dying on lonely battlefields.

He thought of friends and their fate, and was mad at the inaction of the Army. He got his chance to fight, though, and was wounded in the Battle of Prairie Grove, Arkansas, December 7, 1862. His wounds kept him hospitalized for months. Upon release, he served for a time as a clerk in the Commissary, and in April 1863, "he was honorably discharged on a special order by General Schofield, to receive a promotion and was immediately re-mustered as a First Lieutenant of Cavalry." He remained with the mounted branch of the service until the end of the war. John Maxwell also served as a Judge Advocate for three months, and was Treasurer of the military post at Fayetteville, Arkansas. He also became a member of a commission drawn up to arbitrate disputes over property rights between soldiers and civilians.

When war ended, and things returned to normal, he became a minister. He resigned this occupation for the opportunity to teach Greek and Latin to college students. Becoming restless with the sedentary life, or perhaps yearning to preach God's word again, he re-entered the ministry, but an offer of a professorship of Latin and History at Purdue called him back to education. He was married in 1871, and still unable to determine upon his calling, he spent the rest of his life going between the ministry and various professorships. He died in 1904.

The following are three of his war letters. They represent one of Lincoln's boys of '62, who came when he was called, and saw the job through.

Otterville, Jan. 8 — 1862

Friend Demp,

Yours of Dec. 29th came last night, and I improve this earliest opportunity to answer it. We have had nice weather for a few days past, and all things looked quite cheerful, but this morning it is wet and disagreeable. How much more life and activity there is about the camp on a pleasant sunshiny day that [than] a wet cloudy one. The health of the camp is very poor now, only about forty four in our Co. fit for duty and we have about as large a company as any. We are in camp on the Learmine River near Otterville, and will probably stay here for some time.

I still enjoy good health. I have had a severe cold but I am better now. You asked me how holidays passed here. Christmas I did some washing. Mrs. Adams is out here now. She arrived here a day or two after Christmas, she is boarding in town. She brought me a cake and can of peaches. On newyears day my mess had a nice feast.

We were reviewed on newyears day by Gen. Pope and some others. I guess one of Gen. McLellans staff was with Gen. Pope (Gen. Van Rensalier.) I received a letter from Binna a few days ago, and she said Caroline had promised them a visit this winter. Ma is visiting now down at Kent. I got a letter from her the other day. Binna nad Dan were visiting the week before Christmas at Chouis Hill. I was glad to hear of Gordons good health. Warren W. Chouis our orderly sergeant, (I do not know wether he went to school when you did or not,) has gone home on recruiting service, to be gone six months unless ordered back by military authorities. He hired me to take his place while he was gone, and I am now staying with the Capt. He has a nice little stove and we keep very comfortable.

Give my love to your Uncle Quillan's family. I close hoping to hear from you soon.

John

To Demp.

Sedalia Mo. May 22 — 1862
Tuesday afternoon

Friend Demp.

Your last was welcomed a few eavnings since. It is very warm weather here at this time. But we who live on the prairie have one advantage, there is most always a good breeze blowing. It makes music on the tents, the ropes flopping around. It is nice to lay in the tent and hear the rain beat upon the canvass. It has such a soothing sound, that it inclines one to sleep. We have had very dry weather here until recently. Yesterday and the night before we had a good rain. It hailed some yesterday. I received a letter last night, and one the night before from home. They are planting corn, broom cord, sorgum, watermelons, muskmelons, cotton seed, & c, & c.

I expected last week before this time we would have been on the road to "Dixie." Our Col., and company officers told us we would leave, and they had reason to believe it, but here we are yet, and I am mad about it. We are here and will never get any name. If we endure the hardships of a soldiers I think we should have some honor along with it, not that we came into service for honor, but while other troops are having a chance to try their pluck, and display their valor, we are kept back, to act constable for Missouri.

Poor Gordon! How sad to think that he had to fall in the prime of life, so far away from dear friends, let us hope for the best. I feel so sorry for his Mother. How much rather would I fall on the battlefield than linger out a few days of sickness, and die. I mean this to apply to the time I am soldiering.

I am still trying to live in that way that will insure me a home in heavin. Remember me when you kneel before your Maker.

Ever your friend
John A.M.

Near Cassiville, Mo.
Monday afternoon, July 21 — 1862

Friend Demp.

Since last I wrote you we have been marching over a good deal of territory. I believe I have not written to you since we left Sedalia. We left on the cars the latter part of June, arrived at Bolla by traveling two days on the 3rd of the month, and marched every day until last Thursday. We got ready to move that day but the orders were contramanded. We passed through Sprinfield one week ago last Saturday. It is a nice town, and larger than I expected to see. Several fine residences in the place. Most every house between here and Springfield has been burned, it is a desolate looking country. Coming here we passed by the Millson Creek battle ground. It still exhibits marks of the dreadful battle fought there. It is about 55 miles from here to Springfield. I hear just now that we have orders to go back to Springfield. I suppose we will leave in the morning. We will probably go right back to Bolla, and be shipped to some field of active operations.

I got your letter Friday night. We have had an unusual amount of rain since we came here, had a thunder shower most every day. We had a nice rain this morning.

Lt. Smith has been going to Bloomington this year. I had not heard a word from him. You speak of the boys being nearly all gone, I think when they fill up the new Regts. that are to be organized in Indiana, the state will be entirely striped.

I can not write more at this time direct your letters to Mo. without designating any town or Post-Office.

Your friend,
John.

(Continued from page 25.)

[38] *OR,* XII, Part 2, p. 650.
[39] *IBID,* p. 72.
[40] *IBID.*
[41] John H. Worsham, *One of Jackson's Foot Cavalry* (Jackson, Tenn.: McCowat-Mercer Press, 1964), p. 69.
[42] *Battles and Leaders of the Civil War* (New York: Castle Books, 1956), p. 502.
[43] William C. Oates, *The War Between the Union and the Confederacy* (Morningside Bookshop, 1974), p. 133.
[44] *Stonewall Jackson,* p. 126.
[45] *OR,* Part 3, p. 654.
[46] *Life And Campaigns,* p. 517.
[47] *I Rode With Stonewall,* p. 134.
[48] Lieut.-Col. W.W. Blackford, *War Years With JEB Stuart* (New York: Charles Scribner's Sons, 1954), pp. 109-10.
[49] *War Between the Union,* p. 134.
[50] *OR,* Part 2, pp. 650-1; *War Between the Union,* p. 135.
[51] *OR,* XII, Part 2, p. 720.
[52] *IBID.*
[53] *IBID,* p. 644.
[54] Mary Anna Jackson, *Life And Letters of General Thomas J. Jackson* (New York: Harper & Bros., 1892. See *Stonewall Jackson* by Chambers, Vol. II, p. 94 for a discussion of this incident.
[55] John O. Casler, *Four Years In the Stonewall Brigade* (Dayton, Ohio: Morningside Bookshop, 1971), pp. 107-8.

(Continued from page 21.)

FOOTNOTES

[1] William Couper, *One Hundred Years at V.M.I.* (Richmond: Garrett and Massie, Inc., 1939), Vol. II, p. 98.
[2] Robert E. Withers, *Autobiography of an Octogenarian,* p. 132.
[3] *One Hundred Years,* Vol. II, p. 97.
[4] George W. Cullum, *Biographical Register of the Officers and Graduates of the U.S. Military Academy* (New York: D. Van Nostrand, 1868), p. 425.
[5] *One Hundred Years,* Vol. I, p. 72.
[6] *IBID,* Vol. II, p. 7; letter of Gen. Richardson to Virginia Gov. Wise, Nov. 17, 1859.
[7] *IBID,* p. 120.
[8] *IBID.*
[9] *IBID,* p. 137.
[10] *IBID;* letter of Gen. Smith to ex-Virginia Gov. Kemper, Dec. 26, 1861.
[11] *IBID,* Vol. II, p. 36.
[12] *IBID,* p. 49.
[13] *IBID,* p. 79.
[14] *Biographical Register,* p. 594.
[15] *IBID.*
[16] Jennings C. Wise, *The Military History of the Virginia Military Institute* (Lynchburg, Va.: J.P. Bell Co., Inc., 1915), p. 90.
[17] *IBID,* p. 93.
[18] *IBID,* P. 90.
[19] *One Hundred Years at V.M.I.,* Vol. II, p. 27; letter of Gen. Richardson to Col. P. St. George Cocke, May 18, 1860.
[20] *IBID,* p. 107.
[21] John H. Worsham, *One of Jackson's Foot Cavalry* (Jackson, Tenn.: McCowat-Mercer Press, Inc., 1964), p. 11.
[22] *IBID,* p. 14.
[23] *One Hundred Yeart at V.M.I.,* Vol. II, p. 135.
[24] *IBID,* p. 136.
[25] C. G. Chamberlayne, *Ham Chamberlayne — Virginian* (Richmond: Dietz Printing Co., 1932), p. 56.
[26] *War of the Rebellion: A Compilation of the Official Records of the Union and Confederate Armies* (Washington: Government Printing Office, Series I, Vol. V, p. 391.
[27] *One of Jackson's Foot Cavalry,* p. 27.
[28] *One Hundred Years at V.M.I.,* Vol. II, p. 181.
[29] *IBID,* P. 147.
[30] *IBID,* Vol. III. p. 1.
[31] *IBID.*
[32] *IBID,* p. 10.
[33] *IBID,* p. 64.
[34] *IBID,* p. 107.
[35] *IBID.*

THE SITUATION IN VIRGINIA.

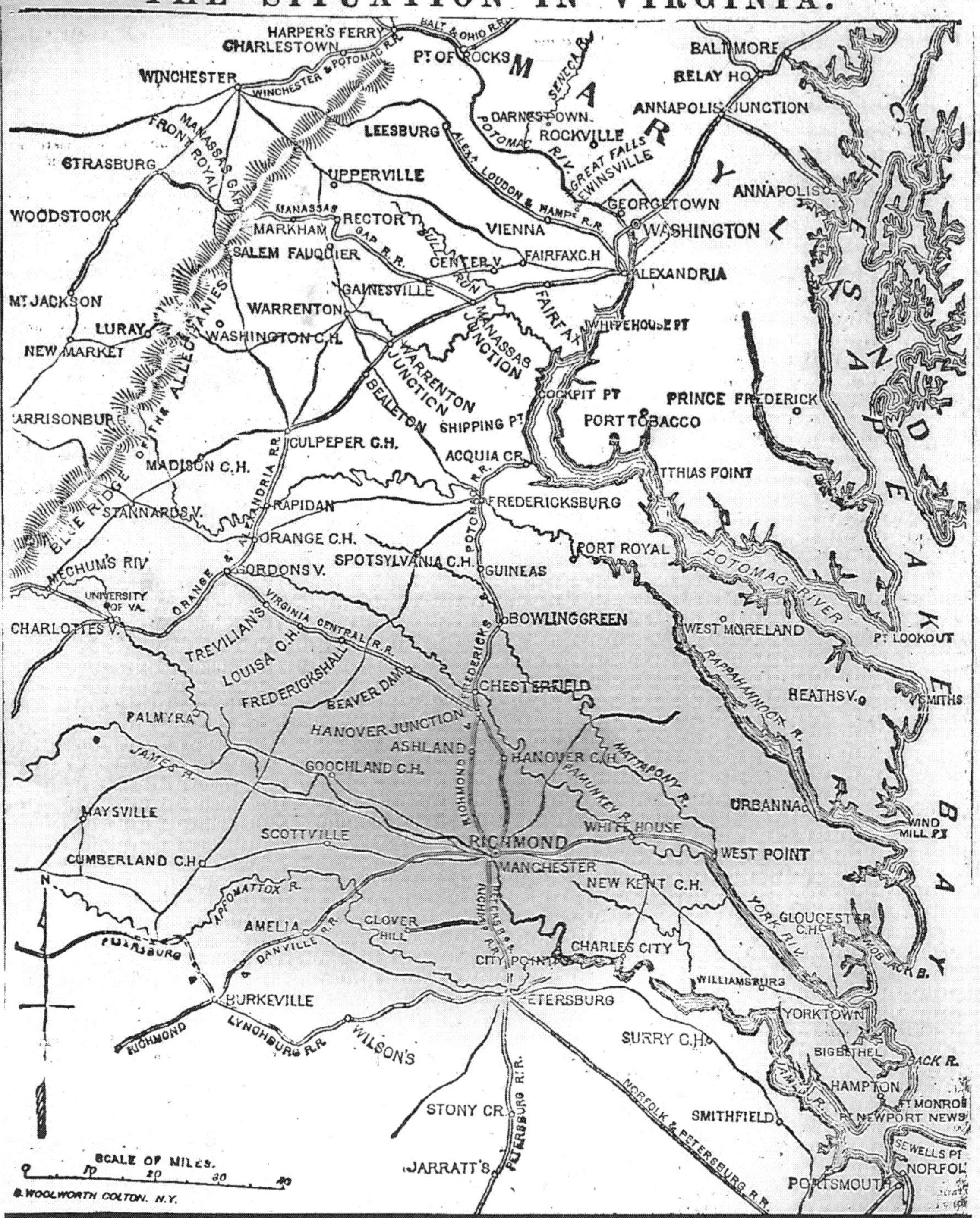

THE KEPI

DECEMBER 1984 — JANUARY 1985 $3.50

"A New Concept in Civil War Reporting"

"A Fire Bell At Night": The Revolt of the East Tennessee Unionists
by Vernon H. Crow

"The Search for Private Tomkins"
by William F. Howard

John Householder
by Wes Rine

A Merry Christmas
and
A Happy New Year!

From the Editors of

THE KEPI

THE KEPI

"A New Concept in Civil War Reporting"

DECEMBER 1984 — JANUARY 1985 VOLUME II, NUMBER 6

THE KEPI is published bi-monthly for $3.50 per issue (Bulk Mail; $4.50 First Class), or $18.00 per year. Canadian single copy, $4.50; subscription $23.00 per year (US funds). United Kingdom single copy, $5.00; subscription $28.00 (US funds) per year (Air Mail). Correspondence to the editors may be directed to: **THE KEPI, P.O. BOX 699, BALDWIN PARK, CA 91706.** Unless expressly stated, **THE KEPI** neither endorses nor is responsible for the content of advertisements appearing in its pages. **THE KEPI** requests authors to query with specific topics before submitting articles, and assumes no responsibility for unsolicited manuscripts not accompanied by return postage. Work not fully documented and footnoted where applicable will not be considered. All inquiries should be directed to the above address. **THE KEPI** reserves the right to edit and copyright all letters. Please allow 6 to 8 weeks for delivery of first issue.

KEPI POLICY

Renewal notices will be sent shortly before the final issue of a subscription is mailed. If we receive no response, the mailing will be repeated once, two months later. Unless otherwise requested, late renewals will automatically start with the issue next in sequence to the last received. This ensures a complete volume and number run of **THE KEPI.** Due to the high cost of postage, these editions will be mailed at the next regularly scheduled bulk mail. Back issues may be purchased directly from the publishers at $3.50 per issue (Bulk Mail), or $4.50 per issue (First Class). Bulk mailings occur every other month, usually the first weeks of February, April, June, August, October and December. Issues sent bulk rate usually require up to three weeks for delivery after mailing. Advertising rates may be requested from the publishers. Unlike many other publications, advertisers with a multiple insertion contract are guaranteed those rates, even if over the course of their contract advertising rates increase. Private inquiries requiring an answer *must* be accompanied by an SASE.

THE REGULARS

THE IRREGULARS

GENERALLY SPEAKING

It is a little-known fact of the Civil War that within the heart of the Confederacy lay a segment of the population so committed to the Union that they wished to break away from their state and form a separate coalition, allied with the United States. Because they were not successful, their story does not speak for itself, like that of the Western Virginians, who carved their state out of the Old Dominion. These men, loyal to the Old Flag, found themselves surrounded by enemy territory: their frustrating struggle for freedom is vividly depicted in our feature article "A Fire Bell At Night: The Revolt of the East Tennessee Unionists." Vernon H. Crow has skillfully woven together the plight of these Tennesseans as they formulated a plan to burn bridges and cut the Confederacy's lines of communication. Desperately, they banded together, and made their raids — only to be abandoned by the Federal forces. No help came and the men that were caught were literally "hung out to dry." Mr. Crow, author of Storm in the Mountains *and two previous KEPI works, "Confederate Indecision and Delay: The Empty Victory at Chickamauga" and "Massacre In Shelton Laurel," has brought to life in words and pictures this important and often ignored subject.*

"The Search for Private Tompkins" began with a faded photograph, and led to the story of a young Alabama soldier. With just a name and a few penciled notations on the back of the time-worn frame, William Howard rescued a Confederate soldier from obscurity.

PRIVATE THOUGHTS explores the diary and letters of John Householder, 2nd Ohio Vol. Inf. The original of this material is in the possession of Wes Rine, a graduate of Wichita State University with a degree in English. Mr. Rine has traced the life of this Union soldier and presents it to KEPI readers as another chapter in our continuing biographies of the unsung but vitally important individual foot soldier.

Tom Evans' two short works, "Pauline Mosby's Kentucky Cousin" and "A Mosby Legend," ring of the irony and humor that often invaded the grim precinct of war. MILITARY REVIEW focuses on two newly published works.

Editors-in-General
S.L. Kotar *J.E. Gessler*

Headquarters of THE KEPI
1st December 1984

GENERAL ORDERS
No. 10
by Command of the Editors-in-General

It was an "infernal machine" that was as likely to kill the soldier carrying it as the enemy he shot at. It's use was either "barbarous" or a legitimate weapon of war, depending on who was discussing it. Both North and South used them at Gettysburg. The loud cracking sound heard overhead in battle was attributed to it, and troops were often unnerved when they suspected their opponents had it. *It* was the Gardiner explosive musket shell. The February-March 1985 issue of THE KEPI explores the history and uses of this highly controversial bullet.

Highlighting this feature article, for the first time anywhere, are the details of the recent discovery that unearthed 8,000 — 10,000 of these extremely rare bullets at a little-known skirmish site in Virginia. This archeological find sheds revealing new light on the Union use of these projectiles and explains why history has been strangely silent about this group of previously unaccounted for Gardiner's.

Accompanying the cover story are the medical-surgical reports made at the time on the effects of the explosive bullet, the Confederate use of the explosive shell and thorough background articles on Civil War smallarms, projectiles and ammunition.

"A FIRE BELL AT NIGHT":
The Revolt of the East Tennessee Unionists

by
Vernon H. Crow

It was nearly midnight when William C. Pickens and his band of fourteen Unionists, following the banks of the Holston River, reached the Strawberry Plains Bridge. The air was cool and frosty, and clouds obscured the moon, making it difficult for the men to see their way. Pickens and a man named Montgomery left their horses with the rest of the party and scrambled to the top of the bridge. Their job was to burn the bridge, called "the most important" on the East Tennessee and Virginia Railroad by Union General Ambrose E. Burnside.[1] Had they known that the sixteen-hundred-foot bridge was guarded by a sentry, or that the sentinel was posted at this, the eastern end of the bridge, they would have been better prepared.

The guard, James Keelan, described as "small in stature," but "brave, and true to his friends," was in his bunk under the end of the bridge when he heard Pickens and Montgomery approaching. As he emerged from his station, he saw Pickens strike a match and realized immediately that these men were the long-expected saboteurs come to burn his bridge. Keelan fired his single-shot pistol at Pickens, who fell wounded in the thigh. Keelan and Montgomery then sprang upon one another and began slashing with their knives. Montgomery, with a twelve-inch blade that he wielded "with blind fury," nearly severed the left hand of Keelan, although the guard did not know it until later. At the sound of the shot, the other arsonists began to climb the embankment, but Miss Mary Stringfield, whose farm adjoined the bridge, heard the firing, too. She lit a lamp in her room and went out to the yard. This stirring in the Stringfield household scared the would-be bridge burners, who, fearing recognition, abandoned their objective. Meanwhile, Keelan, knocked off the embankment, ran to William Elmore's nearby house for help. Aroused from sleep, Elmore, seeing this half-crazed man asked if Keelan had been drinking. "No, Billy; they have killed me, but I've saved the bridge."[2]

Keelan survived, becoming a hero of sorts in East Tennessee. But the attempt on the Strawberry Plains Bridge, along with simultaneous attacks on other East Tennessee bridges, signaled a planned revolt by Union men who rejected Tennessee's secession.

President Abraham Lincoln called East Tennessee the "most valuable stake we have in the South."[3] Early in the war, he recognized the strategic importance of the railroad that ran 220 miles from Chattanooga to Bristol. Lincoln saw, by looking at maps, that if a Federal army planted itself squarely on the road, Virginia would be effectively cut off from the western Confederacy. Nor was the political significance of East Tennessee lost on the Federal president. Opposed to secession, the majority of East Tennesseans had become rebels in a land of rebels.

Isolated from the rest of Tennessee by the Cumberland Mountains and separated from North Carolina by the southern Appalachias, East Tennessee is a great valley drained by the Tennessee and its tributaries, crossed by smaller mountain chains. Its mountainous nature made slavery unprofitable, though the highlanders were not abolitionists. They were, for the most part, poor farmers who had little in common with the Southern planters.

When all Tennesseans were summoned to vote on the question of secession in February 1861, they turned down the proposal, much to the disappointment of Governor Isham G. Harris, who wanted his state out of the Union. Supported in the western and middle portions of the state, Harris was bitterly opposed in East Tennessee by Congressmen Horace Maynard and Thomas A. R. Nelson, Senator Andrew Johnson and Knoxville newspaper publisher William G. Brownlow, a former traveling Methodist minister-turned-politician.

On May 6, Harris pushed an ordinance of secession through the legislature which the people of Tennessee were asked to ratify in June. The Unionists, headed by Nelson and Brownlow, called for an East Tennessee convention which met in Knoxville on May 30. Nearly 400 delegates from twenty-six East Tennessee counties adopted twelve resolutions which smacked of out-and-out defiance of Nashville. The Unionists, despite hard campaigning, were outnumbered two to one in Tennessee, and were soundly defeated at the polls on June 8. East

Tennesseans, casting seventy percent of the vote against separation, felt they had been dragged out of the Federal Union against their will.[4]

As soon the results were known, Congressman Nelson called for another convention to be held in Greeneville, Senator Johnson's home town. Twenty-nine counties (including two from Middle Tennessee) sent delegates who adopted a "Declaration of Grievances" and verbally thrashed the Nashville secessionists. Though it was understood that Nashville would quickly dispense with any action coming out of Greeneville, a petition was drafted urging the formation of a separate state of East Tennessee. Citing "hopeless and irreconcilable difference of opinion and purpose," Maynard and Oliver P. Temple, a Knoxville attorney, prepared the document in the form of a memorial and submitted it to the legislature. To Governor Harris, a separate East Tennessee was unthinkable: his goal was to lead the state into the Confederacy, and he certainly was not giving up the strategic valley.[5]

The Confederate government, too, recognized East Tennessee's importance, not just for the vital railroad, but also because it was a principal source of wheat, copper (used in percussion caps and field artillery) and saltpeter (for gunpowder and lead).[6]

All through the summer of 1861, Harris maintained a conciliatory policy toward the "tories," as the Unionists were called. It was a political decision influenced by the gubernatorial election slated for August, after which Harris began to take a harder line in East Tennessee — the arrests of Union men became frequent. Even Representative Nelson was taken into custody as he attempted to go north and claim his seat in the U.S. Congress.[7] As the ill feelings between the "Lincolnites" and pro-Confederates intensified, concern for the region grew. Confederate Major General Leonidas Polk telegraphed Jefferson Davis, the Southern president, that: "No time is to be lost in East Tennessee." Polk, worried because only 2,000 troops were there, wanted another 8,000 sent in and a separate department created to be commanded by Felix K. Zollicoffer, then a brigadier general in the Tennessee state army.[8] Davis was also being nudged by William G. Swan, a Knoxville attorney and candidate for the Confederate Congress, who was troubled by the way armed Union men were assembling in "minuteman" fashion in Bradley County. In July, Landon C. Haynes, a Tennessee state senator, warned Richmond that "the peril of civil war in East Tennessee" was imminent. Davis finally acted on Polk's suggestion and sent Zollicoffer, recently ap-

Congressman Thomas A.R. Nelson was arrested by the Confederates when he attempted to leave the southern lines and take his seat in the U.S. Congress.

pointed a general in the Southern service, to Knoxville with orders to "preserve peace, protect the railroad, and repel invasion."[9]

At forty-nine, Zollicoffer was one of the leading men in the state of Tennessee. A former member of Congress and editor of an influential Nashville newspaper, he had worked hard for the Bell-Everett ticket in 1860. He was generally viewed as a peaceful man who could reconcile the Unionists while enforcing Confederate policy in his district. Zollicoffer reached his new command by the end of July and found his force too small to effectively resist any movement into East Tennessee. On August 6, he wrote the War Department, expressing concern about an Union invasion from Kentucky and reported that he was constructing a series of infantry posts. These would be anchored by Cumberland Gap on the right flank and extend westward, blocking the logical entry routes. A week later, Richmond sent him three regiments, two from Polk's army at Memphis and Colonel William B. Wood's 16th Alabama from Tuscumbia.[10]

Zollicoffer attempted to appease the "disaffected" populace. He admonished his officers to not "alarm or irritate" those who previously advocated the "National Union," but who now quietly submitted to the authority of the Confederate government. He told one colonel to keep "your men in camp, so as to prevent the soldiers from committing trespasses," and encouraged those who had fled East Tennessee to return and live in peace. Even the irascible "Parson" Brownlow admitted that he felt personally safe with Zollicoffer in command.[11] But, despite the general's friendly and liberal attitude, the tension between pro-Confederates and Unionista grew increasingly worse. Believing a "reign of terror" was nearly upon them, thousands packed their bags and escaped to Kentucky. Colonel Wood of the Alabama regiment, writing to newly appointed Confederate Secretary of War Judah P. Benjamin, exclaimed that the Union men were "as hostile to [the government] as the people of Ohio." Wood was loath to report that "20 to 100 are every day passing through the narrow and unfrequented gaps of the mountain into Kentucky to join the enemy."[12] Piloted by daring men such as Daniel Ellis, the refugees were "a constant stream," silently working their way north by night, often leaving everything behind, including their families, and traveling with only the clothes on their backs. Some found their way into the Federal army, others went on to Washington, (as did Maynard and Johnson) to work for the relief of East Tennessee. Many stayed in eastern Kentucky and waited for the Lincoln government to deliver them. Exactly how many refugees from East Tennessee crossed the border is difficult to ascertain, but it is known that fifty-three regiments of Tennesseans, aggregating over 31,000 men, were mustered into the Federal army.[13]

Library of Congress

General Felix K. Zollicoffer, Confederate commander in East Tennessee, attempted to pacify the Unionists. After the revolt, he wanted to "pursue to extermination" those who had taken up arms.

From Ellis' *Thrilling Adventures*

Daniel Ellis, later captain of Co. A, 13th Tennessee Cavalry, guided hundreds of Union men over the mountains of Kentucky.

One refugee, William Bloun Carter, a Presbyterian minister fron Elizabethton in upper East Tennes see, developed a scheme that eve President Lincoln thought woul wrestle control of Carter's home land from the Confederates.

The Carter family was well known in Washington. Samuel P Carter, William's forty-two-yea old brother and a lieutenant in th U.S. Navy, had proclaimed his allegiance to the Union so loudly an with so much fanfare that he cam to the attention of Andrew Johnson The senator, recognizing Carter' value in promoting the Federal government in East Tennessee, procured Carter's temporary assignment to the War Department in July. Samuel Carter would eventually attain the rank of major general in the Union army and subsequently — after the war — rear admiral in the navy (the only American officer to accomplish this feat). In the fall of 1861, however, Carter was in charge of the "East Tennessee Brigade," assigned to General George H. Thomas's division stationed at Camp Dick Robinson in Kentucky. Thomas's command was part of General William T. Sherman's Department of the Cumberland, headquartered at Louisville.[14]

At the end of September, William B. Carter escaped across the border to see his brother and General Thomas, to whom he outlined an ambitious plan to interrupt Confederate communications in East Tennessee and stop the flow of Southern forces between Virginia and the western Confederacy. Carter's plan, first suggested in a Brownlow editorial back in May, was to recruit local Unionists and burn the nine major bridges between Bristol, Virginia, and Bridgeport, Alabama, thus disabling what Thomas called "the grand trunk railroad" through Tennessee. Thomas heartily recommended the plan to Major General George B. McClellan remarking that "It would be one of the most important services that could be done for the country...."[15] Thomas wanted McClellan to use his influence in Washington to promote Carter's idea.

Carter made his way to the Federal capitol, where he had little trouble obtaining an interview with Lincoln, McClellan, and Secretary of State William H. Seward. During the meeting (which apparently occurred in the first week of October), Carter laid out his plan to simultaneously wreck all the bridges, cut the telegraph lines, and suggested that as part of this attack a Federal army be brought up to the border, ready to invade East Tennessee. With the railroad inoperative and telegraph lines down, the Confederates would be temporarily paralyzed. Local Unionists would then rise up and join the advancing Federal troops. According to Oliver Temple, who spoke with Carter after the war, Lincoln was enthusiastic about the scheme, and the secretary of state agreed to provide $2,500 to recruit saboteurs, while

From Ellis' *Thrilling Adventures*

This scene depicts Ellis leading Unionists out of East Tennessee.

General McClellan "promised to aid in the movement by sending an army into East Tennessee as soon as possible."[16] Carter took this promise to mean that a Federal army would march immediately following the destruction of the bridges, a serious misunderstanding, as events would prove.

Returning to Thomas's camps in Kentucky, Carter informed the general of his success with the Washington authorities. Thomas, anxious to carry out the plan—including the invasion—detailed two army officers, Captains William Cross and David Fry, to accompany Carter. Leaving General Thomas to prepare his invasion force, or so Carter thought, he and the two officers began to make their way back into East Tennessee.

On October 21, while Carter was in East Tennessee recruiting his saboteurs, Zollicoffer's forces skirmished with an advance element of Thomas's command at Rockcastle Hills near London, Kentucky. Sherman, believing a major Confederate drive was about to take place, refused to let Thomas move any farther south.

By the first few days of November, Carter had enlisted six assistants who were, in turn, to recruit an additional five to six men each. The four bridges targeted between Bridgeport and Charleston in lower East Tennessee were assigned to A.M. Cate. The important bridge at Loudon, thirty miles south of Knoxville, was to be burned by Captain William Cross, while William C. Pickens took charge of the attack on the Strawberry Plains Bridge. The seventh target, across Lick Creek, fifteen miles west of Greeneville, was assigned to Captain David Fry. The next two bridges were at Carter's Depot over the Watauga River and the Union Depot Bridge over the Holston River, just a few miles south of Bristol. Both were the responsibility of Daniel Stover, the son-in-law of Andrew Johnson.[17]

Carter, confident of success and believing that the hour of deliverance was at hand in East Tennessee, chose Friday, November 8, as the date of the simultaneous assaults. On that night, Carter's people attacked and achieved partial success: five bridges were burned and the telegraph lines were downed for miles. The Confederates were completely surprised, though there had been rumors for weeks that an attempt might be made to burn the bridges. On the day Carter launched his attack, Landon C. Haynes predicted that "the flames of rebellion" were about to flash throughout East Tennessee and that "the railroad will be destroyed...."[18]

Library of Congress

General Samuel P. Carter, brother of the chief Union conspirator, William B. Carter, commanded a brigade of East Tennesseans, which, poised over the border in Kentucky, was anxious to march on Knoxville.

Library of Congress

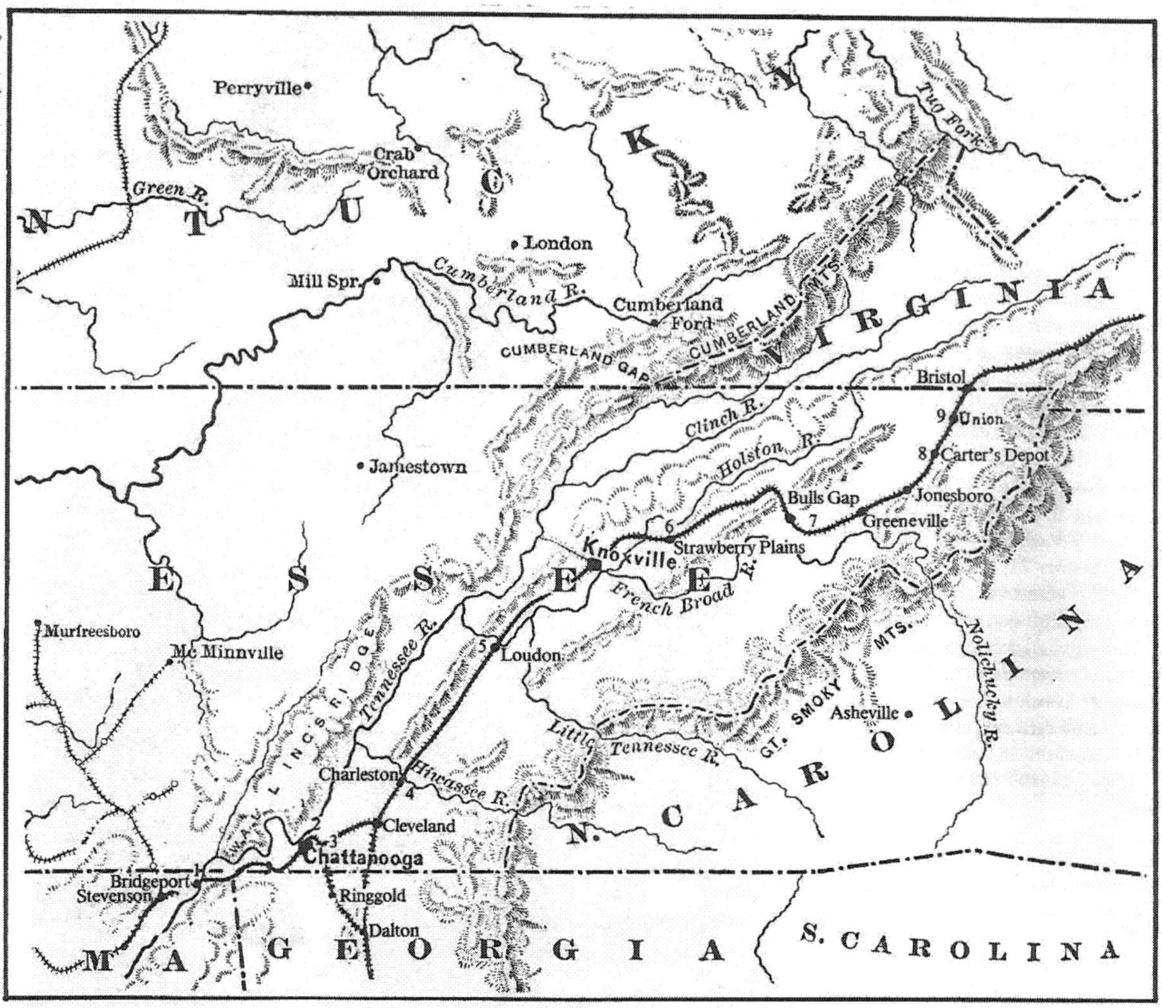

"The grand trunk railroad," as General George H. Thomas called it, ran 220 miles from Bristol to Chattanooga. The nine bridges selected for burning were at: 1) Bridgeport; 2 & 3) two small bridges across Chickamauga Creek, east of Chattanooga; 4) across the Hiwassee River at Charleston; 5) at Loudon; 6) at Strawberry Plains; 7) across Lick Creek, fifteen miles west of Greeneville; 8) at Carter's Depot, and 9) at Union.

Why Zollicoffer did not provide better protection for the bridges is not clear. He had some bridges guarded by full companies; others, such as the one at Strawberry Plains, were patroled by a single sentry. The success of Carter's plan was due, therefore, more to the lack of Southern vigilance than anything else. Andrew Johnson's son-in-law, Daniel Stover, found the bridge at Carter's Depot guarded by Captain David McClellan's cavalry company, but easily overpowered the two sentries at Union Depot. The bridge was set on fire, and the prisoners set loose upon their promise to keep the identities of the bridge burners secret, a promise they did not keep.[19]

Captain Fry and his comrades struck the Lick Creek Bridge between 2:00 and 3:00 in the morning, while five of the six men on guard duty were in their tent near the bridge. Isaac N. Hacker, one of the bridge guards, testified at the trial of a bridge burner that forty to sixty armed men surrounded the tent, and in "less than five minutes the bridge was in flames."[20] Hacker and his fellow prisoners then were required to take the oath of allegiance to the U.S. or face death.

The bravery of the guard and the carelessness of the would-be burners saved the bridge at Strawberry Plains. The Loudon Bridge (below Knoxville) also escaped destruction because two companies of Confederate troops were found posted there; but the one over the Hiwassee River at Charleston was easily burned by A.M. Cate and four others.[21]

East of Chattanooga, two bridges across Chickamauga Creek were burned, one of the East Tennessee and Georgia line, connecting Chattanooga with Knoxville, the other on the Western and Atlantic road which ran south to Dalton, Georgia. Both were burned by two men, one the brother of A.M. Cate, who apparently found them unguarded. The last to be burned—the long important bridge over the Tennessee at Bridgeport—was the responsibility of R.B. Rogan and James D. Keener, who, finding it heavily guarded, abandoned the attempt.[22]

Word of the attacks spread quickly across Confederate-held East Tennessee. Said Oliver Temple: "The news of the burning of the bridges...came upon the country on the morning of November 9, 1861, like the sound of a fire-bell at night."[23] The president of the East Tennessee and Virginia Railroad, John R. Banner, telegraphed Secretary of War Benjamin: "There is great excitement along the whole line," said Banner, "and evidence that the Union party [is] organizing...."[24] Confederate General Sidney Johnston, commanding the western armies, dashed off a note to Governor Harris, urging him to "ascertain the extent, power and organization of this insurrection...."[25] The superintendent of the East Tennessee and Virginia line reported to President Davis that "The country [is] in great excitement and terror."[26]

Tennessee State Library

Confederate Governor of Tennessee, Isham G. Harris, was lenient toward the Unionists before his reelection in August 1861.

By the evening of November 9, the Unionists were in complete revolt. Colonel Wood, commandant of the post at Knoxville, sent word to Richmond that 500 Union men were threatening Strawberry Plains, and 1,500 were assembling in Hamilton County near the bridge at Loudon. Another 500 Unionists were reported gathering at Elizabethton. (Some rumors put their number at 1,500.) Over all of East Tennessee, small parties of twenty to fifty Unionists were congregating, some armed, some not. Said Wood: "The whole country is now in a state of rebellion." Governor Harris urged President Davis to send in troops from western Virginia and told the Confederate chief executive that the "rebellion must be crushed out instantly, the leaders arrested and summarily punished."[27]

Davis and the Confederate high command acted quickly. General Braxton Bragg, then at Pensacola, was ordered to send Colonel S.A.M. Wood's 7th Alabama to Chattanooga. The War Department hurried Brigadier General William H. Carroll, stationed at Memphis with three regiments, to Knoxville, and sent Colonel Danville Leadbetter of the Engineering Corps scampering to East Tennessee with the 3rd Georgia Battalion and a battery of light artillery. Leadbetter was given command of the troops strung along the railroad with instructions to "reconstruct the bridges, repair and keep open the line of communications" between Chattanooga and Bristol. Robert Vance's 29th North Carolina, recruited from the mountains separating Tennessee and western North Carolina, was ordered to join Leadbetter at Jonesborough. Zollicoffer, whose forces were northwest of Knoxville in the Cumberland Mountains watching Thomas's Federals, sent back the 29th Tennessee. By November 13, Benjamin was able to telegraph John Branner: "Troops are now moving to East Tennessee to crush the traitors. You shall be amply protected."[28]

Library of Congress

Colonel S.A.M. Wood of the 7th Alabama put Chattanooga under martial law and dispersed the "Lincolnites" in the lower valley.

Carter and his Unionists had done their job. They destroyed five railroad bridges, interrupted communications, risen up in great numbers over East Tennessee and had thrown the Confederate military and railroad officials into what Oliver Temple called a "wild and unreasonable panic."[29] Now, they wondered, where was General Thomas and his invasion force?

The Rev. Carter had led his conspirators to understand that a Federal army would be waiting over the border in Kentucky. Said Union guide Daniel Ellis who had helped burn the Union Bridge: "It was generally believed by all who had been advised of the project, that if this destruction could be accomplished, there would then be nothing in the way to prevent the Federal troops from coming to East Tennessee to stay...."[30] But the Union army did not come. In fact, it would be two years before one made the trip from Kentucky over the mountains. It was not that Lincoln, Seward or even McClellan had changed their minds—quite the contrary. The Union high command was clearly in favor of it. Nobody, however, could convince General Don C. Buell (recently appointed to succeed Sherman in command of the Department of the Cumberland) of the urgency. Just twenty-four hours before the bridges were torched, McClellan told Buell to "throw the mass of your forces by rapid marches...on Knoxville in order to occupy the railroad at that point and thus enable the loyal citizens of Eastern Tennessee to rise...."[31] Buell apparently objected, pleading that Louisville needed protection. McClellan answered on November 25 that Louisville could still be defended while the movement into the eastern portion of Tennessee was made. Buell then assured his chief that he intended to move on East Tennessee as soon as he concentrated a "sufficient force before Bowling Green to hold [Confederate General Simon S.] Buckner there...."[32] That seemed to satisfy McClellan for a while, but not General Samuel P. Carter, whose East Tennessee regiments were begging to be led against Knoxville. Carter reported to General Thomas

East Tennessee bridges, such as this one across Platt Creek, were constant targets for both sides during the war.

that his brother, William, had burned the bridges, escaped from East Tennessee and returned on November 16. Now a great turmoil existed among the secessionists, and it was time to move.

McClellan, prodded by President Lincoln, again urged Buell on December 3 to march. Buell was also getting pressure from Horace Maynard and Andrew Johnson, who wrote the general on December 7, demanding an invasion. Maynard, exasperated at the Union army's inaction, scolded General Thomas: "You are still farther from East Tennessee than when I left you nearly six weeks ago. There is shameful wrong somewhere...."[33]

By January 1 of 1862, Buell was claiming he needed 120,000 men for his department, and wanted a minimum of three divisions for East Tennessee. He claimed lack of transportation was preventing the invasion, but as soon as he procured sufficient means, the movement would go forth.[34] Transportation never arrived; neither did Buell send a column (under Thomas) across the mountains. Before long, General U.S. Grant's campaign against Forts Henry and Donelson would change the focus of the War Department in Washington, shelving the Unionists in East Tennessee.

While authorities in Washington were pleading with Buell, the Confederates were busy crushing the revolt in East Tennessee. Colonel S.A.M. Wood, with his 7th Alabama regiment, arrived in Chattanooga on Thursday, November 17, and found the city in "a general panic." Taking command, he issued an order which put Chattanooga under martial law, then turned his attention to the "Lincolnites" in camp on Sale Creek, fifteen miles from the village of Washington. He telegraphed Colonel William B. Wood, who had come down from Knoxville to Athens with the 16th Alabama, to make a "simultaneous movement across the Tennessee River" against the insurgents' camp. Wood hired a steamboat to transport the 7th Alabama twenty-seven miles up the river and landed nine miles from the Unionists' camp; but when the column reached the area, it was found deserted.[35]

In upper East Tennessee, the Unionists had not been so quick to flee. On the night of November 10, a Confederate patrol skirmished with their main camp at Taylor's Ford, near Carter's Depot, killing nine and capturing two. The "tories" retreated to Doe River Cove, six miles south of Elizabethton in the mountains of Carter County, to await the appearance of a Federal army.[36]

Colonel Leadbetter, who reached his command by the middle of November, followed the "Lincolnites" to the Doe River. As the Confederates approached, the "tory" pickets fled to their camp, where it was decided that "every man [was to] take care of himself." Leadbetter reported on November 19 that he had dispersed the insurgents and

Northern-born Colonel Danville Leadbetter of the Confederate engineers carried out Secretary of War Judah P. Benjamin's ghoulish orders to hang bridge-burners near the railroad.

had captured "thirty prisoners in the neighborhood; none very prominent." He then moved his headquarters to Greeneville on the 24th, where he learned that another 200 to 300 Unionists were camped at Chimney Top in the northern part of Greene County. He sent Marcellus Stovall and the 3rd Georgia Battalion, along with two companies of the 29th Tennessee and some cavalry, after the insurgents, but they eluded Stovall.[37]

In Sevier County, the other "hot spot" of rebellion, an estimated 400 Unionists marched toward Strawberry Plains, but never got farther than the French Broad, where they failed to cross the river. They remained in camp for a day and a half, skirmished with Confederate cavalry and finally melted away into the Smoky Mountains.[38]

Colonel William Wood announced to Richmond on November 20 that "The rebellion in East Tennessee has been put down in some of the counties, and will be effectively suppressed in less than two weeks in all the counties." By Christmas, Leadbetter was still reporting that "disturbances are frequent," but, for the most part, the revolt had been crushed.[39]

The real tragedy of the rebellion was the effect it had on Confederate authorities in Knoxville; even the conciliatory Zollicoffer turned against the Unionists. The general told one of his subordinates that they "have acted with base duplicity, and should no longer be trusted," called them "fugitive traitors" and now advocated that they be "pursued to extermination."[40]

Hundreds of men all over East Tennessee were arrested, some of suspicion of bridge burning, some as armed insurrectionists, some just for being known Union men. Almost every jail in the eastern end of Tennessee was soon busting at the seams, but the military authorities refused to cooperate with the civilian courts. District Attorney James C. Ramsey informed Judah Benjamin that captured bridge burners were going to be tried by court-martial. Ramsey wanted to know what he should do; Benjamin's answer must have shocked him: "I am very glad to hear of the action of the military authorities, and hope to hear they have hung every bridge-burner at the end of the burned bridge."[41]

Colonel Wood at Knoxville also asked Benjamin for instructions to deal with the jailed "traitors." "All such as can be identified as having been engaged in bridge-burnings are to be tried summarily by drum-head court-martial," answered the secretary of war, "and if found guilty, executed on the spot by hanging." The other "traitors" were nothing less than prisoners of war, and were to be conducted under armed guard to Tuscaloosa, Alabama. By the middle of December, about 400 men had been sent to prison there; others were sent to Nashville.[42]

.rom "Parson" Brownlow's Book

Jacob Harmon and his son, Henry, were convicted of burning the Lick Creek Bridge in 1861, and were hanged side-by-side in Knoxville. The bridge, strategically situated on the East Tennessee and Virginia Railroad between Bull's Gap and Greeneville, was quickly rebuilt by the Confederates.

Benjamin's ghoulish orders to hang convicted bridge burners at the end of the burned structure was stupidly carried out by Colonel Leadbetter, who had captured three of the Lick Creek Bridge conspirators. Jacob M. Hensie and Henry Fry were tried at Greeneville, found guilty, and hung near the railroad from the limb of a tree, though the burned bridge was miles away. Leadbetter, a Northerner by birth, left the bodies dangling from the tree for about twenty-four hours before having them removed. On December 10, A.C. Haun was executed at noon in Knoxville for his role in the Lick Creek affair. A week later, two more insurgents, Jacob Harmon and his son, Henry, were sent to the gallows for taking part in the same bridge burning. Harrison Self was also found guilty and was to be executed at 4:00 p.m. on December 27, but his daughter, Elizabeth (with help from William G. Brownlow) pleaded with President Davis for her father's life: "...he is my only earthly stay I beg you pardon him." Davis took pity on the poor girl and spared Self's life.[43]

Leadbetter told the War Department in December that he had "secured the outward obedience of the people" and that "the executions of the bridge-burners [was] producing the happiest effect." The Confederates had, indeed, quieted the Unionists. Their leaders had fled or been arrested, while those Union men who remained were "taking the Oath [of allegiance to the Confederacy] in pretty large numbers."[44]

Only one sticky problem remained: William G. Brownlow.

Parson Brownlow, a perpetual thorn in the side of the Confederate authorities at Knoxville, ran newspaper editorial after editorial attacking them, the Richmond government and even the Southern soldiers themselves, whom he labeled "riff raff."[45] Brownlow fully expected to have his newspaper shut down and be arrested at any moment because of his open defiance. When rumors began to spread that Brownlow was to be indicted in Nashville by a grand jury, he decided to cease publication on his own. In his final *Whig* editorial, October 24, he charged that the real reason for his contemplated arrest was to "silence, and destroy the last and only Union paper left in the eleven seceded states...."[46] He then went home and waited to be arrested, but when that did not happen, he began to feel uneasy for his family. On November 4, he mounted his horse and rode into Sevier County, where he hid with fellow refugees.

After secluding himself for a month, Brownlow learned from a friend, John Baxter, who had visited Richmond, that the Confederate military would allow him to leave the country under what was thought of as a military pass, issued by the new Confederate commander in East Tennessee, Major General George B. Crittenden, who superseded Zollicoffer on December 1. Crittenden agreed to protect Brownlow from "the military mob," as the Parson called it, but told him that he would be subject to the civilian authorities, which is exactly what Brownlow desired. He was arrested on December 6 when he entered the city, and thrown into

From "Parson" Brownlow's Book

Jacob M. Hensie and Henry Fry were hanged at Greeneville for their part in the Lick Creek Bridge burning. Their bodies were left dangling for twenty-four hours.

William G. "Parson" Brownlow, minister and editor, was arrested by the Confederates for his editorials, one which advocated open rebellion. Later, he was Union governor of Tennessee.

Judge Oliver P. Temple, a Knoxville attorney and key Union leader, urged the formation of a separate state of East Tennessee.

U.S. Congressman Horace Maynard escaped to the North, and spent two years urging the Lincoln government to invade East Tennessee.

prison. Brownlow wanted to be tried by the civilian authorities to achieve his martyrdom, but, on December 27, District Attorney Ramsey entered a "nolle prosequi." Benjamin decided not to prosecute Brownlow, as it appeared he had been entrapped. Instead, because of his illness, he was placed under military protection at his home, where he remained until sent north in March.[47]

Oliver Temple called the revolt of the Unionists in East Tenneessee "most unwise and unfortunate." The burning of the bridges "did but little injury to the enemy [most were quickly rebuilt], while it brought untold calamities and sufferings on the Union people."[48] Still, they could take heart in the knowledge that their insurrection forced the hard-pressed Confederacy to station thousands of troops in East Tennessee when those troops were needed on other fronts.

From "Parson" Brownlow's Book

William G. "Parson" Brownlow was arrested on December 6, 1861, as a "traitor," and thrown into the Knoxville jail, partially achieving his sought-after martyrdom.

FOOTNOTES

[1] *War of the Rebellion, Official Records of the Union and Confederate Armies* (Washington: Government Printing Office, 1881, Series I, Volume XXIII, Part 1, p. 385, hereinafter cited as *OR*. Unless otherwise stated, all references are to Series I; Oliver Temple, *East Tennessee and the Civil War* (Cincinnati: Robert Clark Co., 1899), pp. 381-82, hereinafter cited as *Temple*.
[2] *Sarah Stringfield Butler, Confederate Veteran* (Nashville: Thomas Broadfoot Reprint, 1983), Vol. III, pp. 187-88.
[3] OR, Series II, Vol. I, p. 900.
[4] Mary E. R. Campbell, *The Attitude of Tennessians Toward the Union, 1847-1861* (New York: Vantage Press, 1961), pp.200-208; E. Merton Coulter, *William G. Brownlow* (Knoxville: University of Tennessee press, 1971, pp. 154-55, hereinafter cited as *Coulter*; *Temple*, p. 340; *OR*, Vol. LII, pp. 148-54.
[5] *OR*, op. cit., pp. 168, 178; *Temple*, p. 347; *Coulter*, p. 156.
[6] Thomas L. Connelly, *Army of the Heartland* (Baton Rouge: Louisiana State University Press, 1967), pp. 12-14.
[7] *IBID*, pp. 41-43.
[8] *OR*, Series II, Vol. I, p. 827.
[9] *IBID*, pp. 824, 828; Vol. IV, p. 374.
[10] Ezra J. Warner, *Generals in Gray* (Baton Rouge: Louisiana State University Press, 1959), pp. 349-50; *OR*, Vol. VII, p. 387.
[11] *OR*, Vol. IV, pp. 393, 389; William G. Brownlow, *Sketches of the Rise, Progress and Decline of Secession* (Philadelphia: George W. Childs, 1862), pp. 278-79, hereinafter cited as *Brownlow*.
[12] *Temple*, pp. 369-70; *OR*, Series II, Vol. I, p. 837.
[13] *Tennesseans in the Civil War* (Nashville: Civil War Centennial Commission, Vol. I, p. 1.
[14] Ezra J. Warner, *Generals in Blue* (Baton Rouge: Louisiana State University Press, 1964), p. 74; *OR*, Vol. VII, pp. 257, 461.
[15] *Temple*, pp. 370-71; *OR*, Series II, Vol. I, pp. 889, 911-12.
[16] *Temple*, pp. 370-75.
[17] *Temple*, pp. 379-85.
[18] *OR*, op. cit., pp. 837-38.
[19] *Temple*, pp. 384-85; *OR*, op. cit., p. 839.
[20] *OR*, op. cit., p. 862; *Temple*, p. 384.
[21] *Temple*, pp. 379-81.
[22] *IBID*.
[23] *IBID*, p. 388.
[24] *OR*, Vol. IV, p. 231.
[25] *IBID*, pp. 240-41.
[26] *OR*, Vol. I, p. 839.
[27] *IBID*, pp. 839-41, 843; Vol. IV, pp. 232-36.
[28] *OR*, Vol. IV, pp. 234, 237-38, 559; Series II, Vol. I, p. 841.
[29] *Temple*, p. 388.
[30] Daniel Ellis, *Thrilling Adventures of Daniel Ellis* (New York: Harper and Brothers, 1867), p. 27, hereinafter cited as Ellis.
[31] *OR*, Vol. VII, p. 358; Series II, Vol. I, p. 891.
[32] *OR*, Vol. VII, pp. 450-51.
[33] *OR*, Series II, Vol. I, pp. 892-93, 897-98.
[34] *OR*, Vol. VII, p. 548.
[35] *OR*, Vol. IV, pp. 246-50.
[36] *IBID*, p. 236; *Temple*, p. 386.
[37] *Ellis*, p. 29; *OR*, Series II, Vol. I, p. 844, Vol. Vii, pp. 712-13.
[38] *Temple*, pp. 386-87.
[39] *OR*, Vol. IV, p. 250; Vol. VII, p. 791.
[40] *OR*, Series II, Vol. I, pp. 843, 848.
[41] *OR*, Vol. VII, pp. 700-01.
[42] *IBID*, p. 778.
[43] *OR*, Series II, Vol. I, p. 851; Vol. VII, p. 759; *Temple*, pp. 393-94; *Brownlow*, pp. 326-27.
[44] *OR*, Series II, Vol. I, p. 853; Vol. VII p. 748.
[45] *Coulter*, p. 164.
[46] *Brownlow*, pp. 249-255.
[47] *Coulter*, pp. 181-82, 186-87, 201, 205-7; *OR*, Series II, Vol. I, pp. 902-28.
[48] *Temple*, p. 387.

PAULINE MOSBY'S KENTUCKY COUSIN

By

Thomas J. Evans

In addition to having a husband fighting for the Confederacy in Virginia, Pauline Clarke Mosby also had a cousin similarly engaged in Kentucky. Marcellus Jerome Clark joined the Confederate Army on his birthday, August 25, 1861, at the age of seventeen. He was a member of a distinguished Kentucky family, his uncle being the Democratic candidate for governor during the election of 1855.

Clark served in the cavalry of General John Hunt Morgan, one of the South's great leaders, referred to by his admirers as "The Thunderbolt of the Confederacy." An impetuous commander, always independent, and showing great personal courage, Morgan led his cavalry command in four long raids through Kentucky, and once across the Ohio River into Union territory. On June 12, 1864, Morgan was defeated at the Battle of Cynthiana, Kentucky, by the forces of Federal General Stephen Burbridge and was forced to retreat into Tennessee. On September 3, 1864, Morgan attacked Union forces near Knoxville, and a pursuing force surrounded and killed him at Greenville as he tried to escape on the morning of September 4, 1864.

Clark was not present at Morgan's defeat and death because of a wound he had suffered at Cynthiana. He had been forced to separate from his company to convalesce, and subsequently joined the staff of Colonel Jack Allen, commander of a Confederate cavalry regiment, who was trying to open the way for an invasion of Kentucky by General Hood's forces. In his staff capacity Clark directed various bands of guerillas and was soon recognized as a successful partisan leader.

Courtesy of Author

Marcellus Jerome Clark ["Sue Mundy"]

Through no fault of his own Clark became the unsuspecting player in a political drama that effectively signed his death warrant. George D. Prentice, the editor of the influential newspaper, the Louisville Courier, was fueding with Federal General Burbridge concerning the general's attempt to swing the forthcoming election to the Republican Party. Knowing the small, long-haired Clark was operating successfully throughout central Kentucky, Prentice developed the strategy of printing in his newspaper that this guerilla was actually a woman. He even provided the name, Sue Mundy, actually borrowed from a notorious black madam then operating in Louisville. Soon readers were learning the details of Sue Mundy's exploits and reading Prentice's scathing editorials attacking General Burbridge and his command for not being able to capture a mere woman.

Naturally, these editorials were not well received by General Burbridge and he made a special effort to capture Sue Mundy. On November 17, 1864, Major Sam Martin reported the capture and wounding of several from a party of guerillas he had surprised at Bloomfield, Kentucky: "The notorious Sue Mundy and Berry said to be the ones who were wounded and made their escape. Sue's fine pipe fell a trophy to my men and various other articles."[1] And on February 3, 1865 Brigadier General E.H. Hobson wired the following orders to the command located at the state capital at Frankfort: "Captain Cockrill can remain. No necessity for him in Owen or Henry counties. Troops now there. If he can hear of Sue Mundy's gang, it will be well for him to look after her."[2]

Clark joined forces with Kansas guerilla chief William C. Quantrell, and as reports were received that Sue Mundy and Quantrell were pillaging central Kentucky, the Union forces launched an all-out campaign to secure their capture. On March 12, 1865, Clark and two guerilla leaders were surprised and captured in a barn by a company of Wisconsin cavalry. Clark refused to surrender until assured he would be treated as a prisoner of war rather than a guerilla. It was difficult to identify who did and who did not come under the conditions of the code of war as bonafide soldiers due to Kentuckians being widely divided in their loyalties. Clark would have probably been treated as a prisoner of war had not one of the guerillas identified him as the much sought Sue Mundy.

Clark was tried by Federal authorities in Louisville the day after his capture and sentenced to hang immediately. He was allowed no witnesses in his defense. Resigned to his fate he requested baptism from the Reverend Mr. Talbott of St. John's Episcopal Church.[3] On March 16, 1865, he mounted the gallows, and in a short speech denied he had committed a

Continued on page 35.

ECHOES OF WAR

A MOSBY LEGEND

by
Thomas J. Evans

"It was a fall day, just the kind of warm day makes a fellow think he was a child again. Me and three other boys was doing picket duty way off from the Regiment. Things was quiet and had been for sometime. Right next to us was this stream and it was deep enough in spots for a good swim. I said to the boys we should go swimming since there wasn't anything going on and nobody would be the wiser anyway. We needed a bath for the graybacks (lice) was starting to act up again like they could feel cold weather coming.

"We had one boy to stand guard and the rest of us stripped off and went in.

Continued on page 35.

E | 54 | Pa.

Thomas Evans

Pvt., Co. D, 54 Reg't Pennsylvania Inf.

Appears on

Company Muster Roll

for Sept. & Oct., 1864.

Present or absent Present

Stoppage, $ ..... 100 for

Due Gov't, $ ..... 100 for

Remarks: $15.32 due U.S. for Enfield Rifle & wrist belt lost. Recruit.

Book mark:

Andrews
(358) Copyist.

"The Search For Private Tompkins"

by
William F. Howard

For many young men who volunteered during the Civil War, the photographs they had taken before leaving home were the only record they left. The flower of a troubled nation, a generation of lost promise, they peer out of sepia-toned images across the gulf of a century. Identified or unknown, Union and Confederate, they are the young men of high hopes and promise who fought in great cataclysmic battles or in forgotten backwoods skirmishes. They are the men and boys who never lived to tell their stories of life in uniform, never survived to grow long gray beards or march in Memorial Day parades. All that remains as testimony to their individual sacrifices are scattered bits and pieces, a faded photograph or a name in the official records that suggests one man's role in the great national crisis.

Marching among those who flocked to the military standard in 1861 was James M. Tompkins. An original albumen portrait photograph in the author's collection helped chronicle Tompkin's story. The photograph, purchased at an Atlanta flea market in 1983 by Lee Joyner, was sold to the author in April, 1984. Extremely faded and tattered, but depicting a young man in period civilian clothing, the photo has an old handwritten note glued on the reverse, that reads: "James M. Tompkins, Fell at the Battle of Gaines Mill before Richmond, June 27th 1862, aged 20 years." Originally thought to be a Georgian, the photo was traced to a Thomaston, Georgia estate sale. From there, some excellent genealogical detective work by Ms. Marie Varrelman Melchiori, CGRS, revealed that Tompkins was actually a member of the Fifth Alabama Infantry Battalion.[1]

What meager records exist show that James M. Tompkins joined the Confederate army just two months after Bull Run. We enlisted September 10, 1861, at Manassas, Virginia, in Captain S. Van de Graff's Company A (nicknamed the North Sumter Rifles), and mustered in as a private for the duration of his service. National Archives records list Private Tompkins present for duty from his enlistment until his death in 1862[2]. When Tompkins fell at Gaines Mill, the Battalion was part of General A.P. Hill's Division, General James Longstreet's command, Army of Northern Virginia. The records do not give the details of his death.

A regimental capsule history provided by Civil War researcher, John F. Walter,[3] does add a little more to the story. Walter's research found that the Fifth Alabama Infantry Battalion was organized at Dumfries, Virginia, in December, 1861. Originally composed of six companies, the battalion was assigned to duty in the Department of Northern Virginia. In the spring of 1862, during reorganization, it was absorbed by the Army of Northern Virginia. After Gettysburg, the large number of casualties compelled consolidation into three companies, and in mid-1863, the 5th Alabama was assigned to provost duty. The battalion members surrendered with Lee's army at Appomattox, Virginia, April 9, 1865.[4]

Private Tompkins did not live to see the destruction of the Confederacy: he died at Gaines Mill, the third of a series of battles before Richmond that pitted General Robert E.

Courtesy of Author

Lee's Confederate army against the new Union defensive position established by General George B. McClellan, A.P. Hill's Division, of which the Fifth Alabama Battalion was a part, began the attack along Beaver Dam Creek early on June 27, 1862. After brushing aside Union skirmishers, Hill's advance was stopped in the afternoon by the main Union line along Boatswain's Swamp. A series of attacks by Confederate brigades failed to break the Union position. Arriving on the field and appreciating the difficulty of Confederate infantry advancing through the swamp immediately in front of the Union line, General Longstreet informed Lee of the situation, and the attack was delayed in anticipation of Jackson's arrival.[5]

General Thomas Jackson reached the field late in the afternoon, but did not receive specific orders, and was consequently unable to support Hill and Longstreet. On June 28, the Confederates finally did launch a general attack all along the reinforced Union line shortly before dark. The Federals were driven back. It was a fierce engagement, and General Fitz John Porter, who commanded the Union center, was ultimately obliged to retreat across the

Chickahominy bridge. In all, out of approximately 57,018 Confederates engaged at Gaines Mill, 8,751 were killed or wounded; of the 34,214 Federals engaged, some 6,837 casualties were reported.[6]

It is uncertain at which point in the battle Private Tompkins was killed. The battalion was engaged most of the day, even losing the regimental colors to the Second Maine Infantry in a vicious hand-to-hand fight.[7] No regimental historian ever told his story, and whoever penned the sad notation on the back of the albumen portrait has long since departed this earth. Still, it is important to remember the legacy of all those American soldiers, like James M. Tompkins, who fought and died for ideals and principles. Their collective sacrifice helped to forge a truly American heritage.

FOOTNOTES

[1] Letter from Ms. Marie Varrelman Melchiori, CGRS, 121 Tapawingo Road, SW, Vienna, Virginia, 22180, to the author, 14 April 1984.

[2] National Archives, Compiled Service Records of Confederate Soldiers Who Served from Alabama, 5th BN., Vols., S-Y, Roll 154.

[3] For a fee, John F. Walter will provide a capsule regimental history of any Union or Confederate regiment that served in the Civil War.

[4] Based on the findings of Walter's research.

[5] Mark M. Boatner, *The Civil War Dictionary* (New York: David McKay Company, Inc., 1959), p. 321.

[6] *Ibid.*

[7] Joseph P. Cullen, *The Peninsula Campaign 1862: McClellan and Lee Struggle for Richmond* (New York: Bonanza Books, n.d.), p. 114.

FAMILY BIBLE

Birthdays of Civil War Notables

by

Paul Kallina

(N = North, S = South.)

DECEMBER

Dec. 1, 1826—Gen. William Mahone, Southhampton County, Virginia. (S)

Dec. 3, 1806—Gen. Henry A. Wise, Drummondtown, Virginia. (S)

Dec. 3, 1826—Gen. George B. McClellan, Philadelphia, Pennsylvania. (N)

Dec. 4, 1818—Gen. William W. Loring, Wilmington, North Carolina. (S)

Dec. 5, 1809—Rev. James C. Furman, educator, Charleston, South Carolina. (S)

Dec. 5, 1839—Gen. George A. Custer, New Rumley, Harrison County, Ohio. (N)

Dec. 6, 1802—Commodore Duncan N. Ingraham, Charleston, South Carolina. (S)

Dec. 6, 1830—Sen. George G. Vest, Frankfort, Kentucky. (S)

Dec. 6, 1833—Col. John S. Mosby, Edgemont, Powhatan County, Virginia. (S)

Dec. 7, 1805—George Opdyke, manufacturer and mayor of New York (1862-1863), Hunterdon County, New Jersey. (N)

Dec. 8, 1765—Eli Whitney, inventor, Westboro, Massachusetts.

Dec. 8, 1828—Henry Timrod, poet laureate of the Confederacy, Charleston, South Carolina. (S)

Dec. 9, 1812—Rev. James H. Thornwell, Marlboro District, South Carolina. (S)

Dec. 10, 1794—Gen. James W. Ripley, Windham County, Connecticut. (N)

Dec. 10, 1805—William Lloyd Garrison, Newburyport, Massachusetts. (N)

Dec. 10, 1813—Sen. Zachariah Chandler, Bedford, New Hampshire. (N)

Dec. 11, 1797—Adm. Hiram Paulding, Westchester County, New York. (N)

Dec. 11, 1824—Maj. Jonathan Letterman, surgeon, Canonsburg, Pennsylvania. (N)

Dec. 12, 1806—Gen. Stand Watie, Coosawallee, Cherokee Nation, Georgia. (S)

Dec. 12, 1820—Rep. James L. Pugh, Waynesboro, Georgia. (S)

Dec. 12, 1830—Gen. Joseph O. Shelby, Lexington, Kentucky. (S)

Dec. 13, 1813—Col. John T. Grant, Greene County, Georgia. (S)

Dec. 13, 1816—Sen. Clement C. Clay, Jr., Huntsville, Alabama. (S)

Dec. 13, 1818—Mary Todd Lincoln, Lexington, Kentucky. (N)

Dec. 13, 1842—Sgt. John D. Billings, *Hardtack and Coffee*, Canton, Massachusetts. (N)

Dec. 16, 1819—Gen. Robert S. Garnett, Essex County, Virginia. (S)

Dec. 16, 1824—Sen. Thomas J. Semmes, Georgetown, District of Columbia. (S)

Dec. 16, 1825—Gen. Henry Heth, Chesterfield County, Virginia. (S)

Dec. 17, 1807—John Greenleaf Whittier, Haverhill, Massachusetts. (N)

MILITARY REVIEW

Medals, Military and Civilian of the United States by David Borthick and Jack Britton, MCN Press, P.O. Box 702073, Tulsa, OK, 74170, July 1984; $19.95 (Hardcover); $12.95 (Softcover); 293 pages, line drawings, four pages of color plates, no index.

Medals, Military and Civilian is an honest attempt to present the unique assortment of awards that have been issued in the United States since 1780. The medals are described and each is illustrated by a line drawing. No photographs are used because of the difficulty of depicting fine detail. Short histories of each medal are included, and the four color plates with ribbons can be removed and glued in the book over the appropriate medals. The period covered is 1780 to the present.

Aside from the fact there are no photographs, the primary drawback of the book is the lack of an index. There is a lengthy list of illustrations in the beginning, but no alphabetical index. The drawing are not to scale, which makes it difficult for the reader to accurately visualize the medals.

For identification of United States medals, this book should be useful. For Civil War students, the book's use is limited to several private (?) and several government issue medals.

Sound of Drums, by Spencer B. King, Mercer University Press, Macon, Georgia, 31207; 1984; 543 pages, no index, no illustrations.

This group of short essays ran in the *Macon Telegraph and News* on a weekly basis, beginning on Sunday, December 11, 1960.

Dec. 18, 1816—Gen. Arnold Elzey, Somerset County, Maryland. (S)
Dec. 18, 1825—Gen. Charles Griffin, Granville, Ohio. (N)
Dec. 18, 1826—Commander John M. Brooke, Brooke Cantonment, on Tampa Bay, Florida. (S)
Dec. 19, 1814—Edwin M. Stanton, Secretary of War (1862-1868), Steubenville, Ohio. (N)
Dec. 19, 1817—Gen. James J. Archer, Stafford, Hartford County, Maryland. (S)
Dec. 20, 1807—Gen. Richard L. Page, Clarke County, Virginia. (S)
Dec. 20, 1808—Commodore Thomas T. Craven, Washington, D.C. (N)
Dec. 20, 1833—Dr. Samuel A. Mudd, Charles County, Maryland. (S)
Dec. 21, 1800—Robert Barnwell Rhett, "the Father of Secession," Beaufort, South Carolina. (S)
Dec. 22, 1803—Gen. Joseph K.F. Mansfield, New Haven, Connecticut. (N)
Dec. 22, 1823—Col. Thomas Wentworth Higginson, *Army Life in a Black Regiment*, Cambridge, Massachusetts. (N)
Dec. 23, 1806—Judge John W. Brockenbrough, Hanover County, Virginia. (S)
Dec. 24, 1808—Capt. Henry Walke, U.S.N., Princess Anne County, Virginia. (N)
Dec. 24, 1809—Col. Christopher "Kit" Carson, Madison County, Kentucky. (N)
Dec. 25, 1808—Commodore Stephen C. Rowan, near Dublin, Ireland. (N)
Dec. 25, 1813—Gen. Milledge L. Bonham, governor of South Carolina (1863-1864), Red Bank, South Carolina. (S)
Dec. 25, 1818—William H. Herndon, Lincoln's law partner, Greensburg, Kentucky. (N)
Dec. 25, 1821—Clara Barton, Oxford, Massachusetts. (N)
Dec. 25, 1828—Col. Prince Felix J.N.M. Salm-Salm, Anholt, Prussia. (N)
Dec. 25, 1835—Col. Orville E. Babcock, Franklin, Vermont. (N)
Dec. 25, 1840—Princess Agnes E. Leclercq Joy Salm-Salm, Franklin County, Vermont. (N)
Dec. 26, 1809—Gen. William N. Pendleton, Richmond, Virginia. (S)
Dec. 26, 1815—Gen. Israel B. Richardson, Fairfax, Vermont. (N)
Dec. 26, 1837—Lt. George Dewey, U.S.N., Montpelier, Vermont. (N)
Dec. 27, 1821—Rep. Josiah Turner, Hillsboro, North Carolina. (S)
Dec. 28, 1822—Gen. William B. Taliaferro, Gloucester County, Virginia. (S)
Dec. 29, 1808—President Andrew Johnson, Raleigh, North Carolina. (N)
Dec. 29, 1809—William E. Gladstone, Chancellor of the Exchequer (1859-1865), Liverpool, England.
Dec. 29, 1809—Gen. Albert Pike, Boston, Massachusetts. (S)
Dec. 29, 1810—Sen. Edward Sparrow, Dublin, Ireland. (S)
Dec. 29, 1823—Capt. Horace L. Hunley, inventor, Sumner County, Tennessee. (S)
Dec. 30, 1784—Col. Stephen H. Long, Hopkinton, New Hampshire. (N)
Dec. 30, 1819—Gen. John W. Geary, Westmoreland County, Pennsylvania. (N)
Dec. 31, 1815—Gen. George Gordon Meade, Cadiz, Spain. (N)

JANUARY

Jan. 1, 1810—Col. Charles Ellet, Jr., Bucks County, Pennsylvania. (N)
Jan. 1, 1822—Gen. Gustavus W. Smith, Georgetown, Kentucky. (S)

Jan. 1, 1835—John Yates Beall, guerrilla and spy, Jefferson County, Virginia. (S)
Jan. 1, 1838—Gen. William H. Young, Booneville, Missouri. (S)
Jan. 1, 1839—James R. Randall, poet, "Maryland, My Maryland," Baltimore, Maryland. (S)
Jan. 2, 1822—Sen. William E. Simms, near Cynthiana, Kentucky. (S)
Jan. 3, 1793—Lucretia C. Mott, activist, Nantucket, Massachusetts. (N)
Jan. 3, 1819—Col. Thomas H. Watts, Attorney General (1862-1863) and governor of Alabama (1863-1865), Conecuh, Alabama. (S)
Jan. 4, 1801—David L. Swain, president of the University of North Carolina (1835-1868), Asheville, North Carolina. (S)
Jan. 4, 1808—Commodore Thomas J. Page, Matthews County, Virginia. (S)
Jan. 4, 1824—Rep. Ethelbert Barksdale, Smyrna, Tennessee. (S)
Jan. 4, 1838—Charles S. Stratton (Gen. Tom Thumb), Bridgeport, Connecticut. (N)
Jan. 5, 1794—Edmund Ruffin, secessionist, Prince George County, Virginia. (S)
Jan. 5, 1813—Gen. Thomas N. Waul, Sumter District, South Carolina. (S)
Jan. 5, 1817—Commander Richard Wainwright, Charlestown, Massachusetts. (N)
Jan. 5, 1822—Gen. Joseph B. Kershaw, Camden, South Carolina. (S)
Jan. 6, 1811—Sen. Charles Sumner, Boston, Massachusetts. (N)
Jan. 6, 1827—Gen. John C. Brown, Giles County, Tennessee. (S)
Jan. 8, 1814—Gen. Thomas Green, Amelia County, Virginia. (S)
Jan. 8, 1821—Gen. James Longstreet, Edgefield District, South Carolina. (S)
Jan. 8, 1830—Gen. Gouverneur K. Warren, Cold Spring, New York. (N)
Jan. 9, 1803—Christopher G. Memminger, Secretary of the Treasury (1861-1864), Nayhingen, Duchy of Wurttemberg, Germany. (S)
Jan. 9, 1816—John P. Usher, Secretary of the Interior (1863-1865), Brookfield, New York. (N)
Jan. 9, 1822—Gen. John P. Hatch, Oswego, New York. (N)
Jan. 9, 1840—Col. Samuel B. Young, Pittsburgh, Pennsylvania. (N)
Jan. 10, 1815—Gen. Alexander B. Dyer, Richmond, Virginia. (N)
Jan. 10, 1841—Eng. George W. Melville, U.S.N., New York City. (N)
Jan. 11, 1813—Commander Tunis A.M. Craven, Portsmouth, New Hampshire. (N)
Jan. 11, 1820—Rep. John B. Baldwin, Staunton, Virginia. (S)
Jan. 12, 1814—Gen. Jones M. Withers, Madison County, Alabama. (S)
Jan. 13, 1808—Chief Justice Salmon P. Chase, Secretary of the Treasury (1861-1864), Cornish, New Hampshire. (N)
Jan. 13, 1815—Gen. William H. French, Baltimore, Maryland. (N)
Jan. 14, 1806—Commander Matthew F. Maury, near Fredricksburg, Virginia. (S)
Jan. 14, 1819—Gen. Frederick Steele, Delhi, New York. (N)
Jan. 14, 1836—Gen. Judson Kilpatrick, near Deckertown, New Jersey. (N)
Jan. 15, 1821—Gen. Lafayette McLaws, Augusta, Georgia. (S)
Jan. 15, 1844—Capt. Thomas C. "Cole" Younger, Jackson County, Missouri. (S)
Jan. 16, 1807—Adm. Charles H. Davis, Boston, Massachusetts. (N)
Jan. 16, 1815—Gen. Henry W. Halleck, Westernville, New York. (N)
Jan. 16, 1845—Ens. Charles D. Sigsbee, Albany, New York. (N)

Continued on page 27.

The purpose was to commemorate the centennial of the Civil War. Dr. King was chairman of the department of history at Mercer University at the time his column appeared, and he accepted no salary for his weekly submissions, which ran for over a four year period.

Sound of Drums is a compilation of these columns, put together in seven categories: Secession, The War, A Soldier's Life, Men of War, The Homefront, Meaning and Purpose, and Journeys. With the emphasis on Macon and Georgia, larger issues are also touched on, as Dr. King followed the progression of war. The strength of the book lies in the previously unpublished letters and diaries that are included.

The text is not for use as a research tool, as noted in the forward by Henry Y. Warnock: it is more of a gentle reminder of times gone by, of people and places and how the war affected them. It serves well as a centennial look back on the War, and for the modern reader, as a look back twenty-odd years to the early 1960's when these articles were written. Dr. King commented on the race riots and the Cuban missile crisis: grim reminders that the problems of inequity and war still haunt us today.

S.L. Kotar — J.E. Gessler

PRIVATE THOUGHTS

John Householder

by
Wes Rine

The 2nd Ohio was organized at Camp Dennison, Ohio, in August and September, 1861. On September 20, David Mitchell, it's commander, took up enlistments at Steubinville and Mitchell's Salt Works. Company K was formed out of an overflow from Captain Sarratt's unit. One young man who would answer the Buckeye state's call would be eighteen-year-old John Marshall Householder (also known as Jehu), son of Peter and Mary Householder of Hammondsville. He left a farming background to enter military service.

In February, 1862, Householder's regiment, after moving on Bowling Green and Gallatin, occupied Nashville. One month later, Don Carlos Buell's main army went to Grant's assistance at Shiloh; O.M. Mitchell's Division, to which the 2nd was attached, advanced on Murfreesboro, Shelbyville, Fayetteville, Tennessee and Huntsville, Alabama. After a brief stay at Battle Creek (near Jasper, Tennessee), the unit was again moving via Manchester, Murfreesboro, Nashville, Bowling Green, Green River and Louisville; John Kell commanded Leonard Harris' brigade. Householder recorded this not-so-glorious aspect of army life in September, 1862. His diary (in the author's collection) for September speaks of Company K spending its time marching, countermarching and undergoing frequent inspection. On September 9, for example, they marched between Mitchellville and Franklin. The leg exercise produced its own perils. The constant movement led Householder to complain of "lower limb pain from all this hard marching," a not too uncommon ailment among foot soldiers during the war. It was not judged to be serious, for it resulted in no medical help. He was treated, however, near a place called Edgefield Junction, Tennessee (north of Nashville and near Hendersonville) by the regimental surgeon for a liver disease contracted from impure water and bad diet.

On September 15, Householder mentioned a review of the 1st Brigade by Lovell Rousseau, whose division became part of Alexander McCook's left wing in Buell's army (late 9th Brigade, Benjamin F. Scribner commanding). This was all in keeping with army reorganization at Louisville. Householder's diary closes with Company K marching toward Altamont, and an excellent day of foraging on the 31st.

As the 2nd marched through Tennessee, affairs in neighboring Kentucky were taking shape. Braxton Bragg, in an attempt to swing it toward the Confederacy, invaded the Blue-grass state. On October 8, his army clashed with Buell at Perryville, the 2nd losing 40% of all engaged. Some three weeks later, William Starke Rosecrans, who, on October 30, 1862, replaced Buell as commander of the newly designed Army of the Cumberland, fought Bragg at Stones River. There, they took up position in the center of George Thomas' 14th Corps (1st Division, 1st Brigade), under the leadership of A.G. McCook and Lieutenant Colonel John Kell. In this engagement, the 2nd Ohio, with aid from Guenther's Battery H, captured the 32nd Arkansas' colors (William Cotter commanding). This battle cost the life of Colonel Kell, 1st Lieutenant Lafayette Van Horn and 1st Lieutenant Richard Chambers. After the fighting ended on January 2, Householder sent a poem, entitled "The Wounded Soldier," to his mother in Hammondsville. Written by Joseph McKee, 74th O.V.I., it reads in part:

"In silent converse with his God
  The wounded hero lay
It seemed to him communion sweet
  No agony to pray.

And mild eyed peace may visit soon
  Stones Rivers rocky shore
But Murfrees chiming sabbath bell
  Will never wake him more."

This touching postscript was added below: "Remember me when this you see."

The soldier's life was not just fighting battles or moving from one place to another. Sometimes he had time to write or receive news from home. On August 14, Householder received correspondence from a woman living in Bellaire, Ohio, named Mary Peeples. She expressed concern about an upcoming draft and strong feeling for those who chose to stay at home, while men such as Householder fought the battles. She wondered how they could call themselves "union." Mary spoke from personal experience, having lost a brother in the "Cruell War." Her letter closes with Miss Peeples wishing John good health and expressing regret for not sending a personal picture. On the latter omission, she wrote:

"I think the best is when you return. It wi" be more pleasure for us both to see the original. So please come and see me & we can talk about the picture."

By late August, 1864, the war was winding toward a conclusion. Preparatory to final discharge, the 2nd reported in with the post commandant at Chattanooga. On the 20th, John wrote his mother concerning Company K's activities. One sight which impressed him was the condition of Chickamauga battlefield. He called it the "worst tore up with artilery" he's ever seen. Upon reaching

Jan. 17, 1800—Caleb Cushing, statesman, Salisbury, Essex County, Massachusetts. (N)
Jan. 17, 1815—Rep. James M. Leach, Randolph County, North Carolina. (S)
Jan. 18, 1782—Daniel Webster, Salisbury, New Hampshire. (N)
Jan. 18, 1815—Gen. James Chestnut, Jr., Camden, South Carolina. (S)
Jan. 19, 1807—Gen. Robert E. Lee, Stratford, Westmoreland County, Virginia. (S)
Jan. 19, 1816—Gen. Henry Gray, Laurens District, South Carolina. (S)
Jan. 20, 1814—Sen. David Wilmot, Bethany, Pennsylvania. (N)
Jan. 20 or 21, 1824—Gen. Thomas J. "Stonewall" Jackson, Clarksburg, Virginia. (S)
Jan. 21, 1813—Gen. John C. Fremont, Savannah, Georgia. (N)
Jan. 21, 1815—Gen. Daniel C. McCallum, Johnston, Renfrewshire, Scotland. (N)
Jan. 21, 1821—Gen. John C. Breckinridge, Secretary of War (1865), Cabell's Dale, near Lexington, Kentucky. (S)
Jan. 22, 1827—Col. Stephen Vincent Benet, St. Augustine, Florida. (N)
Jan. 22, 1832—Commander George E. Belknap, Newport, New Hampshire. (N)
Jan. 23, 1823—Commander Stephen P. Quackenbush Albany, New York. (N)
Jan. 23, 1833—Gen. John R. Chambliss, Jr., Hicksford, Greensville County, Virginia. (S)
Jan. 24, 1820—Henry J. Raymond, editor, *New York Times*, Lima, New York. (N)
Jan. 24, 1832—Gen. John Pegram, Petersburg, Virginia. (S)
Jan. 25, 1786—Gouverneur Kemble, ordnance manufacturer, New York City. (N)

Ringgold with some mules, he got word that Rebels had attacked Dalton. Joe Wheeler's cavalry was at this time raiding Union outposts in Northern Georgia. His report from Soring Place, Tennessee, on August 16 confirmed this. The Union garrison, consisting of Colonel Bernard Liaboldts 2nd Missouri, retreated to a small but strongly fortified position outside of town. As a result of this enemy movement, Féderal reinforcements were immediately sent via railroad cars from Ringgold toward the town's relief. Householder witnessed at least four trains loaded with men passing by him near Graysville.

The Confederate presence in Dalton was short-lived, according to Householder's letter. Soon after its occupation, Colonel Thomas J. Morgan's 14th United States Colored Troops, acting in conjunction with the 2nd Missouri, 29th, 51st, 69th Indiana, 108th Ohio and 78th Pennsylvania, drove Confederate cavalry out. General James B. Steedman (Householder spells it "Steadman"*), commanding the Division of the Etowah, put Wheeler's command at no less than 6,000. Steedman wrote this concerning the Union operations at Dalton:

"I moved forward and soon cleared the town of enemy, but being without cavalry could not pursue."

Two days later, Wheeler struck again near Graysville. He remained only long enough to tear up some road between there and Dalton. Once finished, Wheeler retired by way of Parker's Gap.

Company K, before spending the night at Graysville, burned some arms and ammunition on General Steedman's orders. Next morning, Householder, with his company, was busily engaged in repairing Wheeler's sabotage. Once this task had been completed, they returned to Chattanooga. Householder closes his letter by making that most universal soldier's wish: to come home.

Jan. 25, 1814—Francis H. Pierpont, governor of the "restored" state of Virginia (1863-1868), Monongalia County, Virginia. (N)
Jan. 25, 1825—Gen. George E. Pickett, Richmond, Virginia. (S)
Jan. 27, 1821—Col. John M. Chivington, Warren County, Ohio. (N)
Jan. 27, 1826—Gen. Richard Taylor, Springfields, near Louisville, Kentucky. (S)
Jan. 28, 1818—Rep. George S. Boutwell, Brookline, Massachusetts. (N)
Jan. 28, 1841—Henry Morton Stanley, soldier and sailor, Denbigh, Wales. (S) & (N)
Jan. 29, 1827—Maj. Charles G. McCawley, U.S.M.C., Philadelphia, Pennsylvania (N)
Jan. 29, 1843—Capt. William McKinley, U.S. Army, Niles, Ohio. (N)
Jan. 30, 1797—Gen. Edwin V. Sumner, Boston, Massachusetts. (N)
Jan. 30, 1816—Gen. Nathaniel P. Banks, Waltham, Massachusetts. (N)
Jan. 30, 1841—George Alfred "Gath" Townsend, war correspondent, Georgetown, Delaware. (N)
Jan. 31, 1810—Gen. Daniel Ruggles, Barre, Massachusetts. (S)
Jan. 31, 1812—Capt. John R. "Handsome Jack" Tucker, C.S.N., Alexandria, District of Columbia. (S)
Jan. 31, 1820—Robert Ould, prisoner exchange commissioner, Georgetown, District of Columbia. (S)
Jan. 31, 1830—Rep. James G. Blaine, West Brownsville, Pennsylvania. (N)
Jan. 31, 1840—Lt. William Pittenger, *The Great Locomotive Chase*, Jefferson County, Ohio. (N)

BIBLIOGRAPHY

Biographical Directory of the United States Executive Branch, 1774-1971. Ed. Robert Sobel, Westport, Connecticut: Greenwood Publishing Company, 1971.

Dictionary of American Biography. 20 vols. New York: Charles Scribner's Sons, 1928-1936.

The National Cyclopaedia of American Biography. 57 vols. New York and Clifton, New Jersey: James T. White & Company, 1892-1977.

Wakelyn, John L. *Biographical Dictionary of the Confederacy*. Westport, Connecticut: Greenwood Press, 1977.

Webster's American Military Biographies. Ed. Robert McHenry. Springfield, Massachusetts: G. & C. Merriam Company, 1978.

Encyclopedia of American History. Vol. II. Ed. Richard B. Morris. New York: Harper & Brothers, 1953.

Kane, Joseph Nathan. *Facts about the Presidents*. New York: The H. W. Wilson Company, 1981.

The McGraw-Hill Encyclopedia of World Biography. 12 vols. New York: McGraw-Hill Book Company, 1973.

Mlecka, Louis F. *Famous People: Historical, Biographical Book of Birthdays*. Brooksville, Florida: Louis F. Mlecka, Publisher, 1973.

Notable American Women, 1607-1950. 3 vols. Ed. Edward T. James. Cambridge, Massachusetts: The Belknap Press of Harvard University Press, 1971.

Reynolds, Clark G. *Famous American Admirals*. New York: Van Nostrand Reinhold Company, 1978.

Warner, Ezra J. *Generals in Blue*. Baton Rouge: Louisiana State University Press, 1964.

Warner, Ezra J. *Generals in Gray*. Baton Rouge: Louisiana State University Press, 1959.

Warner, Ezra J., and W. Buck Yearns. *Biographical Register of the Confederate Congress*. Baton Rouge: Louisiana State University Press, 1975.

Who Was Who in America. Historical Volume, 1607-1896. Chicago: The A. N. Marquis Company, 1967.

Who Was Who in America. Vol. I. 1897-1942. Chicago: The A. N. Marquis Company, 1942.

*Steedman, who had a horse shot from under him at Chickamauga, was a dogmatic man. Before going back into battle, he instructed a staff officer to see that his name was spelled correctly in obituaries: *Steed*man, not *Stead*man.

On October 10, 1864, John Householder was mustered out with his regiment at Camp Chase, Ohio. Upon going home, he resumed a farming life. He remained in Ohio until 1870, when he married Miss Eva Dickenson and became a permanent resident of the state of Kansas. His later years were spent in Winfield, where Householder belonged to GAR Post 85 (organized July 6, 1882.)

Householder's application for an army pension was rejected in 1889. The War Department based its decision on the fact that records furnished showed no evidence of permanent disability. He tried again 26 years later, and, this time, after appearing once more before Winfield's pension board in 1920, was granted one for $72.00 a month.* Householder died three years later of what doctors described as hydrostatic pneumonia and myocarditis.

Diary of John Householder
1862

Monday Sept - 1
Still laying in camp 15 miles from Manchester.
Tues 2d
Started early in the morning for Manchester and arrived there in the afternoon and camped till morning.
Wed 3d
Then we started on the march and marched 13 miles and camped till the next morning.
Tues [Thurs] 4th
Marched within 3 miles of Murfreesboro and camped near a large spring.
Friday 5th
Started from Murfreesboro and marched 15 miles to Savene by dark and took 2 hours rest and ate supper and then took up our line of march again and marched within 6 miles of Nashville by dark and camped till morning.
Sat. 6
Started after breakfast and marched to our camp, 3-3½ miles of Nashville and stayed there till the next day.
Sund 7
Rec'd orders to move out - 12 O'clock and marched 4 miles from Nashville to our old camp and had orders to get our supplies and be ready to move in 2 hours but were countermanded and did not march.
Sept 9
Ready to march early and marched 5 miles and stopped and got breakfast and then marched to Mitchelleville and got our dinner and rested till 4 o'clock and then marched within 2 miles of Franklin and stayed there till morning, and we crossed the state line about 4½ o'clock.
Wed 10
Marched within 6½ miles of Bowling green and stopped for dinner till 5 o'clock then marched within 2½ miles of town and camped by sinking creek till morning.
Thus 11
Marched to town and camped between Bowling Green and the river, detailed to haul wood.
Fri. 12
Co. Inspection in the afternoon.
Sat 13
24 men detailed to work on the fortifications.
Sund 14
The regiment is on picket one mile from camp near Collige Hill.

* During the years after the war, and well into the 20th Century, many private laws for the relief of individuals whose claims to pensions had been rejected under general laws were enacted. (Title 4, Case 165, Pension Act of 1920.)

Mon 15
General review of the 9th Brigade, Gen. S.H. Rousseau present.
Wed 27.
Still laying in camp. We get orders to be ready to move by daylight on Thur. morn.
Thurs. 28
Ready to march by daylight: Co. K detailed to help the wagon train over the mountain but was ordered back to the foot of the mountain.

Friday 29
Started early for Altamont and got up the hill all safe and go to the division in the Afternoon.
Saturday 30
The regiment marched out 2½ miles South of Altamont and when we came back the division had moved so we marched 9 miles over the mountain to a cave and camped till morning.
Sund 31
We moved in the woods. A great day of foraging.

Bellair Ohio
Aug 14th/64

Soldier friend

With great pleasure I Address you a few lines in Answer to yours which came to hand last Evening. I had given up any Answer to my letter. I was happy to hare from you. I am now in Bellair But going Back to Wellsville in a few days. Since I wrote your home of theWellsville Boys got home. as for your Cousin Mariah was not Enjoying good helth when I came away. But I herd from here a few days ago and she is Better.

Well all the talk here North is the draft and the war going to have a draft the 5th of Next mounth and I hope that Every man that has not Been in the sirvice will be drafted and taken out into the field and learn the hardships of a soldier life. we have some that stay of there homes and still they say they are union.

I myself cannot have the same Enterest in this war that I had when I had a Brother in the Army. But still I hope that soon this war will Close.

You say in your letter your not Mistaken when you did not got my picture. well now I say pleas dont think hard when I dont send it this time I think the best is when you Return home it will Be more pleasure for us Both to see the original so pleas Call and see me and then we can talk about the picture.

Well I must tell you I saw some Reble prisoners going through here the other day. This is Sunday evening and I am going to Church. this place is very dull but we dont have much Enjoyment over the world at this time. But I hope to see the day when we will have peace and happiness once more.

the weather is very dry and warm. But I suppose you have warm weather where you are.

Well I will Close hoping to here from you soon a soldier letter is wellcom at all times so pleas write woon.

As a friend
Mary Peeples

Good By
for the present.

Address - to Wellsville

Chattanooga Tennessee
Friday August 20th, .64

Dear and affectionate Mother

I take pleasure in letting you know that I am well and in good spirit. I received a letter from home since I last wrote and allso the thread and needles that was sent in Daves letter.

Well Mother we have been a running on the rail road since last Tuesday. there has been some raiders through between here and Dalton.

Last Sunday there was a detail to take a drove of mules to Ring gold, and I was one of the detail. we went through the Chickamauga battle ground. it is the worst tore up with artilery than any other place I ever seen before. we had some seven or eight hundred mules, and twenty three Ambulances. we only went as a guard and we got to ride from Rossville in the Ambulances. we arrived in Ring gold a little before sundown and took a train for Chattanooga. we got the news at Ringgold that the Rebels had attacked Dalton. there was a train that had just come from there to telegraph to Chattanooga for reenforcements. We passed four train loads a going at Graysville.

The rebels took Dalton but our men held a fort near the town. the 14th Colored troops charged them out of the town, but they held the town one night and they tore up ________________.

tuesday morning there was four train loads of troops sent to repair the road. but it wasen't tore up to amount to much untill we got to Dalton. We arrive in Dalton about nine o clock at night and stayed till morning and then took alittle Scout out side of town and then took the train for Chattanooga. but the rebs had burnt the road between Dalton and Graysville in two places. there was a house near one place and there was some arms and ammunition found there and it was burnt by order of Gen. Steadman. We camped at Garysville a Wednesday night for the road was burnt on this side about a mile and half.
a Thursday morning we went to work to repairing the road and then returned to Chattanooga, but we had been in camp more than an hour than there was an order for our Regt to go to Keington with a load of Rail Road Iron. We started from here in the afternoon and arrived at Dalton about nine or ten oclock at night and stayed untill morning. The road being open there was several trains went to the front in the night. We started in the fore noon yesterday for Kingston and arrived there all right. Keingston is 79 miles from Chattanooga. we took the train again for camp at two oclock and arrived here in the night.

Dave came back a Sunday he is well. We will tell you about the Scout when we come home. I canno tell you yet wheh we will get to start home but I hope it will not be long.

I will close every remaining
Your Son

J.M. Householder
Co K 2nd Ohio I.

Wellsville, Ohio
Aug. 30th

Dear friend

I received your kind and wellcome letter this evening was happy to hear that you was numbered with the living for so many of our Brave soldiers are falling every day.
Well I am sorry that I cannot tell you any thing about the draft. it has not come off yet.
Well I will say I saw Mariah to day her health is Better than when I wrote you last. I am ________ with the same family that she is with. But she is going home next week to spend a few weeks with her Brothers when they return. I think she ought to Be a happy girl to have a Brother to Return for I know I should if my Brother was leaving an Returning home. But still I am not all the one that must morn the loss of a Dear friend since this Cruell War has been going on which I hope will soon close.

Well I will soon close for I am writing after night. I don't know whether you can read those few imperfect lines or not but I hope you can.

I tought it was my duty to send you an Answer as prompt as possible as I did not know how long you would remain at Chattanooga. so please excuse this poor writing and I will do better Hereafter.

And I will close hoping this will find you enjoying the great Blessings of good health which I do and also hope to see you soon.

So good night
Your friend with
My warmest
Regards
Mary E. Peeples
please write soon.

Tracts

"Justification and Sanctification"

by

Rev. Allen Kelley

There is, perhaps, no such thing as a simple religion. Faith, itself, may be uncomplicated, but the trappings that surround it invariably are not. Murderous, bloody wars have been waged in the name of God; individuals persecuted for their beliefs; nations divided over the interpretations of obscure ancient passages. Moral codes of behavior were drawn up and "religiously" practiced, so the faithful would find their way to Heaven in the afterlife.

In the 19th Century, a great deal of consideration was given to two peculiar words: *justification* and *sanctification*. It was the goal of a good Christian to aim for this cleansing of sin, or *perfection*. Many sermons were preached on the subject, and ministers of the gospel carried tracts and booklets into the battlefields of the Civil War on these very subjects. There, they would expose the crowds of soldiers to their philosophies and guidelines for a holy life. What lasting impressions these men wrought is difficult to determine, but it is sure they put the fear of God into many a young man's heart.

Consequences for non-compliance to their views were obvious — hell fire and eternal damnation. Repentence and a conversion to godliness was encouraged, but even that might not be enough. A typical religious work of the time, "The Christian's Manual," warned that a casual life of religious observance and good works was not enough to ensure "so high a place in heaven as they would have had, if they had chosen a better path." To be of the higher order of Christians, the devout were taught to strive for Christian perfection. To achieve this state of complete absence from sin, ministers taught first of justification. The premise was easy enough: a guilty man must be pardoned. To elaborate, the mystic men of the cloth told of the sinner who suddenly realized his life was filled with corruption. This self-knowledge brought him to despair, and at length, after finding no good thing in himself, he was compelled to turn to the Savior. This first act of submission, followed by the act of acceptance, as he realized he was a child of God, filled his heart with love and thankfulness. When he reached this point, he had achieved justification and victory over the dark powers. He must not be content, however, for to remain in one place was to backslide into doubts and temptations. Bearing in mind there was always room to grow in grace, the true Christian must now model his life on Christ.

It was preached that full sanctification came in a flash and filled the flesh, as well as the spirit, with the positive influence of God. This renewal of self cleaned away whatever pollution remained behind after justification, and left the man full of righteousness, "like a light shining in a dark place."

Christian perfection, they expounded, did not make one perfect in knowledge, nor free from ignorance. It did not reveal the secrets of the Almighty or free the possessor from inward or outward imperfections, such as the want of a ready or retentive memory, slowness of speech, ungracefullness of pronunciation, infirmities or free the person from temptations. What it did mean was a state of *holiness*, which in the Scriptual sense, meant perfection. This, then, allowed the sanctified never to mistake the things essential to salvation. They glorified God in body and spirit and spared no pains to arrive at the summit of Christian holiness. Their strictness with themselves allowed them zealously to pursue good works of every kind, attend to all ordinances of God and emulate the path Christ walked. They trampled on every pleasure which did not prepare them for taking pleasure in God, accepted no praise or flattery, felt no anger and yielded to no temptations.

The witness or testimony of perfection was an impression on the soul that all sins have been washed away, and a feeling that one's life was wholly set apart for God. For those who professed perfection, there were standards to be maintained. They were advised to ascribe all their knowledge to God, watch and pray continually against pride, always be ready to admit any faults, give no place to enthusiasm, beware of bigotry (or sharing friendship only with members of your own religious group), lose no opportunity to do good of any kind, give no place to sloth, observe every rule of society, never be absent from public meetings (as they are the sinews of society), be wary in speaking and keep at the utmost distance from small talk. To grow daily in grace, regular times for prayer should be set aside and adhered to; church meetings should always be attended and it was advised, "be *exemplary* in all things: particularly in outward things as in dress; in little things, as the laying out of your money, avoiding every needless expense; in deep, steady seriousness, and in the solidity and usefullness of your conversation."

This would not guarantee, however, that life on earth would be free from trials. On the contrary, the believer was warned to expect contradiction, opposition and many crosses to bear in daily life. Suffering was not only unavoidable, in the course of Providence, but profitable, even necessary. Trials were to strengthen the soul and help build the person's house in heaven. And though a man be thought a *fool,* he should follow Christ *fully.*

It was commonly believed in the mid-19th Century that a complete devotion to God was not only necessary for the soul, but for the body, as well. A strong belief in Jesus comforted many a man as he went to war, and many a loved one who stayed behind. Attendance at church meetings, devoutness in habits of prayer, attributing all good things to God and thanking Him for his blessings was the accepted, *expected* conduct. It was neither unusual for soldiers to sit around a campfire and discuss morality, nor for officers to gather and debate justification and sanctification. A knowledge of religion was as necessary as an education in the three R's, and there was seldom a man who succeeded without a profession of some faith. And while the old axiom "You can't judge a book by its cover" was probably as true then as now, it was most definitely believed you *could* judge a man by his religion.

• SIGNALS •

Dear Editors:

Please send the following: Vol. I, Nos. 1 - 5 of THE KEPI.

I've enclosed a check.

Clinton F. Firstbrook III
Lanark, IL

P.S. Especially enjoyed the Oct/Nov [1984] issue about "Stonewall."

Dear Editors:

Your article, "Christmas in August," placed before me the facts in a clear, concise manner. I found it very entertaining reading, as well as thought-provoking. The TIMELINE piece cleared up some confusion on the part of a friend of mine, and I suspect that he will soon be subscribing. Keep up the good work.

Randy White
Poughkeepsie, NY

Dear Editors:

My special interest is Gettysburg. I have just subscribed to THE KEPI, and was wondering if your magazine will be running a piece on this battle.

Jim Barnes
Oklahoma City, OK

[*THE KEPI will be presenting an article in our April-May 1984 issue, entitled:* "Joshua Chamberlain: An Amateur Among Professionals" *which concerns his actions at Gettysburg. We hope all our readers will enjoy it.*

E's-in-G.]

WHIMSEY

A London artist passing the shop of M. Hautcoeur, Rue de Rivoli, publisher of engravings, remarked that you would naturally go there for High 'Art.

[From *Harper's Weekly,* New York, Saturday, October 26, 1861.]

The same unfortunate Cockney, having heard that horseflesh is eaten at certain Parisian hotels, evinced the utmost horror when, looking over a *carte* at Vefour's, he came to the *hors d'oeuvres.* He says that a saddle of mutton is the nearest approach he can bear to equestrian viands.

[From *Harper's Weekly,* New York, Saturday, October 26, 1861.]

●

TO BE PARTED WITH, for the veriest trifle, a LONG ESTABLISHED COLD, of a sonorous, deep-toned quality. The advertiser can highly recommend it, as it has been in his possession now for the last three years, and has never left him, either day or night, for a single minute. The only motive for parting with it, is because the owner has recently joined a Choral Society, and he finds that his fellow-students strongly object to his practicing with them. Six dozen boxes of cough lozenges will be thrown in, as a *douceur,* with the above. Immediate possession can be had, and a month's trial allowed for approval. Letters to be addressed to "A. BARKER, ESQ., care of the Secretary of the Tonic Sol-Fallal-de-Riddle-ol Association."

[From *Harper's Weekly,* New York, Saturday, October 26, 1861.]

CONCURRENT EVENTS

SNOW STORM IN SYRIA— A letter just received from Beyrout, under date of the 9th ultimo, states that the country was, on the 27th of February, inundated by a heavy fall of rain which lasted four days, and on the fifth day it turned into a regular snowstorm, which continued for two days, much to the astonishment of the natives of the country, "The Oldest Man" of whom had never witnessed a similar occurrence. This snowstorm was succeeded on the Fifth of March, by a tremendous watery gale, causing heavy damage to shipping, sinking fifty-two merchant vessels along the coast, and a French steamer before Beyrouth. About 600 persons are supposed to have perished. The crew of the steamer were all saved.

[From the *Detroit Advertiser and Tribune,* April 6, 1863. Courtesy of Bill Christen.]

JEFF DAVIS LIKELY TO LOSE HIS EYE — A correspondent of the Montgomery *Advertiser,* writing from Richmond on the 23 ult. says President Davis has been confined to his house for two weeks, and is a great sufferer. An abscess has formed on his temple, next to his sound eye, and it is feared by his friends that he will lose the sight of that eye also.

[From the *Detroit Advertiser and Tribune,* April 6, 1863. Courtesy of Bill Christen.]

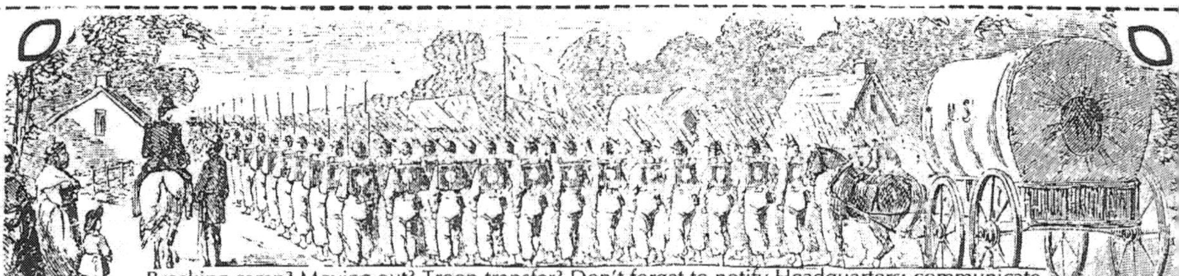

Breaking camp? Moving out? Troop transfer? Don't forget to notify Headquarters: communicate new location to THE KEPI.

NAME ______________________________

ADDRESS ______________________________

Continued from page 21.

tenth of the crimes attributed to him. At 4 p.m. he dropped through the trap door of the gallows, and Pauline Mosby lost a beloved cousin. His final recorded words were: "I believe and die for the Confederate cause."[4]

FOOTNOTES

[1] OR, Series I, Vol. XXXIX, Part I, p. 899.
[2] OR, Series I, Vol. XLIX, Part I, p. 641.
[3] The Courier-Journal, Nov. 20, 1960 edition.
[4] Kentucky Historical Society.

ACKNOWLEDGEMENTS

Kim B. Holien, Historian, U.S. Army, Washington, D.C. Cynthia L. Musselman, Genealogist, Stafford County, Va. Homer B. Musselman, Historian, Stafford County, Va.

Continued from page 21.

We were frolicking around and all and finally called the boy on guard to come on in. Well, he done it, and we were having a good old time when I noticed the other boys got quiet all of a sudden. I looked up and three Rebels in gray uniforms was just sitting on their horses there on the bank enjoying watching us.

'Boys, you are having a good time and I don't want to spoil it for you,' said a wiry little Confederate officer. Says, 'But I am going to take these nice rifles and cartridge boxes you have stacked here for me. If I was a mean man I would take your uniforms, too, but I reckon you will have need for them.'

"And that is what they did do, and then they all just rode off slowly, carrying our rifles and equipment.

'That's Mosby,' said one of the boys. Says, 'We are sure lucky he didn't take us prisoner.'

"Well, there wasn't a thing to do but go on back to camp and confess to being the fool. The Captain took on something awful and the old Colonel got red as a beet and jumped around.

'You idiots are going to pay for those rifles and that equipment,' shouted the Colonel. Says, 'Expected more from you, Tommy Evans, you hard-headed Welshman. Serve you right if some Rebel shoots you with your own gun.'

"Now we had a bad time out of it and had to pay up. I was sure sorry to lose that rifle. It was a good English one I used down in the Shenandoah Valley when I won the Medal of Honor. But we could have been taken prisoner or killed by Mosby while acting the fool, and I still don't enjoy swimming as much as I used to before that day."

(As told to the writer's grandfather by Pvt. Thomas Evans, 54th Pa. Vols.)

THE KEPI
P.O. Box 699
Baldwin Park, Ca. 91706

BULK MAIL

THE KEPI Volume I &II INDEX

Made in the USA
Monee, IL
04 November 2026

67383148R00226